CW00925623

THE LOEB CLASSICAL LIBRARY

FOUNDED BY JAMES LOEB

HIPPOCRATES

IV

LCL 150

HIPPOCRATES

VOLUME IV

HERACLEITUS

ON THE UNIVERSE

WITH AN ENGLISH TRANSLATION BY

W. H. S. JONES

HARVARD UNIVERSITY PRESS
CAMBRIDGE, MASSACHUSETTS
LONDON, ENGLAND

First published 1931
Reprinted 1943, 1953, 1959, 1967, 1979,
1992, 1998

ISBN 0-674-99166-4

Printed in Great Britain by St Edmundsbury Press Ltd,
Bury St Edmunds, Suffolk, on acid-free paper.
Bound by Hunter & Foulis Ltd, Edinburgh, Scotland.

CONTENTS

PREFACE

The work of preparing this volume has taken all my leisure for over five years, the most laborious part being the collation of the manuscripts Urb. 64, A, M, V, *θ*, C′, Holkhamensis and Caius $\frac{50}{27}$. I have not quoted all the variants, perhaps not the greater number of them; the rule I have tried to follow is to record only those readings that are intrinsically interesting and those that seriously affect the meaning. The readings recorded by my predecessors are often wrongly transcribed; knowing by experience the risk of mistakes in collations, however carefully done, I am sure that there are some errors in the notes in this volume. The readings of Urb. 64 are here printed for the first time, as also are many from the manuscripts M, V.

I wish to thank my pupil, Mr. A. W. Poole, for help in preparing the index.

W. H. S. J.

INTRODUCTION

I

INTENTIONAL OBSCURITY IN ANCIENT WRITINGS

To a modern it appears somewhat strange that a writer should be intentionally obscure. An author wishes to be easily understood, knowing that neither critics nor readers will tolerate obscurity of any kind. But in ancient times the public taste was different; the reader, or hearer, was not always averse to being mystified, and authors tried to satisfy this appetite for puzzles.

It was probably the oracles, with their ambiguous or doubtful replies, that set the fashion, which was followed most closely by those writers who affected an oracular style. The difficulties of Pindar and of the choral odes of Aeschylus, who was imitated in this by later dramatists, were not entirely or even mainly due to the struggle of lofty thought seeking to find adequate expression in an as yet inadequate medium. They were to a great extent the result of an effort to create an atmosphere congenial to religion and religious mystery. So Plato, who can when it suits his purpose be transparently clear, affects an almost unnatural obscurity when he wishes to attune his readers' mind to truths that transcend human understanding. Much of the *Phaedrus* and of the *Symposium*, the Number in the *Republic*, and a great part of the *Timaeus*, are oracular utterances rather than reasoned argument, taking their colour from the difficulty of their subject. But prose remained comparatively free from intentional obscurity; lyric poetry, on the other hand, at any rate the choral lyric, seems to have been particularly prone to it. In Alexandrine times obscure writing became one of the fads of literary pedants, and Lycophron is a warning example of its folly when carried to extremes.

There must have been something in Greek mentality to account for the persistence of this curious habit, which appears all the more curious when we remember how fond the Greeks were of clear-cut outlines in all forms of art. The reason is probably to be found in the restless activity of the Greek mind, which never had enough material to occupy it fully. The modern has perhaps too much to think about, but before books and other forms of mental recreation became common men were led into all sorts of abnormalities and extravagances. The unoccupied mind broods, often becoming fanciful, bizarre or morbid. To quote but two instances out of many, the "tradition" condemned by Jesus in the Gospels, and the elaborate dogmas expounded at tedious length by the early Fathers, were to some extent at least caused by active brains being deprived of suitable material. It is a tribute to the genius of the Greeks that they found so much healthy occupation in applying thought to everyday things, thus escaping to a great extent the dangers that come when the mind is insufficiently fed. A tendency to idle speculation is the only serious fault that can be found with Greek mentality; indulgence in intentional obscurity is perhaps a fault, but only a slight and venial one. As has been said above, oracular responses seem to have started the fashion of purposely hiding thought, but it was kept up by the Greeks' love of solving puzzles, of having something really difficult with which to exercise their brains.

It has already been pointed out, in the introduction to *Decorum*, that certain (probably late) tracts in the *Corpus* are intentionally difficult, but the reason for their difficulties may well be due to a desire to keep secret the ritual or liturgy of a guild; *Decorum*, *Precepts* and *Law* are in a class by themselves. This explanation, however, will not apply to the obscure passages in *Humours*. This work has nothing to do with secret societies. It is a series of notes which, however disjointed or unconnected, are severely practical. Their obviously utilitarian purpose makes their obscurity all the more difficult to understand; a text-book, one might suppose, ought at least to be clear. Yet when we have made allowances for hasty writing and for the natural obscurity of all abbreviated notes, there remains in *Humours* a large residue of passages in which the difficulties appear to be intentional. The fact that these passages[1] are sometimes written in a rather lofty style seems to suggest an explanation of them. *Humours* is akin, though not closely so, to *Nutriment*; it is aphoristic after the manner of Heracleitus "the dark." This thinker adopted the oracular style when expounding his philosophical system, and certain later thinkers

[1] I seem to detect the characteristics to which I refer chiefly in Chapter I, and in the various lists of symptoms, etc.

followed his example. Perhaps it was thought that a "dark" subject required a "dark" medium of expression. The writer of *Nutriment,* who was striving to wed Heracleiteanism and physiology, succeeds in producing a not altogether incongruous result. But Heracleitean obscurity is sadly out of place in a work entirely free from philosophy, whether Heracleitean or other, and the modern reader is repelled by it. The ancients, however, appear to have been attracted, for *Humours* is often referred to, and commentators upon it were numerous. It is interesting to note that the author, or compiler, of *Aphorisms,* who was a really great scientific thinker, while adopting the oracular aphorism as a medium of expression, and keeping the lofty style appropriate to it, makes no use of intentional obscurity, realising, consciously or unconsciously, how unsuitable it is in a work intended to instruct medical students and practising physicians.

II

THE FORM AND CONSTRUCTION OF CERTAIN HIPPOCRATIC WORKS

Many books in the Hippocratic Collection are not strictly "books" at all; they consist of separate pieces written continuously without any internal bond of union. Already, in Volume I, we have discussed the curious features presented by *Epidemics* I and III,[1] and by *Airs Waters Places*.[2] The aphoristic works, being at best compilations, exhibit a looseness of texture which makes additions and interpolations not only easy to insert but also difficult to detect. *Nature of Man* and *Regimen in Health* appear as one work in our MSS., and the whole has been variously divided by commentators from Galen onwards. *Humours* has scarcely any texture at all, and the disjointed fragments of which it is composed can in not a few places be traced to other works in the *Corpus*.

The scholars who have devoted themselves to the study of *Nature of Man—Humours*, probably because of its hopeless obscurity, has been very much neglected—seem to make, perhaps unconsciously, a more than doubtful assumption. They suppose the present form of the book to be due to a compiler,

[1] Vol. I. pp. 141, 142. [2] Vol. I. p. 66.

who acted on some definite purpose. It is, however, quite possible that the "conglomerates," as they may be called, are really the result of an accident. A printed book goes through a fixed routine, which fact is apt to make us forget that a papyrus roll may well have been a chance collection of unconnected fragments. In the library of the medical school at Cos there were doubtless many rough drafts of essays, lecture notes, fragments from lost works, and quotations written out merely because a reader happened to find them interesting. Some tidy but not over-intelligent library-keeper might fasten together enough of these to make a roll of convenient size, giving it a title taken perhaps from the subject of the first, or perhaps from that of the longest fragment. Later on, scribes would copy the roll, and the high honour in which the Hippocratic school was held would give it a dignity to which it was not entitled by its intrinsic value.

Of course these remarks are mere guess-work. Positive evidence to support the hypothesis is very slight, but it should be noticed that a work in the *Corpus* often ends with a fragment taken from another work. Take, for instance, *Regimen in Health*. There are seven chapters of good advice on the preservation of health. The subject is treated in an orderly and logical manner, but the reader feels that at the end of the seventh chapter there is an abrupt break in the description of regimen for athletes. For the eighth chapter is a fragment from the beginning of the second book of *Diseases*, and gives some symptoms of "diseases arising from the brain," and the ninth chapter is a fragment from the beginning of *Affections*, which insists on the importance of health and of making efforts to recover from illnesses. Here *Regimen in Health* ends.

Several points need careful consideration :—

(1) *Regimen in Health* proper ends abruptly and is apparently unfinished ;

(2) This unfinished work has two short fragments tacked on to it, the second of which is but slightly connected, and the first quite unconnected, with the subject matter of the first seven chapters ;

(3) These fragments are taken from the beginnings of other works in the *Corpus*.

Is it possible for such a conglomerate to be the result of design? What author or editor could be so stupid as to complete an incomplete work by such unsuitable additions? What particular kind of accident is responsible nobody could say for certain, but it is at least likely that some librarian, and not an author, added the two fragments. It must be remembered that the parts of a book that get detached most easily, whether the books be a roll or composed of leaves, are the beginning and the end. These places are also the most convenient for making additions. Suppose that the end of *Regimen in Health* was lost and the beginnings of copies of *Diseases* II and of *Affections* became detached; surely it is not unreasonable to suppose that a librarian preserved the latter by adding them to the former.

Nature of Man is similar in construction, but the fragments added to the main piece are longer; *Regimen in Health*, in fact, is itself one of them.

First we have seven chapters treating of the four humours, which end with the relation between these humours and the four seasons. The eighth chapter[1] deals with the relation between the seasons and diseases. The ninth chapter[2] begins with the cure of diseases by their opposites. After three sentences a complete break occurs, and a fresh start is made, beginning with *αἱ δὲ νοῦσοι γίνονται*; and the rest of the chapter, about 50 lines, is concerned with a classification of diseases into (1) those arising from regimen and (2) those caused by the atmosphere. Incidentally it may be noted that the first part of this section is paraphrased in Menon's *Iatrica* VII. 15 and attributed to Hippocrates. The tenth chapter briefly postulates a relationship between the virulence of a disease and the "strength" of the part in which it arises. Then comes the famous passage dealing with the veins, which Aristotle in *Historia Animalium* III. 3 attributes to Polybus. The twelfth chapter deals with the cause, in the case of patients of thirty-five years or more, of "pus" in sputa, urine or stools. The thirteenth chapter contains two unconnected remarks, the first to the effect that knowing the cause of a disease enables the physician to forecast better its history, the second insisting upon the necessity of the patient's co-operation in effecting a cure. The fourteenth

[1] There is an unfulfilled promise in *τὴν δὲ περίοδον αὖτις φράσω τῶν ἡμερέων*, which Fredrich would delete as an interpolation.

[2] This chapter has two references to passages that are not extant, *ὥσπερ μοι πέφρασται καὶ ἑτέρωθι*, and *ὥσπερ μοι καὶ πάλαι εἴρηται*. If *Nature of Man* consists of sections taken from works now lost, these cross-references are easily explained.

chapter deals with deposits in urine. The last chapter contains a very brief classification of fevers.

It requires a special pleader, biased by a subconscious conviction that a Greek book must be an artistic whole, to maintain that this aggregate follows any logical plan. Yet Fredrich, an excellent scholar and a keen student of Hippocrates, sums up his opinion in these words: "Vir quidam, medicus videlicet, in usum suum collegit et composuit res memoria dignas: complures de origine morborum et curatione sententias (π. φύσ. ἀνθρ.; c. ix, 1; ix, 2; 10; 13) dissertationes de venis (c. 11) de pure (12) de urina (14) de febribus (15) de diaeta (1–7) de capitis doloribus (8) principium sanandi (9); et haec quidem duo capita addidit fort., quod initia librorum ei carorum erant."[1] There is nothing unreasonable in assigning the collection of extracts to "medicus quidam"; a physician is perhaps as likely a person as a librarian. But "composuit" does not in the least describe the work of the collector. The sections are not "arranged"; if any effort was made to put them in order it was a very unsuccessful effort. It is a far more likely hypothesis to suppose that fragments of papyrus were fastened together by someone, perhaps a physician, perhaps a library attendant, to prevent their getting lost.

A similar problem faces us when we examine *Humours*, but here the *disiecta membra* are even more incongruous and disordered. An analysis of the work may prove useful.

Chapter I. The humours, and how to divert or deal with them when abnormal.

[1] C. Fredrich, *de libro* περὶ φύσιος ἀνθρώπου *pseudippocrateo*, p. 15.

Chapters II–IV. A mass of detail the physician should notice when examining a patient.
Chapter V. How to find the *κατάστασις* of a disease. What should be averted and what encouraged.
Chapter VI. The proper treatment at paroxysms and crises. Various rules about evacuations.
Chapter VII. Abscessions.
Chapter VIII. Humours and constitutions generally; their relation to diseases
Chapter IX. Psychic symptoms and the relation between mind and body.
Chapter X. External remedies.
Chapter XI. The analogy between animals and plants.
Chapter XII. The fashion of diseases, which are congenital, or due to districts, climate, etc.
Chapters XIII–XVIII. Seasons, winds, rains, etc., and their influence on health and disease.
Chapter XIX. Complexions.
Chapter XX. Quotation from *Epidemics* VI. 3, 23, dealing chiefly with abscessions and fluxes.

There are many quotations or paraphrases from various Hippocratic treatises.

Chapter III. *Aph.* IV. 20; *Prognostic* II.
Chapter IV. I. *Prorrhetic* 39; *Joints* 53.
Chapter VI. *Aph.* I. 19; 20; *Epi.* I. 6; *Aph.* I. 22; 21; 23; 24.
Chapter VII. *Aph.* IV. 31; *Epi.* VI. 7, 7; *Aph.* IV. 32; *Epi.* VI. 1, 9; IV. 48; *Aph.* IV. 33; *Epi.* IV. 27 and 50; *Epi.* VI. 1, 9; 3, 8; 7, 7; 7, 1; 7, 7.
Chapter X. *Epi.* II. 1, 7; 5, 9; *Epi.* IV. 61.

Chapter XII. *Aph.* III. 4; *Airs, Waters, Places* 9, 7 and 9.
Chapter XIII. *Epi.* II. 1, 5; *Aph.* III. 8; 6; *Epi.* I. 4.
Chapter XIV. *Aph.* III. 5, 21 and 5.
Chapter XV. *Aph.* III. 1.
Chapter XX. *Epi.* VI. 3, 23, to 4, 3.

In other words the following passages are quoted:—

Aph. I. 19; 20; 21; 22; 23; 24.
Aph. III. 1; 4; 5; 6; 8; 21.
Aph. IV. 20; 31; 32; 33.
Epi. I. 4.
Epi. II. 1, 5; 1, 6; 1, 7; 5, 9.
Epi. IV. 27; 48; 50; 61.
Epi. VI. 1, 9; 3, 8; 3, 23 to 4, 3.
Epi. VII. 1; 7.
Prognostic II.
I. *Prorrhetic* 39.
Joints 53.
Airs, Waters, Places 7 and 9.

In all there are thirty-five borrowed passages.

The analysis of *Humours* given above is by no means adequate; a careful reader will note many omissions of details. It is, in fact, impossible to analyse what is itself in many places an analysis. Some parts of the book read just like lecture notes, or heads of discourse to be expanded orally by a teacher or lecturer. It is indeed hard to believe that the lists in Chapters II, III, IV, V, IX are not either such notes or else memoranda made by a student for his own guidance. How and why the other parts were added it is impossible to say, with the possible exception of the first chapter and the last. As has already been said, the beginning and end of an ancient scrap-book are the places where additions are most easily made. The first chapter, while similar in character to the rest of the book, is separated from it by the words σκεπτέα ταῦτα·, with which the second chapter begins. These words may well have been the title, as it were, of the memoranda which we assume form the basis of the whole work. The last chapter is obviously a fragment added to the end of the roll by somebody who did not wish it to be lost.

Neither *Humours* nor *Nature of Man* must be judged by the canons used in appreciating literature. They are not literary compositions, and only the first chapters of *Nature of Man* are artistically written. *Humours* is not only inartistic but also often ungrammatical. The writer, or writers, wrote down rough notes without thinking of syntactical structure. Not intended for publication, these jottings show us that the Greek writers were sometimes inaccurate or inelegant in speech. The textual critic, deprived of one of his most powerful weapons, that a faulty expression is probably due to the carelessness of a scribe, is forced to pause and think. If the scientists were often slipshod, perhaps the literary writers were occasionally so. A linguistic error in the text of, say, Demosthenes may be due, not to the mistake of a scribe, but to the inaccuracy of Demosthenes himself. Even the greatest artists are not infallible.

In conclusion, it should be remembered that a papyrus roll could contain no foot-notes, and that marginal notes did not come into general use before the age of the scholiasts. No author annotated his own works; he worked any necessary annotations into the text itself, and these might consist of illustrative passages from other works. As one reads *Humours* the conviction grows that many of its apparently irrelevant passages are really notes of this type. A good example occurs in Chapter XIV. The subject is the influence of south winds and of north winds on health, and the author concludes his remarks at μᾶλλον. Some note, however, is required, to deal with a special case. This special case brings in (1) the question of droughts and (2) the humours. So two fresh notes are added, one stating that either wind may accompany drought, and the other that humours vary with season and district. Between the two notes is inserted a remark (διαφέρει γὰρ καὶ τἄλλα οὕτω· μέγα γὰρ καὶ τοῦτο), the connection of which is very obscure. It may refer to the effects of winds (as in the translation), or it may mean that other things beside winds influence the character of diseases. So there are apparently four notes, one at least of which is a note added to the first note.

But this explanation of irrelevant passages must not be pushed too far. It cannot account for the amorphous construction of many Hippocratic treatises, which is almost certainly due to the welding together of detached or separate fragments of various sizes in order to preserve them in book form.

III

SCIENCE AND IMAGINATION

The progress of scientific thought depends upon two factors. One is the collection of facts by observation and experiment; the other is constructive imagination, which frames hypotheses to interpret these facts. The Greek genius, alert and vigorous, was always ready with explanations, but it was too impatient, perhaps because of its very quickness, to collect an adequate amount of evidence for the framing of useful hypotheses. This fault was not altogether a bad thing; the constructive imagination needs to be developed by practice if progress is to be possible. But imagination needs also training and education, and the Greek mind was so exuberant that it shirked this necessary discipline. The drudgery of collecting facts, and of making sure that they square with theory, proved too laborious. Experiment was entirely, or almost entirely, neglected. The hypotheses of early Greek thought are mere guesses, brilliant guesses no doubt, but related to the facts of experience only in the most casual way. Medicine, indeed, did usually insist on the collection and classification of phenomena, but guesses mar all but the very best work in the Hippocratic *Corpus,* and it was not until Aristotle and Theophrastus laid the foundations of biology that the importance of collecting sufficient evidence was fully realised.

It is interesting in this connection to note that the arts were distinguished from the sciences only when Greek thought was past its zenith. The word τέχνη can mean either "art" or "science," though it inclines more towards the former, sometimes in a slightly derogatory sense ("knack"). Σοφία is almost equally ambiguous, and we have to wait until Aristotle, completing the work of Plato, gave a new, specialised meaning to ἐπιστήμη before there is a word approximately equivalent to our "science" without any additional notion of "art." Now the arts demand much more imagination and freedom of thought than do the sciences, and the Greeks' having the same word for both is a sign that the discipline necessary for accurate science was not appreciated.

Greek imagination was not only luxuriant; it was also picturesque, and demanded artistic detail. A Greek philosopher felt bound to paint a complete picture when he formulated a theory, however few were the certain facts that he could use in its construction. So a Greek philosophical system is likely to contain many details, not indeed incongruous, but unscientific in the modern sense of the term. The Greek love of a completed picture is well illustrated by the "myths" of Plato's dialogues. When a theory cannot be finished, because of the limitations of the human intellect, a myth is added to fill up the ugly gap. The reason, for instance, can prove that the soul never really dies if we admit Plato's Ideas. In the *Phaedo* this proof is elaborated, but there is an inevitable hiatus in the account. Granted that the soul exists in the next world, what sort of a life does it pass there? This cannot be told by reason, so that an imaginary story is added for the sake of completeness.

Nobody would interpret *Revelation* as one would Darwin's *Origin of Species.* An important principle of interpretation follows. A Greek theory cannot always be treated like a truly scientific account. Conformity with experience, a *sine qua non* of scientific reasoning, is not to be demanded of works in which imagination plays a large part.

The medical treatises of the Hippocratic Collection sometimes contain a philosophic element. *Nutriment,* translated in Volume I, is an attempt to apply the principles of Heracleitus, using the language of Heracleitus, to the problems of food and its assimilation. Parts of *Regimen,* a treatise translated in the present volume, are similar in character, although following a different philosophic system. We must not expect of them too much consistency, too much conformity with experience, too much scientific method. We must realise that they are in part works of imagination, often figurative, allusive and metaphorical. They portray truth, or what the writers consider to be truth, in an allegorical guise. Like a modern futurist picture, they try to express reality by a mass of detail which does not strictly correspond with objective fact. Provided that he produces the general impression he desires, the writer is not over-careful about the patches of colour that make up the whole.

What is true of *Nutriment* and *Regimen* is *a fortiori* true of the fragments of Heracleitus in this volume.

They attempt to explain the material universe in a style that is largely poetical. Heracleitus, like most Greek writers, failed to confine himself to a single *rôle*. He is a philosopher and a scientist, but he cannot help being at the same time an artist, a prose poet, and a religious reformer.

IV

NATURE OF MAN

Nature of Man and *Regimen in Health* formed one work in ancient times and are joined together in our manuscripts. Galen comments on the whole work, dividing it into three main parts: Chapters I–VIII, IX–XV and XVI to the end.[1] It is clear that in Galen's time the book had the form it has now, but we do not know when that form was first received. Aristotle[2] refers to the description of the veins in Chapter XI, ascribing it to Polybus, the son-in-law of Hippocrates, and to the same Polybus is ascribed in the *Anonymus Londinensis*[3] a part of the first section, which has given a name to the whole composition. On the other hand, the *Anonymus*[4] quotes, or rather paraphrases, the passage in Chapter IX that begins with αἱ δὲ νοῦσοι γίνονται, and prefaces the quotation with ἀλλὰ γὰρ ἔτι φησὶν Ἱπποκράτης κ.τ.ἕ. If *Nature of Man* had been known as a unity, it is strange that there is here ascribed one part to Polybus and another part to Hippocrates.

[1] See Villaret's discussion of Galen xv. 9 foll. (*op. cit.* pp. 4–6).

[2] *Hist. Animal.* III. 3.

[3] *Iatrica*, XIX; Diels, pp. 33, 34. Chapters II, III, and IV are referred to.

[4] *Iatrica*, VII. 15. See Diels, pp. 10, 11.

Galen is convinced that the first section is referred to by Plato in the famous passage in the *Phaedrus*,[1] and that the whole work, in spite of Aristotle, should be assigned to Hippocrates himself.[2] It should be noted in passing that neither the first section nor the second is complete. The former contains an unfulfilled promise,[3] the latter back references [4] to a discussion of regimen no longer extant.

Most of our difficulties disappear if we look upon *Nature of Man* and *Regimen in Health* as a chance collection of fragments, varying in size and completeness, and perhaps put together by a librarian or book-dealer. Aristotle and Menon may be referring to the complete works from which the extant fragments were taken.

We must now consider the internal evidence. In Chapter I Melissus the Eleatic, who flourished about 440 B.C., is mentioned in such a way as to show that his doctrines were not yet forgotten or out of date, and throughout the first eight chapters the influence of Empedocles is strong. We ought then to postulate for the first section a date not earlier than 440 B. C. and not later than (say) 400 B.C. The style is clear and forcible, pointing to a time when prose-form had already received careful attention, some years later, in fact, than the rise of Sophistic rhetoric. Finally, even a superficial reader will notice the general likeness of the first section of *Nature of Man*

[1] 270 C–E. See Vol. I. pp. xxxiii–xxxv.

[2] See Littré, Vol I. pp. 297, 298, 346. Littré himself is convinced that the *Phaedrus* passage refers, not to *Nature of Man*, but to *Ancient Medicine*.

[3] VIII: τὴν δὲ περίοδον αὖτις φράσω τὴν τῶν ἡμερέων.

[4] IX: ὥσπερ μοι πέφρασται καὶ ἑτέρωθι and ὥσπερ μοι καὶ πάλαι εἴρηται.

to *Ancient Medicine*. It is difficult to resist the conclusion that they were written at approximately the same time, that is, during a period of an eclectic revival of the older philosophies.

The smaller fragments that follow show no reliable clues as to their date, except the similarity of the section on veins to *Sacred Disease* (VI. foll.). *Regimen in Health* ends in two fragments from other treatises in the *Corpus*—περὶ νούσων II and περὶ παθῶν—either stray strips of papyrus added by chance or the result of repeated wrong division of works written as though one treatise were the continuation of the preceding. The main portion belongs to that period, referred to by Plato in his polemic against medicine in the *Republic*, when men grew "fussy" about their health and followed elaborate rules in order to ward off diseases and keep themselves fit. It is not unreasonable to suppose that its date falls within the first quarter of the fourth century B.C.

The main interest of *Nature of Man* lies in the Empedoclean doctrine contained in the first eight chapters. The four humours are not the four elements of Empedocles, but they are analogous and perform analogous functions. It is their κρᾶσις that produces a healthy body,[1] and the whole argument implies that they are elemental and in themselves unchangeable. There was something vital in the philosophy of Empedocles, and as a basis of physics it reappears, modified but not essentially changed, in Plato's *Timaeus* and in Aristotle's *Physics*. Modern chemistry, with its theory of "elements," is nearer akin to Empedocles than it is to atomism. The number of elements may be four or four hundred—the number

[1] Chapter IV.

is immaterial—but the essential factor, whether it is called μῖξις, κρᾶσις or combination, remains constant. *Nature of Man* is a striking, though minor, instance of perennial vitality in the thought of Empedocles.[1]

Modern scholars have found the treatise more attractive than most of the others in the Hippocratic Collection. Carl Fredrich[2] wrote a doctoral thesis on its composition, and published further research five years later.[3] More recently an excellent edition was published as a doctoral thesis by Villaret.[4] Galen has given us a full and interesting commentary.

The chief manuscripts are A, M and V. The first shows its usual superiority in most cases where the manuscripts differ, but sometimes MV give a preferable reading. In particular, there are several omissions in A almost certainly due to careless copying.

[1] Between Empedocles and *Nature of Man* came Philistion, who probably exerted some influence upon its author. Villaret, p. 66.

[2] *De libro περὶ φύσιος ἀνθρώπου pseudippocrateo* scripsit Carolus Fredrich, Gottingae, 1894.

[3] *Hippokratische Untersuchungen*, Berlin, 1899 (pp. 13 foll.).

[4] *Hippocratis De Natura Hominis* scripsit Oskar Villaret, Gottingae, 1911.

V

HUMOURS

THIS work is perhaps the most puzzling in the Hippocratic Collection. It is obviously a scrap-book of the crudest sort; it has no literary qualities and it is obscure to a degree. Yet in ancient times *Humours* attracted great and continued attention. Apparently Bacchius worked on it, and it was familiar to Glaucias, Zeuxis and Heracleides of Tarentum. There are three Galenic commentaries, which recent German scholarship maintains are a Byzantine compilation containing, however, certain passages from the commentary, now lost, which Galen actually did write.[1] As the genuine Galenic commentary has been replaced by a forgery, one is tempted to suppose that the ancient *Humours* has suffered a similar fate. But there can be no doubt that our *Humours* was the work known to Erotian by that name.[2]

Humours is then ancient, but only a few of the old critics attributed it, or parts of it, to Hippocrates

[1] See Galen, XIX. 35. One of the passages in the extant commentaries supposed to be genuine contains the mention of Zeuxis and Glaucias.

[2] See *e.g.* under *πεπασμός, ἀνασμός, ἀπαρτί, πινώδεσι, αἴρεται* and *φῦσα* in Nachmanson's edition of Erotian.

himself.[1] Its true genesis is a matter of doubt; I have already suggested that it may be a haphazard collection of fragments put together by a careful but uncritical librarian.

The popularity [2] of *Humours* in ancient times may be due in part to its very difficulty; it was, as it were, a challenge to the ingenuity of an ingenious people. A riddle provokes many answers, and *Humours* is a continuous riddle. But it has merits of its own, in addition to the provoking nature of its problems; it is more utilitarian than many of the treatises in the Hippocratic *Corpus.* Prognosis is for once in the background. If we omit those portions that are identical with other passages in the *Corpus,* the remainder are chiefly concerned with the treatment and the prevention of disease. This is a refreshing change from the somewhat arid but otherwise similar propositions in *Aphorisms.* Particularly interesting are the catalogues or lists which appear in Chapters II–V. Are they heads of discourses, lecture-notes made by a professor to facilitate his instruction of a medical class, or are they analyses made by a student attending such a class? The reader inclines to this view or to that according to his mood at the time, but however doubtful their origin, nobody can doubt the value of such lists at a time when pathology had not yet been systematised and treatment was still lacking in breadth and thoroughness. Catalogues, by enumerating the possibilities, widened the outlook of the practitioner and made it less unlikely that favourable opportunities would be overlooked.

[1] See Littré, Vol. I. pp. 369, 370.

[2] Cf. Littré, I. 369: "En lisant ce livre, on s'explique difficilement la faveur dont il a joué dans l'antiquité."

The title of the book is deceptive. After the first sentences there is little mention of the humours; indeed *Nature of Man* is the only Hippocratic work that deserves to be called περὶ χυμῶν. The true nucleus seems to be the catalogues beginning σκεπτέα ταῦτα (Chapter II), and the name was probably taken from the opening sentence and given to the whole scrap-book by some ignorant librarian.

The first edition came out in 1555, and few modern scholars have paid any serious attention to the work. Ermerins leaves whole chapters untranslated, with a brief remark in Latin that they are hopeless. Littré has very little to say about it, and his translation is often both unintelligible and unfaithful.

The chief manuscripts are A and M. I have collated both of these and also the Caius manuscript $\frac{50}{27}$.

VI

APHORISMS

This is the best known work in the whole Hippocratic Collection. From the earliest times it has been regarded with a reverence almost religious. Its authority was unquestioned until the breakdown of the Hippocratic tradition. The Greek manuscripts are more numerous than those containing any other work, while there are translations into Hebrew, Arabic, Syriac and Latin.[1] Editions abound in almost every modern language. "The titles alone," says Adams, "occupy ten pages in the edition of Littré, and still more in that of Kühn." The most lavish praise has been bestowed upon the collection; Suidas says, *ἀνθρωπίνην ὑπερβαίνουσι σύνεσιν*, and as late as the nineteenth century it has been called "the physicians' Bible."

Yet it must be confessed that a modern reader finds *Aphorisms* disappointing; the promise of its dignified opening is scarcely fulfilled. The propositions are not arranged after any definite system, and the seven "sections" into which, since the time

[1] There are 140 Greek MSS., 232 Latin, 70 Arabic, 40 Hebrew and 1 Syriac. Besides Galen, the ancient commentators include Meletius, Stephanus of Athens and Theophilus. See further Pauly-Wissowa, *s.v.* Hippocrates, 16, p. 1845.

of Galen, they have been grouped, are somewhat arbitrary. While containing much accurate and interesting information, *Aphorisms* is not useful enough to account for its astounding popularity. Why did it thrust *Regimen in Acute Diseases* and *Prognostic* into comparative obscurity? It may be urged that these are treatises, text-books in the shape of essays, and therefore not easily committed to memory by dull medical students. In this argument there is much truth; the aphorism is naturally popular with minds of a certain type at a particular stage of their development. We might therefore expect the aphoristic works to find greater favour with students than monographs, but why should *Aphorisms* be so much preferred before *Coan Prenotions* and *Prorrhetic I*, or even before the Cnidian books, with their short and clear rules for diagnosis and treatment? Moreover, for sheer utility the later compilers of medical works, such as Celsus and Aretaeus, might be supposed far superior in meeting the needs of the general practitioner.

The problem must remain somewhat of a puzzle, but a few reasons may be suggested why *Aphorisms* enjoyed so long a vogue. In the first place it carried all the authority of a great name, and until comparatively modern times authority exerted an overwhelming influence in all regions of thought. The tradition is that Hippocrates composed it in his old age as a summary of his vast experience, and there is no reason to doubt that this tradition, with certain reservations, is essentially true. Then again it is a very comprehensive work, dealing with most sides of medical, if not of surgical practice.

The ancient testimony in favour of the Hippocratic

authorship of *Aphorisms* is overwhelming, and points at least to an intimate connection between the collection and Hippocrates himself. Yet very many of the propositions obviously belong to the mass of medical aphorisms traditionally current in the schools of ancient Greece. These have come down to us in a number of collections, including *Coan Prenotions*, *Prorrhetic I*, *Nutriment*, *Dentition* and parts of *Epidemics*. Sometimes the same aphorism appears twice, with slight differences of form. *Aphorisms*, for instance, has 68 propositions found in *Coan Prenotions*.[1] So it may represent a collection of aphorisms made by Hippocrates from the vast number current either in literature or in tradition. Many new ones were probably added from the store of his personal experience, and several seem to be old aphorisms corrected and improved.

The various propositions are grouped according to subject, those, for instance, dealing with fevers being classed together. One proposition is sometimes a natural sequel to another, and so finds its final place.[2] How the groups of propositions are themselves arranged it is difficult to say. An alphabetical arrangement would be ideal for reference, as a book of aphorisms is more akin to a dictionary than to a text-book, but a close inspection fails to detect any such order in *Aphorisms*. Perhaps the writer did not see any reason for arranging the sections in any particular order, and so contented himself with an arrangement of the propositions.

[1] See my *Hippocrates*, II. pp. xx–xxix.

[2] In the case of one aphoristic book, *Dentition*, it can be shown that the order is an alphabetical one, depending on key-words. See Vol II. pp. 318, 319.

A few details may be noticed here. The first aphorism is certainly from the hand of the "great" Hippocrates, and was placed by him in a position of prominence to mark the importance to the physician of the truths that are contained in it. The first section shows a fondness for the adjective σφαλερός, which occurs in I, III (four times), IV, V (twice) Here we have obviously an author's "pet" word, and, occuring where it does, it may be the favourite of Hippocrates himself. Finally, the section on fevers (IV. xxvii–lxxiii) ends with the sentence ἐν πυρετοῖσι δὲ ταῦτα (γίνεται). This appears to apply, not to the proposition in which it occurs (LXXIII), but to the whole section. It means, "These are points to observe in fevers." We seem to have here the compiler's note to mark the end of a section. Again, IV. xiii begins with πρὸς τοὺς ἐλλεβόρους (which seems to be a title), and V. xxi closes with θέρμη δὲ ταῦτα ῥύεται, possibly a misplaced title belonging to the long aphorism that follows. Perhaps most sections were never finished, and so received no note to mark their beginning or their end.

Aphoristic works invite interpellation, and many such additions are suspected in *Aphorisms*. I have generally noted these, and likewise those passages which occur again in other parts of the Hippocratic *Corpus*.

In an earlier volume I have given reasons for supposing that *Aphorisms* was written about 415 B.C.[1]

Ancient commentaries were numerous and careful,[2] the best now extant being those of Galen and Theophilus. The first edition appeared in 1488,

[1] See Vol. II. pp. xxviii and xxix.
[2] See Pauly-Wissowa, VIII. 2, p. 1845.

the last was Beck's German translation published in 1907. F. Adam's second volume contains a good English translation with an excellent commentary, to which I am very much indebted. The texts of Ermerins and Reinhold I refer to under the abbreviations "Erm." and "Rein."

I have myself collated all the chief manuscripts containing *Aphorisms*. They are C′, V, M and Urbinas 64 (referred to in notes as "Urb."). The last is a tenth or eleventh century manuscript in the Vatican, containing, among other things, the text of *Aphorisms* with the commentary of Theophilus. I do not think that its readings have been noted before,[1] and the same applies to much of V and M. Littré relied on C′ and the Paris manuscripts, many of which are so closely related to either V or M that few of the readings of the latter were unknown to Littré, although he could not know their authority.

[1] They bear a strong resemblance to those of Littré's S, and the two manuscripts are probably closely related.

VII

REGIMEN I

THE long work called *Regimen* attracted little attention in early times. Erotian does not mention it, and Galen, though he makes several references[1] to it, is not an enthusiastic admirer. The second book, he says, might reasonably be considered worthy of Hippocrates, but the first is entirely divorced from his way of thinking.[2]

There were apparently two editions, one beginning with Book I and the other with Book II; of the latter, some copies began with Χωρίων δὲ θέσιν and others with Σιτίων δὲ καὶ πομάτων δύναμιν. The first of these editions was called περὶ φύσεως ἀνθρώπου καὶ διαίτης, the second περὶ διαίτης.[3] In Galen's time the whole work was divided up into three parts, as it is in our manuscript θ, the last section (περὶ ἐνυπνίων) having no separate title in that manuscript.

The three (or four) books are evidently closely connected in subject, though a special pleader might argue that they are not all by the same hand. They deal with what the author calls his "discovery"

[1] The chief passages are: V. 881; VI. 455, 473, 496, 541, 543; XV. 455; XVII A. 214; XVIII. A 8. They are discussed by Fredrich, Diels and the writer in Pauly-Wissowa.

[2] VI. 473.

[3] Galen, VI. 473.

(εὕρημα), how, that is, one may learn from symptoms which of the two factors of health, food and exercise, is in excess, and to take precautions against the diseases that may spring from such excess.[1] This thesis is developed in the third book (with *Dreams*), while the second book gives the characteristics of various foods and exercises. The first book, after setting forth the subject that the author intends to treat, goes on to discuss the nature of man and of the universe of which man forms a part. This attempt to explain physiological processes by the principles of philosophic physics explains why scholars have found περὶ διαίτης I interesting in spite of its amazing difficulties.

It has been pointed out already that the difficulty is partly intentional, being due to the fashion of imitating oracular responses. But it is also partly caused by the author's carelessness; the details are sometimes blurred because they are not regarded as essential to the main argument. There is always a danger of over-systematisation in explaining ancient philosophy; the parts do not in every case fit exactly into their places, for a philosopher was sometimes inconsistent with himself. It is a great mistake for an interpreter to insist on making all the detail harmonise exactly. The work may be thus analysed.

The author complains of want of comprehensiveness in the work of his predecessors (Chap. I).

[1] See especially III. 1 (LXVII); ἀλλὰ γὰρ αἱ διαγνώσιες ἔμοιγε ἐξευρημέναι εἰσὶ τῶν ἐπικρατεόντων ἐν τῷ σώματι, ἤν τε οἱ πόνοι ἐπικρατέωσι τῶν σίτων, ἤν τε τὰ σῖτα τῶν πόνων, καὶ ὡς χρὴ ἕκαστα ἐξακεῖσθαι, προκαταλαμβάνειν τε ὑγείην, ὥστε τὰς νούσους μὴ προσπελάζειν κ.τ.ἕ.

Correct dieting presupposes a knowledge of physiology. Health is due to the correct correspondence between food and exercise (Chap. II).

All things are composed of two different but complementary elements, fire and water. The δύναμις of fire is to cause motion, that of water is to nourish.

These elements are continually encroaching one on the other, but neither ever completely masters the other (Chap. III).

These elements are themselves logically capable of analysis into—

(*a*) the hot and dry (fire);
(*b*) the cold and the moist (water).

Fire, however, has some moisture from water, and water some dryness from fire. It is the mingling and separating of these elements that are inaccurately termed birth, death, decay and change (Chap. IV).

All nature is in a state of constant flux; there is a perpetual swinging of the pendulum, and a swaying from one opposite to the other (Chap. V).

Man, both body and soul, consists of fire and water, and there is a give and take in his case also, like "parts" joining like "parts" and rejecting the unlike (Chap. VI).

Diet must contain all the "parts" of man, otherwise there could be no growth. The taking in of nutriment, and the resulting growth and evacuation, are like the up-and-down motions of sawing a log. One implies the other (Chap. VII).

How the elements behave in the processes of generation and growth; there is no real birth and decay, but only increase and diminution (Chap. VIII–X).

The processes of the arts and crafts are copies of those of the universe and of the nature of man, the apparent opposites are merely different aspects of the same thing (Chap. XI–XXIV).

The soul of man, a blend of fire and water, helps to feed the body, and the body helps to feed the soul (Chap. XXV).

The development of the embryo (Chap. XXVI). Males (inclining to fire) and females (inclining to water) generate offspring that are male or female according to the predominance of the male or female element.

(1) Male from man and male from woman: brilliant men.
(2) Male from man mastering female from woman: brave men.
(3) Male from woman mastering female from man: hermaphrodites.
(4) Female from both man and woman: lovely women.
(5) Female from woman mastering male from man: bold but modest women.
(6) Female from man mastering male from woman: brazen women.

The generation of twins (Chap. XXX).

Superfetation (Chap. XXXI).

The various constitutions of man due to the character of the water and fire of which the body is composed. The following combinations are considered :—

(1) finest fire and rarest water;
(2) strongest fire and densest water;
(3) densest water and finest fire;
(4) moistest fire and densest water;
(5) strongest fire and finest water;
(6) rarest fire and driest water (Chap. XXXII).

The composition of the body at the various ages (Chap. XXXIII).

Sex and the composition of the body (Chap. XXXIV).

The intelligence (sensitiveness) of the soul in relation to the blend of fire and water (Chap. XXXV).

What regimen can, or cannot, do to effect a change in the soul (Chap. XXXVI).

The weakness of the writer's thesis is plain to all. He takes an unproved postulate and builds upon it a detailed theory of health and disease—the very fault attacked by the author of *Ancient Medicine.* This defect tends to vitiate the very sensible observations in the second and third books dealing with foods, drinks, exercise and regimen generally. Had the writer confined himself to these, and worked out his scheme without any bias due to the supposed effects of fire and water, he would have achieved a more useful result without in the least weakening his boasted εὕρημα.

It should be noticed, however, that Dr. Peck maintains that the εὕρημα was just this point—the expression of health-factors in their fire-and-water values enables a man accurately to adjust the proportion of food to exercises. But in Chapter II (Book I) and again in Chapters LXVII and LXIX

(Book III) the "discovery" is clearly identified with προδιάγνωσις, how to tell beforehand, by symptoms, whether food or exercise is in excess, and by so doing προκαταλαμβάνειν τὴν ὑγείην. When the details of the discovery are discussed, in Chapters LXX–LXXXV, fire and water come in only in so far as want of exercise is supplemented by warmth, and want of nourishment by a "moist" diet. The mere equation of exercise with fire and of food with water does not, and could not, carry the author very far.

But in spite of this inherent fault the theory is worked out most cleverly. The philosophic position is that of an intelligent and progressive eclectic, who combines, instead of merely adding together, the results reached by his predecessors. The perpetual flux of Heracleitus and his harmony through opposition; the four "opposites" of Empedocles; the brilliant theory of change elaborated by Anaxagoras—all these are worked up into a system that appears like the creation of a single mind. Recent criticism [1] has shown a close resembance between the account of the soul and certain parts of Plato's *Timaeus*.[2] The latter may be from Pythagorean sources, and it is interesting to note that Chapter VIII, and perhaps other places also, shows strong Pythagorean influence.[3] Yet there is no patchwork effect, so skilfully are the parts woven together.

[1] Especially the doctoral thesis of A. L. Peck, not yet published.

[2] See especially 37 B, C; 71 B–79 B: 81 E–86 A (diseases of the body); 86 B–87 B (diseases of the soul depending on bodily condition); 91 A (the seed). Peck notices also a resemblance between the account of generation and that given in the treatise περὶ γονῆς.

[3] *E.g.* the dualism of fire)(water.

The writer's theory becomes a little plainer if we look upon the universe as the mutual and alternating encroachment upon each other of fire and water. These elements (whether they are regarded as limited in amount is not quite clear) contain the four traditional opposites:—

(*a*) fire contains the hot, the dry and the moist;

(*b*) water contains the cold, the moist and the dry. Fire advances, sets water in motion and turns it to steam; then it retires and the steam condenses to water. But there are limits to this advance and retirement; the water is never completely "mastered," nor is the fire ever completely quenched. The various things of this world, including animals, are all the result of this alternate swaying, and represent, so to say, various stages in a never-ending process. The writer gives a few details, but hastens on to the application of this general theory to living bodies. Both body and soul contain fire and water, but presumably soul is the more "fiery" of the two. The fire is regarded as the cause of the circulation of food, which enters the body, causes growth, and then is (at least partly) evacuated. Here "give and take" is continually and clearly illustrated. If it were not for the entering in of certain nourishment and the going out of *excreta* and *secreta*, the animal would die. Any abnormality, any grit in the machinery, any disproportion between the incomings and the outgoings, results in disease. Life, in fact, is identified with change, and change with biological, organised growth, as distinct from mere quantitative increase or decrease.

So far the picture is fairly clear, but when the writer proceeds to explain growth he becomes obscure. It is obviously not quantitative increase only, as in Chapter VII growth is said to imply the existence of all the "parts" of the body in the foods that nourish it. What are these parts? Are they the blood, flesh and marrow, etc. of Anaxagoras? If so, how do they become fixtures, what differentiates the proportion of fire and water which makes up blood from the same proportion before it is blood? What is it, in fact, that makes blood "breed true," and have a permanent existence as a specific substance? In general terms, what is it that causes specific differences, separating for ever blood from marrow, horse from man, and rose from daisy? No clear answer is given, but in Chapter VIII it seems to be implied that it is all a matter of "attunement."[1] Water and fire, if they attain one attunement, become one thing, if another attunement, another thing. As a modern chemist might say, one attunement of oxygen and hydrogen produces water, another attunement hydrogen peroxide. Exact proportions in favourable conditions produce, not mechanical mixture, but chemical change.[2]

The name of the author will probably never be known to us. Even in Galen's time there was no manner of agreement among students. Some indeed attributed περὶ διαίτης to Hippocrates himself; others, however, considered the writer to be Philistion, or

[1] This doctrine of attunement (ἁρμονίη) was Pythagorean in origin, but was developed by Heracleitus, who made it one of the pillars of his system.

[2] Dr. Peck thinks that the crucial passage is the first part of Chapter VI, where ὅλα ὅλων may refer to the chemical attunements (if I may so call them) that differentiate species from species, and μέρεα μερέων to those that differentiate one "part" of the body from another.

Ariston, or Euryphon, or Phaon, or Philetas.[1] Modern scholars are equally uncertain in their opinions. Littré would attribute it to Hippocrates himself, were it not for the weight of ancient authority against that view. Fredrich assigns it to a "Compilator" who lived at the end of the fifth century B.C.[2] Peck does not assign the treatise to any particular author, but sees close affinities to Philistion, Diocles, Plato and the author of περὶ γονῆς. Teichmüller would assign the work to the period between Heracleitus and Anaxagoras, Zeller to the period between 420 and 380 B.C.

One may be fairly certain that the date of composition is not far from 400 B.C.—all the lines of evidence point to that date—but the author cannot be identified with any certainty or even probability. He must, however, have belonged to that school of "health-faddists" of whom Plato[3] speaks in such disparaging terms. Perhaps the work owes to Herodicus of Selymbria "who killed fever-patients by excessive exercise,"[4] more than is yet generally conceded.[5]

[1] See Galen, VI. 473, and XV. 455. In XVIII. A 9 Pherecydes is mentioned as one to whom the work was sometimes ascribed.

[2] See *Hippokratische Untersuchungen*, p. 223: "Der Verfasser hat zweifellos nach Heraclit, nach Anaxagoras gelebt und ist ein—vielleicht etwas jüngerer—Zeitgenosse eines Archelaos, Kratylos und Herodikos von Selymbria. Das weist auf das Ende des fünften Jahrhunderts."

[3] See *Republic*, 406 B–D.

[4] *Epidemics*, VI. (Littré, V. 302).

[5] But see Fredrich, *op. cit.*, pp. 217–221. I may add that it is somewhat difficult to decide whether the author was a practising physician or not. No passages can be quoted that are really conclusive, but the general conclusion suggested by Books II and III is that the author was a "health expert," and not a professional doctor.

One more point remains to be noticed. The great importance attached to regimen in this treatise is characteristic of all that is best in Greek medicine. Upon it the physician relied, both to preserve health and to heal diseases. Drugs, of which he had only a few, and these chiefly purges, were regarded as of secondary importance only. "Live a healthy life," said the Greek doctor, "and you are not likely to fall ill, unless you have an accident or an epidemic occurs. If you do fall ill, proper regimen will give you the best chance of recovery." It is not surprising that *Regimen* has close affinities to other works in the *Corpus*, notably *Ancient Medicine, Regimen in Acute Diseases* and *Regimen in Health.* However much they may differ in scope and detail, all these works are written under the conviction that medicine is merely a branch of dietetics.

The first book of *Regimen* has attracted many modern scholars. Bywater included Chapters I–XXIV in his *Heracliti Ephesii reliquiae.*[1] Carl Fredrich has fully discussed the work, in many places reconstructing the text, in his *Hippokratische Untersuchungen.*[2] H. Diels has published two interesting papers in *Hermes,*[3] and a great part of the text appeared in his *Herakleitos von Ephesos.*[4] Several other less important contributions are mentioned in the article *Hippokrates* (16) in Pauly-Wissowa.[5] But

[1] Oxford, 1877.

[2] Pp. 81–230.

[3] *Hippokratische Forschungen* I in Band 45, pp. 125–150, and *Hippokratische Forschungen* II and III in Band 46, pp. 267–285.

[4] Berlin, 1909.

[5] *E.g.* Feuchtersleben, Bernays, Schuster, Teichmüller, Zeller and Gomperz.

all these are superseded by a masterly discussion of the whole of the first book, in its relation to Greek philosophic thought, submitted by Dr. A. L. Peck in 1928 for the degree of Ph.D. This work is not yet published, and I must express my gratitude to Dr. Peck for allowing me to read it at my leisure, and for discussing with me difficult points of interpretation.

The chief manuscripts are θ and M, both of which have been specially collated for this edition of the text. There is an old Latin translation, Paris. lat. 7027, which may have been made in the sixth century, although the manuscript itself is of the tenth century.[1] A very interesting manuscript, which unfortunately I have been unable to collate, is the manuscript referred to by Littré as K′. It almost certainly gives the right reading in Chapter XXXVI, where both θ and M go astray.

[1] See Diels, *Hipp. Forschungen*, 1, p. 137. Readings from this manuscript are occasionally given by Littré.

VIII

REGIMEN II—IV

THE last three books of *Regimen* leave the translator very uneasy. It is not that they are full of mysterious puzzles, as are *Regimen I*, *Precepts* and *Decorum*. These stare one in the face, and cannot be overlooked; but the greater part of *Regimen* is full of concealed traps, into which even an experienced translator may fall unawares. The Greek is somewhat curious, and a temptation exists to apply the strict rules of criticism and interpretation that are applied to Plato and Demosthenes. The result is often to force on the original a meaning that makes indifferent sense. Again, the writer is fond of using common words in a semi-technical sense, difficult to apprehend. Even after a study of Dr. A. L. Peck's *Pseudo-Hippocrates Philosophus* one is in great doubt as to the meaning, in *Regimen*, of δύναμις, περίοδος, ἀπόκρισις, and many other words. Synonyms present an equal difficulty. There may be, for instance, a subtle danger in translating both γυμνάσια and πόνοι by "exercises"; but it is just as dangerous to discriminate between them by rendering the former "gymnastics," while to suggest in an English translation the right amount of effort or fatigue implied in πόνος is past the ability of the present translator, at any rate. Even a casual reader will be worried by the author's use of προσάγω, one of his favourite words. Does it always imply, as Littré and Ermerins indicate by their translations, a gradual increase? Such a progression is certainly signified by the phrase ἐκ προσαγωγῆς, but I have felt most disquieting doubts when so rendering the simple verb without the addition of κατὰ μικρὸν or ἐξ ὀλίγου. A similar uncertainty perplexes the mind when our best manuscript presents a reading at variance with the received canons of Greek grammar or of Greek idiom. In the case of a second-rate writer, not over-careful in style, which of the two is to be preferred: (1) a slipshod expression in a very faithful manuscript or (2) a more elegant and accurate expression in manuscripts presenting every appearance of having been emended by zealous editors or scribes? Each case has to be decided on its merits, and into every decision enters a disquieting amount of guess-work.

It is pleasant to turn from these troublesome, if minor, details to the general purpose of the work, which is a justification and exposition of προδιάγνωσις, "the nipping of a disease in the bud." Plato, indeed, attacks with justice the hypochondriacism that turns life into a lingering death, but nothing but praise is due to the man who first conceived the idea of anticipating disease, of meeting it half-way, and of attempting to check it before it can get a fatal hold. The author, in fact, was the father of preventive medicine; ἀλλὰ χρὴ προθυμεῖσθαι is his oft-repeated slogan. His merit is all the greater when we remember that the most famous Hippocratic works know nothing of προδιάγνωσις but only of προγνωσις. They take a fatalistic view, and assume that every disease must take its course. The author of *Regimen* says: "No, the course can be cut short, and the severity of the disease mitigated." In this view there is a large element of truth. By taking care in good time many a patient suffering from a cold has prevented a fatal bronchitis or pneumonia; many a "weak-chested" person has by similar precautions kept away consumption. On the other hand, some diseases must run their course with but slight, if any, modification. Typhoid and measles, for example, can never be completely aborted when once they have been definitely introduced into the human system. It is more than doubtful whether the author of *Regimen*, or any Greek of the classical period, knew the ordinary zymotic diseases, but had he done so he could never have understood (nobody could understand before Pasteur) why the εὕρημα was inapplicable to at least one large class of maladies.

In fact προδιάγνωσις, while marking an advance, does not go far enough. To abort a disease is good; to prevent it altogether is far better. The Greek had experience enough to outline a course of regimen designed to preserve in ordinary circumstances a fair standard of health, but he had not the experience required to prevent an outbreak of epidemic disease.

It would be beyond the scope of the present edition to discuss in detail the qualities assigned in the second book to foods, drinks, exercises and so forth, or to appreciate the value of the prescriptions in the third book for undoing the mischief caused by excess of food or by excess of exercise. A lengthy volume would be required to do even moderate justice to these questions, and even a full discussion could lead only to the unsatisfactory conclusion that the author has twisted facts wholesale to make them square with his theory. The same remarks reply to the fourth book, sometimes called *Dreams*. One or two details, however, call for a passing notice because of their intrinsic interest.

Dreams contains the first occurrence in classical literature—at least I can discover no earlier one—of a supposed connection between the heavenly bodies and the fates of individual human lives. The connection, indeed, is not clearly defined; we are not told that these bodies actually interfere with the course of events. But it is definitely stated that to dream about them, at any rate to see certain dreams in which they behave in certain ways, means health or a risk of illness. To a modern it is indeed strange that dreams of this sort occupy so large a portion of the book. But a modern, unless he be an astronomer, knows or cares little about the stars. Clocks and watches, the compass, calendars and almanacs have made star-lore quite unnecessary for most people. But the ancients were forced, by the very exigencies of existence, to contemplate the heavens carefully and continuously. The sun, moon and stars entered largely into their conscious and subconscious life, and we need not be surprised that celestial phenomena figured largely in their dreams.

Regimen is the only book in the *Hippocratic Corpus* that lays any emphasis on prayer to the gods. There is, indeed, a passing reference in *Prognostic*, Chapter I, deleted by modern editors, to the possibility of there being *τι θεῖον* in certain cases of illness, while Chapter VI of *Decorum* appears to regard the gods as the cause of cures in medicine and surgery, the doctor being only the means. But to the author of *Regimen* prayer seems to be an integral part of many prescriptions.[1]

A mention should be made of the importance attached to walking as a means of attaining health or of preserving it. Even after violent exercise a walk is prescribed in many cases, possibly or perhaps probably to avoid stiffness and to allow the body to cool slowly. Early-morning walks, and walks after dinner, are recommended constantly. It is unnecessary to point out how wise this advice is, and how well it agrees with the best modern methods of training.

There are many features of *Regimen* that strike us as strangely modern. Unconsciously we are in the habit of putting massage among newly-discovered methods of therapeutics. Yet τρῖψις in the fifth century before Christ was both popular and long-established. What can φωνῆς πόνοι represent except breathing exercises and the like? And even modern hydropathy must confess that the Russian bath has a very near relative in the πυρία. The ἀνακούφισμα was certainly not "relief" (new Liddell and Scott), but a raising of the body from the prone position by using the arms, a well-known form of exercise.

I have not tried to distinguish between σιτία and σῖτα; indeed M regularly prefers the former word and θ the latter. While translating both by "food" I am aware that farinaceous foods are usually meant. Similarly I have rendered ὄψα by "meats," although

[1] See *e.g.* Chapters LXXXVIII and XC. It is interesting to note that a reader (possibly a reviser or even the original scribe) of the MS. θ tried to erase the names of heathen deities.

fish is included under the term. Any attempt to be pedantically accurate, besides being awkward, results in more confusion being introduced than that which is removed.

One or two technical terms of the gymnasium present special difficulties. Few expressions in *Regimen* are more common than τρόχος (or τροχός, as it is spelt in our manuscripts) and καμπτοὶ δρόμοι. Yet our dictionaries and books of reference either neglect them or describe them in a most uncertain way. Sometimes τρόχος is assumed to be a mere equivalent of δρόμος, a view perhaps derived from such passages as Euripides *Medea* 46 and *Hippolytus* 1133; the old translators, followed by Littré and Ermerins, make out the τρόχος to be a round track.

The καμπτὸς δρόμος is even more perplexing. It is obviously a "bent" track; but what was the nature of the bend? Was it a zig-zag? Or was it a turning, as the name suggests, round the καμπτήρ to the starting-point? Was the καμπτὸς δρόμος, in other words, the generic word for a type of track of which the δίαυλος was a specific instance? Whichever answer we see fit to give, the puzzle remains that the Greeks placed καμπτοὶ δρόμοι in one class and the straight course in another, although why a straight quarter of a mile should differ essentially from two hundred and twenty yards there and then back is indeed a curious enigma.

Regimen contains many passages in which occurs the same difficulty as that which is to be found so often in *Epidemics I* and *III*. Do the plurals of πολύς and ὀλίγος refer to size or frequency? Does περίπατοι πολλοὶ mean "many walks" or "long walks"? The same answer, it seems to me, should be given as I gave in the *General Introduction* to Vol. I, p. lxi. In the great majority of cases size, not frequency, is referred to, and, unless the general sense is against this interpretation, πολλοὶ and ὀλίγοι should be translated by "long" and "short."

IX

THE MANUSCRIPTS AND DIALECT OF THE HIPPOCRATIC COLLECTION

A CAREFUL reader will observe that whereas I have not materially changed my opinion of the relative value of our manuscripts A, θ, C′ are our primary authorities – I am somewhat dubious about the rules for spelling given by Kühlwein in the *Prolegomena* to the Teubner edition of Hippocrates, Vol. I, pp. lxvi–cxxviii. In my first volume of the Loeb series I accepted without question the following principles for determining the orthography of the Hippocratic *Corpus*:—

(1) That the pronominal forms in ὁκ- should be avoided;
(2) ϵ + ϵ contract, but not ϵ + ο;
(3) γίνεσθαι not γίγνεσθαι;
(4) various rules for ν ἐφελκυστικόν;
(5) the pseudo-ionisms αὐτέῳ, etc., are to be avoided;
(6) σύν not ξύν.

A prolonged study of the manuscripts has made me feel very doubtful about some of these principles, and my doubts appear to be shared by I. L. Heiberg, who edited the first volume of Hippocrates in the *Corpus Medicorum Graecorum*. Heiberg indeed does not follow strictly any of these rules; my own view is that two are correct and the others more or less uncertain. The pseudo-ionisms have very little authority, nor has γίγνεσθαι. The form ξύν is very doubtful, and I have printed in every case σύν, without, however, being confident that no Hippocratic writer ever wrote ξύν. The case is much the same with ε + ε, which I always contract, and with ε + ο, which I rarely contract to ευ. For the pronominal forms I follow usually the best MS. authority in each case. There is a tendency for our earliest manuscripts not to use the ὀκ- forms, but it is only a tendency, and ought not, I think, to be narrowed to a rigid rule. As for ν ἐφελκυστικόν, Kühlewein's "rules" are so complicated that they can scarcely have been followed by the not over-careful writers whose works are contained in the *Corpus*.

I believe, in short, that those scholars are mistaken who attribute strict uniformity to the authors, and indiscriminate carelessness to the scribes and copyists. It is very hard to be convinced that all the writers, of various degrees of ability, and living at various times and (apparently) at various places, were perfectly at home in a dialect obviously artificial, kept up simply out of respect for tradition. Surely a more probable supposition is that our manuscripts exhibit a slight but varying carelessness on the part of the writers, made even more confusing by greater carelessness on the part of many generations of scribes. In brief, we cannot determine exactly the Ionic of the Hippocratic collection; the most we can do is to observe tendencies.

The conviction that I expressed in the preceding volumes, that at some period or periods the manuscripts were copied with but slight regard for verbal accuracy, has grown stronger with prolonged study. In my critical notes I have quoted in full the readings of our chief manuscripts in places which put, I think, my contention beyond all reasonable doubt.

HIPPOCRATES

NATURE OF MAN

ΠΕΡΙ ΦΥΣΙΟΣ ΑΝΘΡΩΠΟΥ

I. Ὅστις μὲν οὖν εἴωθεν ἀκούειν λεγόντων ἀμφὶ τῆς φύσιος τῆς ἀνθρωπείης[1] προσωτέρω ἢ ὅσον αὐτῆς[2] ἐς ἰητρικὴν ἀφήκει,[3] τούτῳ μὲν οὐκ ἐπιτήδειος ὅδε ὁ λόγος ἀκούειν· οὔτε γὰρ τὸ πάμπαν ἠέρα λέγω τὸν ἄνθρωπον εἶναι, οὔτε πῦρ, οὔτε ὕδωρ, οὔτε γῆν, οὔτ' ἄλλο οὐδὲν ὅ τι μὴ φανερόν ἐστιν ἐνεὸν[4] ἐν τῷ ἀνθρώπῳ· ἀλλὰ τοῖσι βουλομένοισι ταῦτα λέγειν παρίημι. δοκέουσι μέντοι[5] μοι οὐκ ὀρθῶς γινώσκειν οἱ 10 ταῦτα[6] λέγοντες· γνώμῃ μὲν γὰρ τῇ αὐτῇ πάντες χρέονται, λέγουσι δὲ οὐ ταὐτά· ἀλλὰ τῆς μὲν γνώμης τὸν ἐπίλογον τὸν αὐτὸν ποιέονται[7] (φασί τε[8] γὰρ ἕν τι[9] εἶναι, ὅ τι ἔστι, καὶ τοῦτο εἶναι τὸ ἕν τε καὶ[10] τὸ πᾶν) κατὰ δὲ τὰ ὀνόματα οὐχ ὁμολογέουσιν· λέγει δ' αὐτῶν ὁ μέν τις φάσκων ἠέρα τοῦτο εἶναι τὸ ἕν τε καὶ τὸ πᾶν, ὁ δὲ πῦρ, ὁ δὲ ὕδωρ,[11] ὁ δὲ γῆν, καὶ ἐπιλέγει ἕκαστος τῷ ἑωυτοῦ λόγῳ μαρτύριά τε καὶ τεκμήρια, ἅ ἐστιν οὐδέν. ὁπότε δὲ γνώμῃ τῇ αὐτῇ[12] προσ-20 χρέονται, λέγουσι δ' οὐ τὰ αὐτά, δῆλον ὅτι οὐδὲ

[1] ἀνθρωπείης A: ἀνθρωπίνης MV.
[2] αὐτῆς A: αὐτέης M: αὐτέη V.
[3] ἀφήκει A: ἀφίκει MV: ἐφήκει Littré.
[4] ἐνεὸν AV: ἓν ἐὸν M. Galen mentions both readings and prefers ἓν ἐόν.
[5] μέντοι A: δὲ MV.

NATURE OF MAN

I. He who is accustomed to hear speakers discuss the nature of man beyond its relations to medicine will not find the present account of any interest. For I do not say at all that a man is air, or fire, or water, or earth, or anything else that is not an obvious constituent of a man; such accounts I leave to those that care to give them. Those, however, who give them have not in my opinion correct knowledge. For while adopting the same idea they do not give the same account. Though they add the same appendix to their idea—saying that "what is" is a unity, and that this is both unity and the all—yet they are not agreed as to its name. One of them asserts that this one and the all is air, another calls it fire, another, water, and another, earth; while each appends to his own account evidence and proofs that amount to nothing. The fact that, while adopting the same idea, they do not give the same account, shows that their knowledge

---

[6] ταῦτα A: τὰ τοιαῦτα MV.
[7] προΐενται A: ποιέονται MV: ποιεῦνται Villaret.
[8] τε A: MV omit. [9] τι MV: τε A.
[10] A omits τὸ ἓν τε καί.
[11] ὁ δὲ ὕδωρ· ὁ δὲ πῦρ A: ὁ δὲ πῦρ· ὁ δὲ ὕδωρ MV.
[12] ὁπότε δὲ γνώμῃ τῇ αὐτῇ A: ὅτι μὲν γὰρ τῆι αὐτεῆι γνώμῆι πάντες M: ὅτι μὲν γὰρ τῇ αὐτέῃ γνώμῃ πάντες V.

γινώσκουσιν αὐτά.[1] γνοίη δ' ἂν τόδε τις[2] μάλιστα παραγενόμενος αὐτοῖσιν ἀντιλέγουσιν· πρὸς γὰρ ἀλλήλους ἀντιλέγοντες οἱ αὐτοὶ ἄνδρες[3] τῶν αὐτῶν ἐναντίον[4] ἀκροατέων οὐδέποτε τρὶς[5] ἐφεξῆς ὁ αὐτὸς περιγίνεται ἐν τῷ λόγῳ, ἀλλὰ ποτὲ μὲν οὗτος ἐπικρατεῖ, ποτὲ δὲ οὗτος, ποτὲ δὲ[6] ᾧ ἂν τύχῃ μάλιστα ἡ γλῶσσα ἐπιρρυεῖσα[7] πρὸς τὸν ὄχλον. καίτοι[8] δίκαιόν ἐστι τὸν φάντα[9] ὀρθῶς γινώσκειν ἀμφὶ τῶν πρηγμάτων παρέχειν 30 αἰεὶ ἐπικρατέοντα τὸν λόγον τὸν ἑωυτοῦ, εἴπερ ἐόντα γινώσκει καὶ ὀρθῶς ἀποφαίνεται. ἀλλ' ἐμοί γε δοκέουσιν οἱ τοιοῦτοι ἄνθρωποι αὐτοὶ ἑωυτοὺς[10] καταβάλλειν ἐν τοῖσιν ὀνόμασι τῶν λόγων αὐτῶν ὑπὸ ἀσυνεσίης, τὸν δὲ Μελίσσου 35 λόγον ὀρθοῦν.

II. Περὶ μὲν οὖν τούτων ἀρκεῖ μοι τὰ εἰρημένα. τῶν δὲ ἰητρῶν οἱ μέν τινες λέγουσιν ὡς ὤνθρωπος αἷμά[11] ἐστιν, οἱ δ' αὐτῶν χολήν φασιν εἶναι τὸν ἄνθρωπον, ἔνιοι δέ τινες φλέγμα· ἐπίλογον δὲ ποιέονται καὶ οὗτοι[12] πάντες τὸν αὐτόν· ἓν γὰρ[13] εἶναί φασιν, ὅ τι ἕκαστος αὐτῶν βούλεται ὀνο-

[1] A omits αὐτά. Wilamowitz and Villaret read οὐδὲν for οὐδὲ and omit αὐτά.

[2] τῶδε (τόδε in another hand) τίς A: τῶιδέ τις M: τόδέ τις V: τις τόδε Littré, with one MS.

[3] ἄνδρες A: ἄνθρωποι MV.

[4] A correcting hand in A has written ω over the ο of ἐναντίον.

[5] Littré says that a later hand in A has emended τρεῖς to τρίς. The rotograph shows τρεῖς. Both M and V have τρεῖς.

[6] A reads τότε μὲν . . . τότε δὲ . . . τότε δέ.

[7] ἐπιρρυεῖσα A: ῥυεῖσα MV.

[8] καίτοι A: καὶ τὸ MV.

too is at fault. The best way to realise this is to be present at their debates. Given the same debaters and the same audience, the same man never wins in the discussion three times in succession, but now one is victor, now another, now he who happens to have the most glib tongue in the face of the crowd. Yet it is right that a man who claims correct knowledge about the facts should maintain his own argument victorious always, if his knowledge be knowledge of reality and if he set it forth correctly. But in my opinion such men by their lack of understanding overthrow themselves in the words of their very discussions, and establish the theory of Melissus.[1]

II. Now about these men I have said enough, and I will turn to physicians. Some of them say that a man is blood, others that he is bile, a few that he is phlegm. Physicians, like the metaphysicians, all add the same appendix. For they say that a man is a unity, giving it the name that severally they

[1] A philosopher of the Eleatic School, who appears to have flourished about 440 B.C. He maintained that Being is eternal, infinite, invariable and a unity. The disputants referred to in the text "established the theory of Melissus" by showing how many difficulties are involved in equating Being with any one of the four elements.

Diels' conjecture would give the meaning "by words opposed to their thesis itself."

---

[9] *τὸν φύσαντα* (altered to *φήσαντα*) A : *τὸν φάντα* M : *τὸ φάντα* V.

[10] *αὐτοὶ ἑωυτοὺς* A : *σφᾶς αὐτοὺς* MV. Diels conjectures (for *ἐν τοῖσιν . . . αὐτῶν*) *ἀντίοισιν ὀνόμασι τῷ λόγῳ αὐτῷ.*

[11] After *αἷμα* V has *μοῦνον.* So M (in margin).

[12] *οὗτοι* A : *αὐτοὶ* MV.

[13] MV have *ἓν γάρ τι.*

μάσας,[1] καὶ τοῦτο[2] μεταλλάσσειν τὴν ἰδέην καὶ τὴν δύναμιν, ἀναγκαζόμενον ὑπό τε τοῦ θερμοῦ καὶ τοῦ ψυχροῦ, καὶ γίνεσθαι[3] γλυκὺ καὶ πικρὸν 10 καὶ λευκὸν καὶ μέλαν καὶ παντοῖον. ἐμοὶ δὲ οὐδὲ ταῦτα δοκεῖ ὧδε ἔχειν.[4] οἱ οὖν[5] πλεῖστοι τοιαῦτά τινα καὶ[6] ἐγγύτατα τούτων ἀποφαίνονται. ἐγὼ δέ φημι, εἰ ἓν ἦν ὥνθρωπος, οὐδέποτ' ἂν ἤλγεεν· οὐδὲ γὰρ ἂν ἦν[7] ὑφ' ὅτου[8] ἀλγήσειεν ἓν ἐών·[9] εἰ δ' οὖν καὶ ἀλγήσειεν, ἀνάγκη καὶ τὸ ἰώμενον ἓν εἶναι· νῦν δὲ πολλά· πολλὰ γάρ ἐστιν ἐν τῷ σώματι ἐνεόντα, ἅ, ὅταν ὑπ' ἀλλήλων παρὰ φύσιν θερμαίνηταί τε καὶ ψύχηται, καὶ ξηραίνηται καὶ ὑγραίνηται, νούσους 20 τίκτει· ὥστε πολλαὶ μὲν ἰδέαι τῶν νοσημάτων, πολλὴ δὲ καὶ ἡ ἴησις ἐστίν. ἀξιῶ δὲ ἔγωγε τὸν φάσκοντα αἷμα εἶναι μοῦνον τὸν ἄνθρωπον, καὶ ἄλλο μηδέν, δεικνύειν αὐτὸν μὴ μεταλλάσσοντα τὴν ἰδέην μηδὲ[10] γίνεσθαι παντοῖον, ἀλλ' ἢ ὥρην τινὰ τοῦ ἐνιαυτοῦ ἢ τῆς ἡλικίης τῆς τοῦ ἀνθρώπου, ἐν ᾗ αἷμα ἐνεὸν φαίνεται μοῦνον ἐν τῷ ἀνθρώπῳ· εἰκὸς γὰρ εἶναι μίαν τινὰ ὥρην,

[1] αὐτῶν βούλεται ὀνομάσας A: ἠθέλησεν ὀνομάσαι αὐτέων M: ονομάσαι ἠθέλησεν αὐτέων V.

[2] After τοῦτο MV have ἓν ἐόν.

[3] After γίνεσθαι MV have καί.

[4] ἐμοὶ δὲ οὐδέν τι (altered to τοι by another hand) δοκέει ταῦτα οὕτως ἔχειν A: ἐμοὶ δ' οὐδὲ ταῦτα δοκέει ὧδε ἔχειν M: ἐμοὶ δ οὐ δοκέει ταῦτα ὧδε ἔχειν V.

[5] οἱ οὖν A: οἱ μὲν οὖν MV.

[6] After καὶ MV have ἔτι. Ermerins reads ἢ ὅτι, perhaps rightly.

[7] ἦν ἂν A: ἂν ἦν MV.

[8] ὑφ' οὗ A: ὑπὸ τοῦ MV: ὑφ' ὅτου Littré after Galen.

[9] ἐόν AMV: ἐών Littré with one MS.

wish to give it; this changes its form and its power,[1] being constrained by the hot and the cold, and becomes sweet, bitter, white, black and so on. But in my opinion these views also are incorrect. Most physicians then maintain views like these, if not identical with them; but I hold that if man were a unity he would never feel pain, as there would be nothing from which a unity could suffer pain. And even if he were to suffer, the cure too would have to be one. But as a matter of fact cures are many. For in the body are many constituents, which, by heating, by cooling, by drying or by wetting one another contrary to nature, engender diseases; so that both the forms [2] of diseases are many and the healing of them is manifold. But I require of him who asserts that man is blood and nothing else, to point out a man when he does not change his form or assume every quality, and to point out a time, a season of the year or a season of human life, in which obviously blood is the only constituent of man. For it is only natural that there should be

[1] By "power" (δύναμις) is probably meant the sum total of a thing's characteristics or qualities. See Vol. I. pp. 338, 339. Recent research, however, makes it likely that in the medical writers δύναμις is often used with ἰδέη or φύσις to form a tautological phrase meaning "real essence."

[2] A. E. Taylor (*Varia Socratica*, p. 229) thinks that this phrase must mean "there are many substances in which disease arises," *i.e.* disease is not necessarily "diseased state of the *blood*."

---

[10] A has αἷμα μόνον εἶναι τὸν ἄνθρωπον καὶ ἄλλο μηδὲν εἶναι δεικνύειν αὐτὸν μήτε ἀλάσσοντα τὴν ἰδέην μήτε. Ermerins reads αὐτὸ (sc. τὸ αἷμα) μὴ μεταλλάσσειν. Villaret has μήτε μεταλλάσσοντα . . . μήτε γινόμενον, probably rightly.

ἐν ᾗ φαίνεται αὐτὸ ἐφ' ἑαυτοῦ ἐνεόν·[1] τὰ αὐτὰ δὲ λέγω καὶ περὶ τοῦ φάσκοντος φλέγμα[2] εἶναι τὸν
30 ἄνθρωπον, καὶ περὶ τοῦ χολὴν φάσκοντος εἶναι. ἐγὼ μὲν γὰρ ἀποδείξω, ἃ ἂν φήσω τὸν ἄνθρωπον εἶναι, καὶ κατὰ τὸν[3] νόμον καὶ κατὰ τὴν[3] φύσιν, ἀεὶ τὰ αὐτὰ ἐόντα ὁμοίως,[4] καὶ νέου ἐόντος καὶ γέροντος, καὶ τῆς ὥρης ψυχρῆς ἐούσης καὶ θερμῆς, καὶ τεκμήρια παρέξω, καὶ ἀνάγκας ἀποφανῶ, δι' ἃς ἕκαστον αὔξεταί τε καὶ φθίνει
37 ἐν τῷ σώματι.

III. Πρῶτον μὲν οὖν ἀνάγκη τὴν γένεσιν γίνεσθαι μὴ ἀφ' ἑνός· πῶς γὰρ ἂν ἕν γ' ἐόν τι γεννήσειεν, εἰ μή τινι μιχθείη; ἔπειτα οὐδ', ἐὰν[5] μὴ ὁμόφυλα ἐόντα μίσγηται καὶ τὴν αὐτὴν ἔχοντα δύναμιν, γεννᾷ,[6] οὐδ' ἂν ταῦτα ἡμῖν συντελέοιτο. καὶ πάλιν, εἰ μὴ τὸ θερμὸν τῷ ψυχρῷ καὶ τὸ ξηρὸν τῷ ὑγρῷ μετρίως πρὸς

[1] εἰκὸς γὰρ ἔς τινα (corrected to ἔστιν τινὰ) ὥρην ἐν ᾗ φαίνηται· αὐτῷ ἐν ἑαυτῷ ἐὸν ὁ ἐστὶν A (with εἰκὸς γὰρ εἶναι ὥρην in margin). εἰκὸς γὰρ εἶναι μίαν τινὰ ὥρην ἐν ἧι φαίνεται αὐτὸ ἐν ἑωυτῶι ἐνεόν, followed by ὅ τι ἐστὶν erased, M, which has μίαν also written over an erasure. V agrees with M, except that it has ἐὸν for ἐνεὸν without ὅτι ἐστίν. Littré with Galen would read μίαν γέ τινα and with Foes ἐφ' ἑωυτοῦ. Villaret reads ἐφ' ἑωυτοῦ ἐόν, ὅ ἐστιν.

[2] After φλέγμα A has μόνον written underneath the line.

[3] Villaret brackets τὸν and τήν. So Van der Linden and Fredrich.

[4] τὰ αὐτὰ ὅμοια ἐόντα A: εἰ ταῦτα ἐόντα ὅμοια MV. The text is Littré's, who follows certain later MSS. in reading ἀεὶ and ὁμοίως.

[5] εἴ ποῦ δ' ἐὰν (with ὅ over εἴ) A: ἔπειτα οὐδὲ ἂν MV: ἔπειτα οὐδ' ἐὰν Littré: ἐπεὶ οὐδ' ἐὰν Wilamowitz.

[6] A has γενναι with αν written over αι. It also omits ταῦτα, for which Galen reads τὰ αὐτά. I give Littré's text, but I suggest that the true reading is ὅπου δ' ἂν μὴ ὁμόφυλα

one season in which blood-in-itself appears as the sole constituent.[1] My remarks apply also to him who says that man is only phlegm, and to him who says that man is bile. I for my part will prove that what I declare to be the constituents of a man are, according to both convention and nature,[2] always alike the same; it makes no difference whether the man be young or old, or whether the season be cold or hot. I will also bring evidence, and set forth the necessary causes why each constituent grows or decreases in the body.

III. Now in the first place generation cannot take place from a unity. How could a unity generate, without copulating? Again, there is no generation unless the copulating partners be of the same kind, and possess the same qualities; nor would there be any offspring.[3] Moreover, generation will not take place if the combination of hot with cold and of dry

[1] Probably Villaret's reading is correct, and we should translate, "in which the real element appears in its proper form."

[2] This strange phrase apparently means "in name as well as in essence," or rather "as much in essence as they are in name." People agree in giving certain names to the constituents of the human body. These names correspond to real entities. Galen explains κατὰ νόμον to mean "according to received opinion."

[3] The translation of the emendation which I propose will be: "And when the copulating partners are not of the same kind, and do not possess the same generating qualities, we shall get no result."

ἐόντα μίσγηται καὶ τὴν αὐτὴν ἔχοντα δύναμιν γεννᾶν, οὐδὲν ἂν ἡμῖν συντελέοιτο. Ermerins would read ἔπειτα δέ, ἐὰν . . . δύναμιν, γέννα οὐδ' ἂν οὕτω ἡμῖν ξυντελέοιτο. Villaret has εἶτ' οὐδ' ἐὰν . . . γέννα οὐδ' ἂν μία συντελέοιτο.

ἄλληλα ἕξει καὶ ἴσως, ἀλλὰ θάτερον θατέρου πολὺ προέξει καὶ τὸ ἰσχυρότερον[1] τοῦ ἀσθε-
10 νεστέρου, ἡ γένεσις οὐκ ἂν γένοιτο. ὥστε πῶς εἰκὸς ἀπὸ ἑνός τι γεννηθῆναι, ὅτε οὐδ' ἀπὸ τῶν πλειόνων γεννᾶται,[2] ἢν μὴ τύχῃ καλῶς ἔχοντα τῆς κρήσιος τῆς πρὸς ἄλληλα; ἀνάγκη τοίνυν, τῆς φύσιος τοιαύτης ὑπαρχούσης καὶ τῶν ἄλλων ἁπάντων καὶ τῆς τοῦ ἀνθρώπου, μὴ ἓν εἶναι τὸν ἄνθρωπον, ἀλλ' ἕκαστον τῶν συμβαλλομένων ἐς τὴν γένεσιν ἔχειν τὴν[3] δύναμιν ἐν τῷ σώματι, οἵην περ συνεβάλετο. καὶ πάλιν γε ἀνάγκη ἀναχωρεῖν[4] ἐς τὴν ἑωυτοῦ φύσιν ἕκαστον,
20 τελευτῶντος τοῦ σώματος τοῦ ἀνθρώπου, τό τε ὑγρὸν πρὸς τὸ ὑγρὸν καὶ τὸ ξηρὸν πρὸς τὸ ξηρὸν καὶ τὸ θερμὸν πρὸς τὸ θερμὸν καὶ τὸ ψυχρὸν πρὸς τὸ ψυχρόν. τοιαύτη δὲ καὶ τῶν ζῴων ἐστὶν ἡ φύσις, καὶ τῶν ἄλλων πάντων· γίνεταί τε ὁμοίως πάντα καὶ τελευτᾷ ὁμοίως πάντα· συνίσταταί τε γὰρ αὐτῶν ἡ φύσις ἀπὸ τούτων τῶν προειρημένων πάντων, καὶ τελευτᾷ κατὰ τὰ εἰρημένα ἐς τὸ αὐτὸ ὅθεν περ συνέστη ἕκαστον.
29 ἐνταῦθα οὖν καὶ ἀπεχώρησεν.[5]

IV. Τὸ δὲ σῶμα τοῦ ἀνθρώπου ἔχει ἐν ἑωυτῷ αἷμα καὶ φλέγμα καὶ χολὴν ξανθὴν καὶ μέλαιναν, καὶ ταῦτ' ἐστὶν αὐτῷ ἡ φύσις τοῦ σώματος, καὶ διὰ ταῦτα ἀλγεῖ καὶ ὑγιαίνει. ὑγιαίνει μὲν οὖν μάλιστα, ὅταν μετρίως ἔχῃ ταῦτα τῆς πρὸς ἄλληλα κρήσιος καὶ[6] δυνάμιος καὶ τοῦ πλήθεος, καὶ μάλιστα[7] μεμιγμένα ᾖ· ἀλγεῖ δὲ ὅταν τού-

[1] For ἰσχυρότερον A reads ἰσχυρόν.
[2] γεννᾶται MV: γίνεται A.
[3] τὴν A: τινὰ MV.

with moist be not tempered and equal—should the one constituent be much in excess of the other, and the stronger be much stronger than the weaker. Wherefore how is it likely for a thing to be generated from one, when generation does not take place from more than one unless they chance to be mutually well-tempered? Therefore, since such is the nature both of all other things and of man, man of necessity is not one, but each of the components contributing to generation has in the body the power it contributed. Again, each component must return to its own nature when the body of a man dies, moist to moist, dry to dry, hot to hot and cold to cold. Such too is the nature of animals, and of all other things. All things are born in a like way, and all things die in a like way. For the nature of them is composed of all those things I have mentioned above, and each thing, according to what has been said, ends in that from which it was composed. So that too is whither it departs.

IV. The body of man has in itself blood, phlegm, yellow bile and black bile; these make up the nature of his body, and through these he feels pain or enjoys health. Now he enjoys the most perfect health when these elements are duly proportioned to one another in respect of compounding, power and bulk, and when they are perfectly mingled. Pain is

---

[4] ἀναχωρέειν A: ἀποχωρέειν MV.

[5] ἐνταῦθα οὖν καὶ ἀπεχώρησεν reads like a gloss, or an alternative reading for τελευτᾷ ἐς τὸ αὐτό.

[6] A omits κρήσιος καί.

[7] After μάλιστα MV have ἦν and A has εἰ above the line in a corrector's hand.

των τι ἔλασσον ἢ πλέον ᾖ ἢ[1] χωρισθῇ ἐν τῷ σώματι καὶ μὴ κεκρημένον ᾖ τοῖσι σύμπασιν.[2]
10 ἀνάγκη γάρ, ὅταν τούτων τι χωρισθῇ καὶ ἐφ' ἑωυτοῦ στῇ, οὐ μόνον τοῦτο τὸ χωρίον ἔνθεν ἐξέστη ἐπίνοσον γίνεσθαι, ἀλλὰ καὶ ἔνθα ἂν στῇ καὶ ἐπιχυθῇ,[3] ὑπερπιμπλάμενον ὀδύνην τε καὶ πόνον παρέχειν. καὶ γὰρ ὅταν τι τούτων ἔξω τοῦ σώματος ἐκρυῇ πλέον τοῦ ἐπιπολάζοντος, ὀδύνην παρέχει ἡ κένωσις. ἤν τ' αὖ πάλιν ἔσω ποιήσηται τὴν κένωσιν καὶ τὴν μετάστασιν καὶ τὴν ἀπόκρισιν ἀπὸ τῶν ἄλλων, πολλὴ αὐτῷ ἀνάγκη διπλῆν τὴν ὀδύνην παρέχειν κατὰ τὰ
20 εἰρημένα, ἔνθεν τε ἐξέστη καὶ ἔνθα ὑπερέβαλεν.

V. Εἶπον δή,[4] ἃ ἂν φήσω τὸν ἄνθρωπον εἶναι, ἀποφανεῖν αἰεὶ[5] ταὐτὰ ἐόντα καὶ κατὰ νόμον καὶ κατὰ φύσιν· φημὶ δὴ εἶναι[6] αἷμα καὶ φλέγμα καὶ χολὴν ξανθὴν καὶ μέλαιναν. καὶ τούτων πρῶτον μὲν κατὰ νόμον τὰ ὀνόματα διωρίσθαι φημὶ καὶ οὐδενὶ αὐτῶν τὸ αὐτὸ ὄνομα εἶναι, ἔπειτα κατὰ φύσιν τὰς ἰδέας κεχωρίσθαι, καὶ οὔτε τὸ φλέγμα οὐδὲν ἐοικέναι τῷ αἵματι, οὔτε τὸ αἷμα τῇ χολῇ,[7] οὔτε τὴν χολὴν τῷ φλέγματι. πῶς
10 γὰρ ἂν ἐοικότα ταῦτα εἴη ἀλλήλοισιν, ὧν οὔτε τὰ χρώματα ὅμοια φαίνεται προσορώμενα, οὔτε τῇ χειρὶ ψαύοντι ὅμοια δοκεῖ εἶναι;[8] οὔτε

[1] A omits ᾖ ἢ, perhaps rightly. M omits, with εἴη ἢ in margin. V has εἴη ἢ in the text.

[2] ξύμπασιν MV: πᾶσιν A.

[3] The reading is that of A. MV have ἔνθεν τε ἐξέστηκεν· οὐ μόνον τοῦτο τὸ χωρίον νοσερὸν γίνεται, and omit στῇ καί.

[4] εἰπὼν δὲ A: εἶπον δὴ MV.

[5] ἀποφανεῖναί οἱ A (Littré says ἀποφανῆναί οἱ out of ἀποφανεῖν αἰεί, but the rotograph only shows that ει is

felt when one of these elements is in defect or excess, or is isolated in the body without being compounded with all the others. For when an element is isolated and stands by itself, not only must the place which it left become diseased, but the place where it stands in a flood must, because of the excess, cause pain and distress. In fact when more of an element flows out of the body than is necessary to get rid of superfluity, the emptying causes pain. If, on the other hand, it be to an inward part that there takes place the emptying, the shifting and the separation from other elements, the man certainly must, according to what has been said, suffer from a double pain, one in the place left, and another in the place flooded.

V. Now I promised to show that what are according to me the constituents of man remain always the same, according to both convention and nature.[1] These constituents are, I hold, blood, phlegm, yellow bile and black bile. First I assert that the names of these according to convention are separated, and that none of them has the same name as the others; furthermore, that according to nature their essential forms are separated, phlegm being quite unlike blood, blood being quite unlike bile, bile being quite unlike phlegm. How could they be like one another, when their colours appear not alike to the sight nor does their touch seem alike to the hand? For they are

[1] See p. 9.

written over some mark, and that *οἱ* is, apparently, on a thorough erasure): *ἀποφαίνειν αἰεὶ* MV.

[6] *δὲ εἶναι* A: *δ' εἶναι* MV: *δὴ εἶναι* Littré.

[7] *τῶ αἵματι ἡ χολή* A: *τῷ αἷμα* (*sic*) M.

[8] *ὅμοια δὲ* (*οὐ* above the line) *δοκεῖ* A.

γὰρ θερμὰ ὁμοίως ἐστίν, οὔτε ψυχρά, οὔτε ξηρά, οὔτε ὑγρά. ἀνάγκη τοίνυν, ὅτε τοσοῦτον διήλλακται ἀλλήλων τὴν ἰδέην τε καὶ τὴν δύναμιν, μὴ ἓν αὐτὰ εἶναι, εἴπερ μὴ πῦρ τε καὶ ὕδωρ ἕν ἐστιν.[1] γνοίης δ' ἂν τοῖσδε, ὅτι οὐχ ἓν ταῦτα πάντα ἐστίν, ἀλλ' ἕκαστον αὐτῶν ἔχει δύναμίν τε καὶ φύσιν τὴν ἑωυτοῦ· ἢν γάρ τινι
20 διδῷς[2] ἀνθρώπῳ φάρμακον ὅ τι φλέγμα ἄγει, ἐμεῖταί σοι φλέγμα, καὶ ἢν διδῷς φάρμακον ὅ τι χολὴν ἄγει, ἐμεῖταί σοι χολή. κατὰ ταὐτὰ δὲ καὶ χολὴ μέλαινα καθαίρεται,[3] ἢν διδῷς φάρμακον ὅ τι χολὴν μέλαιναν ἄγει· καὶ ἢν τρώσῃς αὐτοῦ τοῦ σώματός τι ὥστε ἕλκος[4] γενέσθαι, ῥυήσεται αὐτῷ αἷμα. καὶ ταῦτα ποιήσει σοι πάντα πᾶσαν ἡμέρην καὶ νύκτα καὶ χειμῶνος καὶ θέρεος, μέχρι ἂν δυνατὸς ᾖ τὸ πνεῦμα ἕλκειν ἐς ἑωυτὸν καὶ πάλιν μεθιέναι, ἢ ἔστ' ἄν τινος
30 τούτων στερηθῇ τῶν συγγεγονότων. συγγέγονε δὲ ταῦτα τὰ εἰρημένα· πῶς γὰρ οὐ συγγέγονε; πρῶτον μὲν φανερός ἐστιν ὥνθρωπος ἔχων ἐν ἑωυτῷ ταῦτα πάντα αἰεὶ[5] ἕως ἂν ζῇ, ἔπειτα δὲ γέγονεν ἐξ ἀνθρώπου ταῦτα πάντα ἔχοντος, τέθραπταί τε ἐν ἀνθρώπῳ ταῦτα πάντα ἔχοντι,
36 ὅσα ἐγώ φημί τε καὶ ἀποδείκνυμι.

VI. Οἱ δὲ λέγοντες ὡς ἕν ἐστιν ὥνθρωπος, δοκέουσί μοι ταύτῃ τῇ γνώμῃ χρῆσθαι·[6] ὁρέοντες τοὺς πίνοντας τὰ φάρμακα καὶ ἀπολλυμένους ἐν τῇσιν ὑπερκαθάρσεσι, τοὺς μὲν χολὴν ἐμέ-

[1] ὕδωρ ἕν ἐστι A: ὕδωρ ταυτόν ἐστι MV: ὕδωρ ἕν τε καὶ ταὐτόν ἐστιν Littré after Galen.

[2] εἰ γάρ τι δοίης (not διδοίης, as Littré says) A: ἢν γάρ τινι δίδως MV.

not equally warm, nor cold, nor dry, nor moist. Since then they are so different from one another in essential form and in power, they cannot be one, if fire and water are not one. From the following evidence you may know that these elements are not all one, but that each of them has its own power and its own nature. If you were to give a man a medicine which withdraws phlegm, he will vomit you phlegm; if you give him one which withdraws bile, he will vomit you bile. Similarly too black bile is purged away if you give a medicine which withdraws black bile. And if you wound a man's body so as to cause a wound, blood will flow from him. And you will find all these things happen on any day and on any night, both in winter and in summer, so long as the man can draw breath in and then breathe it out again, or until he is deprived of one of the elements congenital with him. Congenital with him (how should they not be so?) are the elements already mentioned. First, so long as a man lives he manifestly has all these elements always in him; then he is born out of a human being having all these elements, and is nursed in a human being having them all, I mean those elements I have mentioned with proofs.

VI. Those who assert that man is composed of one element seem to me to have been influenced by the following line of thought. They see those who drink drugs and die through excessive purgings vomiting,

---

[3] χολὴ μέλαινα καθαίρεται A : χολὴν μέλαιναν καθαίρει MV.
[4] ἕλκος A : τραῦμα MV (in M over an erasure).
[5] αἰεὶ M : ἀεὶ V : ἰδεῖν A.
[6] χρῆσθαι AV : κεχρῆσθαι M.

οντας, τοὺς δέ τινας φλέγμα, τοῦτο δὲ ἕκαστον αὐτῶν ἐνόμισαν εἶναι τὸν ἄνθρωπον, ὅ τι καθαιρόμενον εἶδον αὐτὸν ἀποθανόντα· καὶ οἱ τὸ αἷμα φάντες εἶναι τὸν ἄνθρωπον τῇ αὐτῇ[1] γνώμῃ χρέονται· ὁρέοντες ἀποσφαζομένους τοὺς ἀν-
10 θρώπους καὶ τὸ αἷμα ῥέον ἐκ τοῦ σώματος, τοῦτο νομίζουσιν εἶναι τὴν ψυχὴν τῷ ἀνθρώπῳ· καὶ μαρτυρίοισι τούτοισι πάντες χρέονται ἐν τοῖσι λόγοισιν. καίτοι τὸ μὲν πρῶτον[2] ἐν τῇσιν ὑπερκαθάρσεσιν οὐδείς πω ἀπέθανε χολὴν μοῦνον καθαρθείς· ἀλλ' ὁπόταν πίῃ τις φάρμακον ὅ τι χολὴν ἄγει, πρῶτον μὲν χολὴν ἐμεῖ, ἔπειτα δὲ καὶ φλέγμα· ἔπειτα δὲ ἐπὶ τούτοισιν ἐμέουσι χολὴν μέλαιναν ἀναγκαζόμενοι,[3] τελευτῶντες δὲ καὶ αἷμα ἐμέουσι καθαρόν. τὰ αὐτὰ δὲ πάσχουσι
20 καὶ ὑπὸ τῶν φαρμάκων τῶν τὸ φλέγμα ἀγόντων· πρῶτον μὲν γὰρ φλέγμα ἐμέουσιν, ἔπειτα δὲ χολὴν ξανθήν, ἔπειτα δὲ μέλαιναν, τελευτῶντες δὲ αἷμα καθαρόν, καὶ ἐν τῷδε ἀποθνήσκουσιν. τὸ γὰρ φάρμακον, ὅταν ἐσέλθῃ ἐς τὸ σῶμα, πρῶτον μὲν ἄγει ὃ ἂν αὐτῷ κατὰ φύσιν μάλιστα ᾖ τῶν ἐν τῷ σώματι ἐνεόντων, ἔπειτα δὲ καὶ τἄλλα ἕλκει τε καὶ καθαίρει. ὡς γὰρ τὰ φυόμενά τε καὶ σπειρόμενα, ὁπόταν ἐς τὴν γῆν ἔλθῃ, ἕλκει ἕκαστον τὸ κατὰ φύσιν αὐτῷ ἐνεὸν ἐν τῇ γῇ, ἔνι δὲ καὶ ὀξὺ
30 καὶ πικρὸν καὶ γλυκὺ καὶ ἁλμυρὸν καὶ παντοῖον· πρῶτον μὲν οὖν πλεῖστον τούτου εἵλκυσεν ἐς ἑωυτό, ὅ τι ἂν ᾖ αὐτῷ κατὰ φύσιν μάλιστα, ἔπειτα δὲ ἕλκει καὶ τἄλλα· τοιοῦτον δέ τι καὶ τὰ φάρμακα ποιεῖ ἐν τῷ σώματι· ὅσα ἂν χολὴν ἄγῃ, πρῶτον μὲν ἀκρητεστάτην ἐκάθηρε χολήν, ἔπειτα δὲ μεμιγμένην· καὶ τὰ τοῦ φλέγματος

in some cases bile, in others phlegm; then they think that the man is composed of that one thing from the purging of which they saw him die. Those too who say that man is composed of blood use the same line of thought. They see men who are cut[1] bleeding from the body, and so they think that blood composes the soul of a man. Such is the evidence they all use in their discussions. Yet first, nobody yet in excessive purgings has vomited bile alone when he died. But when a man has drunk a drug which withdraws bile, he first vomits bile, then phlegm also. Afterwards under stress men vomit after these black bile, and finally they vomit also pure blood. The same experiences happen to those who drink drugs which withdraw phlegm. First they vomit phlegm, then yellow bile, then black, and finally pure blood, whereon they die. For when the drug enters the body, it first withdraws that constituent of the body which is most akin to itself, and then it draws and purges the other constituents. For just as things that are sown and grow in the earth, when they enter it, draw each that constituent of the earth which is nearest akin to it—these are the acid, the bitter, the sweet, the salt and so on—first the plant draws to itself mostly that element which is most akin to it, and then it draws the other constituents also. Such too is the action of drugs in the body. Those that withdraw bile first evacuate absolutely pure bile, then bile that is mixed.

[1] Literally, "have their throat cut."

---

[1] τοιαύτη A.
[2] καίτοι τὸ μὲν πρῶτον A: καὶ πρῶτον μὲν MV.
[3] ἀναγκαζόμενοι MV: A omits.

φάρμακα πρῶτον μὲν ἀκρητέστατον τὸ φλέγμα ἄγει, ἔπειτα δὲ μεμιγμένον· καὶ τοῖσιν ἀποσφαζομένοισι τὸ αἷμα ῥεῖ πρῶτον θερμότατόν[1] τε
40 καὶ ἐρυθρότατον, ἔπειτα δὲ ῥεῖ φλεγματωδέστερον
41 καὶ χολωδέστερον.

VII. Αὔξεται δὲ ἐν τῷ ἀνθρώπῳ τὸ φλέγμα τοῦ χειμῶνος· τοῦτο γὰρ τῷ χειμῶνι κατὰ φύσιν ἐστὶ μάλιστα τῶν ἐν τῷ σώματι ἐνεόντων, ψυχρότατον γάρ ἐστιν.[2] τεκμήριον δὲ τούτου, ὅτι τὸ μὲν φλέγμα ψυχρότατον, εἰ θέλοις[3] ψαῦσαι φλέγματος καὶ χολῆς καὶ αἵματος,[4] τὸ φλέγμα εὑρήσεις ψυχρότατον ἐόν· καίτοι γλισχρότατόν ἐστι καὶ βίῃ μάλιστα ἄγεται μετὰ[5] χολὴν μέλαιναν· ὅσα δὲ βίῃ ἔρχεται, θερμότερα γίνεται, ἀναγκαζόμενα ὑπὸ τῆς βίης· ἀλλ' ὅμως
10 καὶ πρὸς ταῦτα πάντα ψυχρότατον ἐὸν τὸ φλέγμα φαίνεται ὑπὸ τῆς φύσιος τῆς ἑωυτοῦ. ὅτι δὲ ὁ χειμὼν πληροῖ τὸ σῶμα φλέγματος, γνοίης ἂν τοῖσδε· οἱ ἄνθρωποι πτύουσι καὶ ἀπομύσσονται φλεγματωδέστατον τοῦ χειμῶνος, καὶ τὰ οἰδήματα λευκὰ[6] γίνεται μάλιστα ταύτην τὴν ὥρην, καὶ τἄλλα νοσήματα φλεγματώδεα. τοῦ δὲ ἦρος τὸ φλέγμα ἔτι μένει ἰσχυρὸν[7] ἐν τῷ σώματι, καὶ τὸ αἷμα αὔξεται· τά τε γὰρ ψύχεα ἐξανίει,[8] καὶ τὰ ὕδατα ἐπιγίνεται, τὸ δὲ
20 αἷμα κατὰ ταῦτα[9] αὔξεται ὑπό τε τῶν ὄμβρων

[1] τὸ αἷμα ῥεῖ πρῶτον θερμότατον A : τὸ αἷμα ῥέει πρῶτον μὲν θερμότατον MV.

[2] τοῦτο γὰρ τῷ χειμῶνι κατὰ φύσιν μάλιστα τῶν ἐν τῷ σώματι ἐνεόντων ψυχρότατον ἐστίν A : τοῦτο γὰρ τῶι χειμῶνι κατὰ φύσιν μάλιστα τῶν ἐν τῶι σώματι ἐνεόντων· ψυχρότατον γάρ ἐστι M : τοῦτο γὰρ τῷ χειμῶνι μάλιστα κατὰ φυσιν τῶν ἐν τῷ σώματι ἐνεόντων ψυχρότατόν ἐστι V.

Those that withdraw phlegm first withdraw absolutely pure phlegm, and then phlegm that is mixed. And when men are cut,[1] the blood that flows is at first very hot and very red, and then it flows with more phlegm and bile mixed with it.

VII. Phlegm increases in a man in winter; for phlegm, being the coldest constituent of the body, is closest akin to winter. A proof that phlegm is very cold is that if you touch phlegm, bile and blood, you will find phlegm the coldest. And yet it is the most viscid, and after black bile requires most force for its evacuation. But things that are moved by force become hotter under the stress of the force. Yet in spite of all this, phlegm shows itself the coldest element by reason of its own nature. That winter fills the body with phlegm you can learn from the following evidence. It is in winter that the sputum and nasal discharge of men is fullest of phlegm; at this season mostly swellings become white, and diseases generally phlegmatic. And in spring too phlegm still remains strong in the body, while the blood increases. For the cold relaxes, and the rains come on, while the blood accordingly increases

[1] Literally "have their throats cut."

[3] θέλοις A: ἐθέλοις M: ἐθέλεις (-οις?) V.
[4] A omits καὶ αἵματος.
[5] μετὰ MV: μετὰ δὲ A.
[6] λευκὰ A: λευκότατα MV.
[7] ἔτι μὲν ἰσχυρὸν τὸ φλέγμα ἐστὶν A: τὸ φλέγμα ἔτι μὲν ἰσχυρότερον M: τὸ φλέγμα ἔστι μὲν ἰσχυρότερον V: τὸ φλέγμα ἔτι μένει ἰσχυρὸν Littré, from Galen and notes in Foes.
[8] ἐξανείει τε A: ἐξανίει MV.
[9] A omits κατὰ ταῦτα.

καὶ ὑπὸ τῶν θερμημεριῶν· κατὰ φύσιν γὰρ αὐτῷ ταῦτά ἐστι μάλιστα τοῦ ἐνιαυτοῦ· ὑγρόν τε γάρ ἐστι καὶ θερμόν. γνοίης δ' ἂν τοῖσδε· οἱ ἄνθρωποι τοῦ ἦρος καὶ τοῦ θέρεος μάλιστα ὑπό τε τῶν δυσεντεριῶν ἁλίσκονται, καὶ ἐκ τῶν ῥινῶν τὸ αἷμα[1] ῥεῖ αὐτοῖσι, καὶ θερμότατοί εἰσι καὶ ἐρυθροί· τοῦ δὲ θέρεος τό τε αἷμα ἰσχύει ἔτι, καὶ ἡ χολὴ αἴρεται ἐν τῷ σώματι καὶ παρατείνει ἐς τὸ φθινόπωρον· ἐν δὲ τῷ φθινο-
30 πώρῳ τὸ μὲν αἷμα ὀλίγον γίνεται, ἐναντίον γὰρ αὐτοῦ τὸ φθινόπωρον τῇ φύσει ἐστίν· ἡ δὲ χολὴ τὴν θερείην[2] κατέχει τὸ σῶμα καὶ τὸ φθινόπωρον. γνοίης δ' ἂν τοῖσδε· οἱ ἄνθρωποι αὐτόματοι ταύτην τὴν ὥρην χολὴν ἐμέουσι, καὶ ἐν τῇσι φαρμακοποσίῃσι χολωδέστατα καθαίρονται, δῆλον δὲ καὶ τοῖσι πυρετοῖσι καὶ τοῖσι χρώμασι τῶν ἀνθρώπων. τὸ δὲ φλέγμα τῆς θερείης[3] ἀσθενέστατόν ἐστιν αὐτὸ ἑωυτοῦ· ἐναντίη γὰρ αὐτοῦ τῇ φύσει ἐστὶν ἡ ὥρη, ξηρή τε ἐοῦσα[4] καὶ
40 θερμή. τὸ δὲ αἷμα τοῦ φθινοπώρου ἐλάχιστον γίνεται ἐν τῷ ἀνθρώπῳ, ξηρόν τε γάρ ἐστι τὸ φθινόπωρον καὶ ψύχειν ἤδη ἄρχεται τὸν ἄνθρωπον· ἡ δὲ μέλαινα χολὴ τοῦ φθινοπώρου πλείστη τε καὶ ἰσχυροτάτη ἐστίν. ὅταν δὲ ὁ χειμὼν καταλαμβάνῃ, ἥ τε χολὴ ψυχομένη ὀλίγη γίνεται, καὶ τὸ φλέγμα αὔξεται πάλιν ὑπό[5] τε τῶν ὑετῶν τοῦ πλήθεος καὶ[6] τῶν νυκτῶν τοῦ μήκεος. ἔχει μὲν οὖν ταῦτα πάντα αἰεὶ[7] τὸ σῶμα τοῦ ἀνθρώπου, ὑπὸ δὲ τῆς ὥρης περιισ-
50 ταμένης ποτὲ μὲν πλείω γίνεται αὐτὰ ἑωυτῶν, ποτὲ[8] δὲ ἐλάσσω, ἕκαστα κατὰ μέρος καὶ[9] κατὰ

[1] τὰ αἵματα A: αἷμα MV.

through the showers and the hot days. For these conditions of the year are most akin to the nature of blood, spring being moist and warm. You can learn the truth from the following facts. It is chiefly in spring and summer that men are attacked by dysenteries, and by hemorrhage from the nose, and they are then hottest and red. And in summer blood is still strong, and bile rises in the body and extends until autumn. In autumn blood becomes small in quantity, as autumn is opposed to its nature, while bile prevails in the body during the summer season and during autumn. You may learn this truth from the following facts. During this season men vomit bile without an emetic, and when they take purges the discharges are most bilious. It is plain too from fevers and from the complexions of men. But in summer phlegm is at its weakest. For the season is opposed to its nature, being dry and warm. But in autumn blood becomes least in man, for autumn is dry and begins from this point to chill him. It is black bile which in autumn is greatest and strongest. When winter comes on, bile being chilled becomes small in quantity, and phlegm increases again because of the abundance of rain and the length of the nights. All these elements then are always comprised in the body of a man, but as the year goes round they become now greater and now less, each in turn and

---

[2] τοῦ θέρεος A: τὴν θερίην M: τὴν θερείην V.

[3] τοῦ θέρεος A: τῆς θερίης M: τῆς θερείης V. Littré records a reading τῆς θέρεος θερείης.

[4] ἐοῖσα A: γάρ ἐστι MV.

[5] ὑπὸ A: ἀπὸ MV.

[6] MV read ὑπὸ before τῶν νυκτῶν.

[7] ἀΐδια A: ἀεὶ MV.

[8] ποτὲ . . . ποτὲ MV: τότε . . . τότε A.

[9] κατὰ μέρος τε καὶ A.

φύσιν. ὡς γὰρ[1] ὁ ἐνιαυτὸς μετέχει μὲν πᾶς πάντων καὶ τῶν θερμῶν καὶ τῶν ψυχρῶν καὶ τῶν ξηρῶν καὶ τῶν ὑγρῶν, οὐ γὰρ ἂν μείνειε τούτων[2] οὐδὲν οὐδένα χρόνον ἄνευ πάντων τῶν ἐνεόντων ἐν τῷδε τῷ κόσμῳ, ἀλλ' εἰ ἕν τί γε[3] ἐκλίποι, πάντ' ἂν ἀφανισθείη·[4] ἀπὸ γὰρ τῆς αὐτῆς ἀνάγκης πάντα συνέστηκέ τε καὶ τρέφεται ὑπ'[5] ἀλλήλων· οὕτω δὲ καὶ εἴ τι ἐκ τοῦ ἀν-
60 θρώπου ἐκλίποι τούτων τῶν συγγεγονότων, οὐκ ἂν δύναιτο ζῆν ὥνθρωπος. ἰσχύει δ' ἐν τῷ ἐνιαυτῷ τοτὲ μὲν ὁ χειμὼν μάλιστα, τοτὲ δὲ τὸ ἔαρ, τοτὲ δὲ τὸ θέρος, τοτὲ δὲ τὸ φθινόπωρον· οὕτω δὲ καὶ ἐν τῷ ἀνθρώπῳ τοτὲ μὲν τὸ φλέγμα ἰσχύει, τοτὲ δὲ τὸ αἷμα, τοτὲ δὲ ἡ χολή, πρῶτον μὲν ἡ ξανθή, ἔπειτα δ' ἡ μέλαινα καλεομένη. μαρτύριον δὲ σαφέστατον, εἰ θέλοις[6] τῷ αὐτῷ ἀνθρώπῳ δοῦναι τὸ αὐτὸ φάρμακον τετράκις τοῦ ἐνιαυτοῦ, ἐμεῖταί σοι τοῦ μὲν χειμῶνος φλεγμα-
70 τωδέστατα, τοῦ δὲ ἦρος ὑγρότατα, τοῦ δὲ θέρεος
71 χολωδέστατα, τοῦ δὲ φθινοπώρου μελάντατα.

VIII. Ὀφείλει οὖν, τούτων ὧδε ἐχόντων, ὅσα μὲν τῶν νοσημάτων χειμῶνος αὔξεται, θέρεος φθίνειν,[7] ὅσα δὲ θέρεος αὔξεται, χειμῶνος λήγειν, ὅσα μὴ αὐτῶν ἐν περιόδῳ ἡμερέων ἀπαλλάσσεται· τὴν δὲ περίοδον αὖτις φράσω τὴν τῶν ἡμερέων. ὅσα δὲ ἦρος γίνεται νοσήματα, προσδέχεσθαι χρὴ φθινοπώρου τὴν ἀπάλλαξιν ἔσεσθαι αὐτῶν· ὅσα δὲ φθινοπωρινὰ νοσήματα, τούτων τοῦ ἦρος

[1] ὡς γὰρ A: ὥσπερ MV.

[2] μείνειεν A: μενηιεν (ει above) τουτέων M: μενεῖ τουτέων V. Holkhamensis 282 reads μενεῖ, but according to Littré C has μένει.

according to its nature. For just as every year participates in every element, the hot, the cold, the dry and the moist—none in fact of these elements would last for a moment without all the things that exist in this universe, but if one were to fail all would disappear, for by reason of the same necessity all things are constructed and nourished by one another—even so, if any of these congenital elements were to fail, the man could not live. In the year sometimes the winter is most powerful, sometimes the spring, sometimes the summer and sometimes the autumn. So too in man sometimes phlegm is powerful, sometimes blood, sometimes bile, first yellow, and then what is called black bile. The clearest proof is that if you will give the same man to drink the same drug four times in the year, he will vomit, you will find, the most phlegmatic matter in the winter, the moistest in the spring, the most bilious in the summer, and the blackest in the autumn.

VIII. Now, as these things are so, such diseases as increase in the winter ought to cease in the summer, and such as increase in the summer ought to cease in the winter, with the exception of those which do not change in a period of days—the period of days I shall speak of afterwards. When diseases arise in spring, expect their departure in autumn. Such diseases as arise in autumn must have their

---

[3] ἕν τί γε A: ἕν τι MV.
[4] ἀφανισθείη MV: ἀφανισθῇ A.
[5] ὑπ' A: ἀπ' MV.
[6] θέλοις AV: ἐθέλοις M: ἐθέλεις Littré.
[7] φθίνειν A: λήγειν MV.

ἀνάγκη τὴν ἀπάλλαξιν γενέσθαι· ὅ τι δ' ἂν τὰς
10 ὥρας ταύτας ὑπερβάλλῃ[1] νόσημα, εἰδέναι χρὴ ὡς ἐνιαύσιον αὐτὸ[2] ἐσόμενον. καὶ τὸν ἰητρὸν οὕτω χρὴ ἰῆσθαι[3] τὰ νοσήματα ὡς ἑκάστου τούτων ἰσχύοντος ἐν τῷ σώματι κατὰ τὴν ὥρην τὴν αὐτῷ
14 κατὰ φύσιν ἐοῦσαν μάλιστα.

IX. Εἰδέναι δὲ χρὴ καὶ τάδε πρὸς ἐκείνοις· ὅσα πλησμονὴ τίκτει νοσήματα, κένωσις ἰῆται, ὅσα δὲ ἀπὸ κενώσιος γίνεται, πλησμονὴ ἰῆται, ὅσα δὲ ἀπὸ ταλαιπωρίης γίνεται, ἀνάπαυσις ἰῆται, ὅσα δ' ὑπ' ἀργίης τίκτεται, ταλαιπωρίη ἰῆται.[4] τὸ δὲ σύμπαν γνῶναι, δεῖ τὸν ἰητρὸν ἐναντίον ἵστασθαι τοῖσι καθεστεῶσι[5] καὶ νοσήμασι καὶ εἴδεσι[6] καὶ ὥρῃσι καὶ ἡλικίῃσι, καὶ τὰ συντείνοντα λύειν, καὶ τὰ λελυμένα συντείνειν·
10 οὕτω γὰρ ἂν μάλιστα τὸ κάμνον ἀναπαύοιτο, ἥ τε ἴησις τοῦτό μοι δοκεῖ εἶναι. αἱ δὲ νοῦσοι γίνονται, αἱ μὲν ἀπὸ τῶν διαιτημάτων, αἱ δὲ ἀπὸ τοῦ πνεύματος, ὃ ἐσαγόμενοι ζῶμεν. τὴν δὲ διάγνωσιν χρὴ ἑκατέρου ὧδε ποιεῖσθαι· ὅταν μὲν ὑπὸ νοσήματος ἑνὸς πολλοὶ ἄνθρωποι ἁλίσκωνται κατὰ τὸν αὐτὸν χρόνον, τὴν αἰτίην χρὴ ἀνατιθέναι τούτῳ ὅ τι κοινότατόν ἐστι καὶ μάλιστα αὐτῷ πάντες χρεόμεθα· ἔστι δὲ τοῦτο ὃ ἀναπνέομεν. φανερὸν γὰρ δὴ ὅτι τά γε διαιτήματα ἑκάστου
20 ἡμῶν οὐκ αἴτιά ἐστιν, ὅτε γε[7] ἅπτεται πάντων ἡ νοῦσος ἑξῆς καὶ τῶν νεωτέρων καὶ τῶν πρεσβυτέρων, καὶ γυναικῶν καὶ ἀνδρῶν ὁμοίως, καὶ τῶν

[1] ὑπερβάλλῃ A and Holk. 282: ὑπερβάλῃ M: ὑπερβάλη V.
[2] αὐτὸ deleted by Wilamowitz.
[3] οὕτω χρὴ ἰῆσθαι πρὸς MV: χρὴ οὕτως ἰᾶσθαι A.
[4] ὅσα δὲ ὑπερτέρη ἀργίη νοσήματα τίκτει, ταῦτα ταλαιπωρίη

departure in spring. Whenever a disease passes these limits, you may know that it will last a year. The physician too must treat diseases with the conviction that each of them is powerful in the body according to the season which is most conformable to it.

IX. Furthermore, one must know that diseases due to repletion are cured by evacuation, and those due to evacuation are cured by repletion; those due to exercise are cured by rest, and those due to idleness are cured by exercise. To know the whole matter, the physician must set himself against the established character of diseases, of constitutions, of seasons and of ages; he must relax what is tense and make tense what is relaxed. For in this way the diseased part would rest most, and this, in my opinion, constitutes treatment. Diseases[1] arise, in some cases from regimen, in other cases from the air by the inspiration of which we live. The distinction between the two should be made in the following way. Whenever many men are attacked by one disease at the same time, the cause should be assigned to that which is most common, and which we all use most. This it is which we breathe in. For it is clear that the regimen of each of us is not the cause, since the disease attacks all in turn, both younger and older, men as much as women, those who drink wine as much as

[1] This passage is quoted, or rather paraphrased, in the *Anonymus Londinensis* VII. 15.

ἰᾶται A: ὁκόσα δ' ὑπ' ἀργίης νοσήματα τίκτεται ταλαιπωρίη ἰῆται MV.

[5] καθεστεῶσι MV: καθεστηκόσι A.

[6] εἴδεσι MV: ἰδέῃσι A.

[7] ὅτε γε A: ὅτε τε MV.

θωρησσομένων καὶ τῶν ὑδροποτεόντων, καὶ τῶν μάζαν ἐσθιόντων καὶ τῶν ἄρτον σιτευμένων, καὶ τῶν πολλὰ ταλαιπωρεόντων καὶ τῶν ὀλίγα· οὐκ ἂν οὖν τά γε διαιτήματα αἴτια εἴη, ὅταν διαιτώμενοι πάντας τρόπους οἱ ἄνθρωποι ἁλίσκωνται ὑπὸ τῆς αὐτῆς νούσου. ὅταν δὲ αἱ νοῦσοι γίνωνται παντοδαπαὶ κατὰ τὸν αὐτὸν χρόνον,[1] δῆλον 30 ὅτι τὰ διατήματά ἐστιν αἴτια ἕκαστα ἑκάστοισι, καὶ τὴν θεραπείην χρὴ ποιεῖσθαι ἐναντιούμενον τῇ προφάσει τῆς νούσου, ὥσπερ μοι πέφρασται καὶ ἑτέρωθι, καὶ τῇ τῶν διαιτημάτων μεταβολῇ.[2] δῆλον γὰρ ὅτι οἷσί γε χρῆσθαι εἴωθεν[3] ὥνθρωπος διαιτήμασιν, οὐκ ἐπιτήδειά οἵ ἐστιν ἢ πάντα, ἢ τὰ πλείω, ἢ ἕν γέ τι αὐτῶν· ἃ δεῖ καταμαθόντα μεταβάλλειν, καὶ σκεψάμενον τοῦ ἀνθρώπου τὴν φύσιν[4] τήν τε ἡλικίην καὶ τὸ εἶδος καὶ τὴν ὥρην τοῦ ἔτεος καὶ τῆς νούσου τὸν τρόπον, τὴν θερα- 40 πείην ποιεῖσθαι, ποτὲ μὲν ἀφαιρέοντα, ποτὲ δὲ[5] προστιθέντα, ὥσπερ μοι καὶ[6] πάλαι εἴρηται, πρὸς ἕκαστα[7] τῶν ἡλικιῶν καὶ τῶν ὡρέων καὶ τῶν εἰδέων καὶ τῶν νούσων ἔν τε τῇ φαρμακείῃ[8] προστρέπεσθαι καὶ ἐν τῇ διαίτῃ.[9] ὅταν δὲ νοσήματος ἑνὸς ἐπιδημίη καθεστήκῃ, δῆλον ὅτι[10] οὐ τὰ διαιτήματα αἴτιά ἐστιν, ἀλλ᾽ ὃ ἀναπνέομεν, τοῦτο αἴτιόν ἐστι, καὶ δῆλον ὅτι τοῦτο νοσηρήν τινα ἀπόκρισιν ἔχον ἀνίει. τοῦτον χρὴ[11] τὸν

[1] κατὰ τὸν αὐτὸν χρόνον MV: κατὰ τοὺς αὐτοὺς χρόνους A.

[2] ἐκ τῶν διαιτημάτων μεταβάλλειν A: τῶν διαιτημάτων μεταβολῇ MV; τῇ τῶν διαιτημάτων μεταβολῇ Littré with many late MSS. Wilamowitz deletes καὶ . . . . μεταβάλλειν.

[3] χρῆσθαι εἴωθεν MV: εἰώθη χρεῖσθαι (not χρῆσθαι as Littré says) A.

[4] A omits τὴν φύσιν.

[5] ποτὲ μὲν . . . . ποτὲ δὲ MV: τὰ μὲν . . . . τὰ δὲ A.

teetotallers, those who eat barley cake as much as those who live on bread, those who take much exercise as well as those who take little. For regimen could not be the cause, when no matter what regimen they have followed all men are attacked by the same disease. But when diseases of all sorts occur at one and the same time, it is clear that in each case the particular regimen is the cause, and that the treatment carried out should be that opposed to the cause of the disease, as has been set forth by me elsewhere also, and should be by change of regimen. For it is clear that, of the regimen the patient is wont to use, either all, or the greater part, or some one part, is not suited to him. This one should learn and change, and carry out treatment only after examination of the patient's constitution, age, physique, the season of the year and the fashion of the disease, sometimes taking away and sometimes adding, as I have already said, and so making changes in drugging or in regimen to suit the several conditions of age, season, physique and disease. But when an epidemic of one disease is prevalent, it is plain that the cause is not regimen but what we breathe, and that this is charged with some unhealthy exhalation. During this period these

---

[6] MV omit *καὶ*.

[7] *ἕκαστα* A : *ἑκάστας* MV.

[8] *τῇ φαρμακείῃ* A : *τῇσι φαρμακίῃσι* M : *τῇσι φαρμακείῃσι* V.

[9] *τῇ διαίτῃ* A: *τῇσι διαιτήμασιν* M : *τοῖσι διαιτήμασιν* V.

[10] *δῆλον ὅτι* M : *καὶ δῆλον ᾖ ὅτι* A : *δηλονότι* V.

[11] *δῆλον ἔτι τοῦτο νοσηρὴν τὴν ἀπόκρισιν ἔχον ἂν εἴη, τοῦτον χρὴ* A : *καὶ δῆλον ὅτι τοῦτο νοσηρήν τινα ἀπόκρισιν ἔχων ἂν εἴη· τοῦτον χρὴ* M : *καὶ δῆλον ὅτι τοῦτο νοσηρήν τινα ἀπόκρισιν ἔχον ἂν εἴη· τοῦτον δεῖ* V. *καὶ δῆλον ὅτι τοῦτο νοσηρήν τινα ἀπόκρισιν ἔχον ἀνίει* Littré. Villaret keeps the reading of A from *καθεστήκῃ*, putting a full stop at *εἴη*.

χρόνον τὰς παραινέσιας ποιεῖσθαι τοῖσιν ἀν-
50 θρώποισι τοιάσδε· τὰ μὲν διαιτήματα μὴ μεταβάλλειν, ὅτι[1] γε οὐκ αἴτιά ἐστι τῆς νούσου, τὸ δὲ σῶμα ὁρᾶν, ὅπως ἔσται ὡς ἀογκότατον[2] καὶ ἀσθενέστατον, τῶν τε σιτίων ἀφαιρέοντα καὶ τῶν ποτῶν, οἷσιν εἰώθει χρῆσθαι, κατ' ὀλίγον· ἢν γὰρ μεταβάλῃ[3] ταχέως τὴν δίαιταν, κίνδυνος καὶ ἀπὸ τῆς μεταβολῆς νεώτερόν τι γενέσθαι ἐν τῷ σώματι, ἀλλὰ χρὴ τοῖσι μὲν διαιτήμασιν οὕτω χρῆσθαι, ὅτε γε[4] φαίνεται οὐδὲν[5] ἀδικέοντα τὸν ἄνθρωπον·
60 τοῦ δὲ πνεύματος ὅπως ἡ ῥύσις ὡς ἐλαχίστη ἐς τὸ σῶμα ἐσίῃ[6] καὶ ὡς ξενωτάτη,[7] προμηθεῖσθαι, τῶν τε χωρίων τοὺς τόπους μεταβάλλοντα[8] ἐς δύναμιν, ἐν οἷσιν ἂν ἡ νοῦσος καθεστήκῃ, καὶ τὰ σώματα λεπτύνοντα· οὕτω γὰρ ἂν ἥκιστα πολλοῦ τε καὶ πυκνοῦ τοῦ πνεύματος[9] χρῄζοιεν
66 οἱ ἄνθρωποι.

X. Ὅσα δὲ τῶν νοσημάτων γίνεται ἀπὸ τοῦ[10] σώματος τῶν μελέων τοῦ ἰσχυροτάτου, ταῦτα[11] δὲ δεινότατά ἐστιν· καὶ γὰρ ἢν αὐτοῦ μένῃ[12] ἔνθα ἂν ἄρξηται, ἀνάγκη, τοῦ ἰσχυροτάτου τῶν μελέων πονεομένου, ἅπαν τὸ σῶμα πονεῖσθαι· καὶ ἢν ἐπί τι τῶν ἀσθενεστέρων[13] ἀφίκηται ἀπὸ τοῦ ἰσχυροτέρου, χαλεπαὶ αἱ ἀπολύσιες γίνονται. ὅσα δ' ἂν ἀπὸ τῶν ἀσθενεστέρων[14] ἐπὶ τὰ ἰσχυρότερα

[1] ὅτι A: ὅτε MV.

[2] ἀογκότατον A: ὡς ἀογκότατον M (ὡς above the line): ὡς εὐογκότατον V.

[3] μεταβάλῃ M: μεταβάλλειν A: μεταβάλῃ V.

[4] ὅτε γε MV: ὅτε A.

[5] οὐδὲν AMV: μηδὲν Littré.

[6] σῶμα ἐσίῃ A: σῶμα ἐσίοι MV: στόμα ἐσίῃ Littré.

[7] ξενωτάτη ἔσται M: ξεναιτάτη A: ξενοτάτη ἔσται V.

are the recommendations that should be made to patients. They should not change their regimen, as it is not the cause of their disease, but rather take care that their body be as thin and as weak as possible, by diminishing their usual food and drink gradually. For if the change of regimen be sudden, there is a risk that from the change too some disturbance will take place in the body, but regimen should be used in this way when it manifestly does no harm to a patient. Then care should be taken that inspiration be of the lightest, and also from a source as far removed as possible; the place should be moved as far as possible from that in which the disease is epidemic, and the body should be reduced, for such reduction will minimise the need of deep and frequent breathing.

X. Those diseases are most dangerous which arise in the strongest[1] part of the body. For should the disease remain where it began, the whole body, as the strongest limb in it feels pain, must be in pain; while should the disease move from a stronger part to one of the weaker parts, the riddance of it proves difficult. But when diseases move from weaker parts to stronger parts, it is easier to get rid of

[1] I follow Galen and Littré in taking "the strongest parts" to be those which are naturally, *i.e.* constitutionally, the most healthy members of the body.

---

[8] μεταβάλλοντα MV : μεταβάλλοντας A.

[9] τοῦ πνεύματος A : πνεύματος MV.

[10] ἀπὸ τοῦ MV : ἀπ' αὐτοῦ τοῦ A.

[11] ταῦτα MV : ταῦτα δὲ A.

[12] μενεῖ A : μὲν μένῃ M : μὲν μένῃ V.

[13] ἐπὶ τῶν ἀσθενεστέρων τι A : ἐπὶ τῶν ἀσθενεστέρων τι (with τι above the line after ἐπὶ) M : ἐπί τι τῶν ἀσθενεστέρων V.

[14] V omits from ἀφίκηται to ἀσθενεστέρων.

ἔλθῃ, εὐλυτώτερά ἐστιν, ὑπὸ γὰρ τῆς ἰσχύος ἀνα-
10 λώσεται[1] ῥηϊδίως τὰ ἐπιρρέοντα.

XI. Αἱ παχύταται τῶν φλεβῶν ὧδε πεφύκασιν. τέσσαρα ζεύγεά ἐστιν ἐν τῷ σώματι, καὶ ἓν μὲν αὐτῶν ἀπὸ τῆς κεφαλῆς ὄπισθεν διὰ τοῦ αὐχένος, ἔξωθεν παρὰ[2] τὴν ῥάχιν ἔνθεν τε καὶ ἔνθεν παρὰ[3] τὰ ἰσχία ἀφικνεῖται καὶ ἐς τὰ σκέλεα, ἔπειτα διὰ τῶν κνημέων ἐπὶ[4] τῶν σφυρῶν τὰ ἔξω καὶ ἐς τοὺς πόδας ἀφήκει.[5] δεῖ οὖν τὰς φλεβοτομίας τὰς ἐπὶ τῶν ἀλγημάτων τῶν ἐν τῷ νώτῳ καὶ τοῖσιν ἰσχίοισιν ἀπὸ τῶν ἰγνύων ποιεῖσθαι καὶ ἀπὸ τῶν
10 σφυρῶν ἔξωθεν.[6] αἱ δ' ἕτεραι φλέβες ἀπὸ[7] τῆς κεφαλῆς παρὰ τὰ ὦτα διὰ τοῦ αὐχένος, αἱ σφαγίτιδες καλεόμεναι, ἔσωθεν παρὰ τὴν ῥάχιν ἑκατέρωθεν φέρουσι παρὰ τὰς ψόας ἐς τοὺς ὄρχιας καὶ ἐς τοὺς μηρούς, καὶ διὰ τῶν ἰγνύων ἐκ[8] τοῦ ἔσωθεν μέρεος, ἔπειτα διὰ τῶν κνημέων ἐπὶ[9] τὰ σφυρὰ τὰ ἔσωθεν καὶ τοὺς πόδας. δεῖ οὖν τὰς φλεβοτομίας ποιεῖσθαι πρὸς τὰς ὀδύνας τὰς ἀπὸ τῶν ψοῶν καὶ τῶν ὀρχίων, ἀπὸ τῶν ἰγνύων καὶ ἀπὸ τῶν σφυρῶν ἔσωθεν. αἱ δὲ τρίται φλέβες ἐκ τῶν κροτάφων
20 διὰ τοῦ αὐχένος ὑπὸ τὰς ὠμοπλάτας,[10] ἔπειτα συμφέρονται ἐς τὸν πλεύμονα καὶ ἀφικνέονται ἡ μὲν ἀπὸ τῶν δεξιῶν ἐς τὰ ἀριστερά, ἡ δὲ ἀπὸ τῶν ἀριστερῶν ἐς τὰ δεξιά, καὶ ἡ μὲν δεξιὴ ἀφικνεῖται ἐκ τοῦ πλεύμονος[11] ὑπὸ τὸν μαζὸν καὶ ἐς τὸν σπλῆνα καὶ ἐς τὸν νεφρόν, ἡ δὲ ἀπὸ τῶν ἀριστερῶν ἐς τὰ δεξιὰ ἐκ τοῦ πλεύμονος ὑπὸ τὸν μαζὸν

[1] ἀποκλῃίζεται A: ἀπαλλάσσεται MV: ἀναλώσεται Littré from a note of Galen.
[2] παρὰ AM (M has ἐπὶ above the line): ἐπὶ V.
[3] Fredrich reads ἐς for παρά.

them, as the strength of the stronger part will easily consume the humours that flow into them.

XI. The thickest of the veins have the following nature. There are four pairs in the body. One pair extends from behind the head through the neck, and on either side of the spine externally reaches to the loins and legs, and then stretches through the shanks to the outside of the ankles and to the feet. So bleeding for pains in the back and loins should be made on the outside, behind the knee or at the ankle. The other pair of veins extend from the head by the ears through the neck, and are called jugular veins. They stretch right and left by the side of the spine internally along the loins to the testicles and thighs, then on the inside through the hollow of the knee, and finally through the shanks to the ankles on the inside and to the feet. Accordingly, to counteract pains in the loins and testicles, bleeding should be performed in the hollow of the knee and in the ankles on the inner side. The third pair of veins passes from the temples through the neck under the shoulder-blades, then they meet in the lungs and reach, the one on the right the left side, and the one on the left the right. The right one reaches from the lungs under the breast both to the spleen and to the kidneys, and the left one to the right from the lungs under

[4] ἐπὶ A : καὶ MV. [5] διήκει MV : ἀφίκη A.
[6] ἔξωθεν MV : ποιέεσθαι A.
[7] φλέβες ἀπὸ A : φλέβες ἔχουσιν ἐκ MV : φλέβες ἐκ Littré.
[8] Fredrich brackets ἐκ.
[9] παρὰ MV : ἐπὶ A.
[10] ὠμοπλάτας MV : ὠμοπλάτους A.
[11] A omits ἡ δὲ ἀπὸ . . . . πλεύμονος.

καὶ ἐς τὸ ἧπαρ καὶ ἐς τὸν νεφρόν, τελευτῶσι δὲ ἐς τὸν ἀρχὸν αὗται ἀμφότεραι. αἱ δὲ τέταρται ἀπὸ τοῦ ἔμπροσθεν τῆς κεφαλῆς καὶ τῶν ὀφθαλμῶν
30 ὑπὸ τὸν αὐχένα καὶ τὰς κληῖδας, ἔπειτα δὲ ἐπὶ[1] τῶν βραχιόνων ἄνωθεν ἐς τὰς συγκαμπάς, ἔπειτα δὲ διὰ τῶν πήχεων ἐς τοὺς καρποὺς καὶ τοὺς δακτύλους, ἔπειτα ἀπὸ τῶν δακτύλων πάλιν διὰ τῶν στηθέων καὶ τῶν πήχεων ἄνω ἐς τὰς συγκαμπάς, καὶ διὰ τῶν βραχιόνων τοῦ κάτωθεν μέρεος ἐς τὰς μασχάλας, καὶ ἐκ τῶν πλευρέων ἄνωθεν ἡ μὲν ἐς τὸν σπλῆνα ἀφικνεῖται, ἡ δὲ ἐς τὸ ἧπαρ, ἔπειτα δὲ ὑπὲρ τῆς γαστρὸς ἐς τὸ αἰδοῖον τελευτῶσιν ἀμφότεραι. καὶ αἱ μὲν παχέαι[2] τῶν
40 φλεβῶν ὧδε ἔχουσιν.[3] εἰσὶ δὲ καὶ ἀπὸ τῆς κοιλίης φλέβες ἀνὰ τὸ σῶμα πάμπολλαί[4] τε καὶ παντοῖαι, δι᾽ ὧν ἡ τροφὴ τῷ σώματι ἔρχεται. φέρουσι δὲ καὶ ἀπὸ τῶν παχειῶν φλεβῶν ἐς τὴν κοιλίην καὶ τὸ ἄλλο σῶμα καὶ ἀπὸ τῶν ἔξω[5] καὶ ἀπὸ τῶν ἔσω, καὶ ἐς ἀλλήλας διαδιδόασιν[6] αἵ τε ἔσωθεν ἔξω καὶ αἱ ἔξωθεν ἔσω. τὰς οὖν φλεβοτομίας[7] ποιεῖσθαι κατὰ τούτους τοὺς λόγους· ἐπιτηδεύειν δὲ χρὴ τὰς τομὰς ὡς προσωτάτω τάμνειν ἀπὸ τῶν χωρίων, ἔνθα ἂν αἱ
50 ὀδύναι μεμαθήκωσι[8] γίνεσθαι καὶ τὸ αἷμα συλλέγεσθαι· οὕτω γὰρ ἂν ἥ τε μεταβολὴ ἥκιστα γίνοιτο μεγάλη ἐξαπίνης, καὶ τὸ ἔθος μεταστήσαις[9] ἂν ὥστε μηκέτι ἐς τὸ αὐτὸ χωρίον
54 συλλέγεσθαι.

[1] ἔπειτα δὲ ἐπὶ A : ἔπειτα ὑπὲρ MV.
[2] παχέαι A : παχύταται MV.
[3] ὧδε ἔχουσιν MV : οὕτω πεφύκασιν A.
[4] πολαὶ (*sic*) A : παμπολαὶ M : παμπολλαὶ V.

the breast both to the liver and to the kidneys, both of them ending at the anus. The fourth pair begin at the front of the head and eyes, under the neck and collar-bones, passing on the upper part of the arms to the elbows, then through the forearms to the wrists and fingers, then back from the fingers they go through the ball of the hand and the fore-arm upwards to the elbow, and through the upper arm on the under side to the armpit, and from the ribs above one reaches to the spleen and the other to the liver, and finally both pass over the belly to the privy parts Such is the arrangement of the thick veins. From the belly too extend over the body very many veins of all sorts, by which nourishment comes to the body. Veins too lead from the thick veins to the belly and to the rest of the body both from the outside and from the inside; they communicate with one another, the inside ones outside and the outside ones inside. Bleeding then should be practised according to these principles. The habit should be cultivated of cutting as far as possible from the places where the pains are wont to occur and the blood to collect. In this way the change will be least sudden and violent, and you will change the habit so that the blood no longer collects in the same place.

---

[5] *ἐξωτάτων* A: *ἐξωτάτωι* (with *-τάτωι καὶ ἀπὸ τῶν* deleted) M: *ἔξω* V.

[6] *διαδίδουσιν* A: *διαδιδόασιν* (with *δια-* half erased) M: *διδόασιν* V.

[7] After *φλεβοτομίας* A has *χρή*.

[8] *μεμαθητικόσι* with *μεμαθήκασι* in margin A: *μεμαθήκασι* MV: *μεμαθήκωσι* Littré.

[9] *μεταστήσιας* A: *μεταστῆσαι* MV: *μεταστήσαις* Littré.

XII. Ὅσοι πῦον πολλὸν πτύουσιν ἄτερ πυρετοῦ ἐόντες,[1] καὶ οἷσιν ὑπὸ τὸ οὖρον πῦον ὑφίσταται πολλὸν ἄτερ ὀδύνης ἐοῦσι,[2] καὶ ὅσοις τὰ ὑποχωρήματα αἱματώδεα ὥσπερ ἐν τῇσι δυσεντερίῃσι καὶ χρόνιά ἐστιν ἐοῦσι[3] πέντε καὶ τριήκοντα ἐτέων καὶ γεραιτέροισι, τούτοισι πᾶσιν ἀπὸ τοῦ αὐτοῦ τὰ νοσήματα γίνεται· ἀνάγκη γὰρ τούτους ταλαιπώρους τε γενέσθαι[4] καὶ φιλοπόνους τῷ σώματι καὶ ἐργάτας νεηνίσκους ἐόντας, ἔπειτα δὲ ἐξανεθέντας τῶν πόνων σαρκωθῆναι μαλθακῇ σαρκὶ καὶ πολὺ διαφερούσῃ τῆς προτέρης, καὶ πολλὸν διακεκριμένον ἔχειν τὸ σῶμα τό τε προϋπάρχον καὶ τὸ ἐπιτραφέν, ὥστε μὴ ὁμονοεῖν.[5] ὅταν οὖν νόσημά τι καταλάβῃ τοὺς οὕτω διακειμένους, τὸ μὲν παραχρῆμα διαφεύγουσιν, ὕστερον δὲ μετὰ τὴν νοῦσον χρόνῳ τήκεται τὸ σῶμα, καὶ ῥεῖ[6] διὰ τῶν φλεβῶν, ᾗ ἂν εὐρυχωρίης μάλιστα τύχῃ, ἰχωροειδές· ἢν μὲν οὖν ὁρμήσῃ ἐς τὴν κοιλίην τὴν κάτω, σχεδόν τι οἷόν περ ἐν τῷ σώματι ἂν ἐνῇ[7] τοιοῦτον καὶ τὸ διαχώρημα γίνεται· ἅ τε γὰρ τῆς ὁδοῦ κατάντεος ἐούσης, οὐχ ἵσταται[8] πολὺν χρόνον ἐν τῷ ἐντέρῳ. οἷσι δ' ἂν ἐς τὰ στήθεα ἐσρυῇ, ὑπόπυον γίνεται· ἅτε γὰρ τῆς καθάρσιος ἀνάντεος ἐούσης, καὶ χρόνον ἐναυλιζόμενον πολὺν ἐν τῷ στήθει, κατασήπεται καὶ γίνεται πυοειδές. οἷσι δ' ἂν ἐς τὴν κύστιν ἐξερεύγηται, ὑπὸ τῆς θερμότητος τοῦ χωρίου τοῦτο καὶ θερμὸν καὶ[9] λευκὸν γίνεται, καὶ διακρίνεται·

[1] ἄτερ πυρετοῦ ἐόντες AV : ἄτε πυρετοῦ ἐόντος (ες written above) M : ἐόντος C (according to Littré).

[2] ἐοῦσι AV : ἐοῦσιν M : ἐούσης Littré.

XII. Such as expectorate much pus without fever, or have a thick sediment of pus in the urine without pain, or whose stools remain stained with blood, as in dysentery, during a long period, being thirty-five years or older, all these are ill from the same cause. For these patients must have been in their youth hard-working, diligent and industrious; afterwards when delivered from their labours they must have put on soft flesh very different from their former flesh, and there must be a wide difference between the previous condition and the hypertrophied condition of their body, so that there is no longer harmony. Accordingly when a disease seizes men in such a condition, at first they escape, but after the disease the body in time wastes and serous matter flows through the veins wherever it finds the broadest passage. Now if the flux be to the lower bowel, the stools become very like the matter shut up in the body, because as the passage slopes downwards the matter cannot remain long in the intestine. When the flux is to the chest the patients suffer suppuration, because since the purging is along an upward passage and abides a long time in the chest it rots and turns to pus. When the matter empties itself into the bladder, owing to the warmth of the place the matter becomes hot and white, and separates itself

---

[3] *ἐστίν, ἐοῦσι* A : *ἃ* (on *ἐστι*?) *νέοισιν ἐοῦσιν* M : *ἃ νέοισιν ἐοῦσι* V.

[4] *τε γενέσθαι* MV : *γεγενῆσθαι* A.

[5] *ὁμολογέειν* AV : *ὁμονοέειν* M (in margin *ὁμολογέειν*).

[6] *ῥεῖ* A : *ῥέει* M : *διαρρέει* V.

[7] *ἐνῇ* A : *ἐνέη* MV : *ἂν ἐνέῃ* Littré.

[8] *ἵσταται* A : *ἴσχεται* MV.

[9] *τοῦτο καὶ θερμὸν καὶ* omitted by A.

καὶ τὸ μὲν ἀραιότατον ἐφίσταται[1] ἄνω, τὸ δὲ
30 παχύτατον κάτω, ὃ δὴ πῦον καλεῖται. γίνονται
δὲ καὶ οἱ λίθοι τοῖσι παιδίοισι διὰ τὴν θερμότητα
τοῦ χωρίου τε τούτου καὶ τοῦ ὅλου σώματος,
τοῖσι δὲ ἀνδράσιν οὐ γίνονται λίθοι διὰ τὴν
ψυχρότητα τοῦ σώματος. εὖ γὰρ χρὴ εἰδέναι,
ὅτι ὁ ἄνθρωπος τῇ πρώτῃ τῶν ἡμερέων θερμότατός
ἐστιν αὐτὸς ἑωυτοῦ, τῇ δὲ ὑστάτῃ ψυχρότατος·
ἀνάγκη γὰρ αὐξανόμενον καὶ χωρέον τὸ
σῶμα πρὸς βίην θερμὸν εἶναι· ὅταν δὲ ἄρχηται
μαραίνεσθαι[2] τὸ σῶμα, καταρρέον πρὸς εὐπέτειαν,
ψυχρότερον γίνεται· καὶ κατὰ τοῦτον τὸν λόγον,
40 ὅσον τῇ πρώτῃ τῶν ἡμερέων πλεῖστον αὔξεται ὁ
ἄνθρωπος, τοσοῦτον θερμότερος[3] γίνεται, καὶ τῇ
ὑστάτῃ τῶν ἡμερέων, ὅσον πλεῖστον καταμαραίνεται,
τοσοῦτον ἀνάγκη ψυχρότερον[4] εἶναι.
ὑγιέες δὲ γίγνονται αὐτόματοι οἱ οὕτω διακείμενοι,
πλεῖστοι μὲν ἐν τῇ ὥρῃ, ᾗ ἂν ἄρξωνται
τήκεσθαι, πεντεκαιτεσσαρακονθήμεροι·[5] ὅσοι δ'
ἂν τὴν ὥρην ταύτην ὑπερβάλλωσιν, ἐνιαυτῷ
αὐτόματοι ὑγιέες γίνονται,[6] ἢν μή τι ἕτερον κακουργῆται
49 ὤνθρωπος.

XIII. Ὅσα τῶν νοσημάτων ἐξ ὀλίγου γίνεται,
καὶ ὅσων αἱ προφάσιες εὔγνωστοι, ταῦτα δὲ
ἀσφαλέστατά ἐστι προαγορεύεσθαι· τὴν δὲ ἴησιν
χρὴ ποιεῖσθαι αὐτὸν[7] ἐναντιούμενον τῇ προφάσει
τῆς νούσου· οὕτω γὰρ ἂν λύοιτο τὸ τὴν νοῦσον
6 παρασχὸν ἐν τῷ σώματι.[8]

[1] ἐφίσταται MV : ἀμφίσταται A.
[2] ἄρχηται μαραίνεσθαι MV : ἄρξηται θερμαίνεσθαι A.
[3] θερμότερος MV : θερμότατος A.
[4] ἀνάγκη ψυχρότερον MV : ψυχρότατον ἀνάγκη A.

out. The finest part becomes scum on the top, while the thickest sinks to the bottom and is called pus. Stones too form in children because of the heat of this place and of the whole body, but in men stones do not form because of the coldness of the body. For you must know that a man is warmest on the first day of his existence and coldest on the last. For it must be that the body is hot which grows and progresses with force; but when the body begins to decay with an easy decline it grows cooler. It is on account of this that a man, growing most on his first day, is proportionally hotter then; on his last day, decaying most, he is proportionally cooler. Most patients in the condition described above recover their health spontaneously forty-five days from the day on which they began to waste. Such of them as exceed this period, should no other illness occur, recover spontaneously in a year.

XIII. Diseases which arise soon after their origin, and whose cause is clearly known, are those the history of which can be foretold with the greatest certainty. The patient himself must bring about a cure by combating the cause of the disease, for in this way will be removed that which caused the disease in the body.

---

[5] A omits τήκεσθαι and has β̄ τε καὶ μ̄ ἡμερέων. Galen mentions readings with 45 and 40. Villaret reads, δυοῖν καὶ τεσσαράκοντα ἡμερέων.

[6] A omits ὑγιέες γίνονται and M has it in the margin.

[7] αὐτὸν A: αὐτέων MV.

[8] οὕτω γὰρ λύοι τὸ τὴν νοῦσον παρεχον τῶ σώματι A: οὕτω γὰρ ἂν λύοιτο τὸ τὴν νοῦσον παρασχὸν ἐν τῷ σώματι M: οὕτω γὰρ ἂν λύοι το τὴν νοῦσον παρασχὸν ἐν τῶ σώματι V. Littré says that C has παρασχών.

XIV. Οἷσι δὲ ψαμμοειδέα ὑφίσταται ἢ πῶροι ἐν τοῖσιν οὔροισι, τούτοισι τὴν ἀρχὴν φύματα ἐγένετο πρὸς τῇ φλεβὶ τῇ παχείῃ, καὶ διεπύησεν, ἔπειτα δέ, ἅτε οὐ ταχέως ἐκραγέντων τῶν φυμάτων, πῶροι συνετράφησαν ἐκ τοῦ πύου, οἵτινες ἔξω θλίβονται[1] διὰ τῆς φλεβὸς σὺν τῷ οὔρῳ ἐς τὴν κύστιν. οἷσι δὲ μοῦνον αἱματώδεα[2] τὰ οὐρήματα, τούτοισι δὲ αἱ φλέβες πεπονήκασιν· οἷσι δὲ ἐν τῷ οὐρήματι παχεῖ ἐόντι σαρκία
10 σμικρὰ τριχοειδέα συνεξέρχεται, ταῦτα δὲ ἀπὸ τῶν νεφρῶν εἰδέναι χρὴ ἐόντα καὶ ἀπὸ ἀρθριτικῶν·[3] ὅσοισι δὲ καθαρὸν τὸ οὖρον, ἄλλοτε δὲ καὶ ἄλλοτε οἷον πίτυρα ἐμφαίνεται[4] ἐν τῷ οὐρήματι,
14 τούτων δὲ ἡ κύστις ψωριᾷ.

XV. Οἱ πλεῖστοι τῶν πυρετῶν γίνονται ἀπὸ χολῆς· εἴδεα δὲ σφέων ἐστὶ τέσσαρα, χωρὶς τῶν ἐν τῇσιν ὀδύνῃσι γινομένων τῇσιν ἀποκεκριμένῃσιν· ὀνόματα δ' αὐτοῖσίν ἐστι σύνοχος καὶ ἀμφημερινὸς καὶ τριταῖος καὶ τεταρταῖος. ὁ μὲν οὖν σύνοχος καλεόμενος γίνεται ἀπὸ πλείστης χολῆς καὶ ἀκρητεστάτης, καὶ τὰς κρίσιας ἐν ἐλαχίστῳ χρόνῳ ποιεῖται· τὸ γὰρ σῶμα οὐ διαψυχόμενον οὐδένα χρόνον συντήκεται ταχέως,
10 ἅτε ὑπὸ πολλοῦ τοῦ θερμοῦ θερμαινόμενον. ὁ δὲ ἀμφημερινὸς μετὰ τὸν σύνοχον ἀπὸ πλείστης χολῆς γίνεται, καὶ ἀπαλλάσσεται τάχιστα τῶν ἄλλων, μακρότερος δέ ἐστι τοῦ συνόχου, ὅσῳ ἀπὸ ἐλάσσονος γίνεται χολῆς, καὶ ὅτι ἔχει ἀνάπαυσιν τὸ σῶμα, ἐν δὲ τῷ συνόχῳ οὐκ ἀνα-

[1] ἐξ ὧν λείβονται A : ἐκ τοῦ πύου· οἵτινες ἔξω θλίβονται MV.
[2] MV omit μοῦνον and insert μὲν after αἱματώδεα.

XIV. Patients whose urine contains a deposit of sand or chalk suffer at first from tumours near the thick vein, with suppuration; then, since the tumours do not break quickly, from the pus there grow out pieces of chalk, which are pressed outside through the vein into the bladder with the urine. Those whose urine is merely blood-stained have suffered in the veins. When the urine is thick, and there are passed with it small pieces of flesh like hair, you must know that these symptoms result from the kidneys and arthritic complaints. When the urine is clear, but from time to time as it were bran appears in it, the patients suffer from psoriasis of the bladder.

XV. Most fevers come from bile. There are four sorts of them, apart from those that arise in distinctly separate pains.[1] Their names are the continued, the quotidian, the tertian and the quartan. Now what is called the continued fever comes from the most abundant and the purest bile, and its crises occur after the shortest interval. For since the body has no time to cool it wastes away rapidly, being warmed by the great heat. The quotidian next to the continued comes from the most abundant bile, and ceases quicker than any other, though it is longer than the continued, proportionately to the lesser quantity of bile from which it comes; moreover the body has a breathing space, whereas in the continued there is

[1] That is, apart from fevers which accompany certain specific diseases and various wounds. The four kinds of fevers are those now recognised as malarial.

---

[3] A omits *καὶ ἀπὸ ἀρθριτικῶν*.

[4] *ἄλλοτε δὲ καὶ ἄλλοτε οἷον πίτυρα ἐμφαίνεται* A: *ἄλλοτε καὶ ἄλλοτε· ὁκοῖον εἰ πίτυρα ἐπιφαίνεται* MV.

παύεται οὐδένα χρόνον. ὁ δὲ τριταῖος μακρότερός ἐστι τοῦ ἀμφημερινοῦ, καὶ ἀπὸ χολῆς ἐλάσσονος γίνεται· ὅσῳ δὲ πλείονα χρόνον ἐν τῷ τριταίῳ ἢ ἐν τῷ ἀμφημερινῷ τὸ σῶμα ἀναπαύε-
20 ται, τοσούτῳ χρονιώτερος οὗτος ὁ πυρετὸς τοῦ ἀμφημερινοῦ ἐστίν. οἱ δὲ τεταρταῖοι τὰ μὲν ἄλλα κατὰ τὸν αὐτὸν λόγον, χρονιώτεροι δέ εἰσι τῶν τριταίων, ὅσῳ ἔλασσον μετέχουσι μέρος τῆς χολῆς τῆς τὴν θερμασίην παρεχούσης, τοῦ τε[1] διαψύχεσθαι τὸ σῶμα πλέον μετέχουσιν· προσγίνεται δὲ αὐτοῖσιν ἀπὸ μελαίνης χολῆς τὸ περισσὸν[2] τοῦτο καὶ δυσαπάλλακτον· μέλαινα γὰρ χολὴ τῶν ἐν τῷ σώματι ἐνεόντων χυμῶν γλισχρότατον, καὶ τὰς ἕδρας χρονιωτάτας ποιεῖται.
30 γνώσῃ δὲ τῷδε, ὅτι οἱ τεταρταῖοι πυρετοὶ μετέχουσι τοῦ μελαγχολικοῦ· φθινοπώρου μάλιστα οἱ ἄνθρωποι ἁλίσκονται ὑπὸ τῶν τεταρταίων καὶ ἐν τῇ ἡλικίῃ τῇ ἀπὸ πέντε καὶ εἴκοσιν[3] ἐτέων ἐς τὰ πέντε καὶ τεσσαράκοντα,[4] ἡ δὲ ἡλικίη αὕτη ὑπὸ μελαίνης χολῆς κατέχεται μάλιστα πασέων τῶν ἡλικιῶν, ἥ τε φθινοπωρινὴ ὥρη μάλιστα πασέων τῶν ὡρέων. ὅσοι δ' ἂν ἁλῶσιν ἔξω τῆς ὥρης ταύτης καὶ τῆς ἡλικίης ὑπὸ τεταρταίου, εὖ χρὴ εἰδέναι μὴ χρόνιον ἐσόμενον τὸν πυρετόν, ἢν
40 μὴ ἄλλο τι κακουργῆται ὤνθρωπος.

[1] τοῦ τε A: τοῦ δὲ MV.
[2] τό τε περισσὸν A: τὸ περισσὸν MV.
[3] (ε̅· καὶ) κ̅ A: εἴκοσιν M: τριήκοντα V.
[4] β̅ τε καὶ μ̅ A: πέντε καὶ τεσσαράκοντα MV.

no breathing space at all. The tertian is longer than the quotidian and is the result of less bile. The longer the breathing space enjoyed by the body in the case of the tertian than in the case of the quotidian, the longer this fever is than the quotidian. The quartans are in general similar, but they are more protracted than the tertians in so far as their portion is less of the bile that causes heat, while the intervals are greater in which the body cools. It is from black bile that this excessive obstinacy arises. For black bile is the most viscous of the humours in the body, and that which sticks fast the longest. Hereby you will know that quartan fevers participate in the atrabilious element, because it is mostly in autumn that men are attacked by quartans, and between the ages of twenty-five and forty-five.[1] This age is that which of all ages is most under the mastery of black bile, just as autumn is the season of all seasons which is most under its mastery. Such as are attacked by a quartan fever outside this period and this age you may be sure will not suffer from a long fever, unless the patient be the victim of another malady as well.

[1] With the reading of A, "forty-two."

REGIMEN IN HEALTH

ΠΕΡΙ ΔΙΑΙΤΗΣ ΥΓΙΕΙΝΗΣ

I. Τοὺς ἰδιώτας ὧδε χρὴ διαιτᾶσθαι· τοῦ μὲν χειμῶνος ἐσθίειν ὡς πλεῖστα, πίνειν δ᾽ ὡς ἐλάχιστα, εἶναι δὲ τὸ πόμα οἶνον ὡς ἀκρητέστατον, τὰ δὲ σιτία ἄρτον καὶ τὰ ὄψα ὀπτὰ πάντα, λαχάνοισι δὲ ὡς ἐλαχίστοισι χρῆσθαι ταύτην τὴν ὥρην· οὕτω γὰρ ἂν μάλιστα τὸ σῶμα ξηρόν τε εἴη καὶ θερμόν. ὅταν δὲ τὸ ἔαρ ἐπιλαμβάνῃ, τότε πόμα χρὴ πλέον ποιεῖσθαι καὶ ὑδαρέστατον[1] καὶ κατ᾽ ὀλίγον, καὶ τοῖσι σιτίοισι μαλακωτέροισι 10 χρῆσθαι καὶ ἐλάσσοσι, καὶ τὸν ἄρτον ἀφαιρέοντα μάζαν προστιθέναι, καὶ τὰ ὄψα κατὰ τὸν αὐτὸν λόγον ἀφαιρεῖν, καὶ ἐκ τῶν ὀπτῶν πάντα ἑφθὰ ποιεῖσθαι, καὶ λαχάνοισιν ἤδη χρῆσθαι[2] τοῦ ἦρος ὀλίγοις, ὅπως ἐς τὴν θερίην καταστήσεται ὥνθρωπος τοῖσί τε σιτίοισι μαλθακοῖσι πᾶσι χρώμενος[3] καὶ τοῖσιν ὄψοισιν ἑφθοῖσι καὶ λαχάνοις ὠμοῖσι καὶ ἑφθοῖσι· καὶ τοῖσι πόμασιν, ὡς ὑδαρεστάτοισι καὶ πλείστοισιν, ἀλλ᾽ ὅπως μὴ μεγάλη[4] ἡ μεταβολὴ ἔσται κατὰ μικρὸν μὴ 20 ἐξαπίνης χρωμένῳ. τοῦ δὲ θέρεος τῇ τε μάζῃ μαλακῇ[5] καὶ τῷ ποτῷ ὑδαρεῖ καὶ πολλῷ καὶ τοῖσιν ὄψοισι ἑφθοῖσι πᾶσι· δεῖ γὰρ χρῆσθαι

[1] ποιεεσθαι· καὶ ὑδαρέστατον A : ποιέειν καὶ ὑδαρέστερον MV.
[2] ἤδη χρῆσθαι A : διαχρῆσθαι MV.
[3] μαλθακοῖσι πᾶσι χρώμενος A : μαλακωτέροισι χρεόμενος M : μαλακωτέροισι χρώμενος V.

REGIMEN IN HEALTH

I THE layman[1] ought to order his regimen in the following way. In winter eat as much as possible and drink as little as possible; drink should be wine as undiluted as possible, and food should be bread, with all meats roasted; during this season take as few vegetables as possible, for so will the body be most dry and hot. When spring comes, increase drink and make it very diluted, taking a little at a time; use softer foods and less in quantity; substitute for bread barley-cake; on the same principle diminish meats, taking them all boiled instead of roasted, and eating when spring comes a few vegetables, in order that a man may be prepared for summer by taking all foods soft, meats boiled, and vegetables raw or boiled. Drinks should be as diluted and as copious as possible, the change to be slight, gradual and not sudden. In summer the barley-cake to be soft, the drink diluted and copious, and the meats in all cases boiled. For one must use these, when it is

[1] By "layman" (ἰδιώτης) in this passage is meant the ordinary, normal person, whose business does not require, as does that, *e.g.*, of the professional athlete, special diet and exercise.

---

[4] *καὶ μὴ μεγάλη* A: *καὶ ὅκως μὴ μεγάλη* (*ὅκως* above line in another hand) M: *καὶ ὅκως μὴ μεγάλη* V. Villaret omits *ὅπως* and reads *ἔστω* for *ἔσται*.

[5] Here V has *τρέφεσθαι*.

τούτοισιν, ὅταν θέρος ᾖ, ὅπως τὸ σῶμα ψυχρὸν καὶ μαλακὸν γένηται· ἡ γὰρ ὥρη θερμή τε καὶ ξηρή, καὶ παρέχεται τὰ σώματα καυματώδεα καὶ αὐχμηρά· δεῖ οὖν τοῖσιν ἐπιτηδεύμασιν ἀλέξασθαι. κατὰ δὲ τὸν αὐτὸν λόγον, ὥσπερ ἐκ τοῦ χειμῶνος ἐς τὸ ἦρ, οὕτω ἐκ τοῦ ἦρος ἐς τὸ θέρος καταστήσεται,[1] *τῶν μὲν σιτίων ἀφαιρέων, τῷ δὲ ποτῷ* 30 *προστιθείς· οὕτω δὲ καὶ τὰ ἐναντία ποιέοντα καταστῆσαι ἐκ τοῦ θέρεος ἐς τὸν χειμῶνα. ἐν δὲ τῷ φθινοπώρῳ τὰ μὲν σιτία πλέω ποιεύμενον καὶ ξηρότερα καὶ τὰ ὄψα κατὰ λόγον, τὰ δὲ ποτὰ ἐλάσσω καὶ ἀκρητέστερα, ὅπως ὅ τε χειμὼν ἀγαθὸς*[2] *ἔσται καὶ ὥνθρωπος διαχρήσεται τοῖσί τε πόμασιν ἀκρητεστάτοισι καὶ ὀλίγοισι καὶ τοῖσι σιτίοισιν ὡς πλείστοισί τε καὶ ξηροτάτοισιν· οὕτω γὰρ ἂν καὶ ὑγιαίνοι μάλιστα καὶ ῥιγῴη* 39 *ἥκιστα· ἡ γὰρ ὥρη ψυχρή τε καὶ ὑγρή.*

II. Τοῖσι δὲ εἴδεσι τοῖσι σαρκώδεσι καὶ μαλακοῖσι καὶ ἐρυθροῖσι συμφέρει τὸν πλεῖστον χρόνον τοῦ ἐνιαυτοῦ ξηροτέροισι τοῖσι διαιτήμασι χρῆσθαι· ὑγρὴ γὰρ ἡ φύσις τῶν εἰδέων τούτων. τοὺς δὲ στρυφνούς τε καὶ προσεσταλμένους καὶ πυρροὺς καὶ μέλανας τῇ ὑγροτέρῃ διαίτῃ χρῆσθαι τὸ πλεῖον τοῦ χρόνου· τὰ γὰρ σώματα τοιαῦτα ὑπάρχει ξηρὰ ἐόντα. καὶ τοῖσι νέοισι τῶν σωμάτων συμφέρει μαλθακωτέροισί τε καὶ ὑγρο- 10 τέροισι χρῆσθαι τοῖσι διαιτήμασιν· ἡ γὰρ ἡλικίη ξηρή, καὶ τὰ σώματα πέπηγεν.[3] τοὺς δὲ πρεσβυτέρους τῷ ξηροτέρῳ τρόπῳ χρὴ τὸ πλέον τοῦ χρόνου διάγειν· τὰ γὰρ σώματα ἐν ταύτῃ τῇ

[1] ἐκ τοῦ χειμῶνος ἐς τὸ ἦρ, οὕτω (καὶ M) ἐκ τοῦ ἦρος εἰς [ἐς M] τὸ θέρος καταστήσεται MV: ἕως τὸ ἔαρ ἐκ τοῦ χειμῶνος· οὕτως

summer, that the body may become cold and soft. For the season is hot and dry, and makes bodies burning and parched. Accordingly these conditions must be counteracted by way of living. On the same principle the change from spring to summer will be prepared for in like manner to that from winter to spring, by lessening food and increasing drink. Similarly, by opposing opposites prepare for the change from summer to winter. In autumn make food more abundant and drier, and meats too similar, while drinks should be smaller and less diluted, so that the winter may be healthy and a man may take his drink neat and scanty and his food as abundant and as dry as possible. For in this way he will be most healthy and least chilly, as the season is cold and wet.

II. Those with physiques that are fleshy, soft and red, find it beneficial to adopt a rather dry regimen for the greater part of the year. For the nature of these physiques is moist. Those that are lean and sinewy, whether ruddy or dark, should adopt a moister regimen for the greater part of the time, for the bodies of such are constitutionally dry. Young people also do well to adopt a softer and moister regimen, for this age is dry, and young bodies are firm. Older people should have a drier kind of diet for the greater part of the time,

---

ἐς τὸ θέρος καταστῆσαι A. In some respects the reading of A is preferable, except for the use of ἕως.

[2] A omits ἀγαθὸς here. The sentence which follows seems a rather idle repetition of the preceding, and possibly the whole from ὅπως ὅ τε χειμὼν to ξηροτάτοισιν should be deleted as a gloss.

[3] καὶ τὰ σώματα πέπηγεν MV (πέπηγε M): πέπηγεν ἔτι A.

ἡλικίῃ ὑγρὰ καὶ μαλθακὰ καὶ ψυχρά. δεῖ οὖν πρὸς τὴν ἡλικίην καὶ τὴν ὥρην καὶ τὸ ἔθος καὶ τὴν χώρην[1] καὶ τὰ εἴδεα τὰ διαιτήματα ποιεῖσθαι ἐναντιούμενον τοῖσι καθισταμένοισι καὶ θάλπεσι
18 καὶ χειμῶσιν· οὕτω γὰρ ἂν μάλιστα ὑγιαίνοιεν.

III. Καὶ ὁδοιπορεῖν τοῦ μὲν χειμῶνος ταχέως χρή, τοῦ δὲ θέρεος ἡσυχῇ, ἢν μὴ διὰ καύματος ὁδοιπορῇ· δεῖ δὲ τοὺς μὲν σαρκώδεας θᾶσσον ὁδοιπορεῖν,[2] τοὺς δὲ ἰσχνοὺς ἡσυχαίτερον.[3] λουτροῖσι δὲ χρὴ πολλοῖσι χρῆσθαι τοῦ θέρεος, τοῦ δὲ χειμῶνος ἐλάσσοσι, τοὺς στρυφνοὺς χρὴ μᾶλλον λούεσθαι τῶν σαρκωδέων. ἠμφιέσθαι δὲ χρὴ τοῦ μὲν χειμῶνος καθαρὰ ἱμάτια, τοῦ δὲ
9 θέρεος ἐλαιοπινέα.

IV. Τοὺς δὲ παχέας χρή, ὅσοι βούλονται λεπτοὶ γενέσθαι, τὰς ταλαιπωρίας νήστιας ἐόντας ποιεῖσθαι ἁπάσας, καὶ τοῖσι σιτίοισιν ἐπιχειρεῖν ἀσθμαίνοντας καὶ μὴ ἀνεψυγμένους καὶ προπεπωκότας οἶνον κεκρημένον μὴ σφόδρα ψυχρόν, καὶ τὰ ὄψα σκευάζειν σησάμοις ἢ ἡδύσμασι καὶ τοῖσιν ἄλλοισι τοῖς τοιουτοτρόποισι· καὶ πίονα δὲ[4] ἔστω· οὕτω γὰρ ἂν ἀπὸ ἐλαχίστων ἐμπιπλαῖντο· καὶ μονοσιτεῖν καὶ ἀλουτεῖν καὶ
10 σκληροκοιτεῖν καὶ γυμνὸν περιπατεῖν ὅσον οἷόν τε μάλιστ᾽ ἂν ᾖ. ὅσοι δὲ βούλονται λεπτοὶ ἐόντες παχέες γενέσθαι, τά τε ἄλλα ποιεῖν τἀναντία κείνοις, καὶ νήστιας μηδεμίην ταλαιπωρίην
14 ποιεῖσθαι.[5]

[1] καὶ τὸ ἔθος καὶ τὴν χώρην omitted by A.
[2] A reads ἡλίου for καύματος and omits ὁδοιπορῇ to θᾶσσον.
[3] ἡσυχαίτερον AV and Holkhamensis 282 : ἡσυχαίστερον M : ὀλιγέστερον Caius 50 and (according to Littré) C. This

for bodies at this age are moist and soft and cold. So in fixing regimen pay attention to age, season, habit, land, and physique, and counteract the prevailing heat or cold. For in this way will the best health be enjoyed.

III. Walking should be rapid in winter and slow in summer, unless it be under a burning heat. Fleshy people should work faster, thin people slower. Bathe frequently in summer, less in winter, and the lean should bathe more than the fleshy. In winter wear unoiled cloaks, but soak them in oil in summer.

IV. Fat people who wish to become thin should always fast when they undertake exertion, and take their food while they are panting and before they have cooled, drinking beforehand diluted wine that is not very cold. Their meats should be seasoned with sesame, sweet spices, and things of that sort. Let them also be rich. For so the appetite will be satisfied with a minimum. They should take only one full meal a day, refrain from bathing, lie on a hard bed, and walk lightly clad as much as is possible. Thin people who wish to become fat should do the opposite of these things, and in particular they should never undertake exertion when fasting.

---

curious difference between Holkhamensis and C (both copies of V), and agreement of C with a MS. of a totally different class, cannot be due to mere chance. As both C and Holkhamensis were copied by the same scribe from V, it looks as though C had been "edited."

[4] Villaret omits *δέ*.

[5] *νήστιας μηδεμίην ταλαιπωρίην ποιέεσθαι* Littré: *νηστείην μηδεμίην καὶ ταλαιπωρίην ποιέεσθαι* (with *καὶ* above the line) A: *νῆστιν* (M *νηστιν* with final *-ν* on an erasure) *μὴδὲ μίην ποιέεσθαι* MV.

V. Τοῖσι δὲ ἐμέτοισι χρὴ καὶ τοῖσι κατακλύσμασι τοῖσι τῆς κοιλίης ὧδε χρῆσθαι· ἓξ μῆνας τοὺς χειμερινοὺς ἐμεῖν, οὗτος γὰρ ὁ χρόνος φλεγματωδέστερος τοῦ θερινοῦ, καὶ τὰ νοσήματα γίνεται περὶ τὴν κεφαλὴν καὶ τὸ χωρίον τοῦτο τὸ ὑπὲρ τῶν φρενῶν· ὅταν δὲ ᾖ θάλπος, τοῖσι κατακλύσμασι χρῆσθαι, ἡ γὰρ ὥρη καυματώδης, καὶ χολωδέστερον τὸ σῶμα, καὶ βαρύτητες ἐν τῇ ὀσφύϊ καὶ ἐν τοῖσι γούνασι, καὶ θέρμαι γίνονται, καὶ ἐν τῇ 10 γαστρὶ στρόφοι· δεῖ οὖν τὸ σῶμα ψύχειν καὶ τὰ μετεωριζόμενα κάτω ὑπάγειν ἀπὸ[1] τῶν χωρίων τούτων. ἔστω δὲ τὰ κατακλύσματα τοῖσι μὲν παχυτέροισι καὶ ὑγροτέροισιν ἁλμυρώτερα καὶ λεπτότερα, τοῖσι δὲ ξηροτέροισι καὶ προσεσταλμένοισι καὶ ἀσθενεστέροισι λιπαρώτερα καὶ παχύτερα· ἔστι δὲ τῶν κατακλυσμάτων λιπαρὰ καὶ παχέα τὰ ἀπὸ τῶν γαλάκτων καὶ ἀπὸ ἐρεβίνθων ὕδωρ ἑφθὸν καὶ τῶν ἄλλων τοιούτων· λεπτὰ δὲ καὶ ἁλμυρὰ τὰ τοιαῦτα, ἅλμη[2] καὶ 20 θάλασσα. τοὺς δὲ ἐμέτους ὧδε χρὴ ποιεῖσθαι· ὅσοι μὲν τῶν ἀνθρώπων παχέες εἰσὶ καὶ μὴ ἰσχνοί,[3] νήστιες ἐμεόντων δραμόντες ἢ ὁδοιπορήσαντες διὰ τάχεος κατὰ μέσον ἡμέρης· ἔστω δὲ ἡμικοτύλιον ὑσσώπου τετριμμένης ἐν ὕδατος χοέϊ, καὶ τοῦτο ἐκπιέτω, ὄξος παραχέων καὶ ἅλας παραβάλλων, ὅπως ἂν μέλλῃ ἥδιστον ἔσεσθαι, πινέτω δὲ τὸ πρῶτον ἡσυχαίτερον, ἔπειτα δ' ἐπὶ θᾶσσον. οἱ δὲ λεπτότεροι καὶ ἀσθενέσ-

[1] ἀπὸ A : ἐκ MV.

[2] ἅλμη A : κράμβῃ MV and Holkhamensis 282 : κράμβην C (according to Littré). If Littré has correctly collated C, this is a case where this MS. differs from V and the Holkham MS.

V. Emetics and clysters for the bowels should be used thus. Use emetics during the six winter months, for this period engenders more phlegm than does the summer, and in it occur the diseases that attack the head and the region above the diaphragm. But when the weather is hot use clysters, for the season is burning, the body bilious, heaviness is felt in the loins and knees, feverishness comes on and colic in the belly. So the body must be cooled, and the humours that rise must be drawn downwards from these regions. For people inclined to fatness and moistness let the clysters be rather salt and thin; for those inclined to dryness, leanness and weakness let them be rather greasy and thick. Greasy, thick clysters are prepared from milk, or water boiled with chick-peas or similar things. Thin, salt clysters are made of things like brine and sea-water. Emetics should be employed thus. Men who are fat and not thin should take an emetic fasting after running or walking quickly in the middle of the day. Let the emetic consist of half a *cotyle* of hyssop compounded with a *chous*[1] of water, and let the patient drink this, pouring in vinegar and adding salt, in such a way as to make the mixture as agreeable as possible. Let him drink it quietly at first, and then more quickly. Thinner and weaker people should partake of food

[1] If the *chous* contained 12 *cotylae* or 5¾ pints it is plain that the prescription gives the proportions of the mixture rather than the size of the dose. A dose of 6 pints seems heroic.

---

[3] A omits καὶ μὴ ἰσχνοί.

τεροι ἀπὸ σιτίων ποιείσθωσαν τὸν ἔμετον
30 τρόπον τοιόνδε· λουσάμενος θερμῷ προπιέτω
ἀκρήτου κοτύλην, ἔπειτα σιτία παντοδαπὰ
ἐσθιέτω, καὶ μὴ πινέτω ἐπὶ τῷ σιτίῳ μηδ' ἀπὸ
τοῦ σιτίου, ἀλλ' ἐπισχέτω ὅσον δέκα στάδια
διελθεῖν, ἔπειτα δὲ συμμίξας οἴνους τρεῖς πίνειν
διδόναι αὐστηρὸν καὶ γλυκὺν καὶ ὀξύν, πρῶτον μὲν
ἀκρητέστερόν τε καὶ κατ' ὀλίγον καὶ διὰ πολλοῦ
χρόνου, ἔπειτα δὲ ὑδαρέστερόν τε καὶ θᾶσσον καὶ
κατὰ πολλόν. ὅστις δὲ εἴωθε τοῦ μηνὸς δὶς
ἐξεμεῖν, ἄμεινον ἐφεξῆς ποιεῖσθαι τοὺς ἐμέτους ἐν
40 δυσὶν ἡμέρῃσι μᾶλλον, ἢ διὰ πεντεκαίδεκα· οἱ δὲ
πᾶν τοὐναντίον ποιέουσιν. ὅσοις δὲ ἐπιτήδειον
ἀνεμεῖν τὰ σιτία, ἢ ὅσοισιν αἱ κοιλίαι οὐκ
εὐδιέξοδοι, τούτοισι πᾶσι συμφέρει πολλάκις τῆς
ἡμέρης ἐσθίειν, καὶ παντοδαποῖσι βρώμασι
χρῆσθαι καὶ ὄψοισι πάντας τρόπους ἐσκευασμέ-
νοισι, καὶ οἴνους πίνειν δισσοὺς καὶ τρισσούς·
ὅσοι δὲ μὴ ἀνεμέουσι τὰ σιτία, ἢ καὶ κοιλίας
ἔχουσιν ὑγράς, τούτοισι δὲ πᾶσι τοὐναντίον τούτου
49 τοῦ τρόπου συμφέρει ποιεῖν.

VI. Τὰ δὲ[1] παιδία χρὴ τὰ νήπια βρέχειν ἐν θερμῷ
ὕδατι πολὺν χρόνον, καὶ πίνειν διδόναι ὑδαρέα
τὸν οἶνον καὶ μὴ ψυχρὸν παντάπασι, τοῦτον δὲ
διδόναι, ὃς ἥκιστα τὴν γαστέρα μετεωριεῖ καὶ
φῦσαν παρέξει· ταῦτα δὲ ποιεῖν, ὅπως οἵ τε
σπασμοὶ ἧσσον ἐπιλάβωσι, καὶ μείζονα γίνηται
καὶ εὐχροώτερα. τὰς δὲ[2] γυναῖκας χρὴ διαιτᾶσθαι τῷ ξηροτέρῳ τῶν τρόπων· καὶ γὰρ τὰ
σιτία τὰ[3] ξηρὰ ἐπιτηδειότερα πρὸς τὴν μαλθακό-
10 τητα τῶν σαρκῶν, καὶ τὰ πόματα ἀκρητέστερα
11 ἀμείνω πρὸς τὰς ὑστέρας καὶ τὰς κυοτροφίας.[4]

before the emetic in the following way. After bathing in hot water let the patient first drink a cotyle of neat wine; then let him take food of all sorts without drinking either during or after the meal, but after waiting time enough to walk ten stades, give him to drink a mixture of three wines, dry, sweet and acid, first rather neat, and taken in small sips at long intervals, then more diluted, more quickly and in larger quantities.

He who is in the habit of taking an emetic twice a month will find it better to do so on two successive days than once every fortnight, though the usual custom is just the contrary. Those who benefit from vomiting up their food, or whose bowels do not easily excrete, all these profit by eating several times a day, partaking of all sorts of food and of meats prepared in every way, and by drinking two or three sorts of wine. Those who do not vomit up their food, or have loose bowels, all these profit by acting in exactly the opposite way to this.

VI. Infants should be washed in warm water for a long time, and be given to drink their wine well diluted and not altogether cold, and such that will least swell the belly and cause flatulence. This must be done that they may be less subject to convulsions, and that they may become bigger and of a better colour. Women should use a regimen of a rather dry character, for food that is dry is more adapted to the softness of their flesh, and less diluted drinks are better for the womb and for pregnancy.

---

[1] A omits δέ. [2] A omits δέ. [3] A omits τά.
[4] *κυοτροφίας* Littré, slightly altering two inferior MSS. A reads *σκιητροφίας* and MV *σκιατροφίας*. Littré's reading certainly seems correct, but the other one must be very old, going back to the archetype of AM and V.

VII. Τοὺς γυμναζομένους χρὴ τοῦ χειμῶνος καὶ τρέχειν καὶ παλαίειν, τοῦ δὲ θέρεος παλαίειν μὲν ὀλίγα, τρέχειν δὲ μή, περιπατεῖν δὲ πολλὰ κατὰ ψῦχος. ὅσοι κοπιῶσιν ἐκ τῶν δρόμων, τούτους παλαίειν χρή· ὅσοι δὲ παλαίοντες κοπιῶσι, τούτους τρέχειν χρή· οὕτω γὰρ ἂν ταλαιπωρέων τῷ κοπιῶντι τοῦ σώματος διαθερμαίνοιτο καὶ συνιστῷτο[1] καὶ διαναπαύοιτο μάλιστα. ὁπόσους γυμναζομένους διάρροιαι λαμβάνουσι, 10 καὶ τὰ ὑποχωρήματα σιτώδεα καὶ ἄπεπτα, τούτοισί τε τῶν γυμνασίων ἀφαιρεῖν μὴ ἔλασσω τοῦ τρίτου μέρεος, καὶ τῶν σιτίων τοῖσιν ἡμίσεσι χρῆσθαι· δῆλον γὰρ δὴ ὅτι ἡ κοιλίη συνθάλπειν οὐ δύναται ὥστε πέσσεσθαι τὸ πλῆθος τῶν σιτίων·[2] ἔστω δὲ τούτοισι τὰ σιτία ἄρτος ἐξοπτότατος, ἐν οἴνῳ ἐντεθρυμμένος, καὶ τὰ ποτὰ ἀκρητέστατα καὶ ἐλάχιστα, καὶ περιπάτοισι μὴ χρήσθωσαν ἀπὸ τοῦ σιτίου· μονοσιτεῖν δὲ χρὴ ὑπὸ[3] τοῦτον τὸν χρόνον· οὕτω γὰρ ἂν μάλιστα 20 συνθάλποιτο ἡ κοιλίη, καὶ τῶν ἐσιόντων ἐπικρατοίη. γίνεται δὲ ὁ τρόπος οὗτος τῆς διαρροίης τῶν σωμάτων τοῖσι πυκνοσάρκοισι μάλιστα, ὅταν ἀναγκάζηται ὥνθρωπος κρεηφαγεῖν, τῆς φύσιος ὑπαρχούσης τοιαύτης· αἱ γὰρ φλέβες πυκνωθεῖσαι οὐκ ἀντιλαμβάνονται τῶν σιτίων τῶν ἐσιόντων· ἔστι δὲ αὕτη μὲν ἡ φύσις ὀξέη, καὶ τρέπεται ἐφ᾽ ἑκάτερα, καὶ ἀκμάζει ὀλίγον χρόνον ἡ εὐεξίη ἐν τοῖσι τοιουτοτρόποισι τῶν σωμάτων. τὰ δὲ ἀραιότερα τῶν εἰδέων καὶ 30 δασύτερα καὶ τὴν ἀναγκοφαγίην[4] δέχεται, καὶ τὰς ταλαιπωρίας μᾶλλον,[5] καὶ χρονιώτεραι γίνονται

[1] συνιστῷτο is omitted by A.

VII. Athletes in training should in winter both run and wrestle; in summer they should wrestle but little and not run at all, walking instead a good deal in the cool. Such as are fatigued after their running ought to wrestle; such as are fatigued by wrestling ought to run. For by taking exercise in this way they will warm, brace and refresh best the part of the body suffering from fatigue. Such as are attacked by diarrhœa when training, whose stools consist of undigested food, should reduce their training by at least one-third and their food by one-half. For it is plain that their bowels cannot generate the heat necessary to digest the quantity of their food. The food of such should be well-baked bread crumbled into wine, and their drink should be as undiluted and as little as possible, and they ought not to walk after food. At this time they should take only one meal each day, a practice which will give the bowels the greatest heat, and enable them to deal with whatever enters them. This kind of diarrhœa attacks mostly persons of close flesh, when a man of such a constitution is compelled to eat meat, for the veins when closely contracted cannot take in the food that enters. This kind of constitution is apt sharply to turn in either direction, to the good or to the bad, and in bodies of such a sort a good condition is at its best only for a while. Physiques of a less firm flesh and inclined to be hairy are more capable of forcible feeding and of fatigue, and their good condition is of

---

[2] Before σιτίων A has ἐσιόντων. [3] A omits ὑπό.

[4] Littré with slight authority reads κρεηφαγίην, "meat-eating."

[5] For μᾶλλον A has μάλιστα.

αὐτοῖσιν αἱ εὐεξίαι. καὶ ὅσοι τὰ σιτία ἀνερεύγονται τῇ ὑστεραίῃ, καὶ τὰ ὑποχόνδρια μετεωρίζεται αὐτοῖσιν ὡς ἀπέπτων τῶν σιτίων ἐόντων, τούτοισι καθεύδειν μὲν πλείονα χρόνον συμφέρει, τῇ δὲ ἄλλῃ ταλαιπωρίῃ ἀναγκάζειν χρὴ αὐτῶν τὰ σώματα, καὶ τὸν οἶνον ἀκρητέστερον πινόντων καὶ πλείω, καὶ τοῖσι σιτίοισιν ἐλάσσοσι χρῆσθαι ὑπὸ τοῦτον τὸν χρόνον· δῆλον γὰρ δὴ ὅτι ἡ
40 κοιλίη ὑπὸ ἀσθενείης καὶ ψυχρότητος οὐ δύναται τὸ πλῆθος τῶν σιτίων καταπέσσειν. ὅσους δὲ δίψαι λαμβάνουσι, τούτοισι τῶν τε σιτίων καὶ τῶν ταλαιπωριέων ἀφαιρεῖν, καὶ τὸν οἶνον πινόντων ὑδαρέα τε καὶ ὅτι ψυχρότατον. οἷσι δὲ ὀδύναι γίνονται τῶν σπλάγχνων ἢ ἐκ γυμνασίης[1] ἢ ἐξ ἄλλης τινὸς ταλαιπωρίης, τούτοισι συμφέρει ἀναπαύεσθαι ἀσίτοισι, πόματι δὲ χρῆσθαι ὅ τι ἐλάχιστον ἐς τὸ σῶμα ἐσελθὸν πλεῖστον οὖρον διάξει, ὅπως αἱ φλέβες αἱ διὰ
50 τῶν σπλάγχνων πεφυκυῖαι μὴ κατατείνωνται πληρεύμεναι· ἐκ γὰρ τῶν τοιούτων τά τε[2] φύματα
52 γίνονται καὶ οἱ πυρετοί.

VIII. Οἷσιν αἱ νοῦσοι ἀπὸ τοῦ ἐγκεφάλου γίνονται, νάρκη πρῶτον ἴσχει τὴν κεφαλήν, καὶ οὐρεῖ θαμινά, καὶ τἄλλα πάσχει ὅσα ἐπὶ στραγγουρίῃ· οὗτος ἐφ᾽ ἡμέρας ἐννέα τοῦτο πάσχει· καὶ ἢν μὲν[3] ῥαγῇ κατὰ τὰς ῥῖνας ἢ κατὰ τὰ ὦτα ὕδωρ ἢ[4] βλέννα, ἀπαλλάσσεται τῆς νούσου, καὶ τῆς στραγγουρίης παύεται· οὐρεῖ δὲ ἀπόνως πολὺ καὶ λευκόν, ἔστ᾽ ἂν εἴκοσιν ἡμέρας παρέλθῃ· καὶ ἐκ τῆς κεφαλῆς ἡ ὀδύνη ἐκλείπει τῷ ἀνθρώπῳ,
10 ἐσορέοντι δὲ βλάπτεταί οἱ ἡ αὐγή.

[1] A has γυμνασίων.

longer duration. Such as throw up their food the day after, whose hypochondria are swollen because of the undigested food, are benefited by prolonging their sleep, but apart from this their bodies should be subjected to fatigue, and they should drink more wine and less diluted, and at such times partake of less food. For it is plain that their bellies are too weak and cold to digest the quantity of food. When people are attacked by thirst, diminish food and fatigue, and let them drink their wine well diluted and as cold as possible. Those who feel pains in the abdomen after exercise or after other fatigue are benefited by resting without food; they ought also to drink that of which the smallest quantity will cause the maximum of urine to be passed, in order that the veins across the abdomen may not be strained by repletion. For it is in this way that tumours and fevers arise.

VIII.[1] When a disease arises from the brain, at first a numbness seizes the head and there is frequent passing of urine with the other symptoms of strangury; this lasts nine days. Then, if water or mucus break out at the nostrils or ears, the illness ceases and there is no more strangury. The patient passes without pain copious white urine for the next twenty days. His headache disappears, but his vision is impaired.

[1] Chapter VIII is a fragment from the beginning of *περὶ νούσων* II.

---

[2] A omits *τά τε*.
[3] A omits *μέν*.
[4] For *ἢ* A has *καί*.

IX. Ἄνδρα δὲ[1] χρή, ὅς[2] ἐστι συνετός, λογισάμενον ὅτι τοῖσιν ἀνθρώποισι πλείστου ἄξιόν ἐστιν ἡ ὑγιείη, ἐπίστασθαι ἐκ[3] τῆς ἑωυτοῦ γνώμης
4 ἐν τῇσι νούσοισιν ὠφελεῖσθαι.

[1] A omits δέ. [2] ὅς MV : ὅστις A.
[3] ἐκ MV : ἀπὸ A.

IX.[1] A wise man should consider that health is the greatest of human blessings, and learn how by his own thought to derive benefit in his illnesses.

[1] Chapter IX is a fragment from the beginning of *περὶ παθῶν* (Littré, vi. 208).

HUMOURS

ΠΕΡΙ ΧΥΜΩΝ

I. Τὸ χρῶμα τῶν χυμῶν, ὅπου μὴ ἄμπωτίς[1] ἐστι τῶν χυμῶν, ὥσπερ ἀνθέων·[2] ἀκτέα, ᾗ ῥέπει τῶν συμφερόντων χωρίων, πλὴν ὧν οἱ πεπασμοὶ ἐκ τῶν χρόνων· οἱ πεπασμοὶ ἔξω ἢ ἔσω ῥέπουσιν, ἢ ἄλλῃ ὅπῃ δεῖ.[3] εὐλαβείη· ἀπειρίη· δυσπειρίη· μαδαρότης· σπλάγχνων κενότης, τοῖσι κάτω, πλήρωσις, τοῖσιν ἄνω, τροφή· ἀναρροπίη, καταρροπίη· τὰ αὐτόματα ἄνω καὶ κάτω, ἃ ὠφελεῖ

[1] ὃς ἂν ἀιπῶτίς (changed into ἀνθήτω τίς) A: ὅκου μὴ ἄμπωτίς M.

[2] ἀνθέων A: ἀνθέῶν M.

[3] Here A has οὐδεμία εὐλάβεια· οὗ δεῖ.

---

[1] I translate the text, which is Littré's. It goes back to the Galenic commentary, which gives three rather forced explanations. (1) Like flowers, humours have their proper colour; (2) the colour of humours is "florid"; (3) consider the colour of humours when they have not left the surface of the flesh leaving it sapless. The repetition of χυμῶν and the variant in A for ἄμπωτις suggest that the original was either τὸ χρῶμα τῶν χυμῶν ὥσπερ ἀνθέον (the corrector of M wrote ο over ω of ἀνθέων) or τὸ χρῶμα τῶν χυμῶν, ὃς ἂν ἀνθῇ. The verb ἀνθῶ, as in *Sacred Disease* VIII (Vol. II, p. 155), seems to mean ἐξανθῶ, "break out," in sores, etc. The sense would be "judge of the colour of humours from an outbreak." ὥσπερ ἀνθέον and ὃς ἂν ἀνθῇ look like the alternative readings which so many places in the *Corpus* show as a "conflated" whole. The Galenic commentary mentions an ancient reading, τὸ χρῶμα τῶν χυμῶν, ὅπου οὐκ ἔστι ταραχὴ αὐτῶν, ὥσπερ τῶν ἀνθῶν ἐν διαδοχῇ τῶν ἡλικιῶν ὑπαλλάττεται—an obvious paraphrase.

HUMOURS

I. The colour of the humours, where there is no ebb of them, is like that of flowers.[1] They must be drawn along the suitable parts whither they tend,[2] except those whose coction comes in due time. Coction tends outwards or inwards, or in any other necessary direction. Caution.[3] Lack of experience. Difficulty of learning by experience. Falling out of hair. Emptiness of bowels, for the lower, repletion, for the upper, nourishment.[4] Tendency upwards;[5] tendency downwards. Spontaneous movements upwards, downwards; beneficial,

[2] I take ᾗ with τῶν συμφερόντων χωρίων, which is partitive. Littré reads διὰ τῶν συμφερόντων χωρίων, with the footnote "διὰ om. Codd." I find it, however, in the Caius MS. $\frac{50}{27}$.

[3] The meaning is most uncertain, and the variant in A suggests either corruption in the vulgate or an attempt at paraphrase. The Galenic commentary quotes with approval the reading εὐλαβείη ἀπειρίῃ, "caution for inexperience." One is very tempted to think that the original was εὐλαβείη δυσπειρίῃ, "be cautious when it is difficult to judge by experience," and that ἀπειρίη is a gloss.

[4] I leave these extraordinary phrases as they are printed in Littré, our MSS. showing no important variants. However, my own feeling is that we ought to read: πλαδαρότης σπλάγχνων, κενότης τοῖσι κάτω, πλήρωσις τοῖσιν ἄνω· τροφή· κ.τ.λ. "Flabbiness of the bowels means emptiness in the lower bowels, fulness in the upper"—not a bad description of certain forms of dyspepsia.

[5] Apparently of the humours, and similarly with the other nouns in the present context.

καὶ βλάπτει·[1] συγγενὲς εἶδος, χώρη, ἔθος, ἡλικίη,
10 ὥρη, κατάστασις νούσου, ὑπερβολή, ἔλλειψις,
οἷσιν ὁπόσον λείπεται, ἢ οὔ·[2] ἄκη· ἔκκλισις· παροχέτευσις ἐς κεφαλήν, ἐς τὰ πλάγια, ᾗ μάλιστα
ῥέπει· ἢ ἀντίσπασις, ἐπὶ τοῖσιν ἄνω, κάτω, ἄνω,
ἐπὶ τοῖσι κάτω· ἢ ξηρῆναι· ἢ οἷσι τὰ κάτω, ἢ
ἄνω ἐκπλύνεται, ἢ οἷσι παρηγορήσεται·[3] μὴ τὰ
ἐκκεχυμωμένα ἐς τὸ ἔσω ἀπολαμβάνειν, ἀλλὰ
τὰς ἀφόδους ξηραίνειν. τάραξις, κατάκλυσις,
διάνιψις, οἷσιν ἀποστήσεται πρὸς ἕδρην, ὅθεν
ἀθέλγεται,[4] ἢ φάρμακον, ἢ ἕλκος, ἢ χυμός τις
20 συνεστηκώς, ἢ βλάστημα, ἢ φῦσα, ἢ σῖτος, ἢ
21 θηρίον, ἢ καῦμα, ἢ ἄλλο τι πάθος.

II. Σκεπτέα ταῦτα· τὰ αὐτόματα λήγοντα, ἢ
οἷον αἱ ἀπὸ καυμάτων ἐπεγειρόμεναι φλύκτεις,

[1] μαδαρότης· σπλάγχνων κειότης· τοῖσι πλείστοισιν· ἢ τοῖσι κάτω πλήρωσις· τοῖσιν ἄνω τροφή· τὰ αὐτόματα ἄνω κάτω ὠφελέει καὶ βλάπτει A: μαδαρότης· σπλάγχνων κενότης· τοῖσι κάτω πλήρωσις· τοῖσιν ἄνω τροφή· ἀναρροπίη· καταρροπίη· τὰ αὐτόματα ἄνω καὶ κάτω· ἃ ὠφελέει καὶ βλάπτει M.

[2] A reads ἃ for οἷσιν ὅποσον and omits ἢ οὔ. Littré adds κάθαρσις καὶ κένωσις before ἄκη (from the Galenic commentary).

[3] ἢ οἷσιν ἄνω παρηγορήσεται A: ἢ οἷσι παρηγορήσεται M.

[4] ἐξαθέλγεται A.

---

[1] εἶδος here seems equivalent to φύσις. See A. E. Taylor, *Varia Socratica*, p. 228.

[2] For κατάστασις see Vol. I, p. 141.

[3] With the reading of A, "defect, and the nature of the deficiency."

[4] This means apparently that "loose" humours in the body ought not to be confined within it, but dried up as

harmful. Congenital constitution,[1] country, habit, age, season, constitution of the disease,[2] excess, defect, the deficient and the amount of the deficiency, or the contrary.[3] Remedies. Deflection. Deviation, to head, to the sides, along the route to which the chief tendencies are. Or revulsion, downwards when there is an upwards tendency, upwards when there is a downwards tendency. Drying up. Cases in which the upper parts, or the lower, are washed out; cases for soothing remedies. Do not shut up extravasated humours inside, but dry up the evacuations.[4] Disturbance; flooding out, washing through,[5] for those who will have an abscession to the seat, whereby is withdrawn poison,[6] or sore, or solidified humour, or growth, or flatulence,[7] or food, or creature,[8] or inflammation, or any other affection.

II. Observe these things: symptoms which cease of themselves, or for example the blisters that rise

evacuations. Littré has "par des moyens siccatifs faciliter les voies"—a very doubtful rendering. Personally I think that the original was *τῇσιν ἀφόδοις* (*ἀφόδοις* and *ἀφόδους* are very similar) and that the meaning is "dry up by evacuations."

[5] The Caius MS. $\frac{50}{27}$ omits *διάνιψις*, and it is probably a gloss on *κατάκλυσις*, or *vice versa*. The general meaning of this passage I take to be that a *τάραξις* of the humours calls for a clyster, should there be signs that the trouble will be resolved to the seat. *ταραχή* and its cognates are generally used of bowel trouble.

[6] *φάρμακον* in the *Corpus* generally means "purge." The meaning here (*substance délétère*, Littré) seems unique in the Hippocratic *corpus*.

[7] Erotian says (Nachmanson, p. 90) *ἐν δὲ τῷ Περὶ Χυμῶν τὸ ἐμφύσημα λέγει.*

[8] Apparently worms.

ἐφ' οἷσιν οἷα βλάπτει ἢ ὠφελεῖ,[1] σχήματα, κίνησις, μετεωρισμός, παλινίδρυσις, ὕπνος, ἐγρήγορσις,[2] ἅ τε ποιητέα ἢ κωλυτέα φθάσαι.[3] παίδευσις ἐμέτου,[4] κάτω διεξόδου,[5] πτυάλου,[6] μύξης, βηχός,[7] ἐρεύξιος, λυγμοῦ,[8] φύσης,[9] οὔρου, πταρμοῦ, δακρύων,[10] κνησμῶν, τιλμῶν, ψαυσίων,[11] δίψης, λιμοῦ, πλησμονῆς, ὕπνων, πόνων, ἀπονίης, 10 σώματος, γνώμης, μαθήσιος, μνήμης, φωνῆς, 11 σιγῆς.[12]

III. Τῇ ὑστερικῇ καθάρσεις, τὰ ἄνωθεν[13] καταρρηγνύμενα, καὶ στροφέοντα, λιπαρά, ἄκρητα, ἀφρώδεα, θερμά, δάκνοντα, ἰώδεα, ποικίλα, ξυσματώδεα, τρυγώδεα, αἱματώδεα, ἄφυσα,[14] ὠμά, ἐφθά,[15] αὖα, ἅσσα περιρρέοι,[16] εὐφορίην καθορέων ἢ δυσφορίην, πρὶν κίνδυνον εἶναι,[17] οἷα οὐ δεῖ παύειν. πεπασμός, κατάβασις τῶν κάτω, ἐπιπόλασις τῶν ἄνω, καὶ τὰ ἐξ ὑστερέων, καὶ ὁ ἐν ὠσὶ ῥύπος·[18]

[1] τὰ αὐτόματα λήγοντα ἐφ' οἷσί τε· οἷα βλάπτει· ἢ ὠφελέει· A: τὰ αὐτόματα λήγοντα· ἢ οἷον αἱ ἀπὸ καυμάτων ἐπεγειρόμεναι φλύκτεις· ἐφ' οἷσιν οἷα βλάπτηται ἢ ὠφελέει. M.

[2] ἔγερσις A: ἐγρήγορσις M. After ἐγρήγορσις the Galenic text implies ἀλύκη, χάσμη, φρίκη. Littré adds these words to his text.

[3] φῦσα A: φθάσαι M.

[4] παρόδευσις ἐμέτου A: παίδευσις ἐμέτου M.

[5] διέξοδοι A: διεξόδου M.

[6] M has ἢ before πτυάλου, but A omits.

[7] βηχὸς μύξης A: μύξης βηχός M.

[8] A omits λυγμοῦ.

[9] φυσέων (έ apparently on an erasure) A: φύσης M.

[10] δακρύου A: δακρύων M.

[11] ψαύσιος A: ψαυσίων M.

[12] σιγίης A: σιγῆς M.

[13] A reads: καὶ ταῦτα σκεπτέα· ἀφρὸς, ὑστερικὴ κάθαρσις· τὰ ἄνωθεν.

[14] A omits τρυγώδεα, αἱματώδεα, ἄφυσα.

upon burns,[1] what are harmful or beneficial and in what cases, positions, movement, rising,[2] subsidence, sleep, waking,[3] to be quick when something must be done or prevented. Instruction[4] about vomit, evacuation below, sputum, mucus, coughing, belching, hiccoughing, flatulence, urine, sneezing, tears, itching, pluckings, touchings, thirst, hunger, repletion, sleep, pain, absence of pain, body, mind, learning, memory, voice, silence.

III. In affections of the womb, purgations; evacuations from above, with colic, that are greasy, uncompounded, foamy, hot, biting, verdigris-coloured, varied, with shreds, lees or blood, without air, unconcocted, concocted, desiccated, the nature of the liquid part, looking at the comfort or discomfort of the patient before danger comes, and also what ought not to be stopped. Coction, descent of the humours below, rising of the humours above, fluxes from the womb, the wax in the ears. Orgasm, opening,

[1] This phrase should probably be omitted (as by A). It reads like a gloss. A's reading gives "symptoms which cease of themselves, and in what cases; what are harmful or beneficial."

[2] *μετεωρισμός* is here opposed to *παλινίδρυσις*, and means either as in *Prognostic* (with Littré), "lifting up of the body," or (with Foës) "inflation of humours."

[3] The words given here by Littré mean "restlessness, yawning, shivering."

[4] The reading of A, *παρόδευσις* ("passing along"), is attractive, but only a few of the genitives suit it. Perhaps it arose from a gloss on *διεξόδου*.

---

[15] A omits *ἐφθά*.
[16] *ὅσα περριρεῖ* A : *ἄσσα περιρρέοι* M.
[17] *πρὶν ἢ κίνδυνον εἶναι* A : *πρὶν κίνδυνον ἰέναι* M.
[18] *καὶ τὰ ἐν ὠσι· ῥύπος·* A : *καὶ ὁ ἐν ὠσὶ ῥύπος* M.

ὀργασμός, ἄνοιξις, κένωσις, θάλψις, ψύξις, ἔσω-
10 θεν, ἔξωθεν, τῶν μέν, τῶν δ' οὔ. ὅταν ᾗ κάτωθεν ὀμφαλοῦ τὸ στρέφον, βραδύς, μαλθακὸς ὁ
12 στρόφος, ἐς τοὐναντίον.

IV. Τὰ διαχωρέοντα, ᾗ ῥέπει,[1] ἄναφρα, πέπονα, ὠμά, ψυχρά, δυσώδεα, ξηρά, ὑγρά. μὴ καυσώδεσι δίψα[2] πρόσθεν μὴ ἐνεοῦσα, μηδὲ καῦμα, μηδὲ ἄλλη πρόφασις, οὖρον,[3] ῥινὸς ὑγρασμός. τὴν ἔρριψιν, καὶ τὸν αὐασμόν, καὶ τὸ ἀσύμπτωτον, καὶ τὸ θαλερὸν[4] πνεῦμα, ὑποχόνδριον, ἄκρεα, ὄμματα προσκακούμενα, χρωτὸς[5] μεταβολή, σφυγμοί, ψύξιες,[6] παλμοί, σκληρυσμὸς δέρματος, νεύρων, ἄρθρων, φωνῆς, γνώμης, σχῆμα ἑκούσιον, τρί-
10 χες, ὄνυχες, τὸ εὔφορον, ἢ μή, οἷα δεῖ.[7] σημεῖα ταῦτα· ὀδμαὶ χρωτός, στόματος, ὠτός, διαχωρήματος, φύσης, οὔρου, ἕλκεος, ἱδρῶτος, πτυάλου, ῥινός, χρὼς ἁλμυρός, ἢ πτύαλον, ἢ ῥίς, ἢ δάκρυον, ἢ ἄλλοι χυμοί·[8] πάντη ὅμοια τὰ ὠφελέοντα, τὰ βλάπτοντα. ἐνύπνια οἷα ἂν ὁρῇ, καὶ ἐν τοῖσιν ὕπνοισιν οἷα ἂν ποιῇ, ἢν ἀκούῃ ὀξύ, καὶ πυθέσθαι[9] προθυμῆται. ἐν τῷ λογισμῷ μέζω καὶ ἰσχυρότερα

[1] ᾗ ῥέπει ἢ δι' ἃ ῥέπει A.
[2] πέπονα· ἢ ψυχρὰ· ἢ θερμὰ· φυσώδεα· ξηρᾶ, ἢ ὑγρᾶ· μὴ καυσῶδες ἢ δίψα A: πέπονα· ὠμά· ψυχρά· δυσώδεα· ξηρά· ὑγρά· ὀδμὴ κακώδεα· δίψα πρόσθεν μὴ ἐνεοῖσα M
[3] οὔρων A: οὖρον M.
[4] θαλερόν A M: θολερόν Littré.
[5] χρωτός A: χρώματος M.
[6] ψύξιες A: ψύξις M.
[7] ἢν μη οἷα δεῖ (η of μη on an erasure) A: ἢ μή· οἷα δεῖ M.
[8] ἢ ἀλλοῖος χυμός A: ἢ ἄλλοι χυμοί M.
[9] πύθεσθαι A: πείθεσθαι M.

emptying, warming, chilling, within or without, in some cases but not in others. When that which causes the colic is below the navel the colic is slow and mild, and *vice versa*.

IV. The evacuations, whither they tend; without foam, with coction, without coction, cold, fetid, dry, moist. In fevers not ardent, thirst that was not present before, brought about neither by heat nor by any other cause, urine, wetness of the nostrils. Prostration, dryness or fulness of the body; rapid[1] respiration; hypochondrium; extremities; eyes sickly; change of complexion; pulsations; chills; palpitations; hardness of the skin, muscles, joints, voice, mind;[2] voluntary posture; hair; nails; power, or the want of power, to bear easily what is necessary. These are signs:—smell of the skin, mouth, ear, stools, flatulence, urine, sores, sweat, sputum, nose; saltness of skin, sputum, nose, tears, or of the humours generally. In every way similar the things that benefit, the things that harm.[3] The dreams the patient sees, what he does in sleep; if his hearing be sharp, if he be interested in information.[4] In estimating signs take the majority that are more important and more promi-

[1] The word θαλερός is poetic in the sense of "frequent" (θαλερὸς γόος in *Odyssey*). But this is no objection when the style is aphoristic. θολερόν would mean "troubled," poetic again in this sense.

[2] If φωνῆς and γνώμης are not mere slips for φωνή and γνώμη, σκληρυσμός must be used metaphorically with them to signify a rigidity of voice and thought not uncommon in serious cases of illness.

[3] I take this to mean that all good signs show a similarity, and so do all bad signs.

[4] The reading πείθεσθαι would mean "is readily obedient to orders."

τὰ πλείω, ἐπικαιρότερα τὰ σῴζοντα τῶν ἑτέρων·[1] ἣν αἰσθάνωνται πάσῃ αἰσθήσει πάντων, καὶ 20 φέρωσιν,[2] ὁποῖον ὀδμάς, λόγους, ἱμάτια,[3] σχήματα, τοιαῦτα, εὐφόρως. ἅπερ καὶ αὐτόματα ἐπιφαινόμενα ὠφελεῖ, καί ποτε κρίσιν καὶ τὰ τοιαῦτα[4] ἐμποιεῖ, οἷον φῦσαι, οὖρον, οἷον ὅσον καὶ πότε· ὁπόσα δ' ἐναντία, ἀποτρέπειν, μάχεσθαι αὐτοῖσιν. τὰ ἐγγὺς καὶ τὰ κοινὰ τοῖσι παθήμασι πρῶτα καὶ 26 μάλιστα κακοῦται.

V. Κατάστασιν δὲ τῆς νούσου[5] ἐκ τῶν πρώτων ἀρχομένων ὅ τι ἂν ἐκκρίνηται, ἐκ τῶν οὔρων ὁποῖα ἂν ᾖ. καὶ οἵη τις σύμπτωσις, χροιῆς ἐξάλλαξις,[6] πνεύματος μείωσις,[7] καὶ τἄλλα μετὰ τούτων ἐπιθεωρεῖν.[8] τὰ μὴ ὅμοια ἃ δεῖ εἰδέναι,[9] διέξοδοι οὔρων,[10] καθ' ὑστέρας, πτύαλα, κατὰ ῥῖνας, ὄμματα, ἱδρώς, ἐκ φυμάτων, ἢ[11] τρωμάτων, ἢ[12] ἐξανθημάτων, ὅσα αὐτόματα, ὅσα τέχνῃσιν, ὅτι ὅμοια ἀλλήλοισι πάντα τὰ κρίνοντα, καὶ τὰ 10 ὠφελέοντα, καὶ τὰ βλάπτοντα, καὶ τὰ ἀπολλύντα,[13]

[1] μείζω καὶ ἰσχυρώτερα τα πλείω, ἐπικαιρότερα τὰ σώζοντα τῶν ἑτέρων A: μέζω ἰσχυρότερα τὰ πλείω· ἐπίκαιρα τὰ σώζοντα τῶν ἑτέρων M. μέζω καὶ ἰσχυρότερα τὰ πλείω, ἐπίκαιρα, σώζοντα· μὴ ἐπίκαιρα. τῶν ἑτέρων Littré, rewriting the text from the Galenic commentary. Littré does not note that A gives ἐπικαιρότερα, not ἐπίκαιρα.

[2] φέρωσιν A: φέρουσιν M.

[3] ἱμάτια A: εἵματα M.

[4] καὶ ποτὲ κρίσιν· καὶ τὰ τοιαῦτα A: καὶ ὁκότε κρίσιν τὰ τοιαῦτα M. After ἐμπιοεῖ Ms add καὶ τοσαῦτα καὶ τοιαῦτα.

[5] τῆς νούσου A: M omits

[6] ἐξάλλαξις A: ἐπάλλαξις M.

[7] μείωσις A: μινύθησις M.

[8] ἐπιθεωρέειν A: τὰ διαιτήματα M.

[9] τὰ μὴ ὅμοια ἃ δεῖ εἰδέναι A: τὸ μὲν εἰ ὅμοια τὰ ἀπιόντα δεῖ εἰδέναι M.

[10] οὔρων A: οὖρα M.

[11] ἢ A: ἐκ M.

[12] ἢ A: omitted by M.

nent—those that denote recovery are more seasonable than the others.[1] If the patients perceive everything with every sense and bear easily, for example, smells, conversation, clothes, postures and so on. Symptoms which benefit even when they manifest themselves spontaneously (and sometimes these too bring about a crisis), such as flatulence and urine, of the right kind, of the right amount, and at the right time.[2] What is contrary avert; combat it. Parts near and common to affected places suffer lesions first and most.

V. In examining the constitution of a disease look to the excretions in the initial stages, the nature of the urine, the state of collapse, change of colour, diminution of respiration and the other symptoms besides. The abnormal conditions that must be known: passage of urine, menstruation, sputum, nasal discharge, eyes, sweat, discharge from tumours, from wounds, from eruptions, what is spontaneous and what artificial; for all critical symptoms follow a norm, as do those that help, those that harm and those that kill. They must be known, that the bad may be shunned and

[1] I translate the reading of A, but I suspect that a gloss has crept into the text, possibly *τὰ σῴζοντα τῶν ἑτέρων*. Littré's translation, "les plus nombreux, les plus forts et les plus considérables," is surely impossible.

[2] I have punctuated from *σχήματα* to *οἷον* roughly as it is in A. Littré puts a full-stop at *σχήματα* and a comma at *εὐφόρως*, translating, "Il y a bonne tolérance, quand les symptômes survenant spontanément soulagent, et quand ils font crise, et quand ils sont suffisants en qualité et en quantité, comme les gaz, etc." In any case the sentence is broken, but the vulgate, which Littré follows, is intolerable.

---

[13] *ἀπολλῦντα* or *ἀπολλύντα* A: *ἀπολλύοντα* M.

ὡς τὰ μὲν περιφεύγων ἀποτρέπῃ, τὰ δὲ προσκαλῆται καὶ ἄγῃ καὶ δέχηται.[1] καὶ τἄλλα δὲ οὕτω, δέρματος, ἀκρέων, ὑποχονδρίων, ἄρθρων, στόματος, ὄμματος,[2] σχημάτων, ὕπνων, οἷα κρίνει, καὶ ὅτε καὶ τὰ τοιαῦτα[3] μηχανᾶσθαι. καὶ ἔτι[4] ὁπόσαι ἀποστάσιες τοιαῦται γίνονται,[5] οἷαι ὠφελέουσι, βρώμασι, πόμασιν, ὀδμῇσιν, ὁράμασιν, ἀκούσμασιν, ἐννοήμασιν, ἀφόδοισιν, θάλψει, ψύξει, ὑγροῖσι, ξηροῖσιν, ὑγρῆναι, ξηρῆναι, χρίσμασιν, ἐγχρίσ-
20 μασιν, ἐπιπλάστοισιν, ἐμπλάστοισιν, ἐπιπάστοισιν, ἐπιδέτοισιν, ἐπιθέτοισι, σχήματα, ἀνάτριψις, ἔασις,[6] πόνος, ἀργίη, ὕπνος, ἀγρυπνίη, πνεύμασιν ἄνωθεν, κάτωθεν, κοινοῖσιν, ἰδίοισι, τεχνητοῖσιν, ἐν τοῖσι παροξύσμοισι μήτε ἐοῦσι, μήτε μέλλουσι, μήτ' ἐν ποδῶν ψύξει, ἀλλ' ἐν καταρρόπῳ τῇ
26 νούσῳ.

VI. Τοῖσιν ἐν τῇσι περιόδοισι παροξυσμοῖσι τὰ προσάρματα μὴ διδόναι μηδ' ἀναγκάζειν, ἀλλ' ἀφαιρεῖν τῶν προσθεσίων πρὸ τῶν κρισίων.[7] τὰ κρινόμενα καὶ τὰ κεκριμένα[8] ἀπαρτὶ[9] μὴ κινεῖν, μήτε φαρμακείῃσι, μήτε ἄλλοισιν ἐρεθισμοῖσιν, μηδὲ νεωτεροποιεῖν,[10] ἀλλ' ἐᾶν. [τὰ

[1] ὅσα μὲν περιφεύγει ἀποτρέπει· τὰ δὲ προσκαλέηται καὶ ἄγῃ καὶ δέχεται A: ὡς τὰ μὲν περιφεύγων, ἀποτρέπει. τὰ δὲ προσκαλεῖται καὶ ἄγῃ κα δέχηται. M.

[2] στόματος· ὄμματος A: ὄμματος· στόματος M.

[3] καὶ ὅτε καὶ τὰ τοιαῦτα A: καὶ ὅτε τὰ τοιαῦτα M. Littré with Galen adds δεῖ after τοιαῦτα.

[4] A omits καὶ ἔτι.

[5] ὁπόσαι ἀποστάσιες τοιαῦται γίνωνται A: ὅσαι τοιαῦται ἀποτάσιες γίνονται M.

[6] ἐάσις A: ἐασις M: ἴησις vulgate.

[7] πρὸ τῶν κρισίων omitted by A.

[8] καὶ τὰ κεκριμένα omitted by A.

[9] ἀπάρτι A: ἀρτίως M.

averted, and that the good may be invited, encouraged and welcomed. Similarly with other symptoms, of the skin, extremities, hypochondria, joints, mouth, eye, postures, sleep, such as denote a crisis, and when symptoms of this kind must be provoked. Moreover, abscessions of a helpful character must be encouraged by foods, drinks, smells, sights, sounds, ideas, evacuations, warmth, cooling, moist things, dry things, moistening, drying, anointings, ointments, plasters, salves, powders, dressings, applications [postures, massage, leaving alone, exertion, rest, sleep, keeping awake],[1] breaths from above, from below, common, particular, artificial[2]—not, however, when paroxysms are present or imminent, nor when the feet are chilled, but when the disease is declining.

VI. At the periodic paroxysms do not give nourishment; do not force it on the patient, but diminish the quantity before the crisis. Do not disturb a patient either during a crisis or just after one, either by purgings or by other irritants; do not try experiments either, but leave the patient

[1] It is hard to see how these nominatives came to be included among the datives. Perhaps they are an addition to the text from a marginal note of a commentator. The dictionaries do not recognise ἔασις, but, although the editions read ἴησις, the word is correctly formed from ἐάω and makes excellent sense in this passage.

[2] These difficult words I take to mean (*a*) letting the air play upon the patient from different directions; (*b*) taking long or deep breaths, and (*c*) the use of a fan. It seems to be better to take κοινοῖσιν . . . τεχνητοῖσιν as agreeing with πνεύμασιν, and not as separate substantives.

---

[10] μηδὲ νεωτερωποιέειν A. Here M has μήτε νεωτεροποιέειν· μήτε φαρμακίηισι μήτε κ.τ.λ.

κρίνοντα ἐπὶ τὸ βέλτιον μὴ αὐτίκα ἐπιφαίνεσθαι.][1] πέπονα φαρμακεύειν καὶ κινεῖν, μὴ ὠμά,[2] μηδὲ ἐν ἀρχῇσιν, ἢν μὴ ὀργᾷ· τὰ δὲ πολλὰ οὐκ ὀργᾷ.
10 ἃ δεῖ ἄγειν, ὅπῃ ἂν μάλιστα ῥέπῃ διὰ[3] τῶν συμφερόντων χωρίων, ταύτῃ ἄγειν. τὰ χωρέοντα μὴ τῷ πλήθει τεκμαίρεσθαι, ἀλλ' ὡς ἂν χωρῇ οἷα δεῖ, καὶ φέρῃ εὐφόρως· ὅπου δὲ δεῖ, γυιῶσαι, ἢ λειποθυμῆσαι, ἕως ἂν τοῦτο ποιηθῇ, οὕνεκα ποιεῖται·[4] εἴ τι ἄλλο τότε δεῖ, ἐπ' ἄλλο ῥέψαι, ἢ ξηρῆναι, ἢ ὑγρῆναι, ἢ ἀντισπάσαι, ἢν ἐξαρκῇ ὁ νοσέων· τούτοισι τεκμαίρεσθαι· τὰ μὲν ξηρὰ θερμὰ ἔσται, τὰ δὲ ὑγρὰ ψυχρά·[5] διαχωρητικὰ δὲ τἀναντία· ἐπὶ τὸ πολὺ δὲ ταῦτα. ἐν τῇσι[6]
20 περισσῇσιν ἄνω, ἢν καὶ αἱ περίοδοι καὶ ἡ κατάστασις τοιαύτη ᾖ τῶν παροξυσμῶν·[7] γίνεται δὲ τὰ πλεῖστα ἐν τῇσιν ἀρτίῃσι κάτω· οὕτω γὰρ καὶ αὐτόματα ὠφελεῖ, ἢν αἱ[8] περίοδοι τοὺς παροξυσμοὺς ἐν τῇσιν ἀρτίῃσι ποιέωνται· ἐν δὲ τοῖσι μὴ τοιούτοισιν,[9] ἐν μὲν ἀρτίῃσιν ἄνω, ἐν δὲ

[1] τὰ κρίνοντα . . . ἐπιφαίνεσθαι not in MSS. but added by Mack and Littré from the Galenic commentary.

[2] πέπονα φαρμακεύειν καὶ κινέειν, μὴ ὠμά M: πέπονα φαρμακεύειν· καὶ μὴ κινέειν ὠμὰ A.

[3] ῥέπῃ διὰ omitted by M.

[4] ἕως ἂν τοῦτο ποιησθῇ. ἢ τί ἄλλο, τότε δε' ἐπ' ἄλλο τρέψεται· ξηρῆναι· ἢ ἀντισπᾶσαι ἕως οὗ εἵνεκα ποιέεται A: ἕως ἂν τοῦτο ποιήσῃς ἐπάλλα ῥεψαι ἢ ξηρῆναι, ἢ ὑγρῆναι. ἢ ἀντισπάσαι· οὗ ἕνεκα τοῦτο ποιέεται M.

[5] After ψυχρά A has ἔσται.

[6] ἐπὶ πολὺ δὲ ταῦτα· ἐν τῇσι A: ἐπὶ τὸ πολύ· ταῦτα τῇισι M: ὡς ἐπὶ τὸ πολὺ δὲ ταῦτα. ἐν τῇσι Littré.

[7] ἢν καὶ αἱ περιοδικαὶ καταστάσιες τοιαῦται; ἕως τῶν παροξυσμῶν· A: ἢν καὶ αἱ περίοδοι καὶ ἡ κατάστασις τοιαύτη ἔῃι τῶν παροξυσμῶν· M.

[8] ἢν αἱ A: ἢν μὴ αἱ M.

[9] τοῖσιν μὴ τοιούτοισιν A: τοῖσι τοιούτοισι M.

alone. [Critical signs of an improvement ought not to be expected to appear at once.] Purge or otherwise disturb concocted, not crude, humours, and avoid the onset of a disease, unless there be orgasm, a thing which rarely occurs then. Evacuate the humours that have to be evacuated in the direction in which they mostly tend, and by the convenient passages. Judge of evacuations, not by bulk, but by conformity to what is proper, and by the way in which the patient supports them. When occasion calls for it, reduce the patient, if need be, to a fainting condition, until the object in view be attained. If then there be need of anything further, shift your ground; dry up the humours, moisten them, treat by revulsion,[1] if, that is, the strength of the patient permits. Take as your tests[2] the following symptoms: the dry will be hot, and the moist cold; purgatives will produce the opposite effect. This is what usually happens. On odd days evacuations should be upwards if the periods and the constitution of the paroxysms be odd. On even days they are generally downwards, for so they are beneficial even when spontaneous, if the periods cause the paroxysms on the even days. But when the circumstances are not such,[3] evacuations should be upwards on even days, downwards

[1] This apparently means that if there be a flux of the humours to one part of the body, they should be "drawn back" by medicines or applications. *E.g.* a flow of blood to the head should be treated by hot water applied to the feet.

[2] To find out, that is, whether your treatment has been successful.

[3] That is, if the paroxysms and evacuations are neither both odd nor both even.

περισσῇσι κάτω· ὀλίγαι δὲ τοιαῦται,[1] αἱ δὲ τοιαῦται δυσκριτώτεραι καταστάσιες. ἀτὰρ καὶ τὰ πρόσω χρόνου προήκοντα[2] ἀνάγκη οὕτως, οἷον τρισκαιδεκαταῖα, τεσσαρεσκαιδεκαταῖα, 30 τρισκαιδεκάτῃ μὲν κάτω,[3] τεσσαρεσκαιδεκάτῃ δὲ ἄνω[4] (πρὸς γὰρ τὸ κρίσιμον οὕτω συμφέρει), καὶ ὅσα εἰκοσταῖα,[5] πλὴν ὅσα κάτω. πολλὰ δεῖ καθαίρειν, ταῦτα δὲ μὴ[6] ἐγγὺς οὕτω κρίσιος, ἀλλὰ προσωτέρω· δεῖ δὲ ὀλιγάκις ἐν ὀξέσι 35 πολλὰ ἄγειν.

VII. Τοῖσι κοπώδεσι τὸ σύμπαν,[7] ἐν τοῖσι πυρετοῖσιν ἐς ἄρθρα καὶ παρὰ γνάθους μάλιστα ἀποστάσιες γίνονται, ἐγγύς τι τῶν πόνων ἑκάστου, ἐπὶ τὸ ἄνω μᾶλλον καὶ τὸ σύμπαν·[8] ἢν ἀργὸς[9] ἡ νοῦσος ᾖ καὶ κατάρροπος, κάτω καὶ αἱ[10]

[1] ὀλίγαι δὲ τοιαῦται omitted by A.

[2] προσήκοντα (the -σ- apparently added afterwards) A: προήκοντα (with -σ- erased) M.

[3] ἄνω A: κάτω M.

[4] κάτω A: ἄνω M.

[5] ὅσα εἰκοστεῖα καὶ τεσσαρακοστεῖα A: ὁκόσα εἰκοσταῖα M.

[6] μὴ is omitted in M.

[7] ἄγειν· τοῖσι δὲ κοπιώδεσιν τὸ σύμπαν ἐν τοῖσι A: ἄγειν, τοῖσι κοπώδεσι τὸ σύμπαν· ἐν τοῖσι M.

[8] πυρετώδεσι· καὶ ἐς ἄρθρα καὶ παρὰ γνάθους ἐγγύς τι τῶν πόνων ἢ ἕκαστον ἐπὶ τὰ ἄνω μᾶλλον· καὶ τὸ σύμπαν ἐν τοῖσι πυρετοῖσι· καὶ ἐς ἄρθρα καὶ παρὰ γνάθους A: πυρετοῖσιν ἐς ἄρθρα καὶ παρὰ γνάθους μάλιστα ἀποστάσιες γίνονται· ἐγγύς τι τῶν πόνων ἑκάστου, ἐπὶ τὸ ἄνω μᾶλλον καὶ τὸ σύμπαν M.

[9] ἀργὸς M and first hand in A: ἀνάρροπος corrector's hand in A.

[10] καὶ αἱ M: αἱ A, which also has ἢ after κατάρροπος.

---

[1] That is, constitutions when paroxysms are on odd days and purges on even days, or *vice versa*. The cases considered seem to be these:—

on odd days. Such constitutions are rare,[1] and the crises are rather uncertain. Prolonged illnesses must be similarly treated—for example, those which last thirteen or fourteen days; purge on the thirteenth day downwards, on the fourteenth upwards (to do so is beneficial for the crisis), and similarly with diseases of twenty days,[2] except when purging should be downwards. Purging must be copious, and not near the crisis but some time before it. Rarely in acute diseases must evacuation be copious.

VII. Generally,[3] in cases of fever with prostration, abscessions [4] are most likely to occur at the joints and by the jaw, in each case near to the part where the pains are, more often, in fact generally, to an upper part. If the disease be sluggish and incline to the lower parts, the abscessions too collect in a

(*a*) A purge is necessary on an odd day. If paroxysms occur on odd days, purge upwards. If paroxysms occur on even days, purge downwards.

(*b*) A purge is necessary on an even day. If paroxysms occur on even days, purge downwards. If paroxysms occur on odd days, purge upwards.

[2] The readings in the text connect these cases with the rare cases mentioned above, where a necessity for purging occurs on a day when a paroxysm is not due. The usual cases are referred to in πλὴν ὅσα κάτω. This is very strange, or at least awkward, and the reading of A, which transposes κάτω and ἄνω, is certainly more natural, but it makes πλὴν ὅσα κάτω absurd. It is possible that these words are a marginal note which has slipped into the text, and that they should be deleted, κάτω and ἄνω being transposed as in A.

[3] The Galenic commentary joins τοῖσι κοπώδεσι with ἄγειν. Littré points out that *Aphorism* IV. 31 is the source of the present passage, and in it τοῖσι κοπιώδεσιν occurs in close conjunction with ἐν τοῖσι πυρετοῖσιν.

[4] For the meaning of "abscession" see Vol. I (Introduction), p. liii.

ἀποστάσιες· μάλιστα δὲ πόδες θερμοὶ κάτω σημαίνουσι, ψυχροὶ δὲ ἄνω. οἷσι δὲ ἀνισταμένοις ἐκ τῶν νούσων, αὐτίκα δὲ χερσὶν ἢ ποσὶ πονήσασιν, ἐν τούτοις ἀφίστανται· ἀτὰρ καὶ ἤν
10 τι προπεπονηκὸς[1] ᾖ, πρὶν ἢ νοσεῖν, ἐς ταῦτα ἀποστηρίζεται, οἷον καὶ τοῖσιν ἐν Περίνθῳ βηχώδεσι καὶ κυναγχικοῖσιν· ποιέουσι γὰρ καὶ αἱ βῆχες ἀποστάσιας, ὥσπερ οἱ πυρετοί· ταῦτα[2] κατὰ τὸν αὐτὸν λόγον ἢ ἀπὸ[3] χυμῶν, ἢ σώματος
15 συντήξιος καὶ ψυχῆς.

VIII. Τοὺς μὲν οὖν χυμοὺς εἰδέναι, ἐν ᾗσιν[4] ὥρῃσιν ἀνθέουσι, καὶ οἷα ἐν ἑκάστῃ νοσήματα ποιέουσι,[5] καὶ οἷα ἐν ἑκάστῳ νοσήματι[6] παθήματα. τὸ δὲ σῶμα τὸ ἄλλο, ἐς ὅ τι μάλιστα νόσημα ἡ φύσις ῥέπει·[7] † οἷόν τι σπλὴν οἰδέων ποιεῖ,[8] τούτων τι καὶ ἡ[9] φύσις· σχεδόν τι καὶ χρώματα κακοήθη, καὶ σώματα[10] σειρέει,[11] καὶ εἴ
8 τι ἄλλο· ταῦτα διαγεγυμνάσθαι.[12] †

[1] τινὰ προπεπονηκῶς A : τι πεπονηκὼς ἔηι M.
[2] ταῦτα A : τοιαύτας· M.
[3] M omits ἢ before ἀπὸ and before σώματος.
[4] ἐν ησιν A : altered to ἦν εἰσίν. This MS. omits ὥρῃσιν ἀνθέουσι, καί.
[5] A omits ποιέουσι.
[6] σώματι A : νοσήματι M.
[7] ῥέπει M : τρέπει A.
[8] ποιέοι A : ποιέει M.
[9] A omits τούτων τι καὶ ἡ.
[10] A omits σώματα.
[11] σειρεοῖ M : σιναροι A.
[12] διαγεγυμνάσθαι M : ἀγυμναστίη A (with ψυχῆς).

---

[1] The reading of A seems to be an attempt to make the grammar square with ταῦτα later on. But the accusative τινά is a curious "accusative of the part affected," and probably ταῦτα is a simple *anacoluthon*.

[2] A reference to *Epidemics* VI. 7, 7 (Littré v. 341).

[3] This seems to mean that abscessions may be the result of

lower part. Hot feet especially signify a lower abscession, cold feet an upper abscession. When patients, on rising after an illness, suffer immediately pains in arms or feet, abscessions form in these parts. Moreover, if a part suffer pain before the illness,[1] it is in it that the humours settle, as was the case with those who in Perinthus[2] suffered from cough and angina. For coughs, like fevers, cause abscessions. These results are the same, whether they come from humours or from wasting of body and soul.[3]

VIII. Know in what seasons the humours break out, what diseases they cause in each, and what symptoms they cause in each disease. As to the body generally, know to what disease the physical constitution most inclines. For example, a swollen spleen produces a certain effect, to which the constitution contributes something. It is much the same with an evil complexion, or the body is parched, and so on. Be practised in these things.[4]

wasting diseases as well as of those caused by "peccant humours."

[4] This chapter towards the end is full of difficulties, and is so irregular, not to say violent, in grammar that I have printed the text between daggers. The general sense of the whole chapter is that the physician must know (1) the effect of the humours in various seasons and in various diseases, and (2) the disease to which an individual constitution is most inclined. Then it seems to be implied that a bad complexion, or a parched, hot skin may also denote a tendency to a particular disease. The sentence *οἷον . . . φύσις* is strange, both *τι* (before *σπλήνν*) and *τούτων* being irregular. It would perhaps be a slight improvement to punctuate : *οἷον· τί σπλὴν οἰδέων ποιεῖ; τούτων τί καὶ ἡ φύσις;* In the next sentence the variants *σιναροί* of A and *σώματα σειρέοι* of M, for *σώματα σειρέει* (Littré, from the Galenic commentary), seem to show that the text is unsound.

IX. Ψυχῆς, ἀκρασίη ποτῶν καὶ βρωμάτων,[1] ὕπνου, ἐγρηγόρσιος, ἢ δι' ἔρωτάς τινας, οἷον κύβων, ἢ διὰ τέχνας ἢ δι' ἀνάγκας καρτερίη πόνων, καὶ ὥντινων τεταγμένη ἢ ἄτακτος· αἱ μεταβολαὶ ἐξ οἵων[2] ἐς οἷα. ἐκ τῶν ἠθέων, φιλοπονίη ψυχῆς, ἢ ζητῶν,[3] ἢ μελετῶν, ἢ ὁρῶν,[4] ἢ λέγων, ἢ εἴ τι ἄλλο, οἷον[5] λῦπαι, δυσοργησίαι, ἐπιθυμίαι· ἢ τὰ[6] ἀπὸ συγκυρίης λυπήματα γνώμης, ἢ[7] τὰ[8] διὰ τῶν ὀμμάτων, ἢ[9] ἀκοῆς·
10 οἷα τὰ σώματα, μύλης μὲν τριφθείσης πρὸς ἑωυτήν, ὀδόντες ᾑμώδησαν, παρά τε κοῖλον παριόντι σκέλεα τρέμει, ὅταν τε τῇσι χερσί τις,[10] ὧν μὴ δεῖται, αἴρῃ, αὗται τρέμουσιν, ὄφις ἐξαίφνης ὀφθεὶς χλωρότητα ἐποίησεν. οἱ φόβοι, αἰσχύνη,[11] λύπη, ἡδονή,[12] ὀργή, ἄλλα τὰ τοιαῦτα,[13] οὕτως ὑπακούει ἑκάστῳ τὸ προσῆκον τοῦ σώματος τῇ πρήξει, ἐν τούτοισιν[14] ἱδρῶτες, καρδίης
18 παλμός, καὶ τὰ τοιαῦτα.

X. Τῶν δυναμένων[15] τὰ ἔξωθεν ὠφελέοντα ἢ βλάπτοντα, ἄλειψις, κατάχυσις, κατάχρισις,[16]

[1] βρωμάτων M : βροτῶν A.

[2] A omits from ὥντινων to οἵων.

[3] ζητῶν A : ζητησίων M.

[4] A omits ἢ ὁρῶν.

[5] A omits οἷον.

[6] ἢ τὰ A.

[7] A adds ἢ before γνώμης.

[8] M omits τὰ before διά.

[9] M has διὰ τῆς.

[10] M omits τις.

[11] M has οἷον before αἰσχύνη.

[12] A has ἡδονὴ λύπη.

[13] ἄλλα τὰ τοιαῦτα A : τὰ ἄλλα τὰ τοιαῦτα M.

[14] οὕτως ἐνακούη ἑκάστω τὸ προσῆκον τοῦ σώματος τῆ πρήξη· ἐν τούτοις A : οὕτως ὑπακούει· ἑκάστωι δὲ τὸ προσῆκον τοῦ σώματος, τῆι πρήξει ἐν τούτοισιν (-ν in second hand) ὑπακούει· M.

[15] τὰ τοιαῦτα τῶν δυναμένων τά ἔξωθεν· ἢ ὠφελέοντα· ἢ βλαπτοντα A : τὰ τοιαῦτα· τῶν δυναμίων τὰ ἔξωθεν ὠφελέοντα ἢ βλάπτοντα, M.

IX. Among psychical symptoms[1] are intemperance in drink and food, in sleep, and in wakefulness, the endurance of toil either for the sake of certain passions (for example, love of dice) or for the sake of one's craft or through necessity, and the regularity or irregularity of such endurance. States of mind before and after changes. Of moral characteristics: diligence of mind, whether in inquiry or practice or sight or speech;[2] similarly, for example, griefs, passionate outbursts, strong desires. Accidents grieving the mind, either through vision or through hearing. How the body behaves: when a mill grinds the teeth are set on edge; the legs shake when one walks beside a precipice; the hands shake when one lifts a load that one should not lift; the sudden sight of a snake causes pallor. Fears, shame, pain, pleasure, passion and so forth: to each of these the appropriate member of the body responds by its action. Instances are sweats, palpitation of the heart and so forth.

X. Of remedies that may help or harm those applied externally include anointing, affusions,

[1] The genitive ψυχῆς ("belonging to the soul are, etc.") is rather strange, and one is strongly tempted to adopt the reading of A, ἀγυμναστίη ψυχῆς, "lack of self-control." Unfortunately this reading leaves ταῦτα without any construction.

[2] This phrase has no grammatical construction with the rest of the sentence, and the manuscript M, with its cognates, reads ζητησίων or ζητήσεων. Glaucias, an old commentator, not understanding the words, added the negative μή before the participles.

[16] A reads καταχρίσεις· καταχύσεις altered to the singular apparently by the original scribe. So with the next two words.

κατάπλασις, ἐπίδεσις ἐρίων καὶ τῶν τοιούτων, καὶ τὰ ἔνδοθεν ὑπακούει[1] τούτων[2] ὁμοίως[3] ὥσπερ καὶ τὰ ἔξω τῶν ἔσω προσφερομένων· ἀτὰρ καὶ τάδε,[4] ἐν ἐρίοισι κοίτη πινώδεσι, καὶ τὸ παρὰ βασιλεῖ[5] λεγόμενον κύμινον, ὀρῶσιν, ὀσφραινομένοισιν· ὅσα κεφαλῆς ἀγωγά, ταρακτικά· λόγοι, φωνή, καὶ τὰ τοιαῦτα. μαζοί, γονή, ὑστέρη, σημεῖα
10 ταῦτ' ἐν τῇσιν ἡλικίῃσι, καὶ ἐν τοῖσι πνιγμοῖσι,
11 καὶ βηξί, τὰ πρὸς ὄρχιν.

XI. Ὥσπερ τοῖσι δένδρεσιν ἡ γῆ οὕτω τοῖσι ζώοισιν[6] ἡ γαστήρ· καὶ τρέφει, καὶ θερμαίνει, καὶ ψύχει·[7] ὥσπερ γῆ κοπρευομένη χειμῶνος θερμαίνει, οὕτως ἡ γαστὴρ θερμὴ γίνεται.[8] δένδρεα φλοιὸν λεπτὸν ξηρὸν ἔχει, ἔσωθεν δὲ ξηρόσαρκα,[9] ὑγιηρά, ἄσηπτα, χρόνια, καὶ ζῴων, οἷον χελῶναι, καὶ ὅ τι τοιοῦτον.[10] ἡλικίῃσιν, ὥρῃσιν, ἐνιαυτοῖς ὅμοια τὰ ζῶντα· οὐ τρίβεται,[11] χρωμένοισι μετρίως βελτίω·[12] ὥσπερ ὑδρεῖον νέον
10 διαπηδᾷ,[13] παλαιούμενον στέγει, οὕτω καὶ ἡ γαστὴρ διίει[14] τὴν τροφήν, καὶ ὑποστάθμην ἴσχει
12 ὥσπερ ἀγγεῖον.

XII. Οἱ τρόποι τῶν νούσων· τὰ μὲν συγγενικὰ ἔστιν εἰδέναι πυθόμενον,[15] καὶ τὰ ἀπὸ τῆς χώρης

[1] ὑπακούσῃ A. [2] τούτων A : τῶν τοιούτων M.
[3] ὁμοίως A : οὐ μόνον M. [4] τάδε M : τὰ τοιάδε A.
[5] βασιλεῖ M : πᾶσι A.
[6] ζώοισιν (-ν in second hand) M : ζωδίοις A.
[7] So A. The reading of M is ψύχει κενουμένη· πληρουμένη θερμαίνει.
[8] κοπριωμένη χειμῶνος θερμαίνει· οὕτως ἡ γαστὴρ θερμὴ γίνεται A : κοπρευομένη χειμῶνος. θερμὴ ἡ κοιλίη· M.
[9] ἔνδοθεν μὲν σκληρόσαρκα· A : ἔσωθεν δέ. ξηρόσαρκα M.
[10] Littré suggests that after τοιοῦτον there has fallen out some phrase like οὕτω καὶ ἡ κοιλίη, in order to make the text

inunction, cataplasms, bandages of wool and the like; the internal parts of the body react to these remedies just as the external parts react to remedies applied internally. Moreover, a bed made out of unwashed fleeces, and the sight or smell of the cumin called "royal." Things that purge the head are disturbing, conversation, voice and so forth. Breasts, seed, womb are symptomatic at the various ages; in chokings and in coughs, fluxes to the testicles.

XI. As the soil is to trees, so is the stomach to animals. It nourishes, it warms, it cools; as it empties it cools, as it fills it warms. As a soil that is manured warms in winter, so the stomach grows warm. Trees have a slight, dry bark, but inside they are of dry texture, healthy, free from rot, durable; so among animals are tortoises and the like. In their ages animals are like the seasons and the year. They do not wear out, but improve with moderate use. As a water-pot, when new, lets the liquid pass through it, but holds it as time goes on, so the stomach lets nourishment pass, and like a vessel retains a sediment.

XII. The fashions of diseases. Some are congenital and may be learned by inquiry, as also may those

---

conform to the Galenic commentary, which says that there is a comparison implied with the membranes of the stomach. Perhaps it is from here that M got its reading of the preceding sentence.

[11] After τρίβεται A adds τὰ ζῶντα.

[12] A omits βελτίω.

[13] A has διαπεῖ· εἰ δέ, but the εἰ is cramped and was apparently added after the other words had been written.

[14] διίει Littré: ἰδέη A: δίει M.

[15] A has πειθόμενον, with υ written over the -ει-.

(οἰκέονται γὰρ οἱ πολλοί, διὸ πλέονες ἴσασι),[1] τὰ δὲ ἐκ τοῦ σώματος, καὶ τὰ ἀπὸ τῶν διαιτημάτων, καὶ καταστάσιος τῆς νούσου,[2] ἢ ἀπὸ ὡρέων. αἱ δὲ χῶραι πρὸς τὰς ὥρας κακῶς κείμεναι τοιαῦτα τίκτουσι νοσήματα, ὁποίη ἂν ἡ ὥρη, ταύτῃ ὁμοίως,[3] οἷον ἀνώμαλον θάλπος ἢ[4] ψῦχος τῆς αὐτῆς ἡμέρης, ὅταν τοιαῦτα ποιῇ,[5] 10 φθινοπωρινὰ ἐν τῇ χώρῃ τὰ νοσήματα· καὶ ἐν τῇσιν ἄλλῃσιν ὥρῃσι κατὰ λόγον. τὰ μὲν ἀπὸ ὀδμέων βορβορωδέων ἢ ἑλωδέων, τὰ δὲ[6] ἀπὸ ὑδάτων, λιθιῶντα, σπληνώδεα, τὰ τοιαῦτα δ' 14 ἀπὸ πνευμάτων χρηστῶν τε καὶ κακῶν.[7]

XIII. Ὥρης δὲ οἷαι ἔσονται αἱ νοῦσοι καὶ καταστάσιες,[8] ἐκ τῶνδε· ἢν[9] αἱ ὧραι ὡραίως, εὐτάκτως, εὐκρινέας νούσους ποιέουσιν· αἱ δ' ἐπιχώριοι τῇσιν ὥρῃσι νοῦσοι δῆλαι[10] τοὺς τρόπους· ὅ τι δ' ἂν ἐξαλλάξῃ ἡ ὥρη, ὅμοια ἢ ἀνόμοια ἔσται[11] τὰ νοσήματα, οἷα ἐν τῇ ὥρῃ ταύτῃ γίνεται· ἢν δ' ὁμοίως ἄγῃ,[12] τοιουτότροπα καὶ ἐπὶ τοιοῦτο εἱλκυσμένα, οἷον ἴκτερον φθινοπω-

[1] οἰκέονται γὰρ διὰ πλειόνων, καὶ πολλοὶ ἴσασι· M : οἰκέονται γὰρ οἱ πολλοὶ διὸ πλέονες ἴσασι· A.

[2] A has ἢ ἀπὸ before τῆς νούσου.

[3] Possibly ταύτῃ ὁμοίως is a marginal explanation of the preceding words, and should be deleted.

[4] A omits ἤ.

[5] A has ὅταν τοιαῦτα ποιέῃ after νοσήματα.

[6] τε A : δὲ M.

[7] τὰ τοιαῦτα δ' ἀπὸ πνευμάτων χρηστῶν καὶ κακῶν. A : τὰ δέ, ἀπὸ πνευμάτων χρηστῶν τε καὶ κακῶν ἄρχονται M.

[8] καταστάσιες M : καταστασίων A.

[9] A omits ἤν.

[10] δῆλαι A : δηλοῦσι M.

[11] ἡ ὥρη αὕτη καὶ οὕτως ἄγῃ, ὅμοια ἔσται A : ἡ ὥρη, ὅμοια, ἢ ἀνόμοια ἔσται M.

that are due to the district, for most people[1] are permanent residents there, so that those who know are numerous. Some are the result of the physical constitution, others of regimen, of the constitution of the disease, of the seasons. Countries badly situated with respect to the seasons engender diseases analogous to the season. *E.g.* when it produces irregular heat or cold on the same day, diseases in the country are autumnal, and similarly in the case of the other seasons. Some spring from the smells of mud or marshes, others from waters, stone, for example, and diseases of the spleen; of this kind are waters[2] because of winds good or bad.

XIII. What the character of a season's diseases and constitutions will be you must foretell from the following signs. If the seasons proceed normally and regularly, they produce diseases that come easily to a crisis. The diseases that are peculiar to the seasons are clear as to their fashions. According to the alterations in a season, the diseases such as arise in this season will be either like or unlike their usual nature.[3] If the season proceeds normally, similar or somewhat similar to the normal will be the diseases, as, for example, autumnal jaundice;

[1] The difference between A and M suggests corruption, M appearing to be an attempt to improve on A. Perhaps *οἱ* should be omitted before *πολλοί*.

[2] So Littré, who bases his interpretation on *Airs, Waters, Places*, IX, where winds are said to give various characteristics to waters. Possibly, however, we should read with M *τὰ δέ* instead of *τοιαῦτα*.

[3] Or, "unlike the seasons."

---

[12] *εἰδ' ὁμοίως ἄγει* A : *ἣν δ' ὁμοίως ἄγηι, ἡ* M.

ρινόν· ψύχεα γὰρ[1] ἐκ θαλπέων, καὶ θάλπος ἐκ
10 ψύχεος·[2] καὶ ἢν τὸ θερινὸν χολῶδες γένηται, καὶ
αὐξηθὲν ἐγκαταλειφθῇ, καὶ ὑπόσπληνοι.[3] ὅταν
οὖν καὶ ἦρ[4] οὕτως ἀγάγῃ, καὶ ἦρος γίνονται
ἴκτεροι· ἐγγυτάτω γὰρ αὕτη ἡ κίνησις[5] τῇ ὥρῃ
κατὰ τοῦτο τὸ εἶδός ἐστιν. ὅταν δὲ θέρος γένηται
ἦρι ὅμοιον,[6] ἱδρῶτες ἐν τοῖσι πυρετοῖσι, καὶ
εὔτροποι, καὶ οὐ κατοξέες, οὐδὲ κατάξηροι γλώσ-
σῃσιν. ὅταν δὲ[7] χειμέριον γένηται ἦρ[8] καὶ
ὀπισθοχειμών,[9] χειμεριναὶ καὶ αἱ νοῦσοι, καὶ
βηχώδεες, καὶ περιπλευμονικαί, καὶ κυναγχικαί.
20 καὶ[10] φθινοπώρου, † ἢν μὴ[11] ἐν ὥρῃ καὶ ἐξαίφνης
χειμάσῃ, μὴ[12] συνεχέως[13] τοιαύτας[14] νούσους
ποιεῖ † διὰ τὸ μὴ ἐν ὥρῃ ἦρχθαι, ἀλλὰ ἀνώμαλα
γίνεται· διόπερ καὶ αἱ ὧραι ἄκριτοι καὶ ἀκατάστατοι γίνονται, ὥσπερ καὶ[15] αἱ νοῦσοι, ἐὰν προεκ-
ρηγνύωνται, ἢ προκρίνωνται, ἢ ἐγκαταλείπωνται·
φιλυπόστροφοι γὰρ καὶ αἱ ὧραι γίνονται,[16] οὕτω
νοσοποιέουσαι. προσλογιστέον οὖν, ὁποίως[17] ἂν[18]
28 ἔχοντα τὰ σώματα αἱ ὧραι παραλαμβάνωσιν.

XIV. Νότοι βαρυήκοοι, ἀχλυώδεες, καρηβαρικοί, νωθροί,[19] διαλυτικοί· ὅταν οὗτος[20] δυναστεύῃ,

[1] A omits from ἴκτερον to γάρ. [2] ψύχεος M : ψύξιος A.
[3] A adds καὶ ἴκτεροι before καὶ ὑπόσπληνοι.
[4] ἦρ M : ἔαρ A. [5] ἡ κίνησις M : κείνη A.
[6] ἔστι δ' ὅτε τὸ θέρος ἔαρι ὅμοιον. ἱδρῶτας A : ὅταν δὲ θέρος γένηται ἦρι ὅμοιον. ἱδρῶτας M.
[7] δὲ omitted by A. [8] ἔαρ A, without γένηται.
[9] ὀπισθοχειμών M : ὀπίσω οὐ χειμὼν ἢ A.
[10] A omits καί. [11] A omits μή. [12] A omits μή.
[13] ξυνεχέας A : συνεχῶς M.
[14] τοιαύτας M : ταύτας τὰς A. [15] A omits καί.
[16] γίνονται M : γινωνται (ω changed to ο?) A.
[17] ὁμοίως A : ὁκοίως M.

for cold spells succeed to hot spells and heat to cold. If the summer prove bilious, and if the increased bile be left behind, there will also be diseases of the spleen. So when spring too has had a bilious constitution, there occur cases of jaundice in spring also. For this motion[1] is very closely akin to the season when it has this nature. When summer turns out like to spring, sweats occur in fevers; these are mild, not acute, and do not parch the tongue. When the spring turns out wintry, with after-winter storms, the diseases too are wintry, with coughs, pneumonia or angina. So in autumn, should there be sudden and unseasonable wintry weather, symptoms are not continuously autumnal, because they began in their wrong season, but irregularities occur.[2] So seasons, like diseases, can fail to show crisis or to remain true to type, should they break out suddenly, or be determined too soon, or be left behind. For seasons, too, suffer from relapses, and so cause diseases. Accordingly, account must also be taken of the condition of a body when the seasons come upon it.

XIV. South winds cause deafness, dimness of

[1] That is, the disturbance of the humours which causes jaundice.

[2] The sense apparently is that an autumnal disease, beginning in a premature winter, does not show continuously autumnal symptoms. But A omits μή both before ἐν ὥρῃ and before συνεχέως, and the latter negative should be οὐ. The true reading seems to be lost.

---

[18] A omits ἂν and reads παραλαμβάνουσιν.

[19] νότος βαρυήκοον. ἀχλυῶδες. καρηβαρικόν. διαλυτικόν. νωθρὸν· A: νότοι βαρυήκοοι· ἀχλυώδεες· καρηβαρικοί· νωθροί. διαλυτικοί M.

[20] οὗτος M: οὕτως A.

τοιουτότροπα ἐν τῇσι[1] νούσοισι πάσχουσιν· ἕλκεα μαδαρά, μάλιστα στόμα, αἰδοῖον, καὶ τἄλλα. ἢν δὲ βόρειον, βῆχες, φάρυγγες, κοιλίαι σκληρότεραι, δυσουρίαι[2] φρικώδεες, ὀδύναι[3] πλευρέων, στηθέων· ὅταν οὗτος[4] δυναστεύῃ, τοιαῦτα προσδέχεσθαι τὰ νοσήματα μᾶλλον. ἢν μᾶλλον πλεονάζῃ, αὐχμοῖσιν οἱ πυρετοὶ ἕπονται καὶ ὄμβροισιν, 10 ἐξ ὁποίων ἂν οἱ πλεονασμοὶ μεταπέσωσι, καὶ ὅπως ἂν ἔχοντα τὰ σώματα παραλάβωσιν ἐκ τῆς ἑτέρης ὥρης, καὶ ὁποιουτινοσοῦν χυμοῦ δυναστεύοντος ἐν τῷ σώματι. ἀτὰρ ἀνυδρίαι νότιοι, βόρειοι· διαφέρει γὰρ καὶ τἄλλα οὕτω· μέγα γὰρ καὶ τοῦτο· ἄλλος γὰρ ἐν ἄλλῃ ὥρῃ καὶ χώρῃ μέγας, οἷον τὸ θέρος χολοποιόν, ἦρ ἕναιμον, τἄλλα 17 ὡς ἕκαστα.

XV. Αἱ μεταβολαὶ μάλιστα τίκτουσι νοσήματα, καὶ αἱ μέγισται μάλιστα, καὶ ἐν τῇσιν ὥρῃσιν αἱ μεγάλαι μεταλλαγαί, καὶ ἐν τοῖσιν ἄλλοισιν· αἳ δ' ἐκ προσαγωγῆς γίνονται,[5] αἱ ὧραι αὗται ἀσφαλέσταται, ὥσπερ καὶ δίαιται καὶ ψῦχος καὶ θάλπος μάλιστα ἐκ προσαγωγῆς, καὶ 7 αἱ ἡλικίαι οὕτω μεταβαλλόμεναι.

XVI. Φύσιες δὲ ὡς πρὸς τὰς ὥρας, αἱ μὲν πρὸς θέρος, αἱ δὲ πρὸς χειμῶνα εὖ καὶ κακῶς πεφύκασιν, αἱ δὲ πρὸς χώρας καὶ ἡλικίας καὶ διαίτας καὶ τὰς ἄλλας καταστάσιας τῶν νούσων ἄλλαι πρὸς ἄλλας εὖ καὶ κακῶς πεφύκασι, καὶ ἡλικίαι πρὸς χώρας καὶ ὥρας καὶ διαίτας καὶ πρὸς καταστάσιας νούσων· καὶ ἐν τῇσιν ὥρῃσι, δίαιται

[1] τῇσι M : τοῖσι A. [2] δυσουρίαι M : δυσουρότεροι A.
[3] ὀδύναι M : ὀδυνώδεες A. [4] οὗτος M : οὕτως A.
[5] M has γίνεται with τὰ δὲ preceding.

vision, headaches, heaviness, and are relaxing. When such winds prevail, their characteristics extend to sufferers from diseases. Sores are soft, especially in the mouth, the privy parts, and similar places. A north wind causes coughs, sore throats, constipation, difficult micturition accompanied by shivering, pains in the side and chest; such are the diseases that one must be prone to expect when this wind prevails. Should its predominance be greater still, the fevers which follow drought and rain are determined by the conditions that preceded this predominance, by the physical condition produced by the previous season, and by the particular humour that prevails in the body. Droughts accompany both south winds and north winds. Winds cause differences—and this too is important—in all other respects also. For humours vary in strength according to season and district; summer, for instance, produces bile; spring, blood, and so on in each case.

XV. It is changes that are chiefly responsible for diseases, especially the greatest changes, the violent alterations both in the seasons and in other things. But seasons which come on gradually are the safest, as are gradual changes of regimen and temperature, and gradual changes from one period of life to another.

XVI. The constitutions of men are well or ill adapted to the seasons, some to summer, some to winter; others again to districts, to periods of life, to modes of living, to the various constitutions of diseases. Periods of life too are well or ill adapted to districts, seasons, modes of living and constitutions of diseases. So with the seasons vary modes of

καὶ σιτία, καὶ ποτά, ὁ μὲν γὰρ[1] χειμὼν ἀργὸς ἔργων, καὶ πέπονα τὰ ἐσιόντα καὶ[2] ἁπλᾶ, μέγα
10 γὰρ καὶ τοῦτο· αἱ ὀπῶραι δὲ ἐργάσιμοι, ἡλιώσιες, τὰ πινόμενα πυκνά,[3] ἀκατάστατα σιτία, οἶνοι,
12 ἀκρόδρυα.

XVII. Ὥσπερ δὲ[4] ἐκ τῶν ὡρέων τὰς νούσους ἔστι τεκμήρασθαι, ἔστι ποτὲ καὶ ἐκ τῶν νούσων ὕδατα καὶ ἀνέμους καὶ ἀνυδρίας προγινώσκειν, οἷον βόρεια, νότια· ἔστι γὰρ εὖ μαθόντι καὶ ὀρθῶς ὅθεν σκεπτέα, οἷον καὶ λέπραι τινὲς καὶ περὶ τὰ ἄρθρα πόνοι,[5] ὕδατα ὅταν μέλλῃ,
7 κνησμώδεές εἰσι, καὶ ἄλλα τοιαῦτα.

XVIII. Καὶ ὑσμάτων οἷα ἢ διὰ τρίτης, ἢ ἑκάστης, ἢ διὰ περιόδων ἄλλων, καὶ τὰ συνεχέα· καὶ ἀνέμων οἱ μὲν πολυήμεροι πνέουσι, καὶ ἀντιπνέουσι ἀλλήλοισιν, ἄλλοι δὲ διὰ βραχυτέρων, οἱ δὲ καὶ αὐτοὶ κατὰ περίοδον· ταῦτα ἔχει τῇσι καταστάσεσιν ὁμοιότητας, ἐπὶ βραχύτερον δὲ τὰ τοιαῦτα. καὶ εἰ μὲν ἐπὶ πλέον τὸ ἔτος τοιοῦτον ἐὸν τὴν κατάστασιν ἐποίησε τοιαύτην, ἐπὶ πλέον[6] καὶ τὰ νοσήματα τοιαῦτα καὶ
10 μᾶλλον[7] ἰσχυρότερα, καὶ μέγιστα νοσήματα οὕτως ἐγένετο[8] καὶ κοινότατα καὶ ἐπὶ πλεῖστον χρόνον. ἐκ τῶν πρώτων ὑδάτων, ὅταν ἐξ ἀνυδρίης πολλῆς μέλλῃ ὕδωρ ἔσεσθαι, ἔστι περὶ ὑδρώπων προειπεῖν, καὶ ὁπόταν τἄλλα σμικρὰ σημεῖα φανῇ ἐν νηνεμίῃ, ἢ ἐν μεταβολῇ,[9] συνακ-

[1] ὁ μὲν γὰρ A : οἷα· ὅτι ὁ μὲν M.
[2] A omits καί.
[3] πυκνά M : συχνά A.
[4] ὡς δ' A.
[5] οἷσι λέπραι καί τινες περὶ τὰ ἄρθρα A : οἷον καὶ λέπραι τινὲς, καὶ τὰ περὶ τὰ ἄρθρα πόνοι M. Probably πόνοι is a note on τὰ περὶ τὰ ἄρθρα, but I have not altered Littré's text.

living, foods and drinks. In winter no work is done and foods are ripe and simple—an important point; in autumn work is done, exposure to the sun is beneficial, drinks are frequent and foods varied, with wine and fruits.

XVII. As it is possible to infer diseases from the seasons, so occasionally it is possible from diseases to forecast rains, winds and droughts; for example, north winds and south winds. For he who has noticed symptoms carefully and accurately has evidence on which to work; certain skin diseases, for instance, and pains at the joints are irritating when rain threatens, to quote one example out of many.

XVIII. Rains occur every other day, or every day, or at other intervals; some are continuous. Winds sometimes last for many days, and are opposed to one another; others are shorter; some, like rains, are periodic. These have resemblances to the seasonal constitutions, though less marked. If the year, having had a certain character to a marked degree, has given this character to the constitution, the diseases too have this character to a marked degree and are more severe; in this way have arisen very serious diseases, very widespread and lasting a very long period of time. After the first rains, when rain is coming after a long drought, it is possible to predict dropsies; and when the other slight signs appear at a period of calm, or at a change, one must infer

---

[6] καὶ εἰ μὲν ἐπὶ πλεῖον τὸ ἔτος τοιοῦτον ἐὸν, τὴν κατάστασιν ἐποίησεν τοιαύτην; ἐπὶ πλέον A: ἢν μὲν ἐπὶ πλέον τὸ ἔτος τοιοῦτον, οἵην τὴν κατάστασιν ἐποίησε. ἐπι πλεῖον M.

[7] Before μᾶλλον M adds ἦν.

[8] τὰ μέγιστα οὕτω νοσήματα ἐγένετο A: μέγιστα νοσήματα, οὕτως ἐγένετο M.

[9] ἀνεμίη ἢ ἐν μεταβολῇ A: νηνεμίηι ἐν μεταβολῆι. M.

τέον,[1] *ὅσαι μὲν ἐφ' οἵοισιν ὕδασιν ἢ*[2] *ἀνέμοισι*[3] *νοῦσοι ἐπισημαίνουσι, καὶ ἀκουστέον εἴ τις οἶδε, τοιοῦδε*[4] *χειμῶνος προγενομένου, οἷον ἦρ ἢ*
19 *θέρος ἔσται.*

XIX. *Τὰ χρώματα οὐχ ὅμοια ἐν τῇσιν ὥρῃσιν, οὐδὲ ἐν βορείοισι καὶ νοτίοισιν, οὐδ' ἐν τῇσιν ἡλικίῃσιν αὐτὸς πρὸς ἑωυτόν, οὐδ' ἄλλος ἄλλῳ οὐδενί. σκεπτέον δὲ ἐξ ὧν ἴσμεν καὶ παρεόντων καὶ ἀτρεμεόντων περὶ χροιῶν,*[5] *καὶ ὅτι αἱ ἡλικίαι τῇσιν ὥρῃσιν ἐμφερέες εἰσὶ καὶ χροιῇ καὶ*
7 *τρόπῳ.*[6]

XX. *Οἱ αἱμορροΐδας ἔχοντες οὔτε πλευρίτιδι, οὔτε περιπνευμονίῃ, οὔτε φαγεδαίνῃ, οὔτε δοθιῆσιν, οὔτε τερμίνθοισιν ἁλίσκονται,*[7] *ἴσως δὲ οὐδὲ λέπρῃσιν, ἴσως δὲ οὐδὲ ἀλφοῖσιν·*[8] *ἰητρευθέντες γε*[9] *μὴν ἀκαίρως, συχνοὶ τοιούτοισιν οὐ*[10] *βραδέως ἑάλωσαν, καὶ ὀλέθρια οὕτως· καὶ ὅσαι ἄλλαι*[11] *ἀποστάσιες, οἷον σύριγγες, ἑτέρων ἄκος· ὅσα δέ, ἐφ' οἷσι γενόμενα ῥύεται, τούτων προγενόμενα κωλύματα·*[12] *οἱ ὕποπτοι τόποι ὑποδεξάμενοι πόνῳ*
10 *ἢ βάρει ἢ ἄλλῳ τινὶ ῥύονται·*[13] *ἄλλοισιν αἱ*

[1] *συνακτέον οὖν* M : *ξυνακτέον·* A. [2] A omits *ἤ*.

[3] Before *νοῦσοι* M adds *αἱ*. [4] *τοιοῦδε* A : *ὅτι τοιούτου* M

[5] *παρεόντων, καὶ ἀτρεμεόντων, περι χροιῶν* M : *περιόντων· καὶ ἀτρεμεύντων· καὶ περὶ χροιῶν* A.

[6] *καὶ χροιὴν καὶ τρόπον* A : *καὶ χροιῆ καὶ τροπωι·* M.

[7] A omits *ἁλίσκονται*.

[8] *ἀλφοῖσιν* M : *ἄλλοῖσιν·* A (the two accents are significant).

[9] A omits *γε*. [10] *οὐ* M : *οὐδὲ* A.

[11] *ἄλλαι* M : *ἄλλων* A.

[12] *ὅσα δὲ ἐφ' οἷσι γενόμενα αἴρεται, τούτων προγενόμενα κωλύματα·* A: *ὅσα πέφυκεν ἐπιφαινόμενα παύειν. ῥύεται τούτων προσγενόμενα κωλύματα·* M.

what diseases are typical of the various rains or winds, and must listen to anyone who knows the nature of the spring or summer that will follow a winter of such and such a character.

XIX. Complexions vary with the seasons; they are not the same in north winds as in south winds; individuals differ, and the same individual varies in complexion as he grows older. Judge of complexions by their permanent characteristics, realising that ages resemble seasons in colour as in character.

XX. Sufferers from hemorrhoids are attacked neither by pleurisy, nor by pneumonia, nor by spreading ulcer, nor by boils, nor by swellings, nor perhaps by skin-eruptions and skin-diseases. However, unseasonably cured, many have been quickly caught by such diseases, and, moreover, in a fatal manner. All other abscessions, too, such as fistula, are cures of other diseases. So symptoms that relieve complaints if they come after their development, prevent the development if they come before. Suspected places cause relief, by acting as receptacles owing to pain, weight, or any other cause.[1] In other cases

[1] The reading of A is a corruption of the reading of *Epidemics* VI. 3, 23 and means, "Places receiving (peccant humours) from another place, through pain, weight or any other cause, bring relief." A "suspected" place is one in which we might expect a morbid affection to arise, and pain here, or an accumulation of humours, might relieve affections elsewhere. The phenomenon is common enough in certain forms of neuralgia, the pains of which often jump from place to place in such a way that one pain seems to relieve another.

---

[13] *ἄλλου τόπου. οἱ τόποι οὗτοι δοξάμενοι· ἢ πόνω· ἢ βάρει· ἢ ἀλλώ τῶ, ῥύονται* A: *ἄλλοισι αἱ κοινωνίαι· οἱ ὕποπτοι τόποι ὑποδεξάμενοι πόνω ἢ βάρει, ἢ ἄλλωι τινὶ ῥύονται.* M.

κοινωνίαι· διὰ τὴν ῥοπὴν οὐκ ἔτι αἷμα ἔρχεται, ἀλλὰ κατὰ τοῦ χυμοῦ τὴν ξυγγένειαν τοιαῦτα πτύουσιν· ἔστιν οἷσιν αἷμα ἀφίεσθαι ἐν καιρῷ[1] ἐπὶ τοῖσι τοιούτοισιν, ἐπ' ἄλλοισι δὲ ὥσπερ ἐπὶ τούτοισι τοῦτο οὐκ εἰκός, κώλυσις. ἐπὶ τοῖσι δὲ δὴ[2] αἱματώδεα πτύουσιν ὥρη, πλευρῖτις, χολή. τὰ παρὰ τὸ οὖς οἷσιν ἀμφὶ κρίσιν γενόμενα μὴ ἐκπυήσει, τούτου λαπασσομένου, ὑποστροφὴ γίνεται,[3] καὶ[4] κατὰ λόγον τῶν ὑποστροφέων[5]
20 τῆς ὑποστροφῆς γενομένης,[6] αὖθις[7] αἴρεται καὶ παραμένει,[8] ὥσπερ αἱ τῶν πυρετῶν ὑποστροφαί, ἐν ὁμοίῃ περιόδῳ· ἐπὶ τούτοισιν ἐλπὶς ἐς ἄρθρα ἀφίστασθαι. οὖρον παχύ, λευκόν, οἷον τῷ[9] τοῦ Ἀντιγένεος, ἐπὶ τοῖσι κοπιώδεσι τεταρταίοις ἔστιν ὅτε ἔρχεται, καὶ ῥύεται τῆς ἀποστάσιος, ἢν δὲ πρὸς τούτῳ[10] καὶ αἱμορραγήσῃ ἀπὸ ῥινῶν ἱκανῶς, καὶ πάνυ. ᾧ τὸ ἔντερον[11] ἐπὶ δεξιὰ ἀρθριτικῷ[12] ἐγένετο· ἦν ἡσυχαίτερος, ἐπεὶ δὲ
29 τοῦτο ἰητρεύθη, ἐπιπονώτερος.

[1] A adds καὶ before ἐν καιρῷ.
[2] τούτοισιν A: τοῖσι δὲ δὴ M.
[3] τὰ παρ' οὖς οἷς ἀμφὶ κρίσιν γενόμενα μὴ ἐκπυήσῃ, τούτου λαπασσομένου· ὑποστροφὴ γίνεται· A: τὰ παρὰ τὸ οὖς· ὅσοισιν ἂν ἀμφι κρίσιν γινόμενα, ἢν μὴ ἐκπυήσηι, τούτου ἀπαλλασσομένου, ὑπὸ στροφὴ γίνεται· M.
[4] καὶ M: τὰ A.
[5] ὑποστροφέων M: ὑποστροφεόντων A.
[6] κρινόμενα AM: γενομένης Littré from Galen.
[7] αὖθις M: ἄν τις A.
[8] παραμένει M: παραμένῃ A.

there is the sympathetic action. The issue, through the flow, ceases to be one of blood, but the patients spit up matter connected with the humour. In some such cases seasonable blood-letting is possible, but in other cases blood-letting, as sometimes in the former cases, is not suitable but only a hindrance. Blood-spitting may be caused by the season, by pleurisy, or by bile. When swellings by the ear do not suppurate at a crisis,[1] a relapse occurs when the swelling softens; when the relapse follows the normal course of relapses, the swelling rises again and remains, following the same periods as occur when fevers relapse. In such cases expect an abscession to the joints. Thick, white urine, as in the case of the slave of Antigenes, sometimes is passed on the fourth day in prostrating fevers, and saves the patient from the abscession, and this is especially so if in addition there is a copious flow of blood from the nostrils. The patient whose right bowel was painful became easier when arthritis supervened, but when this symptom was cured the pains became worse.[2]

[1] Or, "occur at a crisis but do not suppurate."

[2] Chapter XX is the same as *Epidemics* VI. 3, 24 to 4, 3. The variations of reading are not very important, but we may note that ἑτέρων ἄκος appears in *Epidemics* as ἢ ἕτεραι· σκέψις. See the *Introduction* to the present treatise.

---

[9] A has τῶ with ο above ώ: M has τό.

[10] τούτωι M: τοῦτο A. [11] ἔντερον M: ἕτερον· A.

[12] So both A and M. Query: ἀρθριτικὸς as in *Epidemics*?

APHORISMS

ΑΦΟΡΙΣΜΟΙ

ΤΜΗΜΑ ΠΡΩΤΟΝ

I. Ὁ βίος βραχύς, ἡ δὲ τέχνη μακρή, ὁ δὲ[1] καιρὸς ὀξύς, ἡ δὲ πεῖρα σφαλερή, ἡ δὲ κρίσις χαλεπή. δεῖ δὲ οὐ μόνον ἑωυτὸν[2] παρέχειν τὰ δέοντα ποιέοντα, ἀλλὰ καὶ τὸν νοσέοντα[3] καὶ τοὺς
5 παρεόντας καὶ τὰ ἔξωθεν.

II. Ἐν τῇσι ταραχῇσι τῆς κοιλίης[4] καὶ τοῖσιν ἐμέτοισι τοῖσιν[5] αὐτομάτοισι[6] γινομένοισιν, ἢν μὲν οἷα δεῖ καθαίρεσθαι καθαίρωνται,[7] συμφέρει τε καὶ εὐφόρως φέρουσιν· ἢν δὲ μή, τοὐναντίον. οὕτω[8] καὶ κενεαγγίη,[9] ἢν μὲν οἷα[10] δεῖ γίνεσθαι γίνηται, συμφέρει τε καὶ εὐφόρως φέρουσιν· ἢν δὲ μή, τοὐναντίον. ἐπιβλέπειν οὖν δεῖ καὶ ὥρην καὶ
8 χώρην[11] καὶ ἡλικίην καὶ νούσους, ἐν ᾗσι[12] δεῖ ἢ οὔ.[13]

III. Ἐν τοῖσι γυμναστικοῖσιν αἱ ἐπ' ἄκρον εὐεξίαι σφαλεραί, ἢν ἐν τῷ ἐσχάτῳ ἔωσιν· οὐ γὰρ δύνανται μένειν ἐν τῷ αὐτῷ οὐδὲ[14] ἀτρεμεῖν· ἐπεὶ[15] δὲ οὐκ ἀτρεμέουσιν, οὐκέτι δύνανται[16] ἐπὶ τὸ βέλτιον ἐπιδιδόναι· λείπεται οὖν ἐπὶ τὸ

[1] δὲ omitted by C′. [2] ἑαυτὸν Urb.
[3] τοὺς νοσέοντας V.
[4] τῇισι κοιλίῃισι M : τῇσι κοιλίῃσι V : τῆς κοιλίῃσι Q.
[5] C′ has τοισι before τοῖσιν.
[6] αὐτομάτοισι V : αὐτομάτοις C′: αὐτομάτως Urb. M.
[7] καθαίρηται Rein.
[8] So C′ Urb. M : οὕτω δὴ V : δὲ Littré.
[9] κεναγγίην C′ : κεναγγείη Urb. V : κενεαγγείη M : κενεαγγείῃ Q.
[10] οἵην Rein.

APHORISMS

FIRST SECTION

I. Life is short, the Art long, opportunity fleeting, experiment treacherous,[1] judgment [2] difficult. The physician must be ready, not only to do his duty himself, but also to secure the co-operation of the patient, of the attendants and of externals.

II. In disorders of the bowels, and in vomitings that are spontaneous, if the matters purged be such as should be purged, the patient profits and bears up well. If not, the contrary. So too artificial evacuations, if what takes place is what should take place, profit and are well borne. If not, the contrary. So one ought to have an eye to season, district, age and disease, to see if the treatment is, or is not, proper in the circumstances.

III. In athletes a perfect condition that is at its highest pitch is treacherous.[3] Such conditions cannot remain the same or be at rest, and, change for the better being impossible, the only possible change is

[1] Or, "deceptive."

[2] It is just possible that κρίσις here means the crisis of a disease, and that the aphorism refers to the danger attending a crisis, and to the need for prompt and skilful treatment at such times.

[3] Or, "dangerous."

---

[11] χώρην καὶ ὥρην Q.

[12] οἷσι Q. C′ has ᾗ εἰσὶ καὶ διοῦ. [13] μὴ Ermerins.

[14] Ermerins omits ἢν . . . ἔωσιν· and μένειν . . . οὐδὲ.

[15] ἐπειδὴ C′. [16] V places δύνανται after βέλτιον.

χεῖρον. τούτων οὖν εἵνεκεν τὴν εὐεξίην λύειν συμφέρει μὴ βραδέως, ἵνα πάλιν ἀρχὴν ἀναθρέψιος λαμβάνῃ τὸ σῶμα. μηδὲ τὰς συμπτώσιας ἐς τὸ ἔσχατον ἄγειν, σφαλερὸν[1] γάρ, ἀλλ' ὁκοίη[2] ἂν
10 ἡ φύσις ᾖ τοῦ μέλλοντος ὑπομένειν, ἐς τοῦτο ἄγειν. ὡσαύτως δὲ καὶ αἱ κενώσιες αἱ ἐς τὸ ἔσχατον ἄγουσαι σφαλεραί· καὶ πάλιν αἱ ἀνα-
13 λήψιες[3] αἱ ἐν τῷ ἐσχάτῳ ἐοῦσαι[4] σφαλεραί.[5]

IV. Αἱ λεπταὶ καὶ ἀκριβέες δίαιται, καὶ[6] ἐν τοῖσι μακροῖσιν αἰεὶ πάθεσι,[7] καὶ ἐν τοῖσιν ὀξέσιν,[8] οὗ μὴ ἐπιδέχεται, σφαλεραί.[9] καὶ πάλιν[10] αἱ ἐς τὸ ἔσχατον λεπτότητος ἀφιγμέναι δίαιται χαλεπαί·[11] καὶ γὰρ καὶ[12] αἱ πληρώσιες αἱ
6 ἐς τὸ ἔσχατον ἀφιγμέναι[13] χαλεπαί.[14]

V. Ἐν τῇσι λεπτῇσι διαίτῃσιν ἁμαρτάνουσιν οἱ νοσέοντες, διὸ μᾶλλον βλάπτονται· πᾶν γὰρ[15] ὃ ἂν γίνηται μέγα γίνεται μᾶλλον ἢ ἐν τῇσιν ὀλίγον ἁδροτέρῃσι διαίτῃσιν. διὰ τοῦτο καὶ τοῖσιν ὑγιαίνουσι σφαλεραὶ αἱ πάνυ λεπταὶ καὶ ἀκριβέες καθεστηκυῖαι[16] δίαιται, ὅτι τὰ ἁμαρτανόμενα χαλεπώτερον φέρουσιν. διὰ τοῦτο οὖν[17]

[1] εἰς τὸ ἐσχάτην ἄγειν· σφαλερὰ Urb. M has ξυμπόσιας for συμπτώσιας.
[2] ὅκου Urb.
[3] ἀναθρέψιες M.
[4] ἀναληψιαις αἱ ἐς τὸ ἔσχατον ἄγουσαι C′. Ermerins omits from ὡσαύτως to the end.
[5] Ermerins omits ὡσαύτως . . . σφαλεραί.
[6] καὶ omitted by V. Ermerins omits from καὶ to δίαιται χαλεπαί.
[7] Urb. has ἀεὶ πάθεσι in the margin in another hand.
[8] After ὀξέσιν (spelt ὀξέσι) C′ has νοσίμασιν. So S according to Littré. This suggests that πάθεσι is a gloss.
[9] χαλεπαὶ V Q.
[10] καὶ πάλιν omitted by Urb. V.
[11] λεπταὶ V.
[12] καὶ C′.

for the worse. For this reason it is an advantage to reduce the fine condition quickly, in order that the body may make a fresh beginning of growth. But reduction of flesh must not be carried to extremes, as such action is treacherous[1]; it should be carried to a point compatible with the constitution of the patient. Similarly, too, evacuations carried to extremes are treacherous,[1] and again new growths, when extreme, are treacherous.[1]

IV. A restricted and rigid regimen is treacherous,[1] in chronic diseases always, in acute, where it is not called for. Again, a regimen carried to the extreme of restriction is perilous; and in fact repletion too, carried to extremes, is perilous.

V. In a restricted regimen the patient makes mistakes, and thereby suffers more; for everything that occurs is more serious than with a slightly more liberal regimen. For this reason in health too an established regimen that is rigidly restricted is treacherous,[1] because mistakes are more hardly borne.

[1] Or, "dangerous."

---

13 ἐν τῷ ἐσχάτῳ ἐοῦσαι Urb.

14 σφαλεραί Urb. (and S according to Littré).

15 After γὰρ Littré with E adds τὸ ἁμάρτημα.

16 So C′: ἀκριβέες καὶ καθεστηκυῖαι Urb.: καθεστηκυῖαι καὶ λεπταὶ καὶ ἀκριβεῖς V: λεπταὶ καὶ καθεστηκυῖαι καὶ ἀκριβέες M: λεπταὶ καθεστηκυῖαι καὶ ἀκριβέες Q.

Here V 2r, l. 13 ends:

καὶ λεπταὶ καὶ ἀκριβεῖς δίαιται

l. 14 ends: αἱ λεπταὶ καὶ ἀκριβεῖς δίαιται

l. 15 begins: σφαλεραὶ ἐς τὰ πλεῖστα . . .

C 2v, l. 8 ends: καὶ λεπταὶ καὶ ἀκριβεῖς δίαιται

l. 9 begins: σφαλεραὶ ἐς τὰ πλεῖστα . . .

The scribe of C, who copied V, omitted one entire line.

17 διὰ τοῦτο οὖν omitted by C′.

αἱ λεπταὶ καὶ ἀκριβέες δίαιται σφαλεραὶ[1] ἐς τὰ
9 πλεῖστα τῶν σμικρὸν[2] ἁδροτέρων.

VI. Ἐς δὲ τὰ ἔσχατα νοσήματα αἱ ἔσχαται
2 θεραπεῖαι ἐς ἀκριβείην κράτισται.

VII. Ὅκου μὲν οὖν κάτοξυ τὸ νόσημα, αὐτίκα
καὶ τοὺς ἐσχάτους πόνους ἔχει, καὶ τῇ ἐσχάτως
λεπτοτάτῃ διαίτῃ ἀναγκαῖον χρῆσθαι· ὅκου δὲ
μή, ἀλλ' ἐνδέχεται ἁδροτέρως διαιτᾶν, τοσοῦτον
ὑποκαταβαίνειν, ὁκόσον ἂν ἡ νοῦσος μαλθακω-
6 τέρη[3] τῶν ἐσχάτων ᾖ.

VIII. Ὁκόταν[4] ἀκμάζῃ τὸ νόσημα, τότε λεπτο-
2 τάτῃ διαίτῃ ἀναγκαῖον χρῆσθαι.

IX. Συντεκμαίρεσθαι δὲ χρὴ[5] καὶ τὸν νοσέοντα, εἰ ἐξαρκέσει τῇ διαίτῃ πρὸς τὴν ἀκμὴν τῆς
νούσου,[6] καὶ πότερον ἐκεῖνος ἀπαυδήσει πρότερον,[7] καὶ οὐκ ἐξαρκέσει τῇ διαίτῃ, ἢ ἡ νοῦσος
5 πρότερον ἀπαυδήσει καὶ ἀμβλυνεῖται.[8]

X. Ὁκόσοισι[9] μὲν οὖν αὐτίκα ἡ ἀκμή, αὐτίκα
λεπτῶς διαιτᾶν· ὁκόσοισι[9] δὲ ὕστερον ἡ ἀκμή,
ἐς ἐκεῖνο καὶ πρὸ ἐκείνου σμικρὸν ἀφαιρετέον·
ἔμπροσθεν δέ, πιοτέρως διαιτᾶν ὡς ἂν[10] ἐξαρκέσῃ
5 ὁ νοσέων.

XI. Ἐν δὲ τοῖσι παροξυσμοῖσι ὑποστέλλεσθαι

[1] σφαλερώτεραι Erm. : μᾶλλον σφαλεραὶ Rein.

[2] μικρῶν C′ Q : σμικρὸν Urb. V : σμικρῶν M.

[3] μαλακοτέρη C′.

[4] After ὁκόταν many MSS. have δὲ. It is omitted by Urb., while C′ has γὰρ.

[5] χρὴ omitted by V.

[6] τῇ διαίτῃ πρὸς τὴν ἀκμὴν τῆς νούσου C′ and Urb. : τῇ διαίτῃ καὶ τὴν ἀκμὴν τῆς νούσου V : τῆι νούσωι· καὶ τὴν ἀκμὴν τῆς νούσου M.

[7] καὶ μὴ πρότερον ἐκεῖνος ἀπαυδήσει Urb. and Magnolus in margin.

For this reason, therefore, a rigidly restricted regimen is treacherous[1] generally as compared with one a little more liberal.

VI. For extreme diseases extreme strictness of treatment is most efficacious.

VII. Where the disease is very acute, immediately, not only is the pain extreme, but also it is essential to employ a regimen of extreme strictness. In other cases, where a more liberal regimen is possible, relax the strictness according as the disease is milder than the most extreme type.

VIII. It is when the disease is at its height that it is necessary to use the most restricted regimen.

IX. Take the patient too into account and decide whether he will stand the regimen at the height of the disease; whether his strength will give out first and he will not stand the regimen, or whether the disease will give way first and abate its severity.

X. When the disease reaches its height immediately, regimen must be restricted immediately. When the height comes later, restrict regimen then and a little before then; before, however, use a fuller regimen, in order that the patient may hold out.[2]

XI. Lower diet during exacerbations, for to give

[1] Or, "dangerous."

[2] So Littré; and, as V omits ἂν, it is probable that the ancient interpretation took ὡς to be final. But it is perhaps better to take ὡς as meaning "how" or "in such a way that," in which case the translation will be "restricting it not more than the patient's strength permits."

---

[8] ἀμβλυνεῖται M V and Urb.: ἀπαμβλύνηταὶ C′: Perhaps ἀπαμβλυνεῖται.

[9] V has οἶσι.

[10] V omits ἂν.

χρή· τὸ προστιθέναι γὰρ βλάβη· καὶ ὁκόσα[1] κατὰ περιόδους παροξύνεται ἐν τοῖσι παροξυσ-
4 μοῖσιν ὑποστέλλεσθαι χρή.[2]

XII. Τοὺς δὲ παροξυσμοὺς καὶ τὰς καταστάσιας δηλώσουσιν[3] αἱ νοῦσοι, καὶ αἱ ὧραι τοῦ ἔτεος, καὶ αἱ[4] τῶν περιόδων πρὸς ἀλλήλας ἀνταποδόσιες,[5] ἤν τε καθ' ἡμέρην, ἤν τε παρ' ἡμέρην, ἤν τε καὶ διὰ πλείονος χρόνου γίνωνται· ἀτὰρ καὶ τοῖσιν ἐπιφαινομένοισιν, οἷον ἐν πλευριτικοῖσι πτύαλον ἢν[6] αὐτίκα ἐπιφαίνηται ἀρχομένου, βραχύνει, ἢν δ' ὕστερον ἐπιφαίνηται, μηκύνει· καὶ οὖρα καὶ ὑποχωρήματα καὶ ἱδρῶτες,[7] καὶ
10 δύσκριτα καὶ εὔκριτα, καὶ βραχέα καὶ μακρὰ[8] τὰ
11 νοσήματα, ἐπιφαινόμενα, δηλοῖ.[9]

XIII. Γέροντες εὐφορώτερα νηστείην φέρουσι, δεύτερα οἱ καθεστηκότες, ἥκιστα μειρακία, πάντων δὲ μάλιστα παιδία, τούτων δὲ ἢν[10] τύχῃ αὐτὰ
4 ἑωυτῶν προθυμότερα ἐόντα.

XIV. Τὰ αὐξανόμενα πλεῖστον ἔχει τὸ ἔμφυτον θερμόν· πλείστης οὖν δεῖται τροφῆς· εἰ[11] δὲ μή, τὸ σῶμα ἀναλίσκεται· γέρουσι δὲ ὀλίγον τὸ θερμόν, διὰ τοῦτο ἄρα ὀλίγων ὑπεκκαυμάτων δέονται· ὑπὸ πολλῶν γὰρ ἀποσβέννυται· διὰ τοῦτο καὶ οἱ πυρετοὶ τοῖσι γέρουσιν οὐχ ὁμοίως
7 ὀξέες· ψυχρὸν γὰρ τὸ σῶμα.

XV. Αἱ κοιλίαι χειμῶνος καὶ ἦρος θερμόταται φύσει, καὶ ὕπνοι μακρότατοι· ἐν ταύτῃσιν οὖν τῇσιν ὥρῃσι καὶ τὰ προσάρματα πλείω δοτέον·

[1] V has ὅσα.
[2] τὸ . . . χρή omitted by C′. χρή is omitted by M.
[3] δηλοῦσιν C′ with many later MSS.
[4] αἱ omitted by Urb. and S.

food is harmful; lower diet too during the exacerbations wherever a disease is exacerbated periodically.

XII. Exacerbations and constitutions will be made plain by the diseases, by the seasons of the year, and by the correspondence of periods to one another, whether they come every day, every other day, or at a longer interval. Moreover, there are supervening symptoms; for example, in pleurisy, if expectoration supervene immediately on the commencement of the disease, it means a shorter illness, if afterwards, a longer one. Urine, stools, sweats, by the manner in which they supervene, show whether the disease will have a difficult crisis or an easy one, whether it will be short or long.

XIII. Old men endure fasting most easily, then men of middle age, youths very badly, and worst of all children, especially those of a liveliness greater than the ordinary.

XIV. Growing creatures have most innate heat, and it is for this reason that they need most food, deprived of which their body pines away. Old men have little innate heat, and for this reason they need but little fuel; much fuel puts it out. For this reason too the fevers of old men are less acute than others, for the body is cold.

XV. Bowels are naturally hottest in winter and in spring, and sleep is then longest; so it is in these seasons that more sustenance is necessary. For the

---

[5] ἐπιδόσιες C′ Urb. Galen and many later MSS.

[6] αὐτίκα ἦν V, with μὲν after ἀρχομένου. Some MSS. have μὲν after ἦν·

[7] After ἱδρῶτες V has καὶ χρώματα.

[8] καὶ μακρὰ omitted by C′.

[9] σημαίνει V. [10] ἃ ἂν Erm. and Rein. [11] ἦν C′ Urb.

καὶ γὰρ τὸ ἔμφυτον θερμὸν πολύ·[1] τροφῆς οὖν πλείονος δέονται·[2] σημεῖον, αἱ ἡλικίαι καὶ οἱ
6 ἀθληταί.[3]

XVI. Αἱ[4] ὑγραὶ δίαιται πᾶσι τοῖσι πυρεταίνουσι συμφέρουσι, μάλιστα δὲ παιδίοισι, καὶ τοῖσιν ἄλλοισι τοῖσιν οὕτως εἰθισμένοισι διαι-
4 τᾶσθαι.

XVII. Καὶ† τοῖσιν†[5] ἅπαξ ἢ δίς, ἢ πλείω ἢ ἐλάσσω, καὶ κατὰ μέρος· δοτέον δέ τι καὶ τῇ ὥρῃ,
3 καὶ τῇ χώρῃ,[6] καὶ τῷ ἔθει, καὶ τῇ ἡλικίῃ.

XVIII. Θέρεος καὶ φθινοπώρου[7] σιτία δυσφορώτατα φέρουσι, χειμῶνος ῥήιστα, ἦρος
3 δεύτερον.

XIX. Τοῖσιν ἐν τῇσι[8] περιόδοισι παροξυνομένοισι μηδὲν διδόναι, μηδ' ἀναγκάζειν, ἀλλ'[9]
3 ἀφαιρεῖν τῶν προσθεσίων[10] πρὸ τῶν κρισίων.[11]

XX. Τὰ κρινόμενα καὶ τὰ κεκριμένα ἀρτίως μὴ κινεῖν, μηδὲ νεωτεροποιεῖν, μήτε φαρμακείῃσι,
3 μήτ' ἄλλοισιν ἐρεθισμοῖσιν, ἀλλ' ἐᾶν.

XXI. Ἃ δεῖ ἄγειν, ὅκου ἂν μάλιστα ῥέπῃ,[12]
2 ταύτῃ ἄγειν, διὰ τῶν συμφερόντων χωρίων.

[1] πλεῖόν ἐστι Rein.

[2] δέονται C′ Urb. M. δεῖται V.

[3] Erm. omits καὶ γὰρ . . . ἀθληταί.

[4] M V omit αἱ.

[5] All our good MSS. have τοῖσιν or τοῖσι. Littré with slight authority reads οἷσιν. Littré would also read κατὰ μέρος δοτέον· δοτέοι δέ τι καὶ κ.τ.ἑ. Erm. and Rein. omit καὶ τοῖσιν.

[6] V omits καὶ τῇ χώρῃ.

[7] Before σιτία C′ has τὰ, and before ῥήιστα Urb. has δὲ.

[8] τακτῇσι Rein.

[9] C′ omits ἀλλ'.

[10] προσθεσίων Urb.: προθεσήων V: προθέσεων C′.

innate heat being great, more food is required; witness the young and athletes.

XVI. A sloppy diet is beneficial in all fevers, especially in the case of children and of those used to such a diet.

XVII. To some, food should be given once, to others, twice; in greater quantity or in less quantity; a little at a time.[1] Something too must be conceded to season, district, habit, and age.

XVIII. In summer and in autumn food is most difficult to assimilate, easiest in winter, next easiest in spring.

XIX. When the patient is suffering from a periodic exacerbation, offer nothing and force nothing, but lessen the nourishment before the crisis.[2]

XX. Do not disturb a patient either during or just after a crisis, and try no experiments, neither with purges nor with other irritants, but leave him alone.

XXI. What matters ought to be evacuated, evacuate in the direction to which they tend, through the appropriate passages.

[1] The reading in this aphorism is more than dubious. The strong evidence for τοῖσιν, which makes no possible grammar with the rest of the sentence, is almost proof positive that the true text has been lost. Fortunately the general sense is quite plain.

[2] As Galen says, "crisis" here may mean either the exacerbation, or the summit of the disease, or the crisis in the strict sense of the word. The aphorism is so like XI. that some editors think it is an interpolation, though an early one.

---

[11] τῶν κρίσεων M V Urb.: τῆς κρίσεως C′: τῶν παροξυσμῶν Erm.

[12] After ῥέπῃ C′ has ἡ φύσις.

XXII. Πέπονα φαρμακεύειν καὶ κινεῖν, μὴ ὠμά, μηδὲ ἐν ἀρχῇσιν, ἢν μὴ ὀργᾷ· τὰ δὲ πλεῖστα[1]
3 οὐκ ὀργᾷ.

XXIII. Τὰ χωρέοντα μὴ τῷ πλήθει τεκμαίρεσθαι, ἀλλ' ὡς ἂν χωρῇ οἷα δεῖ, καὶ φέρῃ[2] εὐφόρως· καὶ ὅκου δεῖ μέχρι[3] λειποθυμίης ἄγειν,
4 καὶ τοῦτο ποιεῖν, ἢν ἐξαρκῇ ὁ νοσέων.

XXIV. Ἐν τοῖσιν ὀξέσι πάθεσιν ὀλιγάκις καὶ ἐν ἀρχῇσι τῇσι φαρμακείῃσι χρῆσθαι, καὶ τοῦτο
3 προεξευκρινήσαντα ποιεῖν.

XXV. Ἢν οἷα δεῖ καθαίρεσθαι καθαίρωνται, συμφέρει τε καὶ εὐφόρως φέρουσιν· τὰ δ' ἐναντία,
3 δυσχερῶς.

ΤΜΗΜΑ ΔΕΥΤΕΡΟΝ

I. Ἐν ᾧ νοσήματι ὕπνος πόνον ποιεῖ, θανάσι-
2 μον· ἢν δὲ ὕπνος ὠφελῇ, οὐ θανάσιμον.

II. Ὅκου παραφροσύνην ὕπνος παύει, ἀγαθόν.

III. Ὕπνος, ἀγρυπνίη, ἀμφότερα μᾶλλον τοῦ
2 μετρίου γινόμενα,[4] κακόν.

IV. Οὐ πλησμονή, οὐ λιμός, οὐδ' ἄλλο οὐδὲν
2 ἀγαθόν, ὅ τι ἂν μᾶλλον τῆς φύσιος ᾖ.

V. Κόποι αὐτόματοι φράζουσι νούσους.

[1] πλεῖστα C' Urb. : πολλὰ M V.
[2] Rein. reads ὅσα for ὡς, εἰ before οἷα, and φέρει.
[3] ἄχρι Urb. Q S. [4] C' has γινόμενα before μᾶλλον.

---

[1] An orgasm is literally a state of excitement, and in this aphorism signifies that the humours are "struggling to get out," as Adams says.

XXII. Purge or otherwise disturb concocted, not crude, humours, and avoid the onset of a disease, unless there be orgasm, which in most cases does not occur.[1]

XXIII. Judge evacuations, not by bulk, but by their conformity to what is proper, and by the ease with which the patient bears them. Where occasion calls for purging until the patient faints, do even this, if the patient's strength be sufficient.

XXIV. In acute diseases use purgatives sparingly and at the onset, and then only after a thorough examination.

XXV. If the matters purged be such as should be purged, the patient benefits and bears up well; otherwise, the patient is distressed.[2]

SECOND SECTION

I. A disease in which sleep causes distress is a deadly one; but if sleep is beneficial, the disease is not deadly.[3]

II. When sleep puts an end to delirium it is a good sign.

III. Sleep or sleeplessness, in undue measure, these are both bad symptoms.

IV. Neither repletion, nor fasting, nor anything else is good when it is more than natural.[4]

V. Spontaneous weariness indicates disease.

[2] Most of *Aphorisms* XIX.–XXIV. will be found in *Humours* VI. The order of the propositions is not quite the same, and there are several interesting variant readings, which, however, do not seriously affect the sense.

[3] "Deadly" means here only "very dangerous."

[4] Perhaps, "too great for the constitution."

VI. Ὁκόσοι, πονέοντές τι τοῦ σώματος, τὰ πολλὰ [1] τῶν πόνων μὴ [2] αἰσθάνονται, τούτοισιν ἡ γνώμη νοσεῖ.

VII. Τὰ ἐν πολλῷ χρόνῳ λεπτυνόμενα σώματα νωθρῶς ἐπανατρέφειν, τὰ δὲ ἐν ὀλίγῳ, ὀλίγως.[3]

VIII. Ἢν ἐκ νούσου τροφὴν λαμβάνων τις [4] μὴ ἰσχύῃ, σημαίνει τὸ σῶμα ὅτι πλείονι τροφῇ χρῆται·[5] ἢν δὲ τροφὴν μὴ λαμβάνοντος τοῦτο γίνηται, σημαίνει [6] ὅτι κενώσιος δεῖται.

IX. Τὰ σώματα χρή,[7] ὅκου ἄν τις βούληται [8] καθαίρειν,[9] εὔροα ποιεῖν.

X. Τὰ μὴ καθαρὰ τῶν σωμάτων,[10] ὁκόσον [11] ἂν θρέψῃς μᾶλλον, βλάψεις.

XI. Ῥᾷον πληροῦσθαι ποτοῦ ἢ σιτίου.

XII. Τὰ ἐγκαταλιμπανόμενα ἐν τῇσι [12] νούσοισι μετὰ κρίσιν ὑποστροφὰς ποιεῖν εἴωθεν.[13]

XIII. Ὁκόσοισι κρίσις γίνεται, τούτοισιν ἡ νὺξ δύσφορος ἡ πρὸ τοῦ παροξυσμοῦ, ἡ δὲ ἐπιοῦσα εὐφορωτέρη ὡς ἐπὶ τὸ πολύ.

XIV. Ἐν τῇσι τῆς κοιλίης ῥύσεσιν αἱ μεταβολαὶ

[1] Erm. Rein. place τὰ πολλὰ after τούτοισιν.

[2] μὴ C′ Urb. M : οὐκ V.

[3] ταχέως Erm.: ἀλέως Rein.

[4] τις omitted by M.

[5] ὅτι πλείονι τροφῇ τὸ σῶμα χρέεται M: ὅτι πλείονι τροφῇ χρῆται V : τῷ σώματι ὅτι πλείονι τροφῇ χρέεται C′: τὸ σῶμα ὅτι πλείονι τροφῇ νρέεται Urb.

[6] σημαίνει V C′ Urb.: χρὴ εἰδέναι M.

[7] χρὴ omitted by V.

[8] ὅκου (ὅπου C′) ἄν τις βούληται C′ Urb. : ὅκου τις (τίς V) βούλεται M V.

[9] M has καθαίρεσθαι for καθαίρειν. After this aphorism C′ has καὶ ἢν μεν ἄνω βουλῃ εὔρυα ποιέειν στησαι τὴν κοιλιην· ἢν δε κάτω βουλη εὔρυα ποιέειν, ὑγραιναι τὴν κοιλιην·

[10] τῶν σωμάτων C′ Urb. : σώματα M V.

[11] ὁκόσον C′ Urb.: ὁκόσωι M : ὁκόσω V.

VI. Those who, suffering from a painful affection of the body, for the most part are unconscious of the pains, are disordered in mind.

VII. Bodies that have wasted away slowly should be slowly restored; those that have wasted quickly should be quickly restored.

VIII. If a convalescent while taking nourishment[1] remains weak, it is a sign that the body is being over-nourished; if there be weakness while he takes none,[1] it is a sign that evacuation is required.

IX. Bodies that are to be purged must be rendered fluent.[2]

X. Bodies that are not clean,[3] the more you nourish the more you harm.

XI. It is easier to replenish with drink than with food.

XII. Matters left behind in diseases after the crisis are wont to cause relapses.

XIII. When a crisis occurs, the night before the exacerbation is generally[4] uncomfortable, the night after more comfortable.

XIV. In fluxes of the bowels, changes in the

[1] The commentators from Galen have been worried by this phrase and the apparent inconsequence of the second part of the proposition. It is plain that τροφὴν λαμβάνειν means "to take nourishment readily and with appetite."

[2] That is, ready to evacuate. The ancients gave various prescriptions to make bodies εὔροα. See p. 213.

[3] That is, free from impurities, disordered or redundant humours.

[4] ὡς ἐπὶ τὸ πολὺ goes with the whole sentence and not with εὐφορωτέρη only.

---

[12] ἐν omitted by C′.

[13] Two late MSS. (and Galen) have ὑποστροφώδεα instead of ὑποστροφὰς ποιεῖν εἴωθεν.

τῶν διαχωρημάτων ὠφελέουσιν, ἢν μὴ ἐς πονηρὰ
3 μεταβάλλῃ.

XV. Ὅκου φάρυγξ νοσεῖ, ἢ φύματα ἐν τῷ σώματι[1] ἐκφύεται,[2] σκέπτεσθαι τὰς ἐκκρίσιας· ἢν γὰρ χολώδεες ἔωσι, τὸ σῶμα συννοσεῖ· ἢν δὲ ὅμοιαι τοῖσιν ὑγιαίνουσι γίνωνται, ἀσφαλὲς τὸ
5 σῶμα τρέφειν.

XVI. Ὅκου λιμὸς οὐ δεῖ πονεῖν.

XVII. Ὅκου ἂν τροφὴ πλείων παρὰ φύσιν
2 ἐσέλθῃ, τοῦτο νοῦσον ποιεῖ,[3] δηλοῖ δὲ ἡ ἴησις.

XVIII. Τῶν τρεφόντων ἀθρόως καὶ ταχέως,
2 ταχεῖαι καὶ αἱ διαχωρήσιες γίνονται.

XIX. Τῶν ὀξέων νοσημάτων οὐ πάμπαν ἀσφαλέες αἱ προαγορεύσιες, οὔτε τοῦ θανάτου, οὔτε τῆς
3 ὑγιείης.

XX. Ὁκόσοισι νέοισιν ἐοῦσιν αἱ κοιλίαι ὑγραί εἰσι, τούτοισιν ἀπογηράσκουσι ξηραίνονται· ὁκόσοισι δὲ νέοισιν ἐοῦσι ξηραίνονται, τούτοισιν
4 ἀπογηράσκουσιν ὑγραίνονται.[4]

XXI. Λιμὸν θώρηξις λύει.

XXII. Ἀπὸ πλησμονῆς ὁκόσα ἂν νοσήματα γένηται, κένωσις ἰῆται, καὶ ὁκόσα ἀπὸ κενώσιος,
3 πλησμονή, καὶ τῶν ἄλλων ἡ ὑπεναντίωσις.

XXIII. Τὰ ὀξέα τῶν νοσημάτων κρίνεται ἐν
2 τεσσαρεσκαίδεκα ἡμέρῃσιν.

[1] For σώματι C′ has τραχηλω with σω after it, the MS. being possibly imperfect at this point.

[2] ἐκθύει Rein.

[3] For νοῦσον ποιεῖ MV have νοσοποιέει.

[4] The MSS. show a great variety of readings in this

excreta are beneficial unless they change to what is bad.

XV. When the throat is affected, or tumours rise on the body,[1] examine the evacuations. If they are bilious, the whole body is affected; if they are such as they are in a state of health, it is safe to nourish the body.

XVI. When on a starvation diet a patient should not be fatigued.

XVII. When more nourishment is taken than the constitution can stand, disease is caused, as is shown by the treatment.

XVIII. Of foods that nourish all at once and quickly, the evacuations too come quickly.

XIX. In the case of acute diseases to predict either death or recovery is not quite safe.[2]

XX. Those whose bowels are loose in youth get constipated as they grow old; those whose bowels are constipated in youth have them loose as they grow old.

XXI. Strong drink dispels hunger.

XXII. Diseases caused by repletion are cured by depletion; those caused by depletion are cured by repletion, and in general contraries are cured by contraries.

XXIII. Acute diseases come to a crisis in fourteen days.

[1] The reading of C′ seems to show that σῶμα means here "the part of the body about the throat," that is, the neck. Swellings here may denote either a local or a general disorder. Possibly φύματα here means "eruptions."

[2] Or, "not at all safe."

---

aphorism, and it is by some regarded as an interpolation. It is here printed as given by C′.

XXIV. Τῶν ἑπτὰ ἡ τετάρτη ἐπίδηλος· ἑτέρης ἑβδομάδος ἡ ὀγδόη ἀρχή, θεωρητὴ δὲ ἡ ἑνδεκάτη· αὕτη γάρ ἐστι τετάρτη τῆς δευτέρης[1] ἑβδομάδος· θεωρητὴ δὲ πάλιν ἡ ἑπτακαιδεκάτη, αὕτη γάρ ἐστι τετάρτη μὲν ἀπὸ τῆς τεσσαρεσκαιδεκάτης, 6 ἑβδόμη δὲ ἀπὸ τῆς ἑνδεκάτης.

XXV. Οἱ θερινοὶ τεταρταῖοι τὰ[2] πολλὰ γίνονται βραχέες, οἱ δὲ φθινοπωρινοί, μακροί, 3 καὶ μᾶλλον[3] οἱ πρὸς τὸν χειμῶνα συνάπτοντες.

XXVI. Πυρετὸν ἐπὶ σπασμῷ βέλτιον γενέσθαι ἢ σπασμὸν ἐπὶ πυρετῷ.

XXVII. Τοῖσι μὴ κατὰ λόγον κουφίζουσιν οὐ δεῖ πιστεύειν, οὐδὲ φοβεῖσθαι λίην τὰ μοχθηρὰ γινόμενα παραλόγως· τὰ γὰρ πολλὰ τῶν τοιούτων ἐστὶν ἀβέβαια, καὶ οὐ πάνυ διαμένειν, οὐδὲ 5 χρονίζειν[4] εἴωθεν.

XXVIII. Τῶν πυρεσσόντων μὴ παντάπασιν ἐπιπολαίως, τὸ διαμένειν καὶ μηδὲν ἐνδιδόναι τὸ σῶμα, ἢ καὶ συντήκεσθαι μᾶλλον τοῦ κατὰ λόγον, μοχθηρόν· τὸ μὲν γὰρ μῆκος νούσου σημαίνει, τὸ 5 δέ, ἀσθένειαν.

XXIX. Ἀρχομένων τῶν νούσων, ἤν τι δοκῇ κινεῖν, κίνει· ἀκμαζουσῶν δέ, ἡσυχίην ἔχειν βέλτιόν 3 ἐστιν.

XXX. Περὶ τὰς ἀρχὰς καὶ τὰ τέλη, πάντα 2 ἀσθενέστερα,[5] περὶ δὲ τὰς ἀκμάς, ἰσχυρότερα.[6]

[1] δευτέρης all important MSS.: ἑτέρης Littré.
[2] Urb. and several Paris MSS. have ὡς before τὰ πολλά.
[3] μᾶλλον C′ V: μάλιστα Urb. M.
[4] χρονίζειν C′ Urb. M: ἐγχρονίζειν V.
[5] ἀσθενέστερα C′ V: ἀσθενέστατα Urb. M.
[6] ἰσχυρότερα C′ V: ἰσχυρότατα Urb. M.

XXIV. The fourth day is indicative[1] of the seven;[2] the eighth is the beginning of another week; the eleventh is to be watched, as being the fourth day of the second week; again the seventeenth is to be watched, being the fourth from the fourteenth and the seventh from the eleventh.

XXV. Summer quartans generally prove short, but those of autumn are long, especially those that are nigh to winter.

XXVI. It is better for a fever to supervene on a convulsion than a convulsion on a fever.

XXVII. One must not trust improvements that are irregular, nor yet fear overmuch bad symptoms that occur irregularly; for such are generally uncertain and are not at all wont to last or grow chronic.

XXVIII. When fevers are not altogether slight, for the body to remain without any wasting, and also for it to become unduly emaciated, is a bad symptom; the former signifies a long disease, the latter signifies weakness.

XXIX. At the beginning of diseases, if strong medicines[3] seem called for, use them; when they are at their height it is better to let the patient rest.

XXX. At the beginning and at the end all symptoms are weaker, at the height they are stronger.

[1] ἐπίδηλος means much the same as θεωρητός, and signifies that a day indicates beforehand whether the usual critical days will be normal or abnormal. See Littré, iv. p. 479.

[2] The translators say "of the seventh day," though how they get this meaning from τῶν ἕπτα is difficult to say. Does the phrase mean "of the sevens," *i.e.* 7, 14, 21, etc.?

[3] κινεῖν often means to administer a purge, an enema, or an emetic.

XXXI. Τῷ ἐξ ἀρρωστίης εὐσιτέοντι, μηδὲν
2 ἐπιδιδόναι τὸ σῶμα, μοχθηρόν.

XXXII. Ὡς τὰ πολλὰ πάντες οἱ φαύλως ἔχοντες, κατ' ἀρχὰς εὐσιτέοντες, καὶ μηδὲν ἐπιδιδόντες, πρὸς τῷ τέλει πάλιν ἀσιτέουσιν· οἱ δὲ κατ' ἀρχὰς μὲν ἀσιτέοντες ἰσχυρῶς, ὕστερον
5 δὲ εὐσιτέοντες, βέλτιον ἀπαλλάσσουσιν.

XXXIII. Ἐν πάσῃ νούσῳ τὸ ἐρρῶσθαι τὴν διάνοιαν καὶ εὖ ἔχειν πρὸς τὰς προσφοράς, ἀγαθόν·
3 τὸ δὲ ἐναντίον, κακόν.

XXXIV. Ἐν[1] τῇσι νούσοισιν ἧσσον κινδυνεύουσιν,[2] οἷς ἂν οἰκείη τῆς φύσιος, καὶ τῆς ἕξιος, καὶ τῆς ἡλικίης, καὶ τῆς ὥρης[3] ἡ νοῦσος ὑπάρχῃ[4] μᾶλλον, ἢ οἷσιν ἂν μὴ οἰκείη κατά τι τού-
5 των ᾖ.

XXXV. Ἐν πάσῃσι τῇσι νούσοισι, τὰ περὶ τὸν ὀμφαλὸν καὶ τὸ ἦτρον πάχος ἔχειν βέλτιόν ἐστι, τὸ δὲ σφόδρα λεπτὸν καὶ ἐκτετηκός, μοχθηρόν· ἐπισφαλὲς δὲ τὸ τοιοῦτο καὶ πρὸς τὰς κάτω
5 καθάρσιας.

XXXVI. Οἱ ὑγιεινῶς ἔχοντες τὰ σώματα, ἐν τῇσι φαρμακείῃσι καθαιρόμενοι[5] ἐκλύονται
3 ταχέως καὶ οἱ πονηρῇ τροφῇ χρεόμενοι.

XXXVII. Οἱ εὖ τὰ σώματα ἔχοντες φαρμα-
2 κεύεσθαι ἐργώδεες.

XXXVIII. Τὸ σμικρῷ[6] χεῖρον καὶ πόμα καὶ

[1] After ἐν V has πάσῃσι.

[2] After κινδυνεύουσιν many MSS. (including C′) have οἱ νοσέοντες.

[3] The MSS. differ considerably in the order of the genitives. I follow Littré.

[4] ὑπάρχει C′ Urb. V: ὑπάρχῃ several Paris MSS.: ᾖ vulgate.

XXXI. When a convalescent has a good appetite without improving his bodily condition it is a bad sign.

XXXII. Generally all sickly persons with a good appetite at the beginning, who do not improve, have no appetite at the end. But those get off better who at the beginning have a very bad appetite but later on have a good one.[1]

XXXIII. In every disease it is a good sign when the patient's intellect is sound and he enjoys his food[2]; the opposite is a bad sign.

XXXIV. In diseases there is less danger when the disease is more nearly related to the patient in respect of constitution, habit, age and season, than when there is no such relationship.

XXXV. In all diseases it is better for the parts about the navel and the abdomen to keep their fulness, while excessive thinness and emaciation is a bad sign. The latter condition makes it risky to administer purgatives.

XXXVI. Those with healthy bodies quickly lose strength when they take purges, as do those who use a bad diet.

XXXVII. Those who are in a good physical condition are troublesome to purge.

XXXVIII. Food or drink which, though slightly

[1] This aphorism is said by the commentators to apply to convalescents. The explanation seems to do some violence to *οἱ φαύλως ἔχοντες*, however much it may suit the sense of the passage. Perhaps the phrase applies to all who, whether convalescent or not, are neither ill nor well. If so, *πάντες* has more point.

[2] Possibly *προσφοραί* includes treatment of all kinds, and it certainly does not exclude drink.

---

[5] *καθαιρόμενοι* omitted by C′. [6] *σμικρὸν* Urb. V.

σιτίον, ἥδιον δέ, τῶν βελτιόνων μέν, ἀηδεστέρων
3 δέ, μᾶλλον αἱρετέον.

XXXIX. Οἱ πρεσβῦται τῶν νέων τὰ μὲν πολλὰ νοσέουσιν ἧσσον· ὅσα δ' ἂν αὐτοῖσι χρόνια νοσήματα γένηται, τὰ πολλὰ συναπο-
4 θνήσκει.

XL. Βράγχοι καὶ κόρυζαι τοῖσι σφόδρα πρεσ-
2 βυτέροισι[1] οὐ πεπαίνονται.

XLI. Οἱ ἐκλυόμενοι πολλάκις καὶ ἰσχυρῶς,
2 ἄνευ φανερῆς προφάσιος, ἐξαπίνης τελευτῶσιν.

XLII. Λύειν ἀποπληξίην ἰσχυρὴν μὲν ἀδύνα-
2 τον, ἀσθενέα δέ, οὐ ῥηίδιον.[2]

XLIII. Τῶν ἀπαγχομένων καὶ καταλυομένων, μηδέπω δὲ τεθνηκότων, οὐκ ἀναφέρουσιν, οἷσιν
3 ἂν ἀφρὸς ᾖ περὶ τὸ στόμα.

XLIV. Οἱ παχέες σφόδρα κατὰ φύσιν,[3] ταχυ-
2 θάνατοι γίνονται μᾶλλον τῶν ἰσχνῶν.

XLV. Τῶν ἐπιληπτικῶν τοῖσι νέοισιν ἀπαλλαγὴν αἱ μεταβολαὶ μάλιστα τῆς ἡλικίης, καὶ τῶν ὡρέων καὶ τῶν τόπων,[4] καὶ τῶν βίων
4 ποιέουσιν.

XLVI. Δύο πόνων ἅμα γινόμενων μὴ κατὰ τὸν αὐτὸν τόπον, ὁ σφοδρότερος ἀμαυροῖ τὸν
3 ἕτερον.

XLVII. Περὶ τὰς γενέσιας τοῦ πύου οἱ πόνοι
2 καὶ οἱ πυρετοὶ συμβαίνουσι μᾶλλον[5] ἢ γενομένου.

[1] πρεσβυτέροισι C′ Urb.: πρεσβύτῃσι M V.

[2] For ῥηίδιον C′ has ῥαδίως. [3] κατὰ φύσιν omitted by V.

[4] So Urb. (with ὡραίων): C′ M V omit καὶ τῶν τόπων. Littré with one MS. reads χωρίων, omitting καὶ τῶν τόπων. The variants seem due to the unusual meaning of μεταβολαὶ τῶν ὡρέων, not "changes of the seasons" but "change of climate." χωρέων Rein. omitting καὶ τῶν τόπων.

inferior, is more palatable, is preferable to that which is superior but less palatable.

XXXIX. Old men generally have less illness than young men, but such complaints as become chronic in old men generally last until death.

XL. Sore throats and colds of the very old are not concocted.

XLI. Those who suffer from a frequent and extreme prostration without any manifest cause die suddenly.

XLII. It is impossible to cure a violent attack of apoplexy, and not easy to cure a slight one.

XLIII. Those who are hanged and cut down[1] before death do not recover if they foam at the mouth.

XLIV. Those who are constitutionally very fat are more apt to die quickly[2] than those who are thin.

XLV. Epilepsy among the young is cured chiefly by change—change of age, of climate, of place, of mode of life.

XLVI. When two pains occur together, but not in the same place, the more violent obscures the other.

XLVII. Pains and fevers occur when pus is forming rather than when it has been formed.

[1] Or, "are in a fainting condition." A clever emendation is *καταδυομένων*, with which reading the aphorism would refer to persons immersed in water until nearly suffocated.

[2] That is, have less power successfully to resist a severe disease. Adams' translation, "are apt to die earlier," would (wrongly) make *ταχυθάνατοι* refer to the average length of life.

---

[5] *μᾶλλον συμβαίνουσι* C′.

XLVIII. Ἐν πάσῃ κινήσει τοῦ σώματος, ὁκόταν ἄρχηται πονεῖν, τὸ διαναπαύειν εὐθύς, 3 ἄκοπον.

XLIX. Οἱ εἰθισμένοι τοὺς συνήθεας πόνους φέρειν, κῆν ὦσιν ἀσθενέες ἢ γέροντες, τῶν ἀσυνηθέων ἰσχυρῶν τε καὶ νέων ῥᾷον φέρου- 4 σιν.

L. Τὰ ἐκ πολλοῦ χρόνου συνήθεα, κἂν ᾖ χείρω τῶν ἀσυνηθέων,[1] ἧσσον ἐνοχλεῖν εἴωθεν· 3 δεῖ δὲ καὶ ἐς τὰ ἀσυνήθεα μεταβάλλειν.

LI. Τὸ κατὰ πολὺ καὶ ἐξαπίνης κενοῦν, ἢ πληροῦν, ἢ θερμαίνειν, ἢ ψύχειν, ἢ ἄλλως ὁκωσοῦν τὸ σῶμα κινεῖν, σφαλερόν, καὶ πᾶν τὸ πολὺ τῇ φύσει πολέμιον· τὸ δὲ κατὰ μικρόν, ἀσφαλές, καὶ ἄλλως τὸ ἐξ ἑτέρου μεταβαίνειν 6 ἐφ᾽ ἕτερον.[2]

LII. Πάντα κατὰ λόγον ποιέοντι, μὴ γινομένων τῶν κατὰ λόγον, μὴ μεταβαίνειν ἐφ᾽ 3 ἕτερον, μένοντος τοῦ δόξαντος ἐξ ἀρχῆς.

LIII. Ὁκόσοι τὰς κοιλίας ὑγρὰς ἔχουσιν, νέοι μὲν ἐόντες, βέλτιον ἀπαλλάσσουσι τῶν ξηρὰς ἐχόντων, ἐς δὲ τὸ γῆρας χεῖρον ἀπαλλάσσουσιν· ξηραίνονται γὰρ ὡς ἐπὶ τὸ πολὺ τοῖσιν 5 ἀπογηράσκουσιν.

LIV. Μεγέθει δὲ σώματος, ἐννεάσαι μέν, ἐλευθέριον καὶ οὐκ ἀηδές[3] ἐστιν· ἐγγηρᾶσαι δέ, 3 δύσχρηστον καὶ χεῖρον τῶν ἐλασσόνων.

[1] Rein. puts the comma after χείρω.

[2] The text differs considerably from that of Littré. I have followed C′ V Urb., except that the last has κατ᾽ ὀλίγον and εἰς for ἐφ᾽ before ἕτερον.

XLVIII. In every movement[1] of the body, to rest at once when pain begins relieves the suffering.

XLIX. Those who are wont to bear accustomed labours, even if they be weak or old, bear them better than strong and young people who are not used to them.

L. Things to which one has been used a long time, even though they be more severe than unaccustomed things, usually cause less distress. Nevertheless, change to unaccustomed things may be necessary.

LI. Excess and suddenness in evacuating the body, or in replenishing, warming, cooling or in any other way disturbing it, is dangerous; in fact all excess is hostile to nature. But "little by little" is a safe rule, especially in cases of change from one thing to another.

LII. When acting in all things according to rule, do not, when results are not according to rule, change to another course of treatment if the original opinion remains.

LIII. Those who when young have relaxed bowels come off better than those who have hard; but in old age they come off worse, the bowels of the old being generally hard.

LIV. Size of body in youth is noble and not unpleasing; in old age it is inconvenient and less desirable than a smaller stature.

[1] See p. 115, Aph. XXIX

---

[3] καὶ οὐκ ἀηδές omitted by Urb. Erm. reads ἀειδές after Galen.

ΤΜΗΜΑ ΤΡΙΤΟΝ.

I. Αἱ μεταβολαὶ τῶν ὡρέων μάλιστα τίκτουσι νοσήματα, καὶ ἐν τῇσιν ὥρῃσιν αἱ μεγάλαι μεταλλαγαὶ ἢ ψύξιος ἢ θάλψιος, καὶ τἄλλα κατὰ 4 λόγον οὕτως.

II. Τῶν φυσίων αἱ μὲν πρὸς θέρος, αἱ δὲ πρὸς 2 χειμῶνα εὖ ἢ κακῶς πεφύκασιν.

III. Τῶν νούσων ἄλλαι πρὸς ἄλλας εὖ ἢ κακῶς πεφύκασι, καὶ ἡλικίαι τινὲς πρὸς ὥρας, καὶ χώρας, 3 καὶ διαίτας.[1]

IV. Ἐν τῇσιν ὥρῃσιν, ὅταν[2] τῆς αὐτῆς ἡμέρης ποτὲ μὲν θάλπος, ποτὲ δὲ ψῦχος γίνηται,[3] 3 φθινοπωρινὰ τὰ νοσήματα προσδέχεσθαι χρή.[4]

V. Νότοι βαρυήκοοι, ἀχλυώδεες, καρηβαρικοί, νωθροί, διαλυτικοί· ὁκόταν οὗτος δυναστεύῃ, τοιαῦτα ἐν τῇσιν ἀρρωστίῃσι πάσχουσιν. ἢν δὲ βόρειον ᾖ,[5] βῆχες, φάρυγγες, κοιλίαι σκληραί, δυσουρίαι φρικώδεες, ὀδύναι πλευρέων, στηθέων· ὁκόταν οὗτος δυναστεύῃ, τοιαῦτα ἐν τῇσιν 7 ἀρρωστίῃσι προσδέχεσθαι χρή.[6]

VI. Ὁκόταν θέρος γένηται ἦρι ὅμοιον, ἱδρῶτας 2 ἐν τοῖσι πυρετοῖσι πολλοὺς προσδέχεσθαι χρή.[7]

VII. Ἐν τοῖσιν αὐχμοῖσι πυρετοὶ ὀξέες γίνονται· καὶ ἢν μὲν ἐπὶ πλέον ᾖ τὸ ἔτος τοιοῦτον,[8] ὁκοίην καὶ[9] τὴν κατάστασιν ἐποίησεν, ὡς ἐπὶ τὸ

[1] Rein. joins II and III, adding before III καὶ δὲ πρὸς χώρας καὶ διαίτας, καὶ τὰς ἄλλας καταστάσιας. After ὥρας he adds τινὰς and καὶ καταστάσιας νούσων after διαίτας.

[2] ὅταν M : ὁκόταν V : C′ omits.

[3] γίνεται C′ : ποιέει M V.

[4] V omits χρή, but has δεῖ before προσδέχεσθαι.

THIRD SECTION

I. It is chiefly the changes of the seasons which produce diseases, and in the seasons the great changes from cold or heat, and so on according to the same rule.

II. Of constitutions some are well or ill adapted to summer, others are well or ill adapted to winter.

III. Certain diseases and certain ages are well or ill adapted to certain seasons, districts and kinds of regimen.

IV. During the seasons, when on the same day occurs now heat and now cold, you must expect diseases to be autumnal.

V. South winds cause deafness, dimness of vision, heaviness of the head, torpor, and are relaxing. When such winds prevail, their characteristics extend to sufferers from illnesses. A north wind causes coughs, sore throats, constipation, difficult micturition accompanied by shivering, pains in the sides and chest; such are the symptoms one must expect in illnesses when this wind prevails.

VI. When summer proves similar to spring you must expect copious sweats to occur in fevers.

VII. In droughts occur acute fevers; and if the year be particularly dry, according to the constitu-

---

[5] ἢ is omitted by two inferior MSS. as it is in *Humours* XIV.

[6] προσδέχεσθαι χρή is omitted by V, which has δεῖ προσδέχεσθαι after τοιαῦτα.

[7] For χρή Urb. and several other MSS. have δεῖ.

[8] Littré reads with many MSS. τοιουτέον ἐὸν; neither C′ nor Urb. has ἐόν, which appears to be a case of dittography.

[9] δκοίην καὶ C′ Urb. V; Littré with some authority οἵην.

πολὺ καὶ τὰ νοσήματα τοιαῦτα δεῖ προσδέ-
5 χεσθαι.

VIII. Ἐν τοῖσι καθεστεῶσι καιροῖσι, καὶ ὡραίως τὰ ὡραῖα ἀποδιδοῦσιν,[1] εὐσταθέες καὶ εὐκρινέες[2] αἱ νοῦσοι γίνονται, ἐν δὲ τοῖσιν
4 ἀκαταστάτοισιν ἀκατάστατοι καὶ δύσκριτοι.[3]

IX. Ἐν φθινοπώρῳ ὀξύταται αἱ[4] νοῦσοι, καὶ θανατωδέσταται τοὐπίπαν, ἦρ δὲ ὑγιεινότατον,
3 καὶ ἥκιστα θανατῶδες.

X. Τὸ φθινόπωρον τοῖσι φθίνουσι κακόν.

XI. Περὶ δὲ τῶν ὡρέων, ἢν μὲν ὁ χειμὼν αὐχμηρὸς καὶ βόρειος γένηται, τὸ δὲ ἔαρ ἔπομβρον καὶ νότιον, ἀνάγκη τοῦ θέρεος πυρετοὺς ὀξέας, καὶ ὀφθαλμίας, καὶ δυσεντερίας γίνεσθαι, μάλιστα τῇσι γυναιξὶ καὶ τοῖς ὑγρὰς ἔχουσι τὰς
6 φύσιας.[5]

XII. Ἢν δὲ νότιος ὁ χειμὼν καὶ ἔπομβρος καὶ εὔδιος[6] γένηται, τὸ δὲ ἔαρ αὐχμηρὸν καὶ βόρειον, αἱ μὲν γυναῖκες, ᾗσιν οἱ τόκοι[7] πρὸς τὸ ἔαρ, ἐκ πάσης προφάσιος ἐκτιτρώσκουσιν· αἳ δ᾿ ἂν τέκωσιν, ἀκρατέα καὶ νοσώδεα τὰ παιδία τίκτουσιν, ὥστε ἢ παραυτίκα ἀπόλλυσθαι, ἢ λεπτὰ καὶ νοσώδεα ζῆν ἐόντα· τοῖσι δὲ ἄλλοισι δυσεντερίαι καὶ ὀφθαλμίαι ξηραὶ γίνονται, τοῖσι δὲ πρεσβυτέροισι κατάρροοι συντόμως ἀπολ-
10 λύντες.

[1] The vulgate text (with M and Urb.) has ἢν ὡραίως . . . ἀποδιδῶσιν. Erm. omits καὶ . . . . . . ἀποδιδοῦσιν. Rein. has ἡνίκα ὡραίως αἱ ὧραι τὰ ὡραῖα ἀποδιδόασιν.

[2] εὐκρινέστατοι C′V ; εὐκρινέσταται Urb. ; εὐκρινέες Littré.

[3] Urb. has ἀκατάστατα καὶ δύσκριτα καὶ τὰ νοσήματα γίγνονται.

[4] MV omit αἱ.

tion it has produced, such for the most part will be the diseases that must be expected.

VIII. In seasons that are normal,[1] and bring seasonable things at seasonable times, diseases prove normal and have an easy crisis; in abnormal seasons diseases are abnormal and have a difficult crisis.

IX. It is in autumn that diseases are most acute and, in general, most deadly; spring is most healthy and least deadly.

X. Autumn is bad for consumptives.

XI. As for the seasons, if the winter be dry and northerly and the spring wet and southerly, of necessity occur in the summer acute fevers, eye diseases and dysentery, especially among women and those with moist constitutions.[2]

XII. But if the winter prove southerly, rainy and calm, and the spring dry and northerly, women whose confinement is due in the spring suffer abortion on the slightest provocation, or, if they do bear children, have weak and unhealthy offspring, so that they either die at once or live with puny and unhealthy bodies. Among the rest prevail dysentery and dry diseases of the eyes, and, in the case of the old, catarrhs that quickly prove fatal.[3]

[1] *καθεστεῶσι* is difficult to translate. It means "having a regular *κατάστασις* (constitution)," just as *ἀκατάστατος* means "having no regular constitution." "Fixed," "established," "regular," are partial but imperfect equivalents.

[2] See *Airs, Waters, Places*, X. (I. p. 98).

[3] See *Airs, Waters, Places*, X. (I. p. 100).

---

[5] So practically all the good MSS. M, however, reads *καὶ τοῖσιν ὑγροῖσι τὰς φύσιας*.

[6] *εὔδιος*. So C′V. *εὐδινὴs* M; *εὐδιεινὸς* Littré.

[7] *τόκοι* most MSS.; *τοκετοὶ* C′.

XIII. Ἢν δὲ τὸ θέρος αὐχμηρὸν καὶ βόρειον γένηται, τὸ δὲ φθινόπωρον ἔπομβρον καὶ νότιον, κεφαλαλγίαι ἐς τὸν χειμῶνα καὶ βῆχες, καὶ
4 βράγχοι, καὶ κόρυζαι, ἐνίοισι δὲ καὶ φθίσιες.

XIV. Ἢν δὲ βόρειον ᾖ[1] καὶ ἄνυδρον, τοῖσι μὲν ὑγροῖσι τὰς φύσιας[2] καὶ τῇσι γυναιξὶ σύμφορον· τοῖσι δὲ λοιποῖσιν ὀφθαλμίαι ἔσονται ξηραί, καὶ πυρετοὶ ὀξέες, καὶ κόρυζαι,[3] ἐνίοισι
5 δὲ καὶ μελαγχολίαι.

XV. Τῶν δὲ καταστασίων τοῦ ἐνιαυτοῦ τὸ μὲν ὅλον οἱ αὐχμοὶ τῶν ἐπομβριῶν εἰσιν ὑγιει-
3 νότεροι, καὶ ἧσσον[4] θανατώδεες.

XVI. Νοσήματα δὲ ἐν μὲν[5] τῇσιν ἐπομβρίῃσιν ὡς τὰ πολλὰ γίνεται, πυρετοί τε μακροί, καὶ κοιλίης ῥύσιες, καὶ σηπεδόνες, καὶ ἐπίληπτοι, καὶ ἀπόπληκτοι, καὶ κυνάγχαι· ἐν δὲ τοῖσιν αὐχμοῖσι, φθινάδες, ὀφθαλμίαι, ἀρθρίτιδες,
6 στραγγουρίαι, καὶ[6] δυσεντερίαι.

XVII. Αἱ δὲ καθ' ἡμέρην καταστάσιες, αἱ μὲν βόρειοι τά τε σώματα συνιστᾶσι, καὶ εὔτονα καὶ εὐκίνητα καὶ εὔχροα[7] καὶ εὐηκοώτερα ποιέουσι, καὶ τὰς κοιλίας ξηραίνουσι, καὶ τὰ ὄμματα δάκνουσι,[8] καὶ περὶ τὸν θώρηκα ἄλγημα ἤν τι

[1] ᾖ is omitted by C′ Urb.

[2] Some good MSS., including C′, have τὴν φύσιν.

[3] After κόρυζαι V Urb. and many other MSS. have χρόνιαι: some have πολυχρόνιοι. As the parallel passage in *Airs, Waters, Places* has πολυχρόνιοι instead of κόρυζαι, some editors would adopt that reading here. But, as Littré points out, the commentary of Theophilus implies κόρυζαι. Evidently there have been efforts to assimilate the text of *Aphorisms* to that of *Airs, Waters, Places.* Rein. has ὀξέες καὶ χρόνιοι, καὶ κόρυζαι χρόνιαι.

[4] For ἧσσον C′ has ἥκιστα.

XIII. If the summer prove dry and northerly, and the autumn rainy and southerly, headaches are common in the winter, with coughs, sore throats, colds and, in some cases, consumption.[1]

XIV. But if ⟨the autumn⟩ be northerly and rainless it is beneficial to those with moist constitutions and to women. To the others will come dry eye diseases, acute fevers, colds and, in some cases, melancholia.[2]

XV. Of the constitutions[3] of the year droughts are, in general, more healthy and less deadly than wet weather.

XVI. The diseases which generally arise in rainy weather are protracted fevers, fluxes of the bowels, mortifications, epilepsy, apoplexy and angina. In dry weather occur consumption,[4] eye diseases, diseases of the joints, strangury and dysentery.

XVII. Of daily constitutions, such as are northerly brace the body, giving it tone and agility, and improving the complexion and the sense of hearing, dry up the bowels and make the eyes tingle, besides

[1] *Airs, Waters, Places*, I. p. 102.

[2] *Airs, Waters, Places*, I. p. 102. "Melancholia" includes all forms of depression, from true melancholia to mere nervousness.

[3] The καταστάσιες of a year are those periods which exhibit definite, well-marked characteristics.

[4] Galen and Theophilus tell us that many commentators took φθινάδες as an adjective qualifying ὀφθαλμίαι, "eye diseases resulting in destruction of the eyes." This is linguistically better than making φθινάδες equivalent to φθίσιες. M has φθινάδεα.

---

[5] μὲν is omitted by many MSS., including Urb.

[6] καὶ is omitted by C′M.

[7] For εὔχροα V has εὐχροώτερα.

[8] For δάκνουσι C′ has δακρύουσι.

προϋπάρχῃ, μᾶλλον πονέουσιν· αἱ δὲ νότιοι διαλύουσι τὰ σώματα καὶ ὑγραίνουσι, καὶ καρηβαρίας καὶ βαρυηκοΐας καὶ ἰλίγγους ἐμποιέουσιν, ἐν δὲ[1] τοῖσιν ὀφθαλμοῖσι καὶ τοῖσι σώμασι
10 δυσκινησίην, καὶ τὰς κοιλίας ὑγραίνουσιν.

XVIII. Κατὰ δὲ τὰς ὥρας, τοῦ μὲν ἦρος καὶ ἄκρου τοῦ θέρεος, οἱ παῖδες καὶ οἱ τούτων ἐχόμενοι τῇσιν ἡλικίῃσιν ἄριστά τε διάγουσι καὶ ὑγιαίνουσι μάλιστα· τοῦ δὲ θέρεος καὶ τοῦ φθινοπώρου, μέχρι μέν τινος οἱ γέροντες· τὸ δὲ λοιπόν,[2] καὶ τοῦ χειμῶνος, οἱ μέσοι τῇσιν
7 ἡλικίῃσιν.

XIX. Νοσήματα δὲ πάντα μὲν ἐν πάσῃσι τῇσιν ὥρῃσι γίνεται, μᾶλλον δ' ἔνια κατ' ἐνίας
3 αὐτέων καὶ γίνεται καὶ παροξύνεται.

XX. Τοῦ μὲν γὰρ ἦρος, τὰ μελαγχολικά, καὶ τὰ μανικά,[3] καὶ τὰ ἐπιληπτικά, καὶ αἵματος ῥύσιες, καὶ κυνάγχαι, καὶ κορύζαι, καὶ βράγχοι, καὶ βῆχες, καὶ λέπραι, καὶ λειχῆνες, καὶ ἀλφοί, καὶ ἐξανθήσιες ἑλκώδεες πλεῖσται, καὶ φύματα,
6 καὶ ἀρθριτικά.

XXI. Τοῦ δὲ θέρεος, ἔνιά τε τούτων, καὶ πυρετοὶ συνεχέες, καὶ καῦσοι, καὶ τριταῖοι πυρετοί,[4] καὶ ἔμετοι, καὶ διάρροιαι, καὶ ὀφθαλμίαι, καὶ ὤτων πόνοι, καὶ στομάτων ἑλκώσιες,
5 καὶ σηπεδόνες αἰδοίων, καὶ ἵδρωα.[5]

[1] C′ omits δὲ, and many MSS., including Urb. V, read τε. Rein. has τε δὲ.

[2] Rein. adds τοῦ φθινοπώρου.

[3] In M. μελαγχολικά and μανικά are transposed, and there are several minor variants in the less important MSS., the order of the diseases showing considerable confusion.

[4] This is the reading of C′. M adds καὶ τεταρταῖοι, which words, apparently, were not in the texts known to Galen.

aggravating any pre-existing pain in the chest; southerly constitutions relax and moisten the body, bring on heaviness of the head, hardness of hearing and giddiness, make the eyes and the whole body slow to move, and the bowels watery.

XVIII. As for the seasons, in spring and early summer children and young people enjoy the greatest well-being and good health; in summer and part of autumn, the aged; for the remainder of autumn and in winter, the middle-aged.

XIX. All diseases occur at all seasons, but some diseases are more apt to occur and to be aggravated at certain seasons.

XX. In spring occur melancholia, madness, epilepsy, bloody flux, angina, colds, sore throats, coughs, skin eruptions and diseases,[1] eruptions turning generally to ulcers, tumours and affections of the joints.

XXI. In summer occur some of the diseases just mentioned, and also continued fevers, ardent fevers, tertians,[2] vomiting, diarrhœa, eye diseases, pains of the ears, ulcerations of the mouth, mortification of the genitals, sweats.

[1] It is not possible to translate the Greek terms for the various skin diseases, as the modern classification is so different from the ancient. We may be sure, however, that λέπρα included many diseases besides leprosy.

[2] With the reading of V, "very many tertians."

See Littré's note. V and many other MSS. read πλεῖστοι, which Littré adopts.

[5] There are many interesting variants in the latter part of this aphorism. C′ has καὶ αἰδύων ἠδρῶτες, and Urb. καὶ αἰδοίων ἴδρωα, with a colon at σηπεδόνες. It gives quite good sense to take these words together, but Galen took αἰδοίων with σηπεδόνες. V reads ἰδρῶτες for ἴδρωα. M omits καὶ before both σηπεδόνες and ἴδρωα, and so supports the other strong testimony that αἰδοίων should go with ἴδρωα.

XXII. Τοῦ δὲ φθινοπώρου, καὶ τῶν θερινῶν τὰ[1] πολλά, καὶ πυρετοὶ τεταρταῖοι, καὶ πλανῆτες, καὶ σπλῆνες, καὶ ὕδρωπες, καὶ φθίσιες, καὶ στραγγουρίαι, καὶ λειεντερίαι, καὶ δυσεντερίαι,[2] καὶ ἰσχιάδες, καὶ κυνάγχαι,[3] καὶ ἄσθματα, καὶ εἰλεοί, καὶ ἐπιληψίαι, καὶ τὰ μανικά, 7 καὶ τὰ μελαγχολικά.

XXIII. Τοῦ δὲ χειμῶνος, πλευρίτιδες, περιπλευμονίαι, λήθαργοι,[4] κόρυζαι, βράγχοι, βῆχες, πόνοι[5] πλευρέων, στηθέων, ὀσφύος, κεφαλαλγίαι, 4 ἴλιγγοι, ἀποπληξίαι.

XXIV. Ἐν δὲ τῇσιν ἡλικίῃσι τοιάδε συμβαίνει·[6] τοῖσι μὲν σμικροῖσι καὶ νεογνοῖσι παιδίοισιν, ἄφθαι, ἔμετοι, βῆχες, ἀγρυπνίαι, 4 φόβοι, ὀμφαλοῦ φλεγμοναί, ὤτων ὑγρότητες.

XXV. Πρὸς δὲ τὸ ὀδοντοφυεῖν προσάγουσιν, οὔλων ὀδαξησμοί, πυρετοί, σπασμοί, διάρροιαι, μάλιστα ὅταν ἀνάγωσι τοὺς κυνόδοντας,[7] καὶ τοῖσι παχυτάτοισι τῶν παίδων, καὶ[8] τοῖσι τὰς 5 κοιλίας σκληρὰς ἔχουσιν.[9]

XXVI. Πρεσβυτέροισι δὲ γενομένοισι, παρίσθμια, σπονδύλου τοῦ κατὰ τὸ ἰνίον εἴσω ὤσιες,[10] ἄσθματα, λιθιάσιες, ἕλμινθες[11] στρογγύλαι, ἀσ-

[1] τὰ is omitted by V.

[2] Urb. omits καὶ λειεντερίαι καὶ δυσεντερίαι, and V omits καὶ δυσεντερίαι.

[3] For κυνάγχαι C′ has βράγχαι.

[4] Many MSS. omit λήθαργοι, and it is not commented on by Galen. It is placed by some MSS. before κόρυζαι, by others after, while a few omit κόρυζαι.

[5] πόνοι πλευρέων στηθέων C′V: πόνοι στηθέων· πλευρέων Urb. and M. M has a colon at πόνων.

[6] For τοιάδε συμβαίνει V has συμβαίνει τὰ τοιαῦτα.

[7] For κυνόδοντας C′ has καλουμένους κοινοδόντας.

XXII. In autumn occur most[1] summer diseases, with quartans, irregular fevers, enlarged spleen, dropsy, consumption, strangury, lientery, dysentery, sciatica, angina, asthma, ileus, epilepsy, madness, melancholia.

XXIII. In winter occur pleurisy, pneumonia, lethargus, colds, sore throat, coughs, pains in the sides, chest and loins, headache, dizziness, apoplexy.

XXIV. In the different ages the following complaints occur: to little children and babies, aphthae, vomiting, coughs, sleeplessness, terrors, inflammation of the navel, watery discharges from the ears.

XXV. At the approach of dentition, irritation of the gums, fevers, convulsions, diarrhœa, especially when cutting the canine teeth, and in the case of very fat children, and if the bowels are hard.[2]

XXVI. Among those who are older occur affections of the tonsils, curvature at the vertebra by the neck, asthma, stone, round worms, ascarides, warts,

[1] With the reading of V, "many."

[2] That is, have a tendency to constipation. The reading of C′ is very interesting. It obviously arose from the apparent inconsistency of saying that diarrhœa occurs in children naturally constipated. So some scribe or commentator changed σκληρὰς ("hard") to ὑγρὰς ("relaxed," "watery"). But the point is that children usually constipated become very relaxed in certain circumstances.

---

[8] καὶ is omitted by M. This reading would give the sense: "fat babies with a tendency to constipation."

[9] So V and many other MSS. M has σκληρὰς before τὰς. Littré says that C′ has τὰς κοιλίας σκληρὰς ἔχουσιν; it actually has τὰς κοιλίας ὑγρὰς ἔχουσιν.

[10] For εἴσω ὤσιες M has ἰσώσιες and V ἰσωώσηες.

[11] For ἕλμινθες C′ has ἕλμινθαι.

καρίδες, ἀκροχορδόνες, σατυριασμοί, χοιράδες,[1]
5 καὶ τἄλλα φύματα.[2]

XXVII. Τοῖσι δὲ[3] πρεσβυτέροισι καὶ πρὸς τὴν ἥβην προσάγουσι,[4] τούτων τὰ πολλά, καὶ πυρετοὶ χρόνιοι μᾶλλον, καὶ ἐκ ῥινῶν αἵματος
4 ῥύσιες.

XXVIII. Τὰ δὲ πλεῖστα τοῖσι παιδίοισι πάθεα κρίνεται, τὰ μὲν ἐν τεσσαράκοντα ἡμέρῃσι, τὰ δὲ ἐν ἑπτὰ μησί, τὰ δὲ ἐν ἑπτὰ ἔτεσι, τὰ δὲ[5] πρὸς τὴν ἥβην προσάγουσιν· ὁκόσα[6] δ' ἂν διαμείνῃ τοῖσι παιδίοισι,[7] καὶ μὴ ἀπολυθῇ περὶ τὸ ἡβάσκειν, ἢ τοῖσι θήλεσι[8] περὶ τὰς τῶν
7 καταμηνίων ῥήξιας, χρονίζειν εἴωθεν.

XXIX. Τοῖσι δὲ νεηνίσκοισιν, αἵματος πτύσιες, φθίσιες, πυρετοὶ ὀξέες, ἐπιληψίαι, καὶ τἄλλα
3 νοσήματα, μάλιστα δὲ τὰ προειρημένα.[9]

XXX. Τοῖσι δὲ ὑπὲρ τὴν ἡλικίην ταύτην, ἄσθματα, πλευρίτιδες, περιπλευμονίαι, λήθαργοι, φρενίτιδες, καῦσοι, διάρροιαι χρόνιαι, χολέραι,
4 δυσεντερίαι, λειεντερίαι, αἱμορροΐδες.

XXXI. Τοῖσι δὲ πρεσβύτῃσι,[10] δύσπνοιαι, κατάρροιαι[11] βηχώδεες, στραγγουρίαι, δυσουρίαι, ἄρθρων πόνοι, νεφρίτιδες, ἴλιγγοι, ἀποπληξίαι,

[1] Before χοιράδες M has στραγγουρίαι.

[2] After φύματα Littré has μάλιστα δὲ τὰ προειρημένα (from aphorism XXIX?).

[3] After δὲ M has ἔτι (and, after τούτων, τε).

[4] Erm. omits καὶ . . . προσάγουσι.

[5] After δὲ C′ Urb. add καὶ.

[6] For ὁκόσα M has ὅσα.

[7] After παιδίοισι V adds πάθεα.

[8] τοῖσι θήλεσι many MSS., including Urb. and V (with καὶ for ἢ): τῇσι θηλείῃσι Littré with two MSS.

[9] For προειρημένα V has εἰρημένα.

swellings by the ears,[1] scrofula and tumours generally.

XXVII. Older children and those approaching puberty suffer from most of the preceding maladies, from fevers of the more protracted type and from bleeding at the nose.

XXVIII. Most diseases of children reach a crisis in forty days, in seven months, in seven years, at the approach of puberty. But such as persist among boys without ceasing at puberty, or, in the case of girls, at the commencement of menstruation, are wont to become chronic.

XXIX. Young men suffer from spitting of blood, phthisis, acute fevers, epilepsy and the other diseases, especially those mentioned above.

XXX. Those who are beyond this age suffer from asthma, pleurisy, pneumonia, lethargus, phrenitis, ardent fevers, chronic diarrhœa, cholera, dysentery, lientery, hemorrhoids.

XXXI. Old men suffer from difficulty of breathing, catarrh accompanied by coughing, strangury, difficult micturition, pains at the joints, kidney

[1] σατυριασμός, the word given by all our MSS., is very difficult. None of the ancient commentators, with the exception of one scholiast, refer to it. Littré thinks that it means the same as σατυρισμοί, a word explained in the Galenic *Glossary* as meaning tumours by the ears. I have adopted this explanation, but at the same time I am not at all sure that satyriasis is not referred to. So Lallemand and Pappas, who would transpose σατυριασμοί and place it next to ἀσκαρίδες, on the ground that the latter often cause the former.

---

[10] For πρεσβύτῃσι many MSS., including C′, read πρεσβυτέροισι.

[11] κατάρροιαι C′V: κατάρροοι or κατάρροι most MSS.

καχεξίαι,[1] ξυσμοὶ τοῦ σώματος ὅλου, ἀγρυπνίαι, κοιλίης καὶ ὀφθαλμῶν καὶ ῥινῶν ὑγρότητες,
6 ἀμβλυωπίαι, γλαυκώσιες, βαρυηκοΐαι.

ΤΜΗΜΑ ΤΕΤΑΡΤΟΝ

I. Τὰς κυούσας φαρμακεύειν, ἢν ὀργᾷ, τετράμηνα καὶ ἄχρι ἑπτὰ μηνῶν, ἧσσον δὲ[2] ταύτας· τὰ δὲ νήπια καὶ τὰ[3] πρεσβύτερα εὐλαβεῖσθαι
4 χρή.[4]

II. Ἐν τῇσι φαρμακείῃσι τοιαῦτα ἄγειν ἐκ τοῦ σώματος, οἷα[5] καὶ αὐτόματα ἰόντα χρήσιμα, τὰ
3 δὲ ἐναντίως ἰόντα παύειν.

III. Ἢν μὲν[6] οἷα δεῖ καθαίρεσθαι καθαίρωνται, συμφέρει τε καὶ εὐφόρως φέρουσι, τὰ δὲ ἐναντία
3 δυσχερῶς.

IV. Φαρμακεύειν θέρεος μὲν[7] μᾶλλον[8] τὰς
2 ἄνω,[9] χειμῶνος δὲ τὰς κάτω.

V. Ὑπὸ κύνα καὶ πρὸ κυνὸς ἐργώδεες αἱ
2 φαρμακεῖαι.

VI. Τοὺς ἰσχνοὺς καὶ[10] εὐημέας ἄνω φαρμα-
2 κεύειν, ὑποστελλομένους χειμῶνα.[11]

VII. Τοὺς δὲ δυσημέας καὶ μέσως εὐσάρκους,
2 κάτω, ὑποστελλομένους θέρος.

[1] Rein. places καχεξίαι after ὅλου.
[2] Rein. has παρὰ before ταύτας.
[3] τὰ is omitted by C′M, but appears in several Paris MSS.
[4] χρὴ is omitted by C′V and by many Paris MSS.
[5] οἷα C′ and several MSS. : ὁκόσα or ὁκοῖα most MSS.
[6] μὲν is omitted by V and several other MSS.
[7] μὲν is omitted by MV.
[8] μᾶλλον is omitted by C′ and by several other MSS.

disease, dizziness, apoplexy, cachexia, pruritus of the whole body, sleeplessness, watery discharges from bowels, eyes and nostrils, dullness of sight, cataract, hardness of hearing.

FOURTH SECTION

I. Purge pregnant women, should there be orgasm,[1] from the fourth to the seventh month, but these last less freely; the unborn child, in the first and last stages of pregnancy, should be treated very cautiously.

II. In purging, bring away from the body such matters as would leave spontaneously with advantage; matters of an opposite character should be stopped.

III. If matters purged be such as should be purged, the patient benefits and bears up well; otherwise, the patient is distressed.[2]

IV. In summer purge by preference upwards, in winter downwards.

V. At and just before the dog-star, purging is troublesome.[3]

VI. Purge upwards thin people who easily vomit, but be careful in winter.

VII. Purge downwards those who vomit with difficulty and are moderately stout, but be careful in summer.

[1] See note on *Aphorisms*, I. XXII.

[2] See *Aphorisms*, I. XXV.

[3] Heat causes prostration, and ancient purges were violent in action.

---

[9] After ἄνω Urb. and some Paris MSS. add κοιλίας, a word which Galen says must certainly be understood.

[10] καὶ C′V and many other MSS. : καὶ τοὺς M : τοὺς Littré.

[11] Erm. Rein. read χειμῶνος and θέρεος in the next aphorism.

VIII. Τοὺς δὲ φθινώδεας ὑποστέλλεσθαι.[1]

IX. Τοὺς δὲ μελαγχολικοὺς ἁδροτέρως τὰς
2 κάτω, τῷ αὐτῷ λογισμῷ τἀναντία προστιθείς.

X. Φαρμακεύειν ἐν τοῖσι λίην ὀξέσιν, ἢν ὀργᾷ,
αὐθημερόν· χρονίζειν γὰρ ἐν τοῖσι τοιούτοισι
3 κακόν.

XI. Ὁκόσοισι[2] στρόφοι, καὶ πόνοι περὶ τὸν
ὀμφαλόν,[3] καὶ ὀσφύος ἄλγημα μὴ λυόμενον μήτε
ὑπὸ φαρμακείης, μήτ' ἄλλως,[4] εἰς ὕδρωπα ξηρὸν
4 ἱδρύεται.

XII. Ὁκόσοισι κοιλίαι λειεντεριώδεες, χειμῶνος
2 φαρμακεύειν ἄνω κακόν.

XIII. Πρὸς τοὺς ἐλλεβόρους[5] τοῖσι μὴ ῥηιδίως
ἄνω καθαιρομένοισι, πρὸ τῆς πόσιος προϋγραίνειν
3 τὰ σώματα πλείονι τροφῇ καὶ ἀναπαύσει.

XIV. Ἐπὴν πίῃ τις ἐλλέβορον, πρὸς μὲν τὰς
κινήσιας τῶν σωμάτων μᾶλλον ἄγειν, πρὸς δὲ
τοὺς ὕπνους καὶ τὰς ἀκινήσιας,[6] ἧσσον·[7] δηλοῖ
δὲ καὶ ἡ ναυτιλίη,[8] ὅτι κίνησις τὰ σώματα
5 ταράσσει.[9]

XV. Ἐπὴν βούλῃ μᾶλλον ἄγειν τὸν ἐλλέβορον,

[1] ὑποστέλλεσθαι is the reading of C′. Most MSS. have some form of the participle, and Littré follows slight MS. authority, supported, however, by Galen's comment, in adding τὰς ἄνω after ὑποστελλομένους. The authority against τὰς ἄνω is overwhelming; it is omitted by C′MV and most less important MSS. Urb. has κάτω θέρεος· τοὺς φθινώδεας ὑποστελλομένους. Rein. reads τὰς ἄνω with a comma at ὑποστελλομένους.

[2] For ὁκόσοισι V has οἷσι. This variation is very common in *Aphorisms* and need not be noticed again.

[3] V has καὶ οἱ περὶ ὀμφαλὸν πόνοι.

[4] V has πως after ἄλλως, and C′ reads μήτε ὑπὸ ἄλλων.

[5] τοῖσι δι' ἐλλεβόρου Erm. : πρὸς τοὺς δι' ἐλλεβόρου Rein.

VIII. Be careful in purging those with a tendency to consumption.

IX. By the same method of reasoning apply the opposite procedure to those who are of a melancholic temperament, and purge downwards freely.

X. In very acute cases purge on the first day should there be orgasm, for in such cases delay causes harm.

XI. Those who suffer from colic, pains about the navel, and ache in the loins, removed neither by purging nor in any other way, finish with a dry dropsy.[1]

XII. It is bad to purge upwards in winter those whose bowels are in a state of lientery.

XIII. In giving the hellebores, those who are not easily purged upwards should, before the draught, have their bodies moistened by increased food and rest.

XIV. When one has taken hellebore, one should be made to increase the movements of the body, and to indulge less in sleep and rest. Sailing on the sea too proves that movement disturbs the body.

XV. When you wish hellebore to be more efficacious, move the body; when you wish the

[1] See *Coan Prenotions*, 298. A "dry dropsy" is, apparently, the dropsy called "tympanites," so named "because in it the belly, when struck, sounds like a drum (tympanum)" (Adams.)

---

[6] *τὰς ἀκινήσιας* C′V and many other MSS. The accent is sometimes written *-ίας*: *μὴ κινήσιας* Littré and M.

[7] For *πρὸς μὲν . . . ἧσσον*. Rein. has *ἢν μὲν βούλῃ μᾶλλον ἄγειν τὸν ἐλλέβορον, κινεῖ τὸ σῶμα.*

[8] Littré's L has *ναυτίη* (sea-sickness), a reading noted by Galen.

[9] C′ has *κίνησις πλείω τὸ σῶμα ταράσσει.*

κίνει τὸ σῶμα· ἐπὴν δὲ παῦσαι,[1] ὕπνον ποίει, καὶ
3 μὴ κίνει.

XVI. Ἐλλέβορος ἐπικίνδυνος τοῖσι τὰς σάρκας
2 ὑγιέας ἔχουσι, σπασμὸν γὰρ ἐμποιεῖ.

XVII. Ἀπυρέτῳ ἐόντι, ἀποσιτίη, καὶ καρδιωγμός, καὶ σκοτόδινος, καὶ στόμα ἐκπικρούμενον,
3 ἄνω φαρμακείης δεῖσθαι σημαίνει.

XVIII. Τὰ ὑπὲρ τῶν φρενῶν ὀδυνήματα ἄνω φαρμακείης δεῖσθαι[2] σημαίνει· ὁκόσα δὲ κάτω,
3 κάτω.

XIX. Ὁκόσοι ἐν τῇσι φαρμακοποσίῃσι μὴ διψῶσι,[3] καθαιρόμενοι οὐ παύονται πρὶν ἢ διψή-
3 σωσιν.[4]

XX. Ἀπυρέτοισιν ἐοῦσιν, ἢν γένηται[5] στρόφος, καὶ γονάτων βάρος, καὶ ὀσφύος ἄλγημα, κάτω
3 φαρμακείης δεῖσθαι σημαίνει.

XXI. Ὑποχωρήματα μέλανα, ὁκοῖον αἷμα,[6] ἀπὸ ταυτομάτου ἰόντα, καὶ σὺν πυρετῷ, καὶ ἄνευ πυρετοῦ, κάκιστα·[7] καὶ ὁκόσῳ ἂν χρώματα[8] πλείω καὶ[9] πονηρότερα ᾖ,[10] μᾶλλον κάκιον· σὺν φαρμάκῳ δὲ ἄμεινον, καὶ ὁκόσῳ ἂν πλείω[11] χρώ-
6 ματα ᾖ, οὐ πονηρόν.[12]

[1] For παῦσαι C′ has παύειν. Rein. has ἢν δὲ παύεσθαι θούλῃ for ἐπὴν . . . παῦσαι.

[2] φαρμακείης (or φαρμακίης) δέεσθαι (or δεῖσθαι) C′V and many other MSS. : φαρμακίην (without δεῖσθαι) M.

[3] διψῶσι MV : διψήσωσι C′.

[4] For διψήσωσιν V has διψήσουσιν

[5] γένηται most MSS., including C′ and Urb. : γίγνηται V : γίνηται M.

[6] After αἷμα some MSS., with Urb., add μέλαν.

[7] Erm. Rein. mark a hiatus at κάκιστα.

[8] After χρώματα some MSS. add τῶν ὑποχωρημάτων.

[9] πλείω καὶ V : C′ Urb. M omit.

effects to stop, make the patient sleep and do not move him.

XVI. Hellebore is dangerous to those who have healthy flesh, as it produces convulsions.

XVII. When there is no fever, loss of appetite, heartburn, vertigo, and a bitter taste in the mouth indicate that there should be upward purging.

XVIII. Pains above the diaphragm indicate a need for upward purging; pains below indicate a need for downward purging.

XIX. Those who suffer no thirst while under the action of a purgative, do not cease from being purged until they have become thirsty.

XX. In cases where there is no fever, should colic come on, with heaviness of the knees and pains in the loins, need is indicated of purging downwards.

XXI. Stools that are black like (black) blood,[1] coming spontaneously, either with or without fever, are a very bad sign, and the more numerous and the more evil the colours, the worse the sign. When caused by a purge the sign is better, and it is not a bad one when the colours are numerous.[2]

[1] Even though μέλαν is omitted from the text, it is clear that it must be understood.

[2] Littré suggests that in this aphorism χρώματα does not mean "colours," but "shades of black." Such an interpretation makes the aphorism more homogeneous, but no ancient commentator mentions it.

---

[10] ἢ omitted by C′. For κάκιον Rein. has καὶ κακόν.

[11] πλείω omitted by C′ (χρώματα πλείονα V).

[12] For πονηρόν M has πονηρά. Littré thinks that οὐ πονηρόν is a gloss.

XXII. Νοσημάτων ὁκόσων ἀρχομένων, ἢν[1]
2 χολὴ μέλαινα ἢ ἄνω ἢ κάτω[2] ὑπέλθῃ, θανάσιμον.

XXIII. [3]Ὁκόσοισιν ἐκ νοσημάτων ὀξέων ἢ[4] πολυχρονίων, ἢ ἐκ τραυμάτων, ἢ ἄλλως[5] λελεπτυσμένοισι[6] χολὴ μέλαινα ἢ[7] ὁκοῖον αἷμα
4 μέλαν[8] ὑπέλθῃ, τῇ ὑστεραίῃ ἀποθνήσκουσιν.

XXIV. Δυσεντερίη ἢν ἀπὸ χολῆς μελαίνης[9]
2 ἄρξηται, θανάσιμον.

XXV. Αἷμα ἄνω μὲν ὁκοῖον ἂν ᾖ,[10] κακόν, κάτω
2 δέ, ἀγαθόν, καὶ[11] τὰ μέλανα ὑποχωρέοντα.[12]

XXVI. Ἢν ὑπὸ δυσεντερίης ἐχομένῳ ὁκοῖον[13]
2 σάρκες ὑποχωρήσωσι,[14] θανάσιμον.[15]

XXVII. Ὁκόσοισιν ἐν τοῖσι πυρετοῖσιν αἱμορραγεῖ πλῆθος[16] ὁκοθενοῦν, ἐν τῇσιν ἀναλήψεσι
3 τούτοισιν αἱ κοιλίαι καθυγραίνονται.

XXVIII. Ὁκόσοισι[17] χολώδεα τὰ[18] διαχωρήματα, κωφώσιος ἐπιγενομένης[19] παύεται,[20] καὶ

[1] ἢν omitted by MV.
[2] C′ has ἢ κάτω ἢ ἄνω.
[3] V has καὶ before ὁκόσοισιν.
[4] Rein. omits ὀξέων ἤ.
[5] C′ has πονηρῶν for ἢ ἄλλως.
[6] λελεπτυσμένοισι Littré and Dietz, with many MSS. and and Galen: λελεπτυμένοισι C′: λελεπτυσμένων V.
[7] ἢ omitted by M.
[8] M has μέλαν αἷμα. Rein. reads μέλαν ἄν.
[9] μελαίνης χολῆς V Urb S.
[10] εἴη C′V and many other MSS.
[11] καὶ is omitted by M and many other MSS. C′ has τὰ δὲ μέλαινα. Rein. has ἢν ᾖ μέλανα τὰ.
[12] For ὑποχωρέοντα C′ has ὑποχωρήματα.
[13] Dietz (from the reading ὁκοῖαι, which Littré has) suggests ὁκοῖον αἱ. C′ with many other MSS., including S and Q, have ὁκοῖον.
[14] C′ has ὑποχωρέουσι. Other readings are ὑποχωρῶσι and ὑποχωρέωσι.
[15] After θανάσιμον C′ adds: οἱ ὑπὸ τεταρταίων ἐχόμενοι ὑπὸ

XXII. Should black bile be evacuated at the beginning of any disease, whether upwards or downwards, it is a mortal symptom.

XXIII. When patients have become reduced[1] through disease, acute or chronic, or through wounds, or through any other cause, a discharge of black bile, or as it were of black blood, means death on the following day.[2]

XXIV. A dysentery beginning with black bile is mortal.

XXV. Blood evacuated upwards, whatever be its nature, is a bad sign; but evacuated downwards it is a good sign, and so also black stools.[3]

XXVI. If a patient suffering from dysentery discharge from the bowels as it were pieces of flesh, it is a mortal sign.

XXVII. When in fevers from whatsoever source there is copious hemorrhage, during convalescence the patients suffer from loose bowels.

XXVIII. When the stools are bilious, they cease

[1] "Attenuated" (Adams).

[2] There does not seem to be any reference, as Adams apparently thinks there is, to the "black vomit" of yellow fever, a disease unknown to Hippocrates.

[3] Galen, seeing the inconsistency of this aphorism with No. XXI, would interpret the latter half as referring to "bleeding piles." It is, however, quite possible that the two aphorisms come from different sources, and that the inconsistency is a real one.

σπασμῶν οὐ πάνυ τι ἁλίσκονται εἰ δὲ καὶ ἁλίσκονται πρότερον καὶ ἐπιγένηται τεταρταῖος, παύονται.

[16] After *πλῆθος* C′ adds *αἵματος*.

[17] For *ὁκόσοισι* C′ (and other MSS.) have *οἷσι*.

[18] C′Q and many other MSS. omit *τὰ*.

[19] C′ has *κωφοσίως ἐπιχολωδίων ἐπιγινομένης*.

[20] V has *παύονται*.

ὁκόσοισι κώφωσις, χολωδέων ἐπιγενομένων
4 παύεται.[1]

XXIX. Ὁκόσοισιν ἐν τοῖσι πυρετοῖσιν ἑκταίοι-
2 σιν ἐοῦσι ῥίγεα γίνεται,[2] δύσκριτα.

XXX. Ὁκόσοισι παροξυσμοὶ γίνονται, ἣν ἂν[3]
ὥρην ἀφῇ, ἐς τὴν αὔριον τὴν αὐτὴν ὥρην ἢν λάβῃ,
3 δύσκριτα.

XXXI. Τοῖσι κοπιώδεσιν ἐν τοῖσι πυρετοῖσιν,
ἐς ἄρθρα καὶ παρὰ τὰς γνάθους μάλιστα αἱ[4]
3 ἀποστάσιες γίνονται.

XXXII. [5]Ὁκόσοισι δὲ ἀνισταμένοισιν[6] ἐκ
τῶν νούσων τι πονέσει,[7] ἐνταῦθα αἱ[8] ἀποστάσιες
3 γίνονται.

XXXIII. Ἀτὰρ ἢν καὶ προπεπονηκός τι[9] ᾖ
2 πρὸ τοῦ νοσεῖν, ἐνταῦθα στηρίζει ἡ νοῦσος.[10]

XXXIV. Ἢν ὑπὸ πυρετοῦ ἐχομένῳ, οἰδήματος
μὴ ἐόντος ἐν τῇ φάρυγγι, πνὶξ ἐξαίφνης ἐπιγέ-
3 νηται,[11] θανάσιμον.

XXXV. Ἢν ὑπὸ πυρετοῦ ἐχομένῳ ὁ τράχηλος
ἐξαίφνης[12] ἐπιστραφῇ,[13] καὶ μόλις καταπίνειν
3 δύνηται, οἰδήματος μὴ ἐόντος,[14] θανάσιμον.[15]

[1] C′ omits καὶ to παύονται. V before παύεται has διαχωρημάτων.

[2] For γίνεται Urb. has γίγνεται δείκνυται.

[3] Urb. has (with Magnolus *in margine*) ἢν ἣν ἂν.

[4] αἱ C′ Urb. Q. Most MSS. omit.

[5] Two MSS. at least omit this aphorism.

[6] V has (for δὲ ἀνισταμένοισιν) διανισταμένοισιν.

[7] The MSS. show a great variety of readings. V has πονέει τι, M τί πονέσηι, C′ ἤν τι πονήσῃ, Q ἤν τι πονήσωσιν.

[8] αἱ C′Q and many other MSS. But many omit.

[9] προπεπονηκός τι Urb. προπεπονηκώς τί MV The reading of C′ and of several other MSS., προπεπονηκώς τις, is very attractive, and may be right.

if deafness supervenes; when there is deafness, it ceases when bilious stools supervene.

XXIX. When rigors occur in fevers on the sixth day the crisis is difficult.

XXX. Diseases with paroxysms, if at the same time as the paroxysm ceases on one day it returns on the next, have a difficult crisis.[1]

XXXI. When in fevers the patient is prostrated with fatigue, the abscessions form at the joints, especially at those of the jaws.

XXXII. If convalescents from diseases have pain in any part, the abscessions form in that part.

XXXIII. But if previous to an illness a part be in a state of pain, the disease settles in that part.

XXXIV. If a patient suffering from fever, with no swelling in the throat, be suddenly seized with suffocation, it is a deadly symptom.

XXXV. If the neck of a fever patient suddenly become distorted, and to swallow be a matter of difficulty, there being no swelling, it is a deadly symptom.

[1] Galen adopts a different interpretation. He explains: "the crisis is difficult if the paroxysm comes on regularly at the same hour, whatever be the hour at which it left off on the preceding day (ἣν ἂν ὥρην ἀφῇ)."

---

[10] V has ἡ νοῦσος στηρίζει.
[11] For ἐπιγένηται M has ἐπιστῆι.
[12] ἐξαίφνης is omitted by V and many other MSS.
[13] ἀποστραφῇ Rein.
[14] After ἐόντος most MSS. have ἐν τῷ τραχήλῳ; Galen's commentary implies that he did not know this reading.
[15] C′ omits this aphorism.

XXXVI. Ἱδρῶτες πυρεταίνοντι ἢν ἄρξωνται, ἀγαθοὶ τριταῖοι, καὶ πεμπταῖοι, καὶ ἑβδομαῖοι, καὶ ἐναταῖοι, καὶ ἑνδεκαταῖοι, καὶ τεσσαρεσκαιδεκαταῖοι, καὶ ἑπτακαιδεκαταῖοι, καὶ μιῇ καὶ εἰκοστῇ, καὶ ἑβδόμῃ καὶ εἰκοστῇ, καὶ τριηκοστῇ πρώτῃ, καὶ τριηκοστῇ τετάρτῃ·[1] οὗτοι γὰρ οἱ ἱδρῶτες νούσους κρίνουσιν· οἱ δὲ μὴ οὕτως γινόμενοι πόνον σημαίνουσι καὶ μῆκος νούσου
9 καὶ ὑποτροπιασμούς.[2]

XXXVII. Οἱ ψυχροὶ ἱδρῶτες, σὺν μὲν ὀξεῖ πυρετῷ γινόμενοι, θάνατον, σὺν πρηϋτέρῳ δέ,[3]
3 μῆκος νούσου σημαίνουσιν.

XXXVIII Καὶ ὅκου ἔνι τοῦ σώματος ἱδρώς,
2 ἐνταῦθα φράζει τὴν νοῦσον.

XXXIX. Καὶ ὅκου[4] ἔνι τοῦ σώματος θερμὸν
2 ἢ ψυχρόν, ἐνταῦθα ἡ νοῦσος.

XL. Καὶ ὅκου ἐν ὅλῳ τῷ σώματι μεταβολαί,[5] καὶ ἢν τὸ σῶμα ψύχηται, ἢ[6] αὖθις θερμαίνηται, ἢ χρῶμα ἕτερον ἐξ ἑτέρου γίνηται,[7] μῆκος νούσου
4 σημαίνει.

XLI. Ἱδρὼς πολὺς ἐξ ὕπνου ἄνευ τινὸς αἰτίης φανερῆς γινόμενος,[8] τὸ σῶμα σημαίνει ὅτι πλείονι τροφῇ χρῆται· ἢν δὲ τροφὴν μὴ λαμβάνοντι
4 τοῦτο γίνηται, σημαίνει ὅτι κενώσιος δεῖται.

[1] The MSS. show several slight variations in the numbers, but no MS. mentions the fourth day, an important omission, as Galen notices. Q has καὶ εἰκοστοὶ ἕβδομοι καὶ τριακοστοὶ πρῶτοι καὶ τριακοστοὶ ἕβδομοι καὶ τεσσαρακοστοί.

[2] ὑποτροπιασμόν Urb. and many other MSS. ὑποστροφὴν τοῦ νοσήματος καὶ ἀνατροπιασμόν C′.

[3] Urb. and many other MSS. place δὲ after σύν.

[4] For ὅκου V has ὅπη.

[5] For μεταβολαί C′ has διαφοραί and omits καί.

[6] For ἢ VQ have καί. C′ reads εἰ.

XXXVI. Sweats in a fever case are beneficial if they begin on the third day, the fifth, the seventh, the ninth, the eleventh, the fourteenth, the seventeenth, the twenty-first, the twenty-seventh, the thirty-first and the thirty-fourth, for these sweats bring diseases to a crisis. Sweats occurring on other days indicate pain, a long disease and relapses.

XXXVII. Cold sweats, occurring with high fever, indicate death; with a milder fever they indicate a protracted disease.

XXXVIII. And on whatever part of the body there is sweat, it means that the disease has settled there.

XXXIX. And in whatever part of the body there is heat or cold, in that part is the disease.

XL. And where there are changes in the whole body, for instance, if the body grow cold, or, again, grow hot, or if one colour follow on another, it signifies a protracted disease.[1]

XLI. Copious sweat, occurring after sleep without any obvious cause, indicates that the body has a surfeit of food. But should it occur to one who is not taking food, it indicates need of evacuation.

[1] The sense is a little clearer if, with C′, we omit *καί*, and with VQ read *καὶ* for *ἢ* before *αὖθις*: "if the body grow cold and hot by turns."

---

[7] *γένηται* Urb.: *γίγνηται* V. Other MSS. have *γίγνοιτο* or *γίνοιτο*.

[8] The MSS. show many slight variations, some reading *φανερῆς*: others, among them M, *ἑτερῆς*: while V has *ἄνευ τινὸς αἰτίου γινομένου*. Galen notes the variants *φανερῆς*, *ἑτερῆς* Urb. has *ἄνευ* *φανερῆς* / *τινὸς* *αἰτίης ἑτέρης γινόμενος*—*φανερῆς* (in another hand) over *τινός*. Rein. has *τροφὴν λαμβάνοντι* after *γινόμενος*.

XLII. Ἱδρὼς πολὺς θερμὸς ἢ ψυχρὸς αἰεὶ
ῥέων, ὁ μὲν[1] ψυχρός, μέζω,[2] ὁ δὲ[3] θερμός,
3 ἐλάσσω[4] νοῦσον σημαίνει.

XLIII. Οἱ πυρετοὶ ὁκόσοι, μὴ διαλείποντες,
διὰ τρίτης ἰσχυρότεροι γίνονται,[5] ἐπικίνδυνοι·
ὅτῳ δ' ἂν τρόπῳ διαλείπωσι, σημαίνει ὅτι
4 ἀκίνδυνοι.

XLIV. Ὁκόσοισι[6] πυρετοὶ μακροί, τούτοισι
2 φύματα ἢ ἐς τὰ ἄρθρα πόνοι ἐγγίνονται.

XLV. Ὁκόσοισι ἢ φύματα ἐς τὰ ἄρθρα ἢ
πόνοι ἐγγίνονται ἐκ πυρετῶν,[7] οὗτοι σιτίοισι
3 πλείοσι χρέονται.

XLVI Ἢν ῥῖγος ἐμπίπτῃ[8] πυρετῷ μὴ δια-
2 λείποντι,[9] ἤδη ἀσθενεῖ ἐόντι,[10] θανάσιμον.

XLVII. Αἱ ἀποχρέμψιες ἐν τοῖσι πυρετοῖσι
τοῖσι μὴ διαλείπουσιν, αἱ πελιδναί, καὶ αἱματώδεες, καὶ δυσώδεες, καὶ χολώδεες,[11] πᾶσαι[12]
κακαί· ἀποχωρέουσαι δὲ καλῶς, ἀγαθαί· καὶ
κατὰ τὴν διαχώρησιν,[13] καὶ κατὰ τὰ οὖρα· ἢν

[1] MV omit μὲν.

[2] For μέζω the MSS. have μείζων, μείζω, πλείω, πλέον, πλείων.

[3] δὲ is omitted by V.

[4] For ἐλάσσω a very great number of MSS., including MVQ, read ἐλάσσων, the -ν coming from νοῦσον which follows. C′ has ἔλασσον.

[5] Before ἐπικίνδυνοι C′Q have καί.

[6] For ὁκόσοισι Urb. Q have ὁκόσοι.

[7] So C′. V has ἢ φύματα ἢ εἰς τὰ ἄρθρα πόνοι, Urb. φύματα ἢ πόνοι . . . . . γίνονται, M φύματα ἐς τὰ ἄρθρα ἢ πόνοι . . . γίγνονται. After πυρετῶν a few MSS. add μακρῶν (from Galen's commentary).

[8] ἐμπίπτῃ Littré's A¹L¹ : ἐμπίπτει C′V : ἐπιπίπτηι M : ἐπιπίπτῃ Urb.

[9] C′ has ἐν πυρετῶ μὴ διαλίποντι.

XLII. Copious sweat, hot or cold, continually running, indicates, when cold, a more serious disease, and when hot, a less serious one.

XLIII. Such fevers as, without intermitting, grow worse every other day,[1] are dangerous; intermittence of any kind[2] indicates that there is no danger.

XLIV. Sufferers from protracted fevers are attacked by tumours or by pains at the joints.

XLV. Those who, after fevers, are attacked either by tumours or pains at the joints, are taking too much food.

XLVI. If rigor attack[3] a sufferer from a continued fever, while the body is already weak,[4] it is a fatal sign.

XLVII. In continued fevers, expectorations that are livid, bloody, fetid, or bilious are all bad, but if properly evacuated they are favourable. It is the same with stools and urine; for if some suitable

[1] These are malignant tertians, "semitertians," as they were called in ancient times.

[2] That is, malaria of the mild, intermittent type.

[3] Littré's view, that Galen's distinction between ἐπιπέσῃ and ἐμπίπτῃ refers to the tenses rather than to the prefixes ἐπι- and ἐμ-, is probably right.

[4] I have printed the harder reading, though the more regular reading of C′ may be correct. The sense is the same in either case.

---

[10] ἀσθενέως ἐόντος τοῦ σώματος C′: ἀσθενεῖ ἐόντι τῶι σώματι Urb.: ἀσθενεῖ ἐόντι M.

[11] C′ inverts the order of δυσώδεες and χολώδεες. M omits καὶ δυσώδεες.

[12] ἐπιστᾶσαι μὲν Erm.: στᾶσαι Rein.

[13] τὴν διαχώρησιν C′V: τὰς διαχωρήσιας M.

δὲ[1] μή[2] τι τῶν συμφερόντων ἐκκρίνηται διὰ
7 τῶν τόπων τούτων, κακόν.

XLVIII. Ἐν τοῖσι μὴ διαλείπουσι πυρετοῖσιν, ἢν τὰ μὲν ἔξω ψυχρὰ ᾖ,[3] τὰ δὲ ἔνδον καίηται,
3 καὶ δίψαν ἔχῃ, θανάσιμον.

XLIX. Ἐν μὴ διαλείποντι πυρετῷ,[4] ἢν χεῖλος, ἢ ὀφθαλμός, ἢ ὀφρύς,[5] ἢ ῥὶς διαστραφῇ, ἢν μὴ βλέπῃ, ἢν μὴ ἀκούῃ,[6] ἤδη[7] ἀσθενέος ἐόντος τοῦ σώματος,[8] ὅ τι ἂν τούτων γένηται, ἐγγὺς ὁ
5 θάνατος.

L. Ὅκου ἐν πυρετῷ μὴ διαλείποντι δύσπνοια
2 γίνεται[9] καὶ παραφροσύνη, θανάσιμον.

LI. Ἐν τοῖσι πυρετοῖσιν ἀποστήματα μὴ λυόμενα πρὸς τὰς πρώτας κρίσιας, μῆκος νούσου
3 σημαίνει.[10]

LII. Ὁκόσοισιν ἐν τοῖσι πυρετοῖσιν, ἢ ἐν τῇσιν ἄλλῃσιν ἀρρωστίῃσι κατὰ προαίρεσιν οἱ ὀφθαλμοὶ δακρύουσιν, οὐδὲν ἄτοπον· ὁκόσοισι
4 δὲ μὴ κατὰ προαίρεσιν, ἀτοπώτερον.[11]

LIII. Ὁκόσοισιν[12] ἐπὶ τῶν ὀδόντων ἐν τοῖσι πυρετοῖσι[13] περίγλισχρα[14] γίνεται, ἰσχυρότεροι
3 γίνονται οἱ πυρετοί.

[1] δὲ is omitted by M.

[2] μή. Galen says that there were in his days some MSS. omitting the negative. It is in all our MSS.

[3] ᾖ omitted by C′ Urb.

[4] Urb. has ἐν τοῖσι μὴ διαλείπουσι πυρετοῖσιν.

[5] M transposes ὀφθαλμὸς and ὀφρύς.

[6] C′ has ἢ μὴ βλέπει ἢ μὴ ἀκούει.

[7] MV omit ἤδη. Many MSS., including C′, have it.

[8] τοῦ σώματος C′V : τοῦ κάμνοντος Littré, with slight authority. Most MSS., including M, omit.

[9] γίνεται a few Paris MSS. : γίνηται C′MV. Rein. reads ὅκου δ' ἂν τῳ.

excretion does not take place through these channels it is a bad sign.[1]

XLVIII. In continued fevers, if the external parts be cold but the internal parts burning hot, while the patient suffers from thirst, it is a fatal sign.[2]

XLIX. In a continued fever, if the patient's lip, eye, eye-brow or nose be distorted, if sight or hearing fail, while the body is already in a weak state—whatever of these symptoms show themselves, death is near.

L. When in a continued fever occur difficulty of breathing and delirium, it is a fatal sign.

LI. In fevers, abscesses that are not resolved at the first crisis indicate a protracted disease.

LII. When in fevers or in other diseases patients weep of their own will, it is nothing out of the common; but it is rather so when they weep involuntarily.

LIII. When in fevers very viscous matter forms on the teeth, the fevers become more severe.

[1] The reading noticed by Galen, which omits *μή*, would mean that if the secretions be substances that the body requires for health (*τῶν συμφερόντων*) evacuation will only do harm. The emendations of Ermerins and Reinhold remove the difficulties of meaning from this aphorism, and one or other is probably right. See VII. lxx. The meaning would be : "are bad if suppressed, but if properly evacuated, etc."

[2] See *Coan Prenotions*, 115.

---

[10] Urb. V have *σημαίνουσι* (V -*ν*).

[11] M differs from the other good MSS. in omitting *οἱ ὀφθαλμοὶ* and reading *ὀκόσοι* (twice).

[12] *ὀκόσοι* M.

[13] V transposes *ἐπὶ τῶν ὀδόντων* and *ἐν τοῖσι πυρετοῖσι*. *περὶ τοὺς ὀδόντας* C′.

[14] *γλισχράσματα* V.

LIV. Ὁκόσοισιν ἐπὶ πολὺ βῆχες ξηραί, βραχέα[1] ἐρεθίζουσαι, ἐν πυρετοῖσι καυσώδεσιν, οὐ
3 πάνυ τι διψώδεές εἰσιν.

LV. Οἱ ἐπὶ βουβῶσι πυρετοί, πάντες[2] κακοί,
2 πλὴν τῶν ἐφημέρων.[3]

LVI. Πυρέσσοντι ἱδρὼς ἐπιγενόμενος, μὴ ἐκλείποντος[4] τοῦ πυρετοῦ, κακόν· μηκύνει γὰρ ἡ
3 νοῦσος, καὶ ὑγρασίην πλείω σημαίνει.

LVII. Ὑπὸ σπασμοῦ ἢ τετάνου ἐχομένῳ[5]
2 πυρετὸς ἐπιγενόμενος λύει τὸ νόσημα.

LVIII. Ὑπὸ καύσου ἐχομένῳ, ῥίγεος ἐπιγε-
2 νομένου, λύσις.

LIX. Τριταῖος ἀκριβὴς κρίνεται ἐν ἑπτὰ περιό-
2 δοισι τὸ μακρότατον.

LX. Ὁκόσοισιν ἂν[6] ἐν τοῖσι[7] πυρετοῖσι τὰ ὦτα κωφωθῇ, αἷμα ἐκ τῶν ῥινῶν ῥυέν, ἢ κοιλίη
3 ἐκταραχθεῖσα, λύει τὸ νόσημα.[8]

LXI. Πυρέσσοντι[9] ἢν μὴ ἐν περισσῇσιν[10] ἡμέρῃσιν ἀφῇ ὁ πυρετός, ὑποτροπιάζειν[11]
3 εἴωθεν.

LXII. Ὁκόσοισιν ἐν τοῖσι πυρετοῖσιν ἴκτεροι ἐπιγίνονται πρὸ τῶν ἑπτὰ ἡμερῶν, κακόν, ἢν[12] μὴ συνδόσιες ὑγρῶν κατὰ τὴν κοιλίην γένων-
4 ται.[13]

[1] βραχέαι C′ : βραχεῖα Urb. : βραχεῖαι S.

[2] πάντες omitted by Urb.

[3] ἐφ' ἡμερῶν M.

[4] ἐκλίποντος Galen.

[5] ἐνοχλουμένω MV. M places this aphorism after LVIII.

[6] V omits ἄν. C′ has ἤν.

[7] τοῖσι omitted by C′Q.

[8] After νόσημα V has τὰ ἐν ἀρτίῃσιν ἡμέραις κρινόμεναι δύσκριτα καὶ φιλυπόστροφα. M τὰ ἐναρτίῃσι κρινόμενα

LIV. Whenever in ardent[1] fevers dry coughs persist, causing slight irritation,[2] there is not much thirst.

LV. Fevers following buboes are all bad except ephemerals.[3]

LVI. Sweat supervening on fever, without the fever's intermitting, is a bad sign; for the disease is protracted, and it is a sign of excessive moisture.

LVII. Fever supervening on a patient's suffering from convulsion or tetanus, removes the disease.

LVIII. A sufferer from ardent fever is cured by the supervening of a rigor.

LIX. An exact tertian reaches a crisis in seven periods at most.[4]

LX. When in fevers there is deafness, if there be a flow of blood from the nose, or the bowels become disordered, it cures the disease.

LXI. If a fever does not leave the patient on the odd days it is usual for it to relapse.

LXII. When jaundice supervenes in fevers before seven days it is a bad sign, unless there be watery discharges by the bowels.

[1] "Ardent" fevers were a kind of remittent malaria.

[2] Adams translates: "with a tickling nature with slight expectoration."

[3] "Ephemerals" are fevers lasting only about a day.

[4] The "exact" tertian is malaria with an access every other day. So the aphorism means that the tertian does not last more than a fortnight.

δύσκριτα καὶ φιλυπόστροφα. These words C′ and some other MSS. place after the next aphorism.

[9] πυρέσσοντι C′ Urb : πυρέσσοντα MV.

[10] περισσῇσιν Urb. MV. : κρισήμησι (*sic*) C′.

[11] ἐπιτροπιάζειν Urb.

[12] εἰ Urb.

[13] C′V omit ἢν . . . γένωνται.

LXIII. Ὁκόσοισιν ἂν ἐν τοῖσι πυρετοῖσι καθ' ἡμέρην ῥίγεα[1] γίνηται, καθ' ἡμέρην οἱ πυρετοὶ
3 λύονται.

LXIV. Ὁκόσοισιν ἐν τοῖσι[2] πυρετοῖσι τῇ ἑβδόμῃ ἢ τῇ ἐνάτῃ ἢ τῇ ἑνδεκάτῃ[3] ἢ τῇ τεσσαρεσκαιδεκάτῃ ἴκτεροι ἐπιγίνονται, ἀγαθόν, ἢν μὴ τὸ ὑποχόνδριον τὸ δεξιὸν[4] σκληρὸν γένηται.[5]
5 ἢν δὲ μή, οὐκ ἀγαθόν.[6]

LXV. Ἐν τοῖσι πυρετοῖσι περὶ τὴν κοιλίην
2 καῦμα ἰσχυρὸν καὶ καρδιωγμός, κακόν.[7]

LXVI. Ἐν τοῖσι πυρετοῖσι τοῖσιν ὀξέσιν οἱ σπασμοὶ καὶ οἱ περὶ τὰ σπλάγχνα πόνοι ἰσχυροί,
3 κακόν.[8]

LXVII. Ἐν τοῖσι πυρετοῖσιν[9] οἱ ἐκ τῶν
2 ὕπνων φόβοι,[10] ἢ σπασμοί, κακόν.[11]

LXVIII. Ἐν τοῖσι πυρετοῖσι τὸ πνεῦμα
2 προσκόπτον, κακόν· σπασμὸν γὰρ σημαίνει.

LXIX. Ὁκόσοισιν οὖρα παχέα,[12] θρομβώδεα, ὀλίγα, οὐκ ἀπυρέτοισι, πλῆθος ἐλθὸν ἐκ τούτων λεπτὸν[13] ὠφελεῖ· μάλιστα δὲ τὰ τοιαῦτα ἔρχεται οἷσιν[14] ἐξ ἀρχῆς ἢ διὰ ταχέων[15] ὑπόστασιν[16]
5 ἴσχει.

LXX. Ὁκόσοισι δὲ[17] ἐν[18] πυρετοῖσι τὰ οὖρα ἀνατεταραγμένα[19] οἷον ὑποζυγίου, τούτοισι κε-
3 φαλαλγίαι ἢ[20] πάρεισιν ἢ παρέσονται.

[1] ῥῖγος C′. [2] τοῖσι omitted by C′.
[3] ἢ τῇ ἑνδεκάτῃ omitted by M.
[4] τὸ δεξιὸν ὑποχόνδριον M.
[5] γένηται C′ Urb. V. : ᾖι M.
[6] For οὐκ ἀγαθόν M has κακόν.
[7] This aphorism is omitted by C′.
[8] This aphorism in Urb. and several other MSS. comes after LXVII.

LXIII. Fevers in which a rigor occurs each day are resolved each day.

LXIV. In fevers, when jaundice supervenes on the seventh day, on the ninth, on the eleventh, or on the fourteenth, it is a good sign, unless the right hypochondrium become hard. Otherwise it is not a good sign.

LXV. In fevers, great heat about the bowels and heartburn are a bad sign.

LXVI. In acute fevers, convulsions and violent pains in the bowels are a bad sign.

LXVII. In fevers, terrors after sleep, or con vulsions, are a bad sign.

LXVIII. In fevers, stoppage of the breath is a bad sign, as it indicates a convulsion.

LXIX When the urine is thick, full of clots, and scanty, fever being present, a copious discharge of ⟨comparatively⟩ thin urine coming afterwards gives relief. This usually happens in the case of those whose urine contains a sediment from the onset or shortly after it.

LXX. In cases of fever, when the urine is turbid, like that of cattle, headaches either are, or will be, present.

---

[9] After πυρετοῖσιν C′ adds τοῖς ὀξέσι.

[10] For φόβοι ἢ C′ has πόνοι καὶ σπασμοί. Galen mentions πόνοι as a variant of φόβοι, adding that either reading makes good sense.

[11] This aphorism in M comes after LXV.

[12] παχέα omitted by Urb.

[13] ἐκ τούτου λεπτῶν Rein.

[14] οἷς ἂν C′.

[15] παχέων M.

[16] ὑπόστασις V.

[17] δὲ omitted by Urb.

[18] After ἐν C′ has τοῖσι.

[19] τεταραγμένα V.

[20] ἢ omitted by M.

LXXI. Ὁκόσοισιν ἑβδομαῖα κρίνεται, τούτοισιν ἐπινέφελον ἴσχει τὸ οὖρον τῇ τετάρτῃ
3 ἐρυθρόν, καὶ τὰ ἄλλα[1] κατὰ λόγον.

LXXII. Ὁκόσοισιν οὖρα[2] διαφανέα[3] λευκά, πονηρά· μάλιστα δὲ ἐν τοῖσι φρενιτικοῖσιν
3 ἐπιφαίνεται.[4]

LXXIII. Ὁκόσοισιν ὑποχόνδρια μετέωρα, διαβορβορύζοντα, ὀσφύος ἀλγήματος ἐπιγενομένου, αἱ[5] κοιλίαι τούτοισι[6] καθυγραίνονται, ἢν μὴ φῦσαι καταρραγέωσιν, ἢ οὔρου πλῆθος
5 ὑπέλθῃ·[7] ἐν πυρετοῖσι δὲ ταῦτα.[8]

LXXIV. Ὁκόσοισιν ἐλπὶς ἐς[9] ἄρθρα ἀφίστασθαι, ῥύεται τῆς ἀποστάσιος οὖρον πολὺ καὶ παχὺ[10] καὶ λευκὸν γινόμενον, οἷον ἐν τοῖσι κοπιώδεσι πυρετοῖσι τεταρταίοισιν ἐνίοισιν ἄρχεται γίνεσθαι· ἢν δὲ καὶ ἐκ τῶν ῥινῶν αἱ-
6 μορραγήσῃ, καὶ πάνυ ταχὺ λύεται.

LXXV. Ἢν αἷμα ἢ[11] πῦον οὐρῇ, τῶν νεφρῶν
2 ἢ τῆς κύστιος ἕλκωσιν σημαίνει.

LXXVI. Ὁκόσοισιν ἐν τῷ οὔρῳ παχεῖ ἐόντι σαρκία σμικρὰ ὥσπερ τρίχες συνεξέρχονται,
3 τούτοισιν ἀπὸ τῶν νεφρῶν ἐκκρίνεται.

LXXVII. Ὁκόσοισιν ἐν τῷ οὔρῳ παχεῖ ἐόντι

[1] V reads τῇ τετάρτῃ ἴσχει ἐπινέφελον καὶ ἐρυθρὸν τὸ οὖρον καὶ τὰ ἄλλα. Urb. puts τῇ τετάρτῃ before ἐπινέφελον, and instead of τῇ τετάρτῃ C′ has ᾗ.

[2] Before οὖρα Urb. has τά.

[3] After διαφανέα Urb. has ᾖ.

[4] ἐπιγίγνεται, M: ἢν ἐπιφαίνηται C′: ἢν ἐπιφαίνεται Urb. Galen notices a reading ἐπιφαίνεται τὰ τοιαῦτα.

[5] αἱ omitted by Urb. [6] τοῖσι τουτέοισι Urb.

[7] ἐπέλθῃ M (and Littré). C′ has οὔρων and Urb. πλῆθος οὔρων.

[8] C′ has γίνεται after ταῦτα. See Introduction, p xxxvi.

LXXI. In cases that come to a crisis on the seventh day, the patient's urine on the fourth day has a red cloud in it, and other symptoms accordingly.

LXXII. Transparent, colourless[1] urine is bad. It appears mostly in cases of phrenitis.[2]

LXXIII. When there are swelling and rumbling in the hypochondria, should pain in the loins supervene, the bowels become watery, unless there be breaking of wind or a copious discharge of urine. These symptoms occur in fevers.

LXXIV. When an abscession to the joints is to be expected, the abscession may be averted by an abundant flow of thick, white urine, like that which in certain prostrating fevers begins on the fourth day.[3] And if there is also nasal hemorrhage the disease is very quickly resolved.

LXXV. Blood or pus in the urine indicates ulceration of the kidneys or bladder.

LXXVI. When the urine is thick, and small pieces of flesh-like hairs pass with it, it means a secretion from the kidneys.[4]

LXXVII. When the urine is thick, and with it is

[1] So Littré from the commentary of Galen. Perhaps, however, λευκά does mean "white."

[2] The reading ἦν ἐπιφαίνηται would mean "bad, especially when it appears in cases of phrenitis."

[3] Adams translates τεταρταίοισιν "quartans." The other meaning seems more probable here. Adams takes γινόμενον with λευκόν, "becoming white."

[4] Similar propositions occur in *Nature of Man*, XIV.

---

[9] V has τὰ ἄρθρα and C′ τἄρθρα.

[10] V has πολὺ παχὺ and M πολὺ κάρτα παχὺ. Urb. has παχὺ καὶ πολὺ.

[11] καὶ C′ Urb.

πιτυρώδεα συνεξουρεῖται, τούτοισιν ἡ κύστις
3 ψωριᾷ.

LXXVIII. Ὁκόσοι ἀπὸ ταὐτομάτου αἷμα οὐρέουσι, τούτοισιν ἀπὸ τῶν νεφρῶν φλεβίου
3 ῥῆξιν σημαίνει.

LXXIX. Ὁκόσοισιν ἐν τῷ οὔρῳ[1] ψαμμώδεα
2 ὑφίσταται, τούτοισιν ἡ κύστις λιθιᾷ.[2]

LXXX. Ἢν αἷμα οὐρῇ καὶ θρόμβους, καὶ στραγγουρίην ἔχῃ, καὶ ὀδύνη ἐμπίπτῃ ἐς[3] τὸ ὑπογάστριον καὶ ἐς τὸν περίνεον, τὰ περὶ τὴν
4 κύστιν πονεῖ.

LXXXI. Ἢν αἷμα καὶ πῦον οὐρῇ καὶ λεπίδας, καὶ ὀσμὴ βαρέη[4] ᾖ, τῆς κύστιος ἕλκωσιν
3 σημαίνει.

LXXXII. Ὁκόσοισιν ἐν τῇ οὐρήθρῃ φύματα φύεται,[5] τούτοισι, διαπυήσαντος καὶ ἐκραγέντος,
3 λύσις.

LXXXIII. Οὔρησις νύκτωρ[6] πολλὴ γινομένη,
2 σμικρὴν τὴν ὑποχώρησιν[7] σημαίνει.

[1] τοῖσιν οὔροισι Urb.

[2] After λιθιᾷ C′ Urb. add καὶ οἱ νεφροί.

[3] Before τὸ Urb. has τὸν κτένα καί. C′ has καὶ τὸν κτένα καὶ τὸν after ὑπογάστριον.

[4] All our good MSS., including C′ Urb. MV, have βαρεῖα.

[5] ἐκφύεται C′.

passed as it were bran, this means psoriasis of the bladder.[1]

LXXVIII. When a patient has a spontaneous discharge of blood and urine, it indicates the breaking of a small vein in the kidneys.

LXXIX. When the urine contains a sandy sediment there is stone in the bladder.

LXXX. If there be blood and clots in the urine, and strangury be present, should pain attack the hypogastrium and the perineum, the parts about the bladder are affected.[2]

LXXXI. If the urine contain blood, pus and scales, and its odour be strong, it means ulceration of the bladder.

LXXXII. When tumours form in the urethra, should these suppurate and burst, there is relief.[3]

LXXXIII. When much urine is passed in the night, in means that the bowel-discharges are scanty.

[1] Similar propositions occur in *Nature of Man*, XIV.
[2] See *Aphorisms*, VII. xxxix.
[3] Or, "it means a cure."

---

[6] C′ has ἐκ νύκτορ (perhaps as one word), Urb. ἐκ νύκτωρ, a few MSS. ἐκ νυκτός.
[7] Before σημαίνει Urb. has ἴσεσθαι.

ΤΜΗΜΑ ΠΕΜΠΤΟΝ

I. Σπασμὸς ἐξ ἐλλεβόρου, θανάσιμον.

II. Ἐπὶ τρώματι σπασμὸς[1] ἐπιγενόμενος,
2 θανάσιμον.[2]

III. Αἵματος πολλοῦ ῥυέντος σπασμὸς ἢ
2 λυγμὸς ἐπιγενόμενος, κακόν.

IV. Ἐπὶ ὑπερκαθάρσει σπασμὸς ἢ λυγμὸς
2 ἐπιγενόμενος, κακόν.

V. Ἢν μεθύων ἐξαίφνης ἄφωνός τις[3] γένηται, σπασθεὶς ἀποθνῄσκει, ἢν μὴ πυρετὸς ἐπιλάβῃ, ἢ ἐς τὴν ὥρην ἐλθών, καθ' ἣν αἱ κραιπάλαι
4 λύονται, φθέγξηται.

VI. Ὁκόσοι ὑπὸ τετάνου ἁλίσκονται, ἐν τέσσαρσιν[4] ἡμέρῃσιν ἀπόλλυνται· ἢν δὲ ταύτας
3 διαφύγωσιν, ὑγιέες γίνονται.

VII. Τὰ ἐπιληπτικὰ ὁκόσοισι πρὸ τῆς ἥβης γίνεται, μετάστασιν ἴσχει· ὁκόσοισι δὲ πέντε καὶ εἴκοσιν ἐτέων γίνεται, τὰ πολλὰ[5] συναπο-
4 θνῄσκει.

VIII. Ὁκόσοι πλευριτικοὶ γενόμενοι οὐκ ἀνακαθαίρονται ἐν τεσσαρεσκαίδεκα ἡμέρῃσι, τού-
3 τοισιν ἐς ἐμπύημα μεθίσταται.[6]

IX. Φθίσιες γίνονται[7] μάλιστα ἡλικίῃσι τῇσιν ἀπὸ ὀκτωκαίδεκα ἐτέων μέχρι τριήκοντα
3 πέντε.

X. Ὁκόσοι κυνάγχην διαφεύγουσι, καὶ ἐς τὸν

[1] After σπασμὸς C′ adds ἢ λυγμός.
[2] For θανάσιμον C′ has κακόν.
[3] τις is placed here by Urb. M, but after μεθύων by C′V.
[4] τέτρασιν C′ Urb.
[5] Several inferior MSS. omit τὰ πολλὰ, an omission noticed by Galen.

FIFTH SECTION

I. Convulsion after hellebore is deadly.[1]

II. A convulsion supervening upon a wound is deadly.[1]

III. Convulsion or hiccough, supervening on a copious flux of blood, is a bad sign.

IV. Convulsion or hiccough supervening on excessive purging, is a bad sign.

V. If a drunken man suddenly become dumb, he dies after convulsions, unless he falls into a fever, or unless he lives to the time when the effects of intoxication disappear, and recovers his voice.

VI. Those who are attacked by tetanus either die in four days or, if they survive these, recover.

VII. Fits that occur before puberty admit of cure,[2] but if they occur after the age of twenty-five they usually last until death.

VIII. Pleurisy that does not clear up in fourteen days results in empyema.

IX. Consumption[3] occurs chiefly between the ages of eighteen and thirty-five.

X. Those who survive angina, should the disease

[1] The word θανάσιμον is said by the commentators to mean here "dangerous." In the next aphorism tetanus is obviously referred to, and θανάσιμον must mean at least "very often fatal."

[2] Or "change," "modification."

[3] Aphorisms IX.–XV. have close parallels in *Coan Prenotions*.

---

[6] Our MSS. show various readings—μεθίσταται, μεθίστανται, περιίσταται, περιίστανται. Littré reads καθίσταται.

[7] φθίσις γίνεται M: φθίσηες μάλιστα γίγνονται V.

πλεύμονα αὐτοῖσι τρέπεται,[1] ἐν ἑπτὰ ἡμέρῃσιν ἀποθνήσκουσιν· ἢν δὲ ταύτας διαφύγωσιν, ἔμπυοι 4 γίνονται.

XI. Τοῖσιν ὑπὸ τῶν φθισίων ἐνοχλουμένοισιν, ἢν τὸ πτύσμα, ὅ τι ἂν ἀποβήσσωσι, βαρὺ ὄζῃ ἐπὶ τοὺς ἄνθρακας ἐπιχεόμενον, καὶ αἱ τρίχες 4 ἀπὸ[2] τῆς κεφαλῆς ῥέωσι, θανατῶδες.[3]

XII. Ὁκόσοισι φθισιῶσιν αἱ τρίχες ἀπὸ τῆς κεφαλῆς ῥέουσιν, οὗτοι, διαρροίης ἐπιγενομένης, 3 ἀποθνήσκουσιν.

XIII. Ὁκόσοι αἷμα ἀφρῶδες ἀναπτύουσι,[4] 2 τούτοισιν ἐκ τοῦ πλεύμονος ἡ ἀναγωγὴ γίνεται.[5]

XIV. Ὑπὸ φθίσιος ἐχομένῳ διάρροια ἐπιγενο- 2 μένη, θανατῶδες.

XV. Ὁκόσοι ἐκ πλευρίτιδος ἔμπυοι γίνονται, ἢν ἀνακαθαρθῶσιν ἐν τεσσαράκοντα ἡμέρῃσιν, ἀφ᾽ ἧς ἂν ἡ ῥῆξις γένηται, παύονται· ἢν[6] δὲ 4 μή, ἐς φθίσιν μεθίστανται.

XVI. Τὸ θερμὸν βλάπτει ταῦτα[7] πλεονάκις χρεομένοισι, σαρκῶν ἐκθήλυνσιν, νεύρων ἀκράτειαν, γνώμης νάρκωσιν, αἱμορραγίας, λειπο- 4 θυμίας, ταῦτα οἷσι θάνατος.[8]

XVII. Τὸ δὲ ψυχρόν, σπασμούς, τετάνους, 2 μελασμούς, ῥίγεα πυρετώδεα.

[1] ἐς τὸν πνεύμονα τρέπεται αὐτέοισι καὶ C′: εἰς τὸν πνεύμονα τουτέοισι τρέπεται καὶ Urb.: εἰς τὸν πλεύμονα αὐτέων τρέπεται καὶ V: ἐς τὸν πλεύμονα αὐτέοισι τρέπεται καὶ M. Littré with two inferior MSS. transposes καὶ to before ἐς. Theophilus says that this alteration is necessary to the sense, and it seems to be the reading of Galen.

[2] C′ has ἐκ.

[3] C′ has θανάσιμον.

[4] For ἀναπτύουσι V has ἀνεμέουσι and M ἀνεμέωσι.

[5] V reads τουτέοισιν ἡ ἀναγωγὴ γίνεται ἐκ τοῦ πλεύμονος.

[6] For ἢν V has εἰ.

turn to the lungs, die within seven days, or, should they survive these, develop empyema.[1]

XI. In patients troubled with consumption, should the sputa they cough up have a strong[2] smell when poured over hot coals, and should the hair fall off from the head, it is a fatal symptom.

XII. Consumptive patients whose hair falls off from the head are attacked by diarrhoea and die.[3]

XIII. When patients spit up frothy blood, the discharge comes from the lungs.

XIV. If diarrhoea attack a consumptive patient it is a fatal symptom.

XV. When empyema follows on pleurisy, should the lungs clear up within forty days from the breaking, the illness ends; otherwise the disease passes into consumption.

XVI. Heat produces the following harmful results in those who use it too frequently: softening of the flesh, impotence of the muscles, dullness of the intelligence, hemorrhages and fainting, death ensuing in certain of these cases.

XVII. Cold produces convulsions, tetanus, blackening, feverish rigors.[4]

[1] Or "become purulent." So Adams (in notes).

[2] Or "offensive," "fetid."

[3] So Littré, who says that to translate "if diarrhoea supervenes" is inconsistent with XIV.

[4] "Blackening" will include "mortification," but is not to be limited to it.

---

[7] After ταῦτα C′ has τοῖσι, which Urb. places over πολλάκις, read by it for πλεονάκις. Rein. has a colon at ταῦτα.

[8] Urb. has γνώμης νάρκωσιν and ταῦτα οἷσι θάνατος in the margin. Galen notices four variants for the end of this aphorism: τούτοισι θάνατος, ταῦτα, ἐφ' οἷς ὁ θάνατος, ταῦτα οἷσι θάνατος, ταῦτα εἰς θάνατον. Rein. ἐς θάνατον. Query: ἔστιν οἷσι.

XVIII. Τὸ ψυχρὸν πολέμιον ὀστέοισιν, ὀδοῦσι, νεύροισιν, ἐγκεφάλῳ, νωτιαίῳ μυελῷ· τὸ δὲ
3 θερμὸν ὠφέλιμον.

XIX. Ὁκόσα κατέψυκται, ἐκθερμαίνειν,[1] πλὴν
2 ὅσα αἱμορραγεῖν μέλλει.[2]

XX. Ἕλκεσι τὸ μὲν ψυχρὸν δακνῶδες, δέρμα περισκληρύνει, ὀδύνην ἀνεκπύητον ποιεῖ, με-
3 λαίνει,[3] ῥίγεα πυρετώδεα,[4] σπασμούς, τετάνους.

XXI. Ἔστι δὲ ὅκου ἐπὶ τετάνου ἄνευ ἕλκεος νέῳ εὐσάρκῳ, θέρεος μέσου, ψυχροῦ πολλοῦ κατάχυσις ἐπανάκλησιν θέρμης ποιεῖται· θέρμη
4 δὲ ταῦτα[5] ῥύεται.

XXII. Τὸ θερμὸν ἐκπυητικόν, οὐκ ἐπὶ παντὶ ἕλκεϊ, μέγιστον σημεῖον ἐς ἀσφαλείην, δέρμα μαλάσσει, ἰσχναίνει, ἀνώδυνον, ῥιγέων, σπασμῶν, τετάνων παρηγορικόν· τῶν δὲ ἐν κεφαλῇ καρηβαρίην λύει·[6] πλεῖστον δὲ διαφέρει ὀστέων κατήγμασι, μᾶλλον δὲ[7] τοῖσιν ἐψιλωμένοισι, τούτων δὲ μάλιστα, τοῖσιν ἐν κεφαλῇ ἕλκεα ἔχουσι· καὶ ὁκόσα ὑπὸ ψύξιος θνῄσκει, ἢ ἑλκοῦται, καὶ ἕρπησιν ἐσθιομένοισιν, ἕδρῃ, αἰδοίῳ, ὑστέρῃ, κύστει, τούτοισι τὸ θερμὸν φίλιον
11 καὶ κρῖνον, τὸ δὲ ψυχρὸν πολέμιον καὶ κτεῖνον.

[1] ἐκθερμαίνει M.

[2] αἱμορραγέει ἢ μέλλει Littré and several Paris MSS.—perhaps rightly, as this is probably the correct reading in aphorism XXIII.

[3] μελασμούς has been suggested for μελαίνει.

[4] Littré reads ποιέει here, without quoting any authority for it, and I have not seen the word in any MS. I have collated. Dietz would place it after τετάνους. Though the meaning is clear, the exact reading has apparently been lost. The text, though ungrammatical, is the reading of all our good MSS.

[5] Two MSS. read τοῦτον. Rein. τετάνων. See *Intr.* p. xxxi.

XVIII. Cold is harmful to bones, teeth, sinews, brain, and spinal marrow, but heat is beneficial.

XIX. Heat parts that are chilled, except where hemorrhage threatens.[1]

XX. Cold makes sores to smart, hardens the skin, causes pain unattended with suppuration; it blackens, and causes feverish rigors, convulsions, tetanus.

XXI. Sometimes in a case of tetanus without a wound, the patient being a muscular young man, and the time the middle of summer, a copious affusion of cold water brings a recovery of heat. Heat relieves these symptoms.[2]

XXII. When heat causes suppuration, which it does not do in the case of every sore, it is the surest sign of recovery; it softens the skin, makes it[3] thin, removes pain and soothes rigors, convulsions and tetanus. It relieves heaviness of the head. It is particularly useful in fractures of the bones, especially when they are exposed, and most especially in cases of wounds in the head. Also in cases of mortification and sores from cold, of corroding herpes, for the seat, the privy parts, the womb, the bladder—for all these heat is beneficial and conduces to a crisis, while cold is harmful and tends to a fatal issue.

[1] With Littré's reading: "Where there is, or threatens to be, hemorrhage."

[2] The emendation τοῦτον is an attempt to get rid of the awkward plural. Perhaps the sentence is a misplaced "title" of the next aphorism.

[3] Perhaps, "the body generally."

---

[6] τὸ δὲ, ἐν κεφαλῇ· καὶ καρηβαρίην λύει, Urb. The MSS. vary very much here, and Littré (combining the readings of several) has τὰ δὲ ἐν τῇ κεφαλῇ, καὶ καρηβαρίην λύει. The text represents C′MV.

[7] μᾶλλον δὲ V : μᾶλλον C′ : μάλιστα δὲ M.

XXIII. Ἐν τούτοισι δεῖ τῷ ψυχρῷ χρῆσθαι, ὁκόθεν αἱμορραγεῖ, ἢ μέλλει,[1] μὴ ἐπ' αὐτά, ἀλλὰ περὶ αὐτά, ὁκόθεν ἐπιρρεῖ· καὶ ὁκόσαι φλεγμοναὶ ἢ ἐπιφλογίσματα ἐς τὸ ἐρυθρὸν καὶ ὕφαιμον ῥέποντα νεαρῷ αἵματι, ἐπὶ ταῦτα,[2] ἐπεὶ τά γε παλαιὰ μελαίνει· καὶ ἐρυσίπελας τὸ μὴ ἑλκούμενον, ἐπεὶ τό γε ἑλκούμενον βλάπτει. 7

XXIV. Τὰ ψυχρά, οἷον χιὼν κρύσταλλος,[3] στήθεϊ[4] πολέμια, βηχέων κινητικά, αἱμορροϊκά, καταρροϊκά. 3

XXV. Τὰ ἐν ἄρθροισιν οἰδήματα καὶ ἀλγήματα, ἄτερ ἕλκεος, καὶ ποδαγρικά, καὶ σπάσματα, τούτων τὰ πλεῖστα ψυχρὸν καταχεόμενον πολὺ[5] ῥηΐζει τε καὶ ἰσχναίνει, καὶ ὀδύνην λύει·[6] νάρκη δὲ[7] μετρίη ὀδύνης λυτική. 5

XXVI. Ὕδωρ τὸ ταχέως θερμαινόμενον καὶ ταχέως ψυχόμενον, κουφότατον. 2

XXVII. Ὁκόσοισι πιεῖν ὀρέξις νύκτωρ τοῖσι πάνυ διψῶσιν,[8] ἢν ἐπικοιμηθῶσιν, ἀγαθόν. 2

XXVIII. Γυναικείων ἀγωγόν, ἡ ἐν ἀρώμασι πυρίη, πολλαχῇ[9] δὲ καὶ ἐς ἄλλα χρησίμη ἂν[10] ἦν, εἰ μὴ καρηβαρίας ἐνεποίει. 3

XXIX. Τὰς κυούσας φαρμακεύειν, ἢν ὀργᾷ,

[1] MV have αἱμορραγέειν μέλλει. So C′, with μὴ before μέλλει. Galen apparently had ἢ μέλλει with the indicative before it. So Littré, following several MSS. Compare aphorism XIX.

[2] ἐπὶ ταῦτα omitted by Urb. V.

[3] κρύσταλλος χιὼν V.

[4] στηθέων C′.

[5] καταχεόμενον πολὺ C′ : πολλὸν καταχεόμενον MV.

[6] νάρκην γὰρ ποιέει Rein. for καὶ . . . λύει.

[7] δὲ MV : γὰρ C′ and many other MSS. But δὲ often has the force of γάρ.

XXIII. Cold should be used in the following cases: when there is, or is likely to be, hemorrhage, but it should be applied, not to the parts whence blood flows, but around them; in inflammations, and in inflamed pustules inclining to a red and bloodshot colour that is due to fresh blood; in these cases apply cold (but it blackens old inflammations), and when there is erysipelas without sores (but it does harm when there are sores).

XXIV. Cold things, such as snow or ice, are harmful to the chest, and provoke coughing, discharges of blood and catarrhs.

XXV. Swellings and pains in the joints, without sores, whether from gout or from sprains, in most cases are relieved by a copious affusion of cold water, which reduces the swelling and removes the pain. For numbness in moderation removes pain.

XXVI. That water is lightest which quickly gets hot and quickly gets cold.

XXVII. When there is a desire, caused by intense thirst, to drink during the night, should sleep follow, it is a good sign.

XXVIII. Aromatic vapour baths promote menstruation, and in many ways would be useful for other purposes if they did not cause heaviness of the head.

XXIX. Purge pregnant women, if there be orgasm, from the fourth month to the seventh, but

---

[8] *ὁκόσοισι πιεῖν ὄρεξις ὕδωρ ἐκ νυκτῶν τούτοισι διψώδεσιν* V. Urb. has *ὁκόσοισιν ὕδωρ πιεῖν ὄρεξις νύκτωρ. τουτέοισι πάνι διψώδεσιν.*

[9] C′ Urb. have πολλαχοῦ, perhaps rightly.

[10] C′ omits ἂν, and Urb. ἂν ἦν.

τετράμηνα, καὶ ἄχρι ἑπτὰ μηνῶν ἧσσον· τὰ δὲ
3 νήπια καὶ πρεσβύτερα εὐλαβεῖσθαι.

XXX. Γυναικὶ ἐν γαστρὶ ἐχούσῃ ὑπό τινος
2 τῶν ὀξέων νοσημάτων ληφθῆναι,[1] θανατῶδες.

XXXI. Γυνὴ ἐν γαστρὶ ἔχουσα, φλεβοτομηθεῖσα, ἐκτιτρώσκει· καὶ μᾶλλον ᾖσι μεῖζον τὸ
3 ἔμβρυον.

XXXII. Γυναικὶ αἷμα ἐμεούσῃ, τῶν κατα-
2 μηνίων ῥαγέντων, λύσις.[2]

XXXIII. Τῶν καταμηνίων ἐκλειπόντων, αἷμα
2 ἐκ τῶν ῥινῶν ῥυέν,[3] ἀγαθόν.

XXXIV. Γυναικὶ ἐν γαστρὶ ἐχούσῃ, ἢν ἡ
2 κοιλίη ῥυῇ πολλάκις,[4] κίνδυνος ἐκτρῶσαι.

XXXV. Γυναικὶ ὑπὸ ὑστερικῶν ἐνοχλουμένῃ,
2 ἢ δυστοκούσῃ, πταρμὸς ἐπιγινόμενος,[5] ἀγαθόν.

XXXVI. Γυναικὶ τὰ[6] καταμήνια ἄχροα, καὶ
μὴ κατὰ τὰ αὐτὰ ἀεὶ[7] γινόμενα, καθάρσιος
3 δεῖσθαι σημαίνει.

XXXVII. Γυναικὶ ἐν γαστρὶ ἐχούσῃ, ἢν
2 ἐξαίφνης[8] μασθοὶ ἰσχνοὶ γένωνται, ἐκτιτρώσκει.

XXXVIII. Γυναικὶ ἐν γαστρὶ ἐχούσῃ[9] ἢν
ὁ ἕτερος μασθὸς ἰσχνὸς γένηται, δίδυμα ἐχούσῃ,
θάτερον ἐκτιτρώσκει· καὶ ἢν μὲν ὁ δεξιὸς ἰσχνὸς
4 γένηται,[10] τὸ ἄρσεν· ἢν δὲ ὁ ἀριστερός, τὸ θῆλυ.

[1] C′ has συλληθῆναι.

[2] Urb. joins together this aphorism and the next, thus: λύσις γίνεται· τῶν δὲ καταμηνίων ἐκλειπόντων· αἷμα ἐκ τῶν ῥινῶν ῥυὲν ἀγαθόν. This reading explains the insertion of γυναικὶ in C′ (which omits γίνεται) before τῶν.

[3] ῥυὲν C′ Urb. : ῥυῆναι MV.

[4] κοιλίη ῥυῇ πολλὰ V: κοιλίη πολλὰ ῥυῆι M: κοιλίη ῥυεῖ πολλάκις C′ : κοιλίη πολλὰ ῥυῇ Urb. C′ has τοῦ ἐκτρῶσαι.

[5] ἐπιγινόμενος C′ : ἐπιγενόμενος Urb. MV.

[6] τὰ omitted by Urb. V.

less in the latter case; care is needed when the unborn child is of less than four months or of more than seven.[1]

XXX. If a woman with child is attacked by one of the acute diseases, it is fatal.[2]

XXXI. A woman with child, if bled, miscarries; the larger the embryo the greater the risk.

XXXII. When a woman vomits blood, menstruation is a cure.

XXXIII. When menstruation is suppressed, a flow of blood from the nose is a good sign.

XXXIV. When a woman with child has frequent diarrhoea there is a danger of a miscarriage.

XXXV. When a woman suffers from hysteria[3] or difficult labour an attack of sneezing is beneficial.

XXXVI. If menstrual discharge is not of the proper colour, and irregular, it indicates that purging[4] is called for.

XXXVII. Should the breasts of a woman with child suddenly become thin, she miscarries.

XXXVIII. When a woman is pregnant with twins, should either breast become thin, she loses one child. If the right breast become thin, she loses the male child; if the left, the female.

[1] This aphorism is omitted by C′V. See *Aphorisms* IV. i.

[2] This aphorism C′V place after XXXI.

[3] Said by some commentators to refer to retention of the placenta. Galen rejects this interpretation, but Littré seems inclined to accept it.

[4] Or, "an emmenagogue."

---

[7] C′ omits ἀεὶ and reads, I think, *κατὰ τὸ αὐτὸ.*

[8] M puts *ἐξαίφνης* after *μασθοὶ.*

[9] C′ has *ἐχούσῃ· δίδυμα,* omitting these words below.

[10] *ἰσχνὸς γένηται* omitted by Urb. (perhaps rightly).

XXXIX. Ἢν γυνὴ μὴ κύουσα, μηδὲ τετοκυῖα,
2 γάλα ἔχῃ, ταύτης[1] τὰ καταμήνια ἐκλέλοιπεν.

XL. Γυναιξὶν ὁκόσῃσιν ἐς τοὺς τιτθοὺς[2] αἷμα
2 συστρέφεται, μανίην σημαίνει.

XLI. Γυναῖκα ἢν θέλῃς εἰδέναι εἰ κύει, ἐπὴν
μέλλῃ[3] καθεύδειν, ἀδείπνῳ ἐούσῃ,[4] μελίκρητον
δίδου[5] πιεῖν· κἢν μὲν στρόφος ἔχῃ περὶ τὴν
4 γαστέρα,[6] κύει· ἢν[7] δὲ μή, οὐ κύει.

XLII. Γυνὴ ἢν[8] μὲν ἄρρεν κύῃ, εὔχροός ἐστιν·
2 ἢν δὲ θῆλυ, δύσχροος.

XLIII. Γυναικὶ κυούσῃ ἐρυσίπελας ἐν τῇ
2 ὑστέρῃ γενόμενον,[9] θανατῶδες.

XLIV. Ὁκόσαι παρὰ φύσιν λεπταὶ ἐοῦσαι
ἐν γαστρὶ ἔχουσιν,[10] ἐκτιτρώσκουσι,[11] πρὶν ἢ
3 παχυνθῆναι.

XLV. Ὁκόσαι τὸ σῶμα μετρίως ἔχουσαι
ἐκτιτρώσκουσι δίμηνα καὶ τρίμηνα ἄτερ προ-

[1] ταύτης Urb.: ταύτηι M: V has τὰ καταμήνια αὐτῇ and C′ τὰ καταμήνια αὐτῆς (followed by ἐξέλειπεν).

[2] Urb. has μασθοὺς τιτθούς.

[3] C′ has ἢν μέλλῃς εἰδέναι ἢ κύει ἢ οὐ, ὅταν μέλλει καθεύδειν.

[4] ἀδείπνωι ἐούσηι M: omitted by C′ Urb. V.

[5] δίδου C′ Urb. V: δοῦναι M: διδόναι Littré (who does not give the authority).

[6] καὶ ἢν μεν στρόφος ἔχει περὶ την κοιλίην C′: καὶ εἰ μὲν στρόφος ἔχει περὶ τὴν κοιλίην Urb.; καὶ ἢν μὲν στροφὰς ἔχῃ περὶ τὴν γαστέρα αὐτῆς V: κἢν μὲν στρόφος ἔχηι περὶ τὴν γαστέρα M.

[7] ἢν Urb. V: εἰ C′M.

[8] Urb. has εἰ μὲν followed by εἰ δὲ.

[9] So C′: ἢν γυναικὶ . . . γένηται Urb. M: γυναικὶ . . . ἢν ἐρυσίπελας . . . γένηται V.

[10] λεπταὶ ἐοῦσιν ἐν γαστρὶ ἔχουσαι C′: λεπται ἐοῦσαι Urb.: λεπταὶ ἐοῦσιν ἐν γαστρὶ ἔχουσιν M: λεπταὶ ἐοῦσαι κύουσιν V.

[11] After ἐκτιτρώσκουσι Urb. has οὐ κύουσι, and M δήμινα (*i.e.* δίμηνα).

XXXIX. If a woman have milk when she neither is with child nor has had a child, her menstruation is suppressed.

XL. When blood collects at the breasts of a woman, it indicates madness.[1]

XLI. If you wish to know whether a woman is with child, give her hydromel to drink [without supper][2] when she is going to sleep. If she has colic in the stomach she is with child, otherwise she is not.

XLII. If a woman be going to have a male child she is of a good complexion; if a female, of a bad complexion.

XLIII. If a pregnant woman be attacked by erysipelas in the womb, it is fatal.

XLIV. Women with child who are unnaturally thin miscarry until they have grown stouter.[3]

XLV. If moderately well-nourished women miscarry without any obvious cause two or three months

[1] Galen says he had never seen such a case, but Adams thinks that the aphorism may refer to rare cases of puerperal mania.

[2] These words are omitted by our best MSS. Littré keeps them, but points out that they are inconsistent with the commentary of Galen, who says that the woman must be well fed (*καὶ πεπληρῶσθαι σιτίων*). He suggests, therefore, that we should either read *οὐκ ἀδείπνῳ* in the text or *μὴ πεπληρῶσθαι* in Galen.

[3] The meaning of this aphorism seems plain enough, though Adams says it is not altogether confirmed by experience. The ancient commentators gave three explanations of the aphorism, and two of *παρὰ φύσιν*. Perhaps the meaning is: "Women, who in pregnancy are unnaturally thin, miscarry before they can recover a better condition."

φάσιος φανερῆς, ταύτῃσιν αἱ κοτυληδόνες[1] μύξης μεσταί εἰσι, καὶ οὐ δύνανται κρατεῖν ὑπὸ τοῦ
5 βάρεος τὸ ἔμβρυον, ἀλλ' ἀπορρήγνυνται.[2]

XLVI. Ὁκόσαι παρὰ φύσιν παχεῖαι ἐοῦσαι μὴ συλλαμβάνουσιν ἐν γαστρί, ταύτῃσι τὸ ἐπίπλοον τὸ στόμα τῶν ὑστερέων ἀποπιέζει,[3] καὶ
4 πρὶν ἢ λεπτυνθῆναι[4] οὐ κύουσιν.

XLVII. Ἢν ὑστέρη ἐν τῷ ἰσχίῳ ἐγκειμένη[5]
2 διαπυήσῃ, ἀνάγκη ἔμμοτον γενέσθαι.

XLVIII. Ἔμβρυα τὰ μὲν ἄρρενα ἐν τοῖσι δεξιοῖσι, τὰ δὲ θήλεα ἐν τοῖσιν ἀριστεροῖσι
3 μᾶλλον.[6]

XLIX. Ὑστέρων[7] ἐκπτώσιες, πταρμικὸν προσθεὶς ἐπιλάμβανε τοὺς μυκτῆρας καὶ τὸ
3 στόμα.[8]

L. Γυναικὶ[9] καταμήνια ἢν βούλῃ ἐπισχεῖν, σικύην ὡς μεγίστην πρὸς τοὺς τιτθοὺς[10]
3 πρόσβαλλε.[11]

LI. Ὁκόσαι ἐν γαστρὶ ἔχουσι, τούτων[12] τὸ
2 στόμα τῶν ὑστερέων συμμύει.[13]

LII. Ἢν γυναικὶ ἐν γαστρὶ ἐχούσῃ γάλα πολὺ ἐκ τῶν μαζῶν ῥυῇ, ἀσθενὲς[14] τὸ ἔμβρυον

[1] C′ has ταύτης αἱ κοτυληδόνες τῆς τῶν ὑστερῶν. After κοτυληδόνες three MSS. add τῆς μήτρας. Urb. omits ταύτῃσιν and adds αὐτῶν after κοτυληδόνες.

[2] ἀπορήγνυνται M.

[3] ὑποπιέζει C′.

[4] After λεπτυνθῆναι C′ Urb. add τοῦτο.

[5] After ἐγκειμένη some MSS. have ἢ (or ἣ) καί.

[6] C′ omits μᾶλλον, and begins the aphorism with ὁκόσα.

[7] ὑστερέων V.

[8] C′ places πταρμικὸν προσθεὶς after στόμα.

[9] C′ reads γυναικὸς and M has τὰ before καταμήνια.

after conception, the cotyledons of the womb are full of mucus, and break, being unable to retain the unborn child because of its weight.

XLVI. When unnaturally fat women cannot conceive, it is because the fat[1] presses the mouth of the womb, and conception is impossible until they grow thinner.

XLVII. If the part of the womb near the hip-joint suppurates, tents[2] must be employed.

XLVIII. The male embryo is usually on the right, the female on the left.

XLIX. To expel the after-birth: apply something to cause sneezing and compress the nostrils and the mouth.

L. If you wish to check menstruation, apply to[3] the breasts a cupping-glass of the largest size.

LI. When women are with child the mouth of the womb is closed.

LII. When milk flows copiously from the breasts of a woman with child, it shows that the unborn

[1] So the commentator Theophilus. *ἐπίπλοον* means literally the fold of the peritoneum.

[2] Plugs of lint to keep the suppurating place open until it is well on the way to heal from the bottom.

[3] Galen would prefer "under," as given by some MSS. in his day.

---

[10] C′ has *ἐν τῷ στήθῃ* for *πρὸς τοὺς τιτθοὺς*, and Galen says that in his time some MSS. read *ὑπὸ τοὺς τιτθοὺς*.

[11] C′ has *πρόσβαλε*.

[12] C′ has *τουτέοισι* for *τούτων*.

[13] *συμμύει* C′V : *ξυμμύει* Urb. : *συμμέμυκεν* M.

[14] *ἀσθενεῖν* V. The aphorism is omitted by C′.

σημαίνει· ἢν δὲ στερεοὶ οἱ μαστοὶ ἔωσιν, ὑγιει-
4 νότερον τὸ ἔμβρυον σημαίνει.

LIII. [1]Ὁκόσαι διαφθείρειν μέλλουσι τὰ ἔμβρυα,[2] ταύτῃσιν οἱ τιτθοὶ ἰσχνοὶ γίνονται· ἢν δὲ πάλιν σκληροὶ γένωνται, ὀδύνη ἔσται[3] ἢ ἐν τοῖσι τιτθοῖσιν, ἢ ἐν τοῖσιν ἰσχίοισιν, ἢ ἐν τοῖσιν ὀφθαλμοῖσιν, ἢ ἐν τοῖσι γούνασι, καὶ οὐ
6 διαφθείρουσιν.[4]

LIV. [5]Ὁκόσῃσι τὸ στόμα τῶν ὑστερέων σκληρόν ἐστι, ταύτῃσιν ἀνάγκη τὸ στόμα τῶν
3 ὑστερέων συμμύειν.

LV. Ὁκόσαι ἐν γαστρὶ ἔχουσαι ὑπὸ πυρετῶν λαμβάνονται, καὶ ἰσχυρῶς ἰσχναίνονται,[6] ἄνευ[7] προφάσιος φανερῆς,[8] τίκτουσι χαλεπῶς καὶ
4 ἐπικινδύνως, ἢ ἐκτιτρώκουσαι κινδυνεύουσιν.

LVI. Ἐπὶ[9] ῥόῳ γυναικείῳ σπασμὸς καὶ
2 λειποθυμίη ἢν ἐπιγένηται,[10] κακόν.

LVII. Καταμηνίων γενομένων πλειόνων,[11] νοῦσοι συμβαίνουσι, καὶ μὴ γενομένων ἀπὸ τῆς
3 ὑστέρης γίνονται νοῦσοι.

LVIII. Ἐπὶ ἀρχῷ φλεγμαίνοντι, καὶ ὑστέρῃ φλεγμαινούσῃ, στραγγουρίη ἐπιγίνεται, καὶ[12] ἐπὶ νεφροῖσιν ἐμπύοισι στραγγουρίη ἐπιγίνεται, ἐπὶ
4 δὲ ἥπατι φλεγμαίνοντι λὺγξ ἐπιγίνεται.

[1] This aphorism is omitted by C′.
[2] Urb. omits τὰ ἔμβρυα. [3] Urb. has γίνεται for ἔσται.
[4] MV have διαφθείρει.
[5] This aphorism is omitted by C′.
[6] For ἰσχναίνονται C′ and several other MSS. have θερμαίνονται.
[7] C′ has καὶ before ἄνευ, for which V reads ἄτερ.
[8] φανερῆς προφάσιος C′. [9] C′ adds ὁκόσαι before ἐπὶ.
[10] ἢν ἐπιγένηται omitted by M.

child is sickly; but if the breasts be hard, it shows that the child is more healthy.[1]

LIII. When women are threatened with miscarriage the breasts become thin. If they become hard again[2] there will be pain, either in the breasts or in the hip-joints, eyes, or knees, and there is no miscarriage.

LIV. When the mouth of the womb is hard it must of necessity be closed.

LV. When women with child catch a fever and become exceedingly thin,[3] without[4] (other) obvious cause, they suffer difficult and dangerous labour, or a dangerous miscarriage.

LVI. If convulsions and fainting supervene upon menstrual flow, it is a bad sign.

LVII. When menstruation is too copious, diseases ensue; when it is suppressed, diseases of the womb occur.

LVIII. On inflammation of the rectum and on that of the womb strangury supervenes; on suppuration of the kidneys strangury supervenes; on inflammation of the liver hiccough supervenes.

[1] Galen takes the sense to be that hard (and not milky) breasts indicate a healthy child. Littré, thinking that this interpretation neglects the comparative ὑγιεινότερον, understands the sense to be that while soft milky breasts indicate a sickly child, hard milky breasts indicate a more healthy one.

[2] Galen says that πάλιν can mean either (1) "again" or (2) "on the other hand." He prefers the second meaning.

[3] Or (with the reading of C′) "feverish."

[4] The phrase "without obvious cause" may also be taken with the preceding clause.

---

[11] πλειόνων γενομένων C′. Some MSS. have γινομένων or γιγνομένων.

[12] καὶ is omitted by C′, and Urb. omits from καὶ to ἐπιγίνεται.

LIX. Γυνὴ ἢν μὴ λαμβάνῃ ἐν γαστρί, βούλῃ δὲ εἰδέναι εἰ λήψεται, περικαλύψας ἱματίοισι, θυμία κάτω·[1] κἢν μὲν πορεύεσθαι[2] δοκῇ ἡ ὀδμὴ διὰ τοῦ σώματος ἐς τὸ στόμα καὶ ἐς τὰς ῥῖνας,[3]
5 γίνωσκε ὅτι αὐτὴ οὐ δι' ἑωυτὴν ἄγονός ἐστιν.[4]

LX. Γυναικὶ ἐν γαστρὶ ἐχούσῃ ἢν αἱ[5] καθάρ-
2 σιες πορεύωνται, ἀδύνατον τὸ ἔμβρυον ὑγιαίνειν.

LXI. Ἢν γυναικὶ[6] αἱ καθάρσιες μὴ[7] πορεύωνται, μήτε φρίκης, μήτε πυρετοῦ ἐπιγινομένου, ἆσαι δὲ[8] αὐτῇ προσπίπτωσι, λογίζου ταύτην ἐν
4 γαστρὶ ἔχειν.[9]

LXII. Ὁκόσαι πυκνὰς[10] καὶ ψυχρὰς τὰς μήτρας ἔχουσιν, οὐ κυΐσκουσιν· καὶ ὁκόσαι καθύγρους ἔχουσι τὰς μήτρας, οὐ κυΐσκουσιν,[11] ἀποσβέννυται γὰρ ὁ γόνος· καὶ ὁκόσαι ξηρὰς μᾶλλον καὶ περικαέας,[12] ἐνδείῃ γὰρ τῆς τροφῆς φθείρεται τὸ σπέρμα· ὁκόσαι δὲ ἐξ ἀμφοτέρων τὴν κρᾶσιν σύμμετρον[13] ἔχουσιν, αἱ τοιαῦται ἐπί-
8 τεκνοι γίνονται.

LXIII. Παραπλησίως δὲ καὶ ἐπὶ τῶν ἀρρένων·[14] ἢ γὰρ διὰ τὴν ἀραιότητα τοῦ σώματος[15]

[1] κάτωθεν V (Urb. has θ above the line).

[2] After πορεύεσθαι MV have σοι, Q and one other MS. οἱ. Urb. C′ omit.

[3] MV transpose τὸ στόμα and τὰς ῥῖνας.

[4] Urb. adds (after ἐστιν) ἀλλὰ διὰ τὸν ἄνδρα.

[5] MV omit αἱ and transpose ἢν to the beginning of the aphorism. Urb. has ἢν at the beginning and retains αἱ.

[6] After γυναικὶ Urb. adds ἐν γαστρὶ ἐχούσῃ.

[7] μὴ is omitted by C′ and three MSS. have παύωνται for μὴ πορεύωνται.

[8] After δὲ three MSS. add ἀλλόκοτοι καὶ ποικίλαι ὀρέξιες.

[9] For ἔχειν C′ has ἴσχειν.

[10] Urb. M. transpose πυκνὰς and ψυχράς.

LIX. If a woman does not conceive, and you wish to know if she will conceive, cover her round with wraps and burn perfumes underneath. If the smell seems to pass through the body to the mouth and nostrils, be assured that the woman is not barren through her own physical fault.

LX. If a woman with child have menstruation, it is impossible for the embryo to be healthy.

LXI. If menstruation be suppressed, and neither shivering nor fever supervenes, but attacks of nausea occur, you may assume the woman to be with child.

LXII. Women do not conceive who have the womb dense and cold; those who have the womb watery do not conceive, for the seed is drowned; those who have the womb over-dry and very hot do not conceive, for the seed perishes through lack of nourishment. But those whose temperament[1] is a just blend of the two[2] extremes prove able to conceive.

LXIII. Similarly with males. Either because of the rarity of the body the breath[3] is borne outwards

[1] Used in the old sense of the word. κρᾶσις really means "blending," "compounding."

[2] As Galen says, four (not two) dispositions have been mentioned; but these can be taken in pairs, and so we get the healthy mean with respect to (1) heat and (2) dryness.

[3] Moving air in the body was called πνεῦμα, which was not confined, as our word "breath" is, to air moving to and from the lungs. The writer of this aphorism was evidently a supporter of the Pneumatists, who tried to explain health and disease by the action of air.

---

[11] καὶ . . . κυΐσκουσιν omitted by C′.

[12] After περικαέας V has ἔχουσιν.

[13] συμμέτρως Urb.

[14] ἀνδρῶν V.

[15] τοῦ σώματος omitted by C′.

τὸ πνεῦμα ἔξω φέρεται πρὸς τὸ μὴ παραπέμπειν[1] τὸ σπέρμα· ἢ διὰ τὴν πυκνύτητα τὸ ὑγρὸν οὐ διαχωρεῖ ἔξω· ἢ διὰ τὴν ψυχρότητα οὐκ ἐκπυροῦται, ὥστε ἀθροίζεσθαι πρὸς τὸν τόπον τοῦτον·
7 ἢ διὰ τὴν θερμασίην τὸ αὐτὸ τοῦτο γίνεται.

LXIV. Γάλα διδόναι κεφαλαλγέουσι κακόν· κακὸν δὲ καὶ τοῖς πυρεταίνουσι, καὶ οἷσιν ὑποχόνδρια μετέωρα καὶ[2] διαβορβορύζοντα, καὶ τοῖσι διψώδεσι· κακὸν δὲ καὶ οἷσι χολώδεες αἱ ὑποχωρήσιες[3] ἐν τοῖσιν ὀξέσι πυρετοῖσιν,[4] καὶ οἷσιν αἵματος διαχώρησις πολλοῦ γέγονεν· ἁρμόζει δὲ φθινώδεσι μὴ λίην πολλῷ πυρέσσουσιν·[5] διδόναι δὲ καὶ ἐν πυρετοῖσι[6] μακροῖσι βληχροῖσι, μηδενὸς τῶν προειρημένων σημείων παρεόντος,[7]
10 παρὰ λόγον[8] δὲ ἐκτετηκότων.

LXV. Ὁκόσοισιν οἰδήματα ἐφ' ἕλκεσι φαίνεται, οὐ μάλα σπῶνται, οὐδὲ μαίνονται· τούτων δὲ ἀφανισθέντων[9] ἐξαίφνης, τοῖσι μὲν ὄπισθεν σπασμοί, τέτανοι, τοῖσι δὲ ἔμπροσθεν[10] μανίαι, ὀδύναι πλευροῦ ὀξεῖαι,[11] ἢ ἐμπύησις, ἢ δυσεν-
6 τερίη, ἢν ἐρυθρὰ μᾶλλον ᾖ τὰ οἰδήματα.

[1] Before τὸ Urb. adds τῷ στόματι reading also παρεμπίπτειν: other MSS. ἐς τὸ στόμα.

[2] Only three (inferior) MSS. have καί. Littré inserts it following the commentary of Galen, which implies it.

[3] After ὑποχωρήσιες Galen thought that a καί should be added for the sake of the sense. One of our MSS. (probably through the influence of Galen) reads καί.

[4] ἐν τοῖσιν ὀξέσι πυρετοῖσιν C′ Urb.: ἐν ὀξέσι πυρετοῖσ(ι) ἐοῦσι MV.

[5] ἢν μὴ λίην πολλῷ πυρέσσωσιν C′. Urb. and some other MSS. omit πολλῷ, which word, as Galen says, seems otiose.

[6] For μακροῖσι C′ has ἤ.

[7] παρεόντων C′V.

[8] παραλόγως Urb.

[9] ἀφανιζομένων C′.

so as not to force along the seed; or because of the density of the body the liquid[1] does not pass out; or through the coldness it is not heated so as to collect at this place;[2] or through the heat this same thing happens.[3]

LXIV. To give milk to sufferers from headache is bad; it is also bad for fever patients, and for those whose hypochondria are swollen and full of rumbling, and for those who are thirsty. Milk is also bad for those whose stools in acute fevers are bilious, and for those who pass much blood. It is beneficial in cases of consumption when there is no very high fever. Give it also in protracted, low fevers, when none of the aforesaid symptoms is present, but when there is excessive emaciation.

LXV. When swellings appear on wounds, there are seldom convulsions or delirium; but when the swellings suddenly disappear, wounds behind are followed by convulsions and tetanus, wounds in front by delirium, severe pains in the side, or suppuration, or dysentery, if the swellings are inclined to be red.[4]

[1] *τὸ ὑγρὸν* here means *τὸ σπέρμα*.

[2] Galen notes that the writer leaves the "place" to be understood by the reader, but *τοῦτον* seems to refer to something already mentioned.

[3] Galen objects to the last clause as inconsistent with the one preceding, and to the whole aphorism as an interpellation.

[4] There are many difficulties of meaning in this aphorism, the chief being that wounds in front do not differ from wounds behind in their probable or possible after-effects, at any rate not to the extent mentioned in the text. See Littré's note.

---

[10] *εἰς τοὔμπροσθεν* C′. Rein. has *ὅτοισι* for *τοῖσι* (twice).

[11] *μανίη ἢ ὀδύνη πλευρέων ὀξείη* Urb.: *μανίη καὶ ὀδύνη πλευρου ὀξία* C′.

LXVI. Ἢν τραυμάτων πονηρῶν ἐόντων[1] 2 οἰδήματα[2] μὴ φαίνηται, μέγα κακόν.

LXVII. Τὰ χαῦνα, χρηστά, τὰ ἔνωμα,[3] 2 κακά.

LXVIII. Τὰ ὄπισθεν τῆς κεφαλῆς ὀδυνωμένα[4] 2 ἡ ἐν μετώπῳ ὀρθίη φλὲψ τμηθεῖσα ὠφελεῖ·

LXIX. Ῥίγεα ἄρχεται, γυναιξὶ μὲν ἐξ ὀσφύος μᾶλλον καὶ διὰ νώτου ἐς τὴν κεφαλήν· ἀτὰρ καὶ ἀνδράσι ὄπισθεν μᾶλλον ἢ ἔμπροσθεν τοῦ σώματος, οἷον πήχεων, μηρῶν· ἀτὰρ καὶ τὸ 5 δέρμα ἀραιόν, δηλοῖ δὲ ἡ θρίξ.[5]

LXX. Οἱ ὑπὸ τεταρταίων ἁλισκόμενοι[6] ὑπὸ σπασμοῦ οὐ πάνυ τι[7] ἁλίσκονται· ἢν δὲ ἁλίσκωνται πρότερον, καὶ ἐπιγένηται τεταρ-4 ταῖος, παύονται.

LXXI. Ὁκόσοισι δέρματα περιτείνεται σκληρὰ[8] καὶ καρφαλέα, ἄνευ ἱδρῶτος τελευτῶσιν· ὁκόσοισι δὲ χαλαρὰ καὶ ἀραιά, σὺν 4 ἱδρῶτι τελευτῶσιν.[9]

LXXII. Οἱ ἰκτεριώδεες οὐ πάνυ τι πνευμα-2 τώδεές εἰσιν.

[1] ἰσχυρῶν καὶ πονηρῶν ἐόντων M.

[2] οἴδημα M.

[3] ἔννομα C′ : δὲ ἔννομα Urb.

[4] ὀδυνωμένωι M.

[5] ἔχουσι, δηλοῖ δὲ τοῦτο ἡ θρίξ Littré without stating his authority. C′ omits all from οἷον to θρίξ, and there are many slight variants in all parts of the aphorism.

[6] ἐχόμενοι Urb.

[7] οὐ πάνυ τι ὑπὸ σπασμῶν C′ : οὐ πάνυ τι ὑπὸ σπασμοῦ Urb. : ὑπὸ σπασμὸν οὐ πάνυ τι M : ὑπὸ σπασμῶν οὐ πάνυ τι V.

LXVI. If swellings do not appear on severe wounds it is a very bad thing.

LXVII. Softness[1] is good, hardness[2] is bad.

LXVIII. Pains at the back of the head are relieved by opening the upright vein in the forehead.

LXIX. Rigors in women tend to begin in the loins and pass through the back to the head. In men too they begin more often in the back of the body than in the front; for example, in the forearms or thighs. The skin too is rare, as is shown by the hair.[3]

LXX. Those who are attacked by quartans are not very liable to be attacked by convulsions. But if they are first attacked by convulsions and a quartan supervenes, the convulsions cease.

LXXI. Those whose skin is stretched, hard and parched, die[4] without sweat. Those whose skin is loose and rare die[4] with sweat.

LXXII. Those subject to jaundice are not very subject to flatulence.

[1] That is, in swellings, etc.

[2] Or "crudity."

[3] Littré thinks that the last sentence is a separate aphorism, contrasting the bodies of women and of men. Commentators mostly think that there is a reference to the fact that the front parts are more hairy than the back; this shows the less rarity of the latter, *i.e.* their greater coldness and liability to rigors.

[4] Perhaps τελευτῶσιν refers to the termination of any disease, not of fatal diseases only. So Theophilus.

---

[8] For σκληρὰ Urb. has ξηρὰ.

[9] ὁκόσοισι δὲ . . . τελευτῶσιν omitted by C′.

ΤΜΗΜΑ ΕΚΤΟΝ

I. Ἐν τῇσι χρονίῃσι λειεντερίῃσιν ὀξυρεγμίη ἐπιγενομένη, μὴ γενομένη[1] πρότερον, σημεῖον
3 ἀγαθόν.

II. Οἷσι ῥῖνες ὑγραὶ φύσει, καὶ ἡ γονὴ ὑγρή,[2] ὑγιαίνουσι νοσηρότερον·[3] οἷσι δὲ τἀν-
3 αντία, ὑγιεινότερον.[4]

III. Ἐν τῇσι μακρῇσι δυσεντερίῃσιν αἱ
2 ἀποσιτίαι, κακόν· καὶ σὺν πυρετῷ, κάκιον.

IV. Τὰ περιμάδαρα ἕλκεα, κακοήθεα.

V. Τῶν ὀδυνέων, καὶ ἐν πλευρῇσι, καὶ ἐν στήθεσι, καὶ ἐν τοῖσιν ἄλλοισι μέρεσιν,[5] εἰ μέγα
3 διαφέρουσι, καταμαθητέον.

VI. Τὰ νεφριτικά, καὶ τὰ κατὰ τὴν κύστιν,[6]
2 ἐργωδῶς[7] ὑγιάζεται τοῖσι πρεσβυτέροισι.

VII. Ἀλγήματα καὶ οἰδήματα[8] κατὰ τὴν κοιλίην γινόμενα, τὰ μὲν μετέωρα κουφότερα,[9]
3 τὰ δὲ μὴ μετέωρα, ἰσχυρότερα.

VIII. Τοῖσιν ὑδρωπικοῖσι τὰ γινόμενα[10] ἕλκεα
2 ἐν τῷ σώματι, οὐ ῥηιδίως ὑγιάζεται.

[1] πρότερον μὴ ἐοῦσα, μὴ γινομένη πρότερον, πρόσθεν μὴ γιγνομένη are other readings.

[2] V has ὑγρὴ (with ὑγρότεραι) and M has ὑγροτέρη with ὑγ ότεραι.

[3] νοσηλότερον Urb., perhaps rightly, as C′ has νοσιλώτερον.

[4] ὑγιεινότεροι M.

[5] For μέρεσιν Rein. has τὰς διαφοράς.

[6] After κύστιν Urb. and many other MSS. add ἀλγήματα.

[7] For ἐργωδῶς C′ Urb. read δυσχερῶς.

[8] οἰδήματα is strongly supported by the MSS. (including C′ and Urb.), and is mentioned by Theophilus. It is not mentioned by Galen, and Littré omits it from his text. τὰ μετὰ ἀλγημάτων ὀδυνήματα, Rein. Perhaps a case of hendiadys.

SIXTH SECTION

I. In cases of chronic lientery, acid eructations supervening which did not occur before are a good sign.

II. Those whose nostrils are naturally watery, and whose seed is watery, are below the average when in health; those of an opposite character are above the average when in health.[1]

III. In cases of prolonged dysentery, loathing for food is bad; if fever be present, it is worse.

IV. Sores, when the hair about them falls off, are malignant.

V. One should observe about pains, in the sides, in the breast and in the other parts, whether they show great differences.[2]

VI. Kidney troubles, and affections of the bladder, are cured with difficulty when the patient is aged.

VII. Pains and swellings[3] of the belly are less serious when superficial, more severe when deep-seated.

VIII. Sores on the body of dropsical persons are not easily healed.

[1] With the reading of M, "are (generally) more healthy."

[2] Littré, relying on *Epidemics*, II, § 7 (end), where this aphorism occurs in an expanded form, would understand τὰς ὥρας after μέρεσιν, making the genitive τῶν ὀδυνέων depend upon it, and would make "the patients" (understood) the subject of διαφέρουσι. He reads ἢν . . . διαφέρωσι.

[3] This word is doubtful, as it does not suit very well the predicates κουφότερα and ἰσχυρότερα.

---

[9] For κουφότερα MV have κοῦφα.

[10] For γινόμενα V has ἐπιγινόμενα.

IX. Τὰ πλατέα ἐξανθήματα, οὐ πάνυ τι
2 κνησμώδεα.

X. Κεφαλὴν πονέοντι καὶ περιωδυνέοντι, πῦον,
ἢ ὕδωρ, ἢ αἷμα[1] ῥυὲν κατὰ τὰς ῥῖνας, ἢ κατὰ
3 τὰ ὦτα,[2] ἢ κατὰ τὸ στόμα, λύει τὸ νόσημα.

XI. Τοῖσι μελαγχολικοῖσι καὶ τοῖσι νεφρι-
2 τικοῖσιν αἱμορροΐδες ἐπιγινόμεναι, ἀγαθόν.

XII. Τῷ ἰηθέντι χρονίας αἱμορροΐδας, ἢν μὴ
μία φυλαχθῇ, κίνδυνος ὕδρωπα ἐπιγενέσθαι ἢ
3 φθίσιν.

XIII. Ὑπὸ λυγμοῦ ἐχομένῳ πταρμός ἐπι-
2 γενόμενος λύει τὸν λυγμόν.

XIV. Ὑπὸ ὕδρωπος ἐχομένῳ,[3] κατὰ τὰς φλέβας
2 ἐς τὴν κοιλίην ὕδατος ῥυέντος, λύσις.

XV. Ὑπὸ διαρροίης ἐχομένῳ μακρῆς ἀπὸ
ταὐτομάτου ἔμετος ἐπιγενόμενος λύει τὴν διάρ-
3 ροιαν.

XVI. Ὑπὸ πλευρίτιδος, ἢ περιπλευμονίης
2 ἐχομένῳ[4] διάρροια ἐπιγενομένη, κακόν.

XVII. Ὀφθαλμιῶντι[5] ὑπὸ διαρροίης ληφ-
2 θῆναι ἀγαθόν.

XVIII. Κυστιν διακοπέντι, ἢ ἐγκέφαλον, ἢ
καρδίην, ἢ φρένας, ἢ τῶν ἐντέρων τι[6] τῶν
3 λεπτῶν, ἢ κοιλίην, ἢ ἧπαρ, θανατῶδες.

XIX. Ἐπὴν διακοπῇ ὀστέον, ἢ χονδρός, ἢ

[1] ἢ αἷμα omitted by MV.

[2] τὰ ὦτα ἢ τὸ στόμα C′V : κατὰ τὰ ὦτα· ἢ κατὰ τὸ στόμα Urb : κατὰ τὸ στόμα· ἢ κατὰ τὰ ὦτα M.

[3] Rein. adds αὐτομάτου before κατὰ.

[4] ἐχομένῳ in V appears before ἢ.

[5] ὀφθαλμιῶντα Urb. and several other MSS. One has ὀφθαλμιῶντας.

IX. Broad exanthemata[1] are not very irritating.

X. When the head aches and the pain is very severe, a flow of pus, water or blood, by the nostrils, ears or mouth, cures the trouble.

XI. Hemorrhoids supervening on melancholic or kidney affections are a good sign.[2]

XII. When a patient has been cured of chronic hemorrhoids, unless one be kept,[3] there is a danger lest dropsy or consumption supervene.

XIII. In the case of a person afflicted with hiccough, sneezing coming on removes the hiccough.

XIV. In the case of a patient suffering from dropsy, a flow of water by the veins into the belly removes the dropsy.

XV. In the case of a patient suffering from prolonged diarrhoea, involuntary vomiting supervening removes the diarrhoea.

XVI. In the case of a patient suffering from pleurisy or pneumonia, diarrhoea supervening is a bad sign.

XVII. It is a good thing when an ophthalmic[4] patient is attacked by diarrhoea.

XVIII. A severe wound of the bladder, brain, heart, midriff, one of the smaller intestines, belly or liver, is deadly.

XIX. When a bone, cartilage, sinew, the slender

[1] It is not known what exanthemata are meant; probably the pustules of scabies.

[2] Hemorrhoids were supposed to be one of Nature's ways of removing impurities.

[3] That is "left." Some MSS. have καταλειφθῇ.

[4] Ancient "ophthalmia" included many eye diseases besides the one now known by this name.

---

[6] τι is placed by C′ before τῶν ἑτέρων and Urb. omits τῶν.

νεῦρον, ἢ γνάθου τὸ λεπτόν, ἢ ἀκροποσθίη, οὔτε
3 αὔξεται, οὔτε συμφύεται.

XX. Ἢν ἐς τὴν[1] κοιλίην αἷμα ἐκχυθῇ παρὰ
2 φύσιν, ἀνάγκη ἐκπυηθῆναι.[2]

XXI. Τοῖσι μαινομένοισι κιρσῶν ἢ αἱμορ-
2 ροΐδων ἐπιγινομένων, μανίης[3] λύσις.

XXII. Ὁκόσα ῥήγματα[4] ἐκ τοῦ νώτου ἐς
2 τοὺς ἀγκῶνας καταβαίνει, φλεβοτομίη λύει.

XXIII. Ἢν φόβος ἢ δυσθυμίη πολὺν χρόνον
2 διατελῇ, μελαγχολικὸν τὸ τοιοῦτον.

XXIV. Ἐντέρων ἢν διακοπῇ τῶν λεπτῶν τι,
2 οὐ συμφύεται.

XXV. Ἐρυσίπελας ἔξωθεν καταχεόμενον[5]
ἔσω τρέπεσθαι οὐκ ἀγαθόν· ἔσωθεν δὲ ἔξω,
3 ἀγαθόν.

XXVI. Ὁκόσοισιν ἂν ἐν τοῖσι καύσοισι
2 τρόμοι γένωνται,[6] παρακοπὴ λύει.

XXVII. Ὁκόσοι ἔμπυοι ἢ ὑδρωπικοὶ τέμνονται
ἢ καίονται,[7] ἐκρυέντος τοῦ πύου ἢ τοῦ ὕδατος
3 ἀθρόου, πάντως[8] ἀπόλλυνται.

[1] τὴν omitted by C′.

[2] ἐκποιηθῆναι ἀνάγκη ἢ διασαπῆναι C′. The alternative is an attempt to express the criticism of Galen, who says that ἐκπυηθῆναι here means, according to several interpreters, not transformation into pus, but "corruption."

[3] Before μανίης Urb. V add τῆς.

[4] For ῥήγματα three MSS. have ἀλγήματα, a reading noticed by Galen. ἀλγήματα ἢ (καὶ) ῥήγματα C′ Urb.

[5] καταχυθὲν V.

[6] γίνονται C′ and V (which omits ἄν).

[7] καίονται ἢ τέμνονται Urb. Before ἐκρυέντος C′ Urb. add τουτέοισι.

[8] πάντως ἀθρόως C′. For πάντως V has μείναντος and καὶ for ἢ.

---

[1] This aphorism has been a puzzle to all commentators from Galen to Littré, as it is difficult to reconcile it with

part of the jaw, or the foreskin is severed, the part neither grows nor unites.[1]

XX. If there be an unnatural flow of blood into the belly,[2] it must suppurate.

XXI. Varicose veins or hemorrhoids supervening on madness[3] remove it.

XXII. Ruptures[4] that descend from the back to the elbows are removed by bleeding.

XXIII. Fear or depression that is prolonged means melancholia.

XXIV. If one of the smaller intestines be severed it does not unite.

XXV. When erysipelas that spreads externally turns inwards it is not a good thing; but it is good when internal erysipelas turns outwards.

XXVI. Whenever tremors occur in ardent fevers, delirium removes these tremors.[5]

XXVII. Whenever cases of empyema or dropsy are treated by the knife or cautery, if the pus or water flow away all at once, a fatal result is certain.

experience. Perhaps all that is meant is that a severe cut (διακοπῇ) is never completely restored, *e.g.* callus is not exactly bone.

[2] If the article τὴν be omitted, "into a cavity."

[3] μανίῃ includes every state when a person is "out of his mind." It is uncertain to which of these many states reference here is made.

[4] Galen notices that some authorities read ἀλγήματα, "pains," a much more appropriate word in the context. Littré thinks that "referred" pains to the elbows are meant: "les brisements dans le dos font sentir dans les coudes." The reading of C′ combines both readings.

[5] Galen thinks that this aphorism is an interpellation, but takes the meaning to be that delirium replaces the fever. It seems more natural to interpret it to mean that delirium replaces the tremors.

XXVIII. Εὐνοῦχοι οὐ ποδαγριῶσιν, οὐδὲ
2 φαλακροὶ γίνονται.

XXIX. Γυνὴ οὐ ποδαγριᾷ, εἰ μὴ τὰ κατα-
2 μήνια ἐκλέλοιπεν αὐτῇ.[1]

XXX. Παῖς οὐ ποδαγριᾷ πρὸ τοῦ ἀφροδι-
2 σιασμοῦ.[2]

XXXI. Ὀδύνας ὀφθαλμῶν ἀκρητοποσίη, ἢ
λουτρόν, ἢ πυρίη, ἢ φλεβοτομίη, ἢ φαρμακοποσίη[3]
3 λύει.

XXXII. Τραυλοὶ ὑπὸ διαρροίης μάλιστα
2 ἁλίσκονται μακρῆς.

XXXIII. Οἱ ὀξυρεγμιώδεες οὐ πάνυ τι
2 πλευριτικοὶ γίνονται.

XXXIV. Ὁκόσοι φαλακροί, τούτοισι κιρσοὶ
μεγάλοι οὐ γίνονται·[4] ὁκόσοις ἂν φαλακροῖσι
3 κιρσοὶ γένωνται πάλιν γίνονται δασέες.[5]

XXXV. Τοῖσιν ὑδρωπικοῖσι βὴξ ἐπιγενομένη,
2 κακόν· τὸ δὲ προγεγονέναι ἀγαθόν.[6]

XXXVI. Δυσουρίην φλεβοτομίη λύει, τάμνειν
2 δὲ τὴν εἴσω φλέβα.[7]

XXXVII. Ὑπὸ κυνάγχης ἐχομένῳ οἴδημα[8]
2 γενέσθαι ἐν τῷ βρόγχῳ[9] ἔξω, ἀγαθόν.[10]

[1] The MSS. offer many readings (ἦν, εἰ, ἐκλίπῃ, ἐπιλέλοιπεν, λέλοιπεν, ἐκλέλοιπεν, ἐκλείπῃ, αὐτῆς, αὐτῇ, αὐτήν), all with approximately the same sense.

[2] ἀφροδισιάζειν Urb. V.

[3] ἢ φαρμακοποσίη omitted by C′: φαρμακείη V.

[4] The reading οὐ γίνονται has poor MS. support but is the one known to Galen. Otherwise one would adopt οὐκ ἐγγίνονται.

[5] In the second part of this aphorism I have adopted the text of V, which seems to be the simple original, altered by various hands to the fuller text found in our other MSS. The variants include δὲ or δ' before ἂν, the omission of ἄν,

XXVIII. Eunuchs neither get gout nor grow bald.

XXIX. A woman does not get gout unless menstruation is suppressed

XXX. A youth does not get gout before sexual intercourse.

XXXI. Pains of the eyes are removed by drinking neat wine, by bathing, by vapour baths, by bleeding or by purging.

XXXII. Those with an impediment in their speech are very likely to be attacked by protracted diarrhoea.

XXXIII. Those suffering from acid eructations are not very likely to be attacked by pleurisy.

XXXIV. Bald people are not subject to large varicose veins; bald people who get varicose veins grow hair again.

XXXV. A cough supervening on dropsy is a bad sign; but if it precede it is a good sign.

XXXVI. Bleeding removes difficulty of micturition; open the internal vein.[1]

XXXVII. In a case of angina it is a good thing when a swelling appears on the outside of the trachea.

[1] Galen suspects that this aphorism is an interpellation. He says that to make good sense καὶ must be understood before φλεβοτομίη: "bleeding, among other things."

ἐουσι(ν) after φαλακροῖσι, μεγάλοι as epithet of κιρσοί, ἐπιγένωνται, οὗτοι before or after πάλιν.

[6] Several MSS., and Littré, omit τὸ δὲ . . . ἀγαθόν.

[7] τὰς ἔσω Littré and Rein., and V adds φλέβας.

[8] οἴδημα C′: οἰδήματα Urb. MV.

[9] τραχήλῳ Urb.

[10] C′ adds ἔξω γὰρ τρέπεται τὸ νόσημα.

XXXVIII. Ὁκόσοισι κρυπτοὶ καρκίνοι γίνονται,[1] μὴ θεραπεύειν βέλτιον· θεραπευόμενοι γὰρ ἀπόλλυνται ταχέως, μὴ θεραπευόμενοι δὲ πολὺν
4 χρόνον διατελέουσιν.

XXXIX. Σπασμοὶ γίνονται[2] ἢ ὑπὸ πληρώ-
2 σιος ἢ κενώσιος·[3] οὕτω δὲ καὶ λυγμός.

XL. Ὁκόσοισι περὶ τὸ ὑποχόνδριον πόνος γίνεται[4] ἄτερ φλεγμονῆς, τούτοισι πυρετὸς
3 ἐπιγενόμενος λύει τὸν πόνον.[5]

XLI. Ὁκόσοισι διάπυόν τι ἐν τῷ σώματι ἐὸν μὴ ἀποσημαίνει, τούτοισι διὰ παχύτητα τοῦ
3 πύου ἢ[6] τοῦ τόπου οὐκ ἀποσημαίνει.

XLII. Ἐν τοῖσιν ἰκτερικοῖσι τὸ ἧπαρ σκληρὸν
2 γενέσθαι,[7] πονηρόν.

XLIII. Ὁκόσοι σπληνώδεες ὑπὸ δυσεντερίης ἁλίσκονται, τούτοισιν, ἐπιγενομένης μακρῆς τῆς δυσεντερίης, ὕδρωψ ἐπιγίνεται ἢ λειεντερίη, καὶ
4 ἀπόλλυνται.

XLIV. Ὁκόσοισιν ἐκ στραγγουρίης εἰλεοὶ γίνονται, ἐν ἑπτὰ ἡμέρῃσιν ἀπόλλυνται,[8] ἢν μὴ
3 πυρετοῦ ἐπιγενομένου ἅλις[9] τὸ οὖρον ῥυῇ.

XLV. Ἕλκεα ὁκόσα ἐνιαύσια γίνεται, ἢ μακρότερον χρόνον ἴσχουσιν,[10] ἀνάγκη ὀστέον ἀφί-
3 στασθαι, καὶ τὰς οὐλὰς κοίλας γίνεσθαι.

XLVI. Ὁκόσοι ὑβοὶ ἐξ ἄσθματος ἢ βηχὸς
2 γίνονται πρὸ τῆς ἥβης, ἀπόλλυνται.

[1] C′ has κρυπτοὶ καρκίνοι γίνονται twice.
[2] σπασμὸς γίνεται C′ Urb. V.
[3] C′ Urb. transpose πληρώσιος and κενώσιος.
[4] Some MSS. have πόνοι γίνονται.
[5] For τὸν πόνον some MSS. have τὸ νόσημα.
[6] Littré omits τοῦ πύου ἢ on the ground that the commentary of Galen implies two readings, one with τοῦ πύου

XXXVIII. It is better to give no treatment in cases of hidden cancer; treatment causes speedy death, but to omit treatment is to prolong life.

XXXIX. Convulsions occur either from repletion or from depletion. So too with hiccough.

XL. When pain in the region of the hypochondrium occurs without inflammation, the pain is removed if fever supervenes.

XLI. When suppurating matter exists in the body without showing itself, this is due to the thickness either of the pus or of the part.

XLII. In jaundice, sclerosis of the liver is bad.

XLIII. When persons with enlarged spleens are attacked by dysentery, if the dysentery that supervenes be prolonged, dropsy or lientery supervenes with fatal results.

XLIV. Those who, after strangury, are attacked by ileus, die in seven days, unless fever supervenes and there is an abundant flow of urine.

XLV. If sores last for a year or longer, it must be that the bone come away and the scars become hollow.

XLVI. Such as become hump-backed before puberty from asthma or cough, do not recover.

---

and the other with τοῦ τόπου. All our MSS. give both phrases, a fact which Littré would explain as an attempt on the part of a scribe to include both of Galen's readings. Reinhold omits ἢ τοῦ τόπου.

[7] For τὸ . . . γενέσθαι Urb. (with many other MSS.) has ἢν τὸ ἧπαρ σκληρὸν γένηται.

[8] οἱ τοιοῦτοι is added after ἀπόλλυνται by V, before ἀπόλλυνται by C′ Urb.

[9] ἁλὲς Rein.

[10] ἴσχει Littré with several MSS. : ἴσχωσιν V.

XLVII. Ὁκόσοισι φλεβοτομίη ἢ φαρμακείη συμφέρει, τούτους τοῦ ἦρος φαρμακεύειν ἢ φλεβο-
3 τομεῖν.[1]

XLVIII. Τοῖσι σπληνώδεσι δυσεντερίη ἐπιγε-
2 νομένη, ἀγαθόν.

XLIX. Ὁκόσα ποδαγρικὰ νοσήματα γίνεται, ταῦτα ἀποφλεγμήναντα ἐν τεσσαράκοντα[2] ἡμέ-
3 ρῃσιν ἀποκαθίσταται.[3]

L. Ὁκόσοισιν ἂν ὁ ἐγκέφαλος διακοπῇ, τούτοισιν ἀνάγκη πυρετὸν καὶ χολῆς ἔμετον
3 ἐπιγίνεσθαι.

LI. Ὁκόσοισιν ὑγιαίνουσιν ἐξαίφνης ὀδύναι γίνονται ἐν τῇ κεφαλῇ, καὶ παραχρῆμα ἄφωνοι γίνονται,[4] καὶ ῥέγκουσιν, ἀπόλλυνται ἐν ἑπτὰ
4 ἡμέρῃσιν,[5] ἢν μὴ πυρετὸς ἐπιλάβῃ.

LII. Σκοπεῖν δὲ χρὴ καὶ τὰς ὑποφάσιας τῶν ὀφθαλμῶν ἐν τοῖσιν ὕπνοισιν· ἢν γάρ τι ὑποφαίνηται τοῦ λευκοῦ,[6] συμβαλλομένων τῶν βλεφάρων, μὴ ἐκ διαρροίης ἐόντι ἢ φαρμακοποσίης, φλαῦρον τὸ σημεῖον καὶ θανατῶδες
6 σφόδρα.

LIII. Αἱ παραφροσύναι αἱ μὲν μετὰ γέλωτος γινόμεναι ἀσφαλέστεραι·[7] αἱ δὲ μετὰ σπουδῆς[8]
3 ἐπισφαλέστεραι.

LIV. Ἐν τοῖσιν ὀξέσι πάθεσι τοῖσι μετὰ
2 πυρετοῦ αἱ κλαυθμώδεες ἀναπνοαὶ κακόν.[9]

[1] After φλεβοτομεῖν C′ has χρή.

[2] After τεσσαράκοντα V has ὀκτώ.

[3] ἀποκάθισται M Dietz, Littré: καθίσταται C′: καθίστανται several MSS.: ἀποκαθίστανται Urb. V.

[4] For γίνονται Littré (without giving authority) has the attractive reading κεῖνται.

XLVII. Such as are benefited by bleeding or purging shall be purged or bled in spring.

XLVIII. In cases of enlarged spleen, dysentery supervening is a good thing.[1]

XLIX. In gouty affections inflammation subsides within forty days.

L. Severe wounds of the brain are necessarily followed by fever and vomiting of bile.

LI. Those who when in health are suddenly seized with pains in the head, becoming[2] forthwith dumb and breathing stertorously, die within seven days unless fever comes on.

LII. One should also consider what is seen of the eyes in sleep; for if, when the lids are closed, a part of the white is visible, it is, should diarrhoea or purging not be responsible, a bad, in fact an absolutely fatal, sign.[3]

LIII. Delirium with laughter is less dangerous, combined with seriousness it is more so.

LIV. In acute affections attended with fever, moaning respiration is a bad sign.

[1] Cf. the forty-third aphorism of this section, where it is said that in such cases a protracted dysentery is followed by fatal results.

[2] The reading κεῖνται would mean "lie prostrate." This word is very appropriate in its context, as apoplectic seizures are referred to.

[3] In Urb. this aphorism is joined to the preceding. It is taken from *Prognostic*.

---

[5] ἀπόλλυνται after ἡμέρῃσιν C′.

[6] Some MSS. place τοῦ λευκοῦ after βλεφάρων.

[7] ἀσφαλέσταται and ἐπισφαλέσταται Urb.

[8] Some MSS. read κλαυθμοῦ (or κλαθμοῦ) for σπουδῆς.

[9] κακαὶ M.

LV. Τὰ ποδαγρικὰ[1] τοῦ ἦρος καὶ τοῦ φθινο-
2 πώρου κινεῖται.

LVI. Τοῖσι μελαγχολικοῖσι νοσήμασιν ἐς τάδε[2] ἐπικίνδυνοι αἱ ἀποσκήψιες· ἀπόπληξιν τοῦ σώματος, ἢ σπασμόν,[3] ἢ μανίην, ἢ τύφλωσιν
4 σημαίνει.[4]

LVII. Ἀπόπληκτοι[5] δὲ μάλιστα γίνονται οἱ[6]
2 ἀπὸ τεσσαράκοντα ἐτέων μέχρις ἑξήκοντα.

LVIII. Ἢν ἐπίπλοον ἐκπέσῃ, ἀνάγκη ἀπο-
2 σαπῆναι.[7]

LIX. Ὁκόσοισιν ὑπὸ ἰσχιάδος ἐνοχλουμένοισιν[8] ἐξίσταται τὸ ἰσχίον, καὶ πάλιν ἐμπίπτει, τούτοισι
3 μύξαι ἐπιγίνονται.[9]

LX. Ὁκόσοισιν ὑπὸ ἰσχιάδος ἐχομένοισι[10] χρονίης τὸ ἰσχίον ἐξίσταται, τούτοισι τήκεται[11]
3 τὸ σκέλος, καὶ χωλοῦνται, ἢν μὴ καυθέωσιν.

ΤΜΗΜΑ ΕΒΔΟΜΟΝ

I. Ἐν τοῖσιν ὀξέσι νοσήμασι ψύξις ἀκρωτηρίων,
2 κακόν.

[1] C′ adds here μᾶλλον and some MSS. ὡς ἐπὶ τὸ πολύ (πουλύ). A few MSS. add καὶ τὰ μανικὰ before τοῦ ἦρος.

[2] ἐς τὰν δὲ some good MSS.

[3] τοῦ σώματος after σπασμόν Urb.

[4] σημαίνουσιν Littré. One MS. has σημείωσιν. One MS. at least omits.

[5] ἀπόπληκτηκοὶ C′.

[6] οἱ C′: οἱ τῇ ἡλικίῃ V: ἡλικίῃ τῇ Littré (ἡλικίη τῆι M): τῇσιν ἡλικίῃσιν or τοῖσιν ἡλικίοισιν many MSS.

[7] After ἀποσαπῆναι Urb. adds καὶ ἀποπεσεῖν.

[8] ὀχλουμένοις V: χρονίης is found after ἰσχιάδος in some old editions.

LV. Gouty affections become active in spring and in autumn.

LVI. In melancholic affections the melancholy humour is likely to be determined in the following ways: apoplexy of the whole body, convulsions, madness[1] or blindness.

LVII. Apoplexy occurs chiefly between the ages of forty and sixty.

LVIII. If the epiploön protrude, it cannot fail to mortify.[2]

LIX. In cases of hip-joint disease, when the hip-joint protrudes and then slips in again, mucus forms.

LX. In cases of chronic disease of the hip-joint, when the hip-joint protrudes, the leg wastes and the patient becomes lame, unless the part be cauterised.

SEVENTH SECTION.

I. In acute diseases chill of the extremities is a bad sign.

[1] See note on p. 185. The word σημαίνει (if the reading be correct) will be almost impersonal "it means."

[2] Galen and all commentators refer this aphorism to abdominal wounds through which the epiploön protrudes. The words added in Urb. mean "and drop off." The epiploön is the membrane enclosing the intestines.

---

[9] μύξα γίγνεται V. C′ omits this aphorism. M has ἐγγίνονται for ἐπιγίνονται.

[10] V omits: some MSS. (and Littré) ἐνοχλουμένοισι.

[11] φθίνει V and many other MSS.

II. Ἐπὶ ὀστέῳ νοσήσαντι σὰρξ πελιδνή, 2 κακόν.

III. Ἐπὶ ἐμέτῳ λὺγξ καὶ ὀφθαλμοὶ ἐρυθροί, 2 κακόν.

IV. Ἐπὶ ἱδρῶτι φρίκη, οὐ χρηστόν.

V. Ἐπὶ μανίῃ δυσεντερίη, ἢ ὕδρωψ, ἢ ἔκστασις, 2 ἀγαθόν.

VI. Ἐν νούσῳ πολυχρονίῃ ἀσιτίη[1] καὶ ἄκρητοι 2 ὑποχωρήσιες, κακόν.

VII. Ἐκ πολυποσίης ῥῖγος καὶ παραφροσύνη, 2 κακόν.

VIII. Ἐπὶ φύματος ἔσω ῥήξει ἔκλυσις, ἔμετος,[2] 2 καὶ λειποψυχίη γίνεται.

IX. Ἐπὶ αἵματος ῥύσει παραφροσύνη ἢ[3] 2 σπασμός, κακόν.

X. Ἐπὶ εἰλεῷ ἔμετος, ἢ λὺγξ, ἢ σπασμὸς, ἢ 2 παραφροσύνη, κακόν.

XI. Ἐπὶ πλευρίτιδι περιπλευμονίη,[4] κακόν.[5]

XII. Ἐπὶ περιπλευμονίῃ φρενῖτις, κακόν.

XIII. Ἐπὶ καύμασιν[6] ἰσχυροῖσι σπασμὸς ἢ 2 τέτανος,[7] κακόν.

[1] ἀποσιτίη V. After ἀσιτίη M has καὶ ἄκρητοι ἔμετοι.

[2] For ἔκλυσις, ἔμετος M has ἔκκρισις αἵματος.

[3] For ἢ some MSS. have καὶ and M has ἢ καὶ.

[4] After περιπλευμονίῃ M has ἐπιγινομένη.

[5] κακόν according to Galen was omitted by certain ancient MSS.

[6] One MS. has τραύμασιν for καύμασιν. Galen mentions both readings.

[7] σπασμοὶ τέτανοι V.

II. In a case of diseased bone, livid flesh on[1] it is a bad sign.

III. For hiccough and redness of the eyes to follow vomiting is a bad sign.

IV. For shivering to follow sweating is not a good sign.

V. For madness to be followed by dysentery, dropsy or raving,[2] is a good sign.

VI. In a protracted disease loss of appetite and uncompounded[3] discharges are bad.

VII. Rigor and delirium after excessive drinking are bad symptoms.

VIII. From the breaking internally of an abscess result prostration, vomiting and fainting.

IX. After a flow of blood delirium or convulsions are a bad sign.

X. In ileus, vomiting, hiccough, convulsions or delirium are a bad sign.

XI. Pneumonia supervening on pleurisy is bad.[4]

XII. Phrenitis[5] supervening on pneumonia is bad.

XIII. Convulsions or tetanus supervening on severe burns are a bad symptom.

[1] It is difficult to decide how far the preposition ἐπὶ in this and the following aphorisms means "after." The common use of ἐπιγίγνεσθαι to signify one symptom supervening on another suggests that ἐπὶ has somewhat of this force in all cases.

[2] By ἔκστασις is meant an increase of the maniacal symptoms, helping to bring the disease to a crisis.

[3] Probably meaning "showing signs that κρᾶσις is absent."

[4] If κακὸν be omitted: "Pneumonia often supervenes on pleurisy."

[5] Phrenitis means here either (*a*) the form of malaria called by this name, or (*b*) some disease with similar symptoms.

XIV. Ἐπὶ πληγῇ ἐς τὴν κεφαλὴν ἔκπληξις ἢ 2 παραφροσύνη, κακόν.[1]

XV. Ἐπὶ αἵματος πτύσει, πύου πτύσις.

XVI. Ἐπὶ πύου πτύσει, φθίσις καὶ ῥύσις· 2 ἐπὴν δὲ[2] τὸ σίελον[3] ἴσχηται, ἀποθνήσκουσιν.

XVII. Ἐπὶ φλεγμονῇ τοῦ ἥπατος λύγξ 2 κακόν.

XVIII. Ἐπὶ ἀγρυπνίῃ σπασμὸς ἢ παραφρο- 2 σύνη κακόν.

XVIII *bis*. Ἐπὶ ληθάργῳ τρόμος κακόν.

XIX. Ἐπὶ ὀστέου ψιλώσει ἐρυσίπελας κακόν.

XX. Ἐπὶ ἐρυσιπέλατι σηπεδὼν ἢ ἐκπύησις.[4]

XXI. Ἐπὶ ἰσχυρῷ σφυγμῷ ἐν τοῖσιν ἕλκεσιν, 2 αἱμορραγίη.

XXII. Ἐπὶ ὀδύνῃ πολυχρονίῳ τῶν περὶ τὴν 2 κοιλίην, ἐκπύησις.

XXIII. Ἐπὶ ἀκρήτῳ ὑποχωρήσει, δυσεν- 2 τερίη.

XXIV. Ἐπὶ ὀστέου διακοπῇ,[5] παραφροσύνη, 2 ἢν κενεὸν λάβῃ.[6]

XXV. Ἐκ φαρμακοποσίης σπασμός, θανα- 2 τῶδες.

XXVI. Ἐπὶ ὀδύνῃ ἰσχυρῇ τῶν περὶ τὴν 2 κοιλίην, ἀκρωτηρίων ψύξις, κακόν.

[1] κακόν omitted (according to Galen) by certain MSS.
[2] καὶ ἐπὴν for ἐπὴν δὲ Urb. V.
[3] πτύελον C′ Urb.
[4] ἢ ἐκπύησις omitted by V. After ἐκπύησις many MSS add κακόν.
[5] Before παραφροσύνη C′ Urb. add ἔκπληξις ἤ.
[6] Rein. puts ἢν κενεὸν λάβῃ with XXV.

---

[1] If κακὸν be omitted: "Stupor or delirium follows a blow on the head."

XIV. Stupor or delirium from a blow on the head is bad.[1]

XV. After spitting of blood, spitting of pus.

XVI. After spitting of pus, consumption and flux;[2] and when the sputum is checked the patients die.

XVII. In inflammation of the liver, hiccough is bad.

XVIII. In sleeplessness, convulsions or delirium is a bad sign.

In lethargus trembling is a bad sign.

XIX. On the laying bare of a bone erysipelas is bad.

XX. On erysipelas, mortification or suppuration ⟨is bad⟩.[3]

XXI. On violent throbbing in wounds, hemorrhage ⟨is bad⟩.[3]

XXII. After protracted pain in the parts about the belly, suppuration ⟨is bad⟩.[3]

XXIII. On uncompounded stools, dysentery ⟨is bad⟩.[3]

XXIV. After the severing of bone, delirium, if the cavity be penetrated.[4]

XXV. Convulsions following on purging are deadly.

XXVI. In violent pain in the parts about the belly, chill of the extremities is a bad sign.

[2] Galen says that ῥύσις means either (*a*) the falling out of the hair or (*b*) diarrhoea.

[3] These words must be understood, as they easily can be in a list of aphorisms giving "bad" symptoms.

[4] Galen states that this aphorism applies, not to any bone, but to severe fractures of the skull piercing the membranes. I have done my best to use the most appropriate prepositions to translate ἐπὶ in aphorisms XVII. to XXIV.

XXVII. Ἐν γαστρὶ ἐχούσῃ τεινεσμὸς ἐπι-
2 γενόμενος ἐκτρῶσαι ποιεῖ.

XXVIII. Ὅ τι ἂν ὀστέον, ἢ χόνδρος, ἢ νεῦρον[1]
2 διακοπῇ[2] ἐν τῷ σώματι, οὐκ αὔξεται.[3]

XXIX. Ἢν ὑπὸ λευκοῦ φλέγματος ἐχομένῳ
2 διάρροια ἐπιγένηται ἰσχυρή, λύει τὴν νοῦσον.

XXX. Ὁκόσοισιν ἀφρώδεα[4] διαχωρήματα
ἐν τῇσι διαρροίῃσι, τούτοισιν ἀπὸ τῆς κεφαλῆς
3 καταρρεῖ.[5]

XXXI. Ὁκόσοισι πυρέσσουσιν[6] κριμνώδεες αἱ
ὑποστάσιες ἐν τοῖσιν οὔροισι γίνονται, μακρὴν
3 τὴν ἀρρωστίην σημαίνουσιν.[7]

XXXII. Ὁκόσοισι χολώδεες αἱ ὑποστάσιες
γίνονται, ἄνωθεν δὲ λεπταί, ὀξείην τὴν ἀρρωστίην
3 σημαίνουσιν.

XXXIII. Ὁκόσοισι δὲ τὰ οὖρα διεστηκότα[8]
γίνεται, τούτοισι ταραχὴ ἰσχυρὴ[9] ἐν τῷ σώματί
3 ἐστιν.

XXXIV. Ὁκόσοισι δὲ ἐν[10] τοῖσιν οὔροισι
πομφόλυγες ὑφίστανται, νεφριτικὰ[11] σημαίνει,
3 καὶ μακρὴν[12] τὴν ἀρρωστίην.

[1] χόνδρος and νεῦρον are transposed by V. Urb. omits νεῦρον.

[2] διακοπῇ C′ Urb. V: ἀποκοπῆι M.

[3] οὐκ αὔξεται C′ Urb.: οὐκ αὔξεται οὔτε συμφύεται V: οὐχ ὑγιάζει οὔτε αὔξεται οὔτε φύεται M.

[4] MV add τὰ before διαχωρήματα.

[5] ἐπικαταρρεεῖ V: φλέγμα καταρρεῖ C′ Urb.: ταῦτα καταρρεῖ M

[6] ἐν τοῖσιν οὔροισι after πυρέσσουσιν MV.

[7] σημαίνει C′. All the best MSS. except Urb. have κρημνώδεες.

[8] διεστηκότα after δὲ MV.

[9] ἰσχυρὴ after σώματι V.

[10] V has ἐπὶ and ἐφίστανται. M has ἐφίστανται but not ἐπὶ. C′ Urb. have ἐν and ὑφίστανται.

[11] νεφριτικὰ MSS.: φρενιτικὰ Dietz. Some MSS. have σημαίνουσι.

XXVII. Tenesmus[1] in the case of a woman with child causes miscarriage.

XXVIII. Whatsoever bone, cartilege or sinew be cut through in the body, it does not grow.[2]

XXIX. When in the case of a white phlegm[3] violent diarrhoea supervenes, it removes the disease.

XXX. In cases where frothy discharges occur in diarrhoea there are fluxes from the head.[4]

XXXI. In fever cases sediments like coarse meal forming in the urine signify that the disease will be protracted.

XXXII. In cases where the urine is thin at the first,[5] and then becomes bilious, an acute illness is indicated.

XXXIII. In cases where the urine becomes divided there is violent disburbance in the body.[6]

XXXIV. When bubbles form in the urine, it is a sign that the kidneys are affected, and that the disease will be protracted.[7]

[1] Straining at evacuations of stools.

[2] A repetition of *Aphorisms* VI. xix.

[3] *I.e.* incipient anasarca.

[4] This medically obscure aphorism should be connected with the doctrines expounded in the latter part of *Sacred Disease*.

[5] Galen and Theophilus give this meaning to ἄνωθεν, and Adams adopts it. Littré translates, "à la partie supérieure," but Galen says he had never seen urine watery above but bilious below.

[6] The word διεστηκότα perplexed Galen, who took it to mean "not homogeneous"; Adams thinks that it refers to a strongly marked line of distinction between the sediment and the watery part.

[7] Adams explains this as referring to albuminuria. Medically ἐφίστανται ("settle on the surface") is the better reading, as albuminous urine is frothy. But the MS. authority for ἐν and ὑφίστανται is strong.

[12] Urb. has ὀξείην for μακρὴν.

XXXV. Ὁκόσοισι δὲ λιπαρὴ ἡ ἐπίστασις[1] καὶ ἀθρόη, τούτοισι νεφριτικὰ[2] καὶ ὀξέα 3 σημαίνει.[3]

XXXVI. Ὁκόσοισι δὲ νεφριτικοῖσιν ἐοῦσι τὰ προειρημένα σημεῖα συμβαίνει, πόνοι τε ὀξέες περὶ τοὺς μύας τοὺς ῥαχιαίους γίνονται, ἢν μὲν περὶ τοὺς ἔξω τόπους γίνωνται,[4] ἀπόστημα προσδέχου ἐσόμενον ἔξω· ἢν δὲ μᾶλλον οἱ πόνοι πρὸς[5] τοὺς ἔσω τόπους,[6] καὶ τὸ ἀπόστημα προσδέχου 7 ἐσόμενον μᾶλλον ἔσω.

XXXVII. Ὁκόσοι[7] αἷμα ἐμέουσιν, ἢν μὲν ἄνευ πυρετοῦ, σωτήριον· ἢν δὲ σὺν πυρετῷ, κακόν· θεραπεύειν[8] δὲ τοῖσι στυπτικοῖσιν ἢ τοῖσι 4 ψυκτικοῖσιν.[9]

XXXVIII. Κατάρροοι ἐς τὴν ἄνω κοιλίην 2 ἐκπυέονται ἐν ἡμέρῃσιν εἴκοσι.

XXXIX. Ἢν οὐρῇ αἷμα καὶ θρόμβους, καὶ στραγγουρίη ἔχῃ, καὶ ὀδύνη ἐμπίπτῃ ἐς τὸν περίνεον καὶ τὸν κτένα, τὰ περὶ τὴν κύστιν νοσεῖν 4 σημαίνει.[10]

[1] ὑπόστασις C′ Urb. V: ἐπίστασις M. Galen mentions both readings, but prefers ἐπίστασις because of the sense.

[2] Galen says that some would read φρενιτικά on the ground that the symptoms mentioned are not confined to nephritis.

[3] After σημαίνει Urb. adds καὶ ὀξείην τὴν ἀρρωστίην ἔσεσθαι, omitting καὶ ὀξέα. C′ has νεφριτικὰ σημαίνει ὀξέα.

[4] ἢν μὲν περὶ . . . γίνωνται omitted by Urb.

[5] For πρὸς Urb. and several MSS have περὶ.

[6] After τόπους many MSS. have γίνωνται.

[7] ὁκόσοισιν C′V: Urb. has the final -σιν erased.

[8] θεραπεύεται Urb. Rein. has ὀλέθριον κάρτα for κακόν (so Urb.), omitting θεραπεύειν κ.τ.ε.

[9] V has τοῖσι στυπτικοῖσι only, adding τὰ ὀξέα τῶν νοσημάτων κρίνεται ἐν τεσσαρεσκαίδεκα ἡμέρῃσι· τριταῖος κρίνεται ἐν

XXXV. When the scum on the urine is greasy and massed together, it indicates acute disease of the kidneys.[1]

XXXVI. When the aforesaid symptoms occur in kidney diseases, and acute pains are experienced in the muscles of the back, if these occur about the external parts, expect an external abscess; if they occur more about the internal parts, expect rather that the abscess too will be internal.

XXXVII. The vomiting of blood, if without fever, may be cured;[2] if with fever, it is bad. Treat it with styptics or refrigerants.

XXXVIII. Catarrhs (fluxes) into the upper cavity[3] suppurate in twenty days.

XXXIX. When a patient passes in the urine blood and clots, suffers strangury and is seized with pain in the perineum and pubes, it indicates disease in the region of the bladder.

[1] The MS. authority for *ὑπόστασις* is very strong, but Galen's comment seems to be decisive. Some ancient commentators, realising that greasy urine is not necessarily a sign of kidney disease, would have altered the reading *νεφριτικά*. Galen would keep *νεφριτικά*, understanding *ἀθρόη* to refer to time, "scum on urine passed at short intervals." But it is the scum, and not the urine, which is called *ἀθρόη*.

[2] This meaning of *σωτήριον* (*θεραπευθῆναι δυνάμενον*) is vouched for by Galen. The word should mean "salutary."

[3] That is, the chest.

ἕπτα περιόδοισι τὸ μακρότατον. M has *τριταῖος . . . μακρότατον· τὰ ὀξέα τῶν νοσημάτων γίνεται κ.τ.ε.*

[10] Galen mentions two readings, *τὰ περὶ τὴν κύστιν νοσέειν σημαίνει*, and *τὴν κύστιν νοσέειν σημαίνει.* M has *κτένα καὶ τὴν κύστιν. νοῦσον σημαίνει.* Urb. omits this aphorism.

XL. Ἢν ἡ γλῶσσα ἐξαίφνης[1] ἀκρατὴς γένηται, ἢ ἀπόπληκτόν τι τοῦ σώματος, μελαγχο-
3 λικὸν τὸ τοιοῦτον.[2]

XLI. Ἢν, ὑπερκαθαιρομένων τῶν πρεσβυ-
2 τέρων,[3] λὺγξ ἐπιγένηται, οὐκ ἀγαθόν.[4]

XLII. Ἢν πυρετὸς μὴ ἀπὸ χολῆς ἔχῃ, ὕδατος[5] πολλοῦ καὶ θερμοῦ κατὰ τῆς κεφαλῆς καταχεο-
3 μένου, λύσις γίνεται τοῦ πυρετοῦ.[6]

XLIII. Γυνὴ ἀμφιδέξιος οὐ γίνεται.

XLIV. Ὁκόσοι ἔμπυοι τέμνονται[7] ἢ καίονται, ἢν μὲν τὸ πῦον καθαρὸν ῥυῇ καὶ λευκόν, περιγίνονται· ἢν δὲ[8] βορβορῶδες καὶ δυσῶδες, ἀπόλ-
4 λυνται.

XLV. Ὁκόσοι[9] ἧπαρ διάπυον καίονται ἢ τέμνονται, ἢν μὲν τὸ πῦον καθαρὸν ῥυῇ καὶ λευκόν, περιγίγνονται (ἐν χιτῶνι γὰρ τὸ πῦον τούτοισίν ἐστιν)· ἢν δὲ οἷον ἀμόργη ῥυῇ, ἀπόλ-
5 λυνται.

XLVI. Ὀδύνας ὀφθαλμῶν,[10] ἄκρητον ποτίσας
2 καὶ λούσας πολλῷ θερμῷ, φλεβοτόμει.

[1] ἐξαίφνης γλῶσσα Urb.

[2] τὸ τοιοῦτο γίγνεται V.

[3] πρεσβυτάτων V. Rein. has datives in -ῳ.

[4] κακὸν C′.

[5] ὕδατος C′ Urb.: ἱδρῶτος MV. Galen mentions both readings, preferring the former.

[6] For τοῦ πυρετοῦ V has τῆς κεφαλῆς. Query τῆς κεφαλαλγίης?

[7] κέονται ἢ τέμνονται Urb.: τέμνονται ἢ καίονται C′: ἢ τέμνονται omitted by MV.

[8] After δὲ Littré has, following slight authority, ὕφαιμον καὶ.

[9] ὁκόσοι ἧπαρ διὰ πύον καίονται V: ὁκόσοις ἧπαρ διὰ πύον καίονται· ἢ τέμνονται· Urb.: ὁκόσοι τὸ ἧπαρ διάπυον καίονται M. C′ omits this aphorism.

[10] ὀφθαλμῶν ὀδύνας λύει λουτρὸν καὶ ἀκρατοποσίη· λούσας πολλῷ θερμῷ φλεβοτόμησον. C′.

XL. If the tongue is suddenly paralysed, or a part of the body suffers a stroke, the affection is melancholic.[1]

XLI. If old people, when violently purged, are seized with hiccough, it is not a good symptom.

XLII. If a patient suffers from a fever not caused by bile, a copious affusion of hot water over the head removes the fever.[2]

XLIII. A woman does not become ambidexterous.[3]

XLIV. Whenever empyema is treated by the knife or cautery, if the pus flow pure and white, the patient recovers: but if muddy and evil-smelling, the patient dies.

XLV. Whenever abscess of the liver is treated by cautery or the knife, if the pus flow pure and white, the patient recovers, for in such cases the pus is in a membrane; but if it flows like as it were lees of oil, the patient dies.

XLVI. In cases of pains in the eyes, give neat wine to drink, bathe in copious hot water, and bleed.

[1] The ancient commentators are at a loss to understand why paralysis is "melancholic," *i.e.* caused by black bile. Perhaps, as μελαγχολία may mean merely "nervousness," the aphorism means that persons of a nervous temperament are peculiarly subject to "strokes."

[2] The reading of V suggests, "relieves the headache."

[3] Some ancient commentators took this aphorism literally; others thought that it referred to the position of the female embryo in the womb; others to the belief that a female is never an hermaphrodite.

ὀφθαλμῶν ὀδύνας· ἄκρατον ποτίσας καὶ λούσας πολλῷ θερμῷ. φλεβοτόμει. Urb. So V, but with *ἄκρητον.*

ὀδύνας ὀφθαλμῶν ἀκρητοποσίη· ἢ λουτρόν· ἢ πυρίη. ἢ φλεβοτομίη λύει· M.

These three readings throw light upon the history of the Hippocratic text. They could not possibly be descendants of a single text copied with the ordinary copyist's blunders.

XLVII. Ὑδρωπιῶντα ἢν βὴξ ἔχῃ,[1] ἀνέλπιστός
2 ἐστιν.[2]

XLVIII. Στραγγουρίην καὶ δυσουρίην θώρηξις
2 καὶ φλεβοτομίη λύει· τέμνειν δὲ τὰς ἔσω.[3]

XLIX. Ὑπὸ κυνάγχης ἐχομένῳ οἴδημα ἢ ἐρύθημα ἐν τῷ στήθει[4] ἐπιγενόμενον, ἀγαθόν· ἔξω
3 γὰρ τρέπεται τὸ νόσημα.

L. Ὁκόσοισιν ἂν σφακελισθῇ ὁ ἐγκέφαλος, ἐν τρισὶν ἡμέρῃσιν ἀπόλλυνται· ἢν δὲ ταύτας διαφύ-
3 γωσιν, ὑγιέες γίνονται.

LI. Πταρμὸς γίνεται[5] ἐκ[6] τῆς κεφαλῆς, διαθερμαινομένου τοῦ ἐγκεφάλου, ἢ διυγραινομένου[7] τοῦ ἐν τῇ κεφαλῇ κενεοῦ·[8] ὑπερχεῖται οὖν ὁ ἀὴρ ὁ ἐνεών,[9] ψοφεῖ δέ, ὅτι διὰ στενοῦ ἡ διέξοδος αὐτῷ
5 ἐστιν.

LII. Ὁκόσοι ἧπαρ περιωδυνέουσι, τούτοισι
2 πυρετὸς[10] ἐπιγενόμενος λύει τὴν ὀδύνην.

LIII. Ὁκόσοισι συμφέρει αἷμα ἀφαιρεῖν ἀπὸ
2 τῶν φλεβῶν, τούτους τοῦ ἦρος χρὴ φλεβοτομεῖν.[11]

LIV. Ὁκόσοισι μεταξὺ τῶν φρενῶν καὶ τῆς γαστρὸς φλέγμα ἀποκλείεται,[12] καὶ ὀδύνην παρέχει, οὐκ[13] ἔχον διέξοδον ἐς οὐδετέρην[14] τῶν κοιλιῶν,

[1] ὑδρωπιέοντι ἢν βῆξ ἐπιγένηται C′.
[2] For ἐστιν V has γίνεται.
[3] τέμνειν δὴ τὴν εἴσω φλέβα C′. Urb. adds φλέβας.
[4] ἐν στήθεσιν C′.
[5] πταρμοὶ γίνονται V.
[6] For ἐκ C′ Urb. have ἀπὸ.
[7] Before τοῦ C′ has ἢ διαψυχραινομένου and V ἢ ψυχομένου.
[8] After κενεοῦ M has πληρουμένου.
[9] After ἐνεών Urb. M have ἔξω. C′ reads ἐὼν ἔσω λεπτὸς ἔξω.
[10] After πυρετὸς V has πρῶτος.
[11] I have followed C′ closely in deciding the text of this aphorism. Urb. omits it and also the preceding. V reads:

XLVII. There is no hope for a dropsical patient should he suffer from cough.

XLVIII. Strangury and dysuria are removed by drinking neat wine and bleeding; you should open the internal veins.

XLIX. In cases of angina, if swelling or redness appear on the breast, it is a good sign, for the disease is being diverted outwards.

L. When the brain is attacked by sphacelus,[1] the patients die in three days; if they outlive these, they recover.

LI. Sneezing arises from the head, owing to the brain being heated, or to the cavity in the head being filled with moisture ⟨or becoming chilled⟩.[2] So the air inside overflows, and makes a noise, because it passes through a narrow place.

LII. When there is severe pain in the liver, if fever supervenes it removes the pain.

LIII. When it is beneficial to practise venesection, one ought to bleed in the spring.

LIV. In cases where phlegm is confined between the midriff and the stomach, causing pain because it has no outlet into either of the cavities,[3] the disease

[1] Sphacelus is incipient mortification, said by some commentators to include *caries* of the bone.

[2] In brackets is a translation of the words found in C′ and V.

[3] *I.e.* chest and bowels.

αἷμα ἀφαιρέεσθαι ἀπὸ τῶν φλεβίων· τουτέους ἔαρι δεῖ φλεβοτομέεσθαι: M αἷμα ἀφαιρέειν ἀπὸ τῶν φλεβῶν. τουτέοισι ξυμφέρει. ἦρος φλεβοτομεῖσθαι.

[12] ἀποκλειίεται C′MV: ἀποκλύεται Urb.: ἀποκεῖται many MSS.

[13] For οὐκ Urb. has μὴ.

[14] οὐδ' ἐς (εἰς M) ἑτέρην MV.

τούτοισι, κατὰ τὰς φλέβας ἐς τὴν κύστιν τρεπο-
5 μένου τοῦ φλέγματος, λύσις γίνεται τῆς νούσου.

LV. Ὁκόσοισι δ' ἂν τὸ ἧπαρ ὕδατος πλησθὲν[1] ἐς τὸν ἐπίπλοον[2] ῥαγῇ, τούτοισιν ἡ κοιλίη ὕδατος
3 ἐμπίπλαται, καὶ ἀποθνήσκουσιν.

LVI. Ἀλύκην, χάσμην, φρίκην,[3] οἶνος ἴσος ἴσῳ[4]
2 πινόμενος λύει.[5]

LVII. Ὁκόσοισιν ἐν τῇ οὐρήθρῃ φύματα γίνεται,[6] τούτοισι, διαπυήσαντος καὶ ἐκρα-
3 γέντος,[7] λύεται ὁ πόνος.[8]

LVIII. Ὁκόσοισιν ἂν[9] ὁ ἐγκέφαλος σεισθῇ ὑπό τινος προφάσιος,[10] ἀνάγκη ἀφώνους[11] γενέσ-
3 θαι[12] παραχρῆμα.

LX. Τοῖσι σώμασι[13] τοῖσιν ὑγρὰς τὰς σάρκας[14] ἔχουσι λιμὸν ἐμποιεῖν· λιμὸς γὰρ ξηραίνει τὰ
3 σώματα.

LIX. Ἢν ὑπὸ πυρετοῦ ἐχομένῳ, οἰδήματος μὴ ἐόντος ἐν τῇ φάρυγγι,[15] πνὶξ ἐξαίφνης ἐπιγένηται, καὶ καταπίνειν μὴ δύνηται, ἀλλ' ἢ μόλις,[16]
4 θανάσιμον.

[1] ὕδατος ἐμπλησθὲν Urb. : ἐμπλησθὲν ὕδατος V : ὕδατος πλησθὲν C'M.

[2] τὸν ἐπίπλουν C'MV : τὸν ἐπίπλοον Urb. (this MS. has ῥαγῇ before εἰς) : Littré (with one MS. cited) has τὸ ἐπίπλοον.

[3] Galen says that some MSS. ungrammatically gave the nominatives ἀλύκη, χάσμη, φρίκη. Littré restores these, against all our MSS. Ungrammatical sentences are not uncommon in the Hippocratic *Corpus*.

[4] ἴσως. εἴσω M. C' too has ἴσως.

[5] After λύει many MSS. add τὴν νοῦσον ; C' has ταῦτα.

[6] For γίνεται C' has ἐκφύεται.

[7] ῥαγέντος C'.

[8] λύσις γίνεται C' : one MS. λύεται ἄνθρωπος ἐκ τοῦ πόνου. Urb. omits this aphorism.

[9] δ' ἂν M.

is removed if the phlegm be diverted by way of the veins into the bladder.

LV. In cases where the liver is filled with water and bursts into the epiploön, the belly fills with water and the patient dies.

LVI. Distress, yawning and shivering are removed by drinking wine mixed with an equal part of water.

LVII. When tumours form in the urethra, if they suppurate and burst, the pain is removed.

LVIII. In cases of concussion of the brain from any cause, the patients of necessity lose at once the power of speech.

LX. Starving should be prescribed for persons with moist flesh; for starving dries the body.

LIX. In the case of a person suffering from fever, there being no swelling in the throat, should suffocation suddenly supervene, and the patient be unable to drink, or drink only with difficulty, it is a mortal symptom.[1]

[1] See *Aphorisms* IV. xxxiv.

---

[10] C′ has *ἀπὸ* for *ὑπὸ*, and Urb. has *ὑπό τινος προφάσιος* in the margin.

[11] V has *ἄφωνον*, a grammatical error said by Galen to be found in some MSS.

[12] *γίνεσθαι* V.

[13] *τοῖς σώμασι* omitted by C′.

[14] *ὑγρὰς ἔχουσι τὰς φύσιας* C′: *ὑγρὰς τὰς σάρκας ἔχουσιν.* Urb.: *ὑγρὰς* after *σάρκας* V.
The numbering of this and of the two next aphorisms is an attempt to reconcile the order in Galen with that of our vulgate, which omits LIX and places LX after LIX *bis.*

[15] For *ἐν τῇ φάρυγγι* C′ has *ἐν τῷ τραχήλῳ*: *ἐν τῷ φάρυγγι πνὶξ ἐξαίφνης ἐγγένηται ἐκ τοῦ φάρυγγος* V.

[16] *ἀλλὰ μόλις* MV: *καὶ καταπίνειν μόγης δύνηται θανάσιμον* C′.

LIX bis. Ἢν ὑπὸ πυρετοῦ ἐχομένῳ ὁ τράχηλος ἐπιστραφῇ, καὶ καταπίνειν μὴ δύνηται, οἰδήματος 3 μὴ ἐόντος ἐν τῷ τραχήλῳ, θανάσιμον.[1]

LXI. Ὅκου ἐν ὅλῳ τῷ σώματι μεταβολαί, καὶ ἢν τὸ σῶμα ψύχηται, καὶ πάλιν θερμαίνηται, ἢ χρῶμα ἕτερον ἐξ ἑτέρου μεταβάλλῃ, μῆκος νούσου 4 σημαίνει.[2]

LXII.[3] Ἱδρὼς πολύς, θερμὸς ἢ ψυχρός, αἰεὶ ῥέων, σημαίνει πλεῖον ὑγρόν· ἀπάγειν οὖν τῷ 3 μὲν ἰσχυρῷ ἄνωθεν, τῷ δὲ ἀσθενεῖ κάτωθεν.[4]

LXIII. Οἱ πυρετοὶ οἱ μὴ διαλείποντες, ἢν ἰσχυρότεροι διὰ τρίτης γίνωνται, ἐπικίνδυνοι· ὅτῳ δ' ἂν τρόπῳ διαλείπωσι, σημαίνει ὅτι 4 ἀκίνδυνοι.[5]

LXIV. Ὁκόσοισι[6] πυρετοὶ μακροί, τούτοισιν 2 φύματα, καὶ[7] ἐς τὰ ἄρθρα πόνοι ἐγγίνονται.[8]

[1] C′ omits this aphorism.

[2] ὅκου ἐν ὅλω τῶ σώματι διαφοραὶ καὶ ἢν τὸ σῶμα ψύχεται ἢ αὖθις θερμαίνηται ἢ χρῶμα ἕτερον ἐξ ἑτέρου μεταβάλλει μῆκος νούσου δηλοῖ C′:

καὶ ὅκου ἐν ὅλω τῶ σώματι μεταβολαί· καὶ τὸ σῶμα ψύχηται καὶ πάλιν θερμαίνηται· ἢ χρῶμα ἕτερον ἐξ ἑτέρου μεταβάλληται. νούσου μῆκος σημαίνει V:

καὶ ὅκου ἢν ἐν ὅλωι τῶι σώματι μεταβολαί· καὶ τὸ σῶμα καταψύχηται καὶ πάλιν θερμαίνηται· ἢ χρῶμα ἕτερον ἐξ ἑτέρου μεταβάλληι. μῆκος νούσου σημαίνει. M.

This is another series of variants that cannot possibly be due to ordinary "corruption."

[3] ἱδρὼς πολὺς ἀεὶ ῥεων θερμὸς ἢ ψυχρός σημαίνει πλεῖον ὑγρὸν ἀπάγειν τῶ μὲν ἰσχυρω ἄνωθεν τῶ ἀσθενῆ κάτωθεν. C′:

ἱδρὼς πολὺς θερμός· ἢ ψυχρὸς ἀεὶ ῥέων. σημαίνει πλεῖον τὸ ὑγρὸν ὑπάγειν· τῶ μὲν ἰσχυρῶ ἄνωθεν. τῶ δὲ ἀσθενεῖ κάτωθεν. Urb., which ends here.

ἱδρὼς πολὺς θερμὸς ἢ ψυχρὸς ἀεὶ ῥέων. σημαίνει πλεῖον ὑγρὸν ὑπάγειν· ἰσχυρῶ μὲν ἄνωθεν· ἀσθενεῖ δὲ κάτωθεν. V:

ἱδρὼς πολὺς ἢ θερμὸς ἢ ψυχρὸς ῥέων αἰεί. σημαίνει πλεῖον

LIX. *bis.* In the case of a person suffering from fever, if the neck be distorted, and the patient cannot drink, there being no swelling in the neck, it is a mortal symptom.[1]

LXI. Where there are changes in the whole body, if the body is chilled, becoming hot again, or the complexion changes from one colour to another, a protracted disease is indicated.[2]

LXII. Much sweat, flowing constantly hot or cold, indicates excess of moisture. So evacuate, in the case of a strong person, upwards, in the case of a weak one, downwards.[3]

LXIII. Fevers that do not intermit, if they become more violent every other day, are dangerous; but if they intermit in any way, it indicates that they are free from danger.[4]

LXIV. In protracted fevers, tumours and pains at the joints come on.[5]

[1] See *Aphorisms* IV. xxxv. [2] See *Aphorisms* IV. xl.

[3] The words added in our best MSS. mean: "Much sweat signifies disease, cold sweat greater disease, hot sweat less."

[4] See IV. xliii. [5] See IV. xliv.

ὑγρόν· ἀπάγειν οὖν τὸ μὲν ἰσχυρὸν. ἄνωθεν· τῷ δὲ ἀσθενεῖ κάτωθεν· M.

Galen is inclined to think this aphorism interpolated.

[4] After LXII C′ adds ἱδρὸς πολὺς νόσον σημαίνει ὁ μεν· ψυχρὸς πολῦν· ὁ δε θερμος ἔλαττο·: V ἱδρὼς πολὺς νοῦσον σημαίνει· ὁ ψυχρὸς πολλὴν. ὁ θερμὸς ἐλάσσω: M ἱδρῶς πουλὺς. νοῦσον σημαίνει· ὁ ψυχρὸς. πολλήν· ὁ θερμὸς. ἐλάσσω.

[5] ὁκόσοι πυρετοὶ μὴ διαλείποντες· διὰ τρίτης ἰσχυρώτεροι γίνονται καὶ ἐπικίνδυνοι· ὁκοίω δ' ἂν τρόπω διαλίπωσιν σημαίνει ὅτι ἀκίνδυνοι C′: πυρετοὶ ὁκόσοι μὴ διαλείποντες διὰ τρίτης ἰσχυρότεροι γίγνονται· καὶ ἐπικίνδυνοι· ὅτω δ' ἂν τρόπω διαλίπωσιν ἀκίνδυνοι ἔσονται V: οἱ πυρετοὶ ὁκόσοι μὴ διαλείποντες. διὰ τρίτης ἰσχυρότεροι γίνονται ἐπικίνδυνοι· ὅτωι δ' ἂν. τρόπωι διαλείπωσι. σημαίνει ὅτι ἀκίνδυνοι· M.

[6] ὁκόσοι C′. [7] ἢ φύματα ἢ MV. [8] γίγνονται V.

LXV. Ὁκόσοισι φύματα καὶ[1] ἐς τὰ ἄρθρα πόνοι ἐγγίνονται[2] ἐκ πυρετῶν, οὗτοι σιτίοισι
3 πλείοσι χρέονται.

LXVI. Ἤν τις πυρέσσοντι τροφὴν διδῷ, ἣν ὑγιεῖ, τῷ μὲν ὑγιαίνοντι ἰσχύς, τῷ δὲ κάμνοντι
3 νοῦσος.[3]

LXVII. Τὰ διὰ τῆς κύστιος διαχωρέοντα ὁρῆν δεῖ,[4] εἰ οἷα τοῖς ὑγιαίνουσιν ὑποχωρεῖται·[5] τὰ[6] ἥκιστα οὖν ὅμοια τούτοισι, ταῦτα νοσερώτερα,[7]
4 τὰ δ' ὅμοια τοῖσιν ὑγιαίνουσιν, ἥκιστα νοσερά.

LXVIII. Καὶ οἷσι τὰ ὑποχωρήματα, ἢν ἐάσῃς στῆναι καὶ μὴ κινήσῃς, ὑφίσταται[8] οἱονεὶ ξύσματα,[9] τούτοισι συμφέρει ὑποκαθῆραι τὴν κοιλίην· ἢν δὲ μὴ καθαρὴν ποιήσας διδῷς τὰ ῥοφήματα, ὁκόσῳ ἂν πλείω διδῷς, μᾶλλον
6 βλάψεις.

LXIX. Ὁκόσοισιν ἂν κάτω ὠμὰ ὑποχωρῇ,[10]

[1] For καὶ MV have μακρὰ ἢ.

[2] γίγνονται V (γίνονται M after πυρετῶν).

[3] ἣν τίς τῷ πυρέσσοντι τροφὴν διδῷ· ἢ τῷ μεν ὑγιαίνοντι ἰσχῦς τῷ κάμνοντι νοῦσος· C′: ἣν τις πυρέσσοντι τροφὴν διδῶι. ἣν ὑγιεῖ· τῶι μὲν ὑγιαίνοντι ἰσχύς· τῶι δὲ κάμνοντι νοῦσος. M: ἤν τις τῷ πυρέσσοντι τροφὴν διδώῃ· ἣν ὑγιεῖ. τῷ μὲν ὑγιαίνοντι ἰσχὺς· τῷ δὲ κάμνοντι νοῦσος. V: Littré with one MS. ἣν ὑγιεῖ. Rein. οἵην ἂν ὑγιέϊ διδῴη.

[4] δῆ C′.

[5] ὑποχωρέει C′.

[6] C′ adds οὖν after τὰ.

[7] νοσηλότερα V.

[8] ὑφίστανται MV.

[9] After ξύσματα V has ἣν ὀλίγα, ὀλίγη ἡ νοῦσος γίγνεται· ἢν δὲ πολλὰ, πολλή· M has καὶ ἢν ὀλίγα ἦι κ.τ.ε.

LXV. In cases where tumours and pains at the joints appear after fevers, the patients are taking too much food.[1]

LXVI. If you give to a fever patient the same food as you would to a healthy person, it is strength to the healthy but disease to the sick.[2]

LXVII. We must examine the evacuations of the bladder, whether they are like those of persons in health; if they are not at all like, they are particularly morbid,[3] but if they are like those of healthy people, they are not at all so.

LXVIII. When the evacuations are allowed to stand and are not shaken, and a sediment of as it were scrapings is formed, in such cases it is beneficial slightly to purge the bowels. But if you give the barley gruel without purging, the more you give the more harm you will do.[4]

LXIX. When the alvine discharges are crude, they are caused by black bile; and the more copious the

[1] See IV. xlv.

[2] Galen says that there were two forms of this aphorism, but gives only one, which omits ἢν ὑγιεῖ, so that we can only guess what the other form was. He blames the way in which the meaning is expressed. This, however, is obvious enough, and is well illustrated in *Regimen in Acute Diseases*.

[3] Galen finds fault with the comparative, and thinks that a superlative is wanted to contrast with ἥκιστα.

[4] Galen criticises this aphorism. The word ξύσματα, he says, is inappropriate to urinary evacuations; while if it applies to stools, the aphorism does not tally with fact. Some old commentators would join this aphorism to the following by means of a καί. As Littré points out, the aphorisms in this part of the work, however just Galen's criticisms may be, were known at least as early as the age of Bacchius.

---

[10] ὑποχωρέῃ C′ (not ἀποχωρέῃ, as Littré says).

ἀπὸ χολῆς μελαίνης ἐστίν, ἢν πλείονα, πλείονος,
3 ἢν ἐλάσσονα, ἐλάσσονος.[1]

LXX. Αἱ ἀποχρέμψιες αἱ ἐν τοῖσι πυρετοῖσι τοῖσι μὴ διαλείπουσι, πελιδναὶ καὶ αἱματώδεες καὶ χολώδεες καὶ[2] δυσώδεες, πᾶσαι[3] κακαί· ἀποχωρέουσαι δὲ καλῶς, ἀγαθαί, καὶ κατὰ κοιλίην καὶ κύστιν· καὶ ὅκου ἄν τι ἀποχωρέον[4]
6 στῇ[5] μὴ κεκαθαρμένῳ, κακόν.

LXXI. Τὰ σώματα χρή, ὅκου τις βούλεται καθαίρειν,[6] εὔροα ποιεῖν· κῆν μὲν ἄνω βούλῃ εὔροα ποιεῖν, στῆσαι τὴν κοιλίην· ἢν δὲ κάτω
4 εὔροα ποιεῖν, ὑγρῆναι τὴν κοιλίην.

LXXII. Ὕπνος, ἀγρυπνίη, ἀμφότερα μᾶλλον
2 τοῦ μετρίου γινόμενα, νοῦσος.[7]

LXXIII. Ἐν τοῖσι μὴ διαλείπουσι πυρετοῖσιν, ἢν τὰ μὲν ἔξω ψυχρὰ ᾖ, τὰ δὲ ἔσω καίηται, καὶ
3 δίψαν[8] ἔχῃ, θανάσιμον.

LXXIV. Ἐν μὴ διαλείποντι πυρετῷ,[9] ἢν χεῖλος ἢ ῥὶς ἢ ὀφθαλμὸς διαστραφῇ, ἢν μὴ βλέπῃ, ἢν μὴ ἀκούῃ, ἤδη ἀσθενέος ἐόντος τοῦ

[1] ἢν πλείονα πλείω· ἢ ἐλάσσονα ἐλάσσω ἡ νοῦσος C′ : πλείω πλείων and ἐλάσσω ἐλάσσων V : πλείονα πλείω and ἐλάσσω ἐλάσσονος (without ἡ νοῦσος) M.

[2] χολώδεες καὶ omitted by M.

[3] στᾶσαι Rein.

[4] τῶ ὑπὸ χωρέοντι C′.

[5] στῇ omitted by C′.

[6] καθαίρεσθαι M. Rein. omits εὔροα ποιεῖν (twice).

[7] μᾶλλον τοῦ μετρίου κακόν. C′V: μᾶλλον τοῦ μετρίου γινόμενα· κακόν. After these words C′ has οὐ πλησμονὴ οὐ λιμὸς οὐδ' ἄλλό τï ἀγαθὸν οὐδέν. ὅτι ἂν μᾶλλον της φύσιος ᾖ. M has οὐδ' ἄλλο οὐδὲν ἀγαθὸν, while V has οὐδὲ λιμὸς and omits τι.

[8] δίψαν C′ V: δίψα M : πυρετὸς Galen, Littré and Reinhold.

[9] ἐν τοῖσι μὴ διαλείπουσι πυρετοῖσ. C′ V.

discharges the more copious the bile, and the less copious the one, the less copious the other.[1]

LXX. In non-intermittent fevers, expectorations that are livid, blood-stained, bilious and fetid are all[2] bad; but if the discharge passes favourably, they are good, as is the case with discharges by the bowels and bladder. And wherever a part of the excreta remains behind without the body being purged, it is bad.[3]

LXXI. When you wish to purge bodies you must make them fluent;[4] if you wish to make them fluent[5] upwards, close the bowels, if downwards, moisten the bowels.[5]

LXXII. Both sleep and sleeplessness, when beyond due measure, constitute disease.[6]

LXXIII. In non-intermittent fevers, if the outside of the body be cold while the inside is burning, and thirst is present, it is a fatal sign.[7]

LXXIV. In a non-intermittent fever, should lip, nostril or eye be distorted, should the patient lose the sense of sight or hearing, the body being

[1] The other reading, more strongly attested by our MSS., ἢν πλείω, πλείων, ἢν ἐλάσσω, ἐλάσσων ἡ νοῦσος, means: "the more copious the discharges the worse the disease."

[2] Or (with Rein.) "are bad if suppressed."

[3] Compare IV. xlvii.

[4] "Bring into a state favourable to evacuations," Adams. The adjective εὔροα is active, but "relaxed" is the nearest single equivalent I can think of. Littré renders by "coulant." See p. 111.

[5] Compare II. ix.

[6] The words added in our best MSS. mean: "neither repletion, nor starvation, nor anything else is good if it be beyond nature." Compare with this aphorism, II. iii.

[7] See IV. xlviii. Galen appears to have known only the reading πυρετὸς ἔχῃ, which is, as he remarks, absurd.

σώματος,[1] ὅ τι ἂν ᾖ τούτων τῶν σημείων,
5 θανάσιμον.

LXXV. Ἐπὶ λευκῷ φλέγματι ὕδρωψ ἐπι-
2 γίνεται.

LXXVI. Ἐπὶ διαρροίῃ δυσεντερίη.

LXXVII. Ἐπὶ δυσεντερίῃ λειεντερίη ἐπι-
2 γίνεται.

LXXVIII. Ἐπὶ σφακέλῳ ἀπόστασις ὀστέου.[2]

LXXIX et LXXX. Ἐπὶ αἵματος ἐμέτῳ φθορὴ[3] καὶ πύου[4] κάθαρσις ἄνω· ἐπὶ φθορῇ[5] ῥεῦμα ἐκ τῆς κεφαλῆς· ἐπὶ ῥεύματι διάρροια· ἐπὶ διαρροίῃ σχέσις τῆς ἄνω καθάρσιος· ἐπὶ τῇ
5 σχέσει[6] θάνατος.

LXXXI. Ὁκοῖα καὶ ἐν τοῖσι κατὰ τὴν κύστιν, καὶ τοῖσι κατὰ τὴν κοιλίην ὑποχωρήμασι, καὶ ἐν τοῖσι κατὰ τὰς σάρκας, καὶ ἤν που ἄλλῃ τῆς φύσιος ἐκβαίνῃ τὸ σῶμα, ἢν ὀλίγον, ὀλίγη ἡ νοῦσος γίνεται,[7] ἢν πολύ, πολλή, ἢν πάνυ
5 πολύ, θανάσιμον τὸ τοιοῦτον.[8]

LXXXII. Ὁκόσοι[9] ὑπὲρ τὰ τεσσαράκοντα ἔτεα φρενιτικοὶ γίνονται, οὐ πάνυ τι ὑγιάζονται· ἧσσον γὰρ κινδυνεύουσιν, οἷσιν ἂν οἰκείη τῆς
4 φύσιος καὶ τῆς ἡλικίης ἡ νοῦσος ᾖ.[10]

LXXXIII. Ὁκόσοισιν ἐν τῇσιν ἀρρωστίῃσιν οἱ ὀφθαλμοὶ δακρύουσιν κατὰ προαίρεσιν, ἀγαθόν·
3 ὁκόσοισι δὲ ἄνευ προαιρέσιος, κακόν.

[1] So C′. ἀσθενέος ἐόντος V : ἀσθενὴς ἐὼν M.

[2] ἀποστάσηες ὀστέων V.

[3] φθόη M. Rein. reads ἐμέτῳ πύου κάθαρσις ἄνω· ἐπὶ τῇ καθ. φθορή· ἐπὶ τῇ φθορῇ κ.τ.ε.

[4] πύου omitted by M.

[5] For φθορῇ M has φθόη, and adds the article before ῥεύματι and διαρροίῃ.

by this time weak, whichever of these symptoms appears, it is a deadly sign.

LXXV. On "white phlegm" supervenes dropsy.

LXXVI. On diarrhoea dysentery.

LXXVII. On dysentery supervenes lientery.

LXXVIII. On sphacelus exfoliation of the bone.

LXXIX and LXXX. On vomiting of blood consumption and purging of pus upwards. On consumption a flux from the head. On a flux diarrhoea. On a diarrhoea stoppage of the purging upwards. On the stoppage death.

LXXXI. In the discharges by the bladder, the belly and the flesh,[1] if the body departs in any way from its natural state, if slightly, the disease proves slight; if considerably, considerable; if very considerably, such a thing is deadly.

LXXXII. If phrenitis attack those beyond forty years of age they rarely recover; for the risk is less when the disease is related to the constitution and to the age.

LXXXIII. When in illnesses tears flow voluntarily from the eyes, it is a good sign, when involuntarily a bad sign.

[1] This probably means "through the skin."

---

[6] διασχέσει M. At the end some MSS. add ἐπὶ αἵματος πτύσει πύου πτύσις καὶ ῥύσις· ἐπὴν δὲ σίαλον ἴσχηται, ἀποθνήσκουσι—Galen's inaccurate quotation of VII. xv. and xvi.

[7] M omits ἡ νοῦσος γίνεται, and goes on, ἢν δὲ πολὺ κ.τ.ἑ.

[8] After τοιοῦτον V adds: ἐντεῦθεν οἱ νόθοι. Galen's commentary ceases here.

[9] ὁκόσοισιν C′, with φρενιτικὰ γίγνεται following.

[10] ἧσσον γὰρ . . . νοῦσος ᾖ omitted by V, which has οὗτοι οὐ πάνυ σώζονται·

LXXXIV. Ὁκόσοισιν ἐν τοῖσι πυρετοῖσι τεταρταίοισιν ἐοῦσιν αἷμα ἐκ τῶν ῥινῶν ῥυῇ,[1]
3 πονηρόν.

LXXXV. Ἱδρῶτες ἐπικίνδυνοι οἱ ἐν τῇσι κρισίμοισιν ἡμέρῃσι μὴ[2] γινόμενοι, σφοδροί τε καὶ ταχέως ὠθούμενοι ἐκ τοῦ μετώπου, ὥσπερ σταλαγμοὶ καὶ κρουνοί,[3] καὶ ψυχροὶ σφόδρα καὶ πολλοί· ἀνάγκη γὰρ τὸν τοιοῦτον ἱδρῶτα[4] πορεύεσθαι[5] μετὰ βίης, καὶ πόνου ὑπερβολῆς,
7 καὶ ἐκθλίψιος[6] πολυχρονίου.

LXXXVI. Ἐπὶ χρονίῳ νοσήματι κοιλίης κατα-
2 φορή, κακόν.

LXXXVII. Ὁκόσα φάρμακα οὐκ ἰῆται, σίδηρος ἰῆται· ὅσα σίδηρος οὐκ ἰῆται, πῦρ ἰῆται· ὅσα δὲ
3 πῦρ οὐκ ἰῆται, ταῦτα χρὴ νομίζειν ἀνίατα.[7]

Φθίσιες μάλιστα γίνονται ἀπὸ ὀκτὼ καὶ δέκα ἐτέων μέχρι τριήκοντα καὶ πέντε.[8] τὰ δὲ κατὰ φύσιν γινόμενα κατὰ φθίσιν πάντα μὲν ἰσχυρά,

[1] ῥυῇ. Query, ῥεῖ?
[2] μὴ omitted by M.
[3] καὶ κροῦνοι καὶ omitted by C′.
[4] τοὺς τοιούτους ἵδρωτας C′.
[5] πονηρεύεσθαι C′ M V.
[6] θλίψεως C′.
[7] C′ omits Aphorisms LXXXVI. and LXXXVII.
[8] C′ omits φθίσιες . . . πέντε.

LXXXIV. When in patients suffering from quartan[1] fevers there is bleeding at the nose, it is a bad symptom.

LXXXV. Sweats are dangerous that do not occur[2] on the critical days, when they are violent and quickly forced out of the forehead, as it were in drops or streams, and are very cold and copious. For such a sweat must be attended with violence, excess of pain and prolonged pressure.

LXXXVI. In a chronic disease excessive flux from the bowels is bad.

LXXXVII. Those diseases that medicines do not cure are cured by the knife. Those that the knife does not cure are cured by fire. Those that fire does not cure must be considered incurable.

In the MSS. C′ and V, before the beginning of *Prognostic*, occur the following fragments, which Littré discusses in Vol. I. pp. 401 and following. He considers that most of the passage belongs to the work *Sevens*. The first sentence, not found in C′, is *Aphorisms* V. ix. The interesting point about the addition of such fragmentary passages to the end of a book is, that compilations like *Nature of Man* and *Humours* may have grown by a repetition of a like process.

Consumption usually occurs between the ages of eighteen and thirty-five. The symptoms that normally[3] occur in consumption are all violent, while

[1] So Adams. Littré takes the Greek to mean: "When in fevers the patient bleeds at the nose on the fourth day," etc.

[2] With the reading of M: "that occur on the critical days," etc.

[3] κατὰ φύσιν may be a mistaken repetition of κατὰ φθίσιν.

τὰ δὲ καὶ θανατώδεα. δεύτερον δέ, ἣν ἐν τῇ ὥρῃ νοσῇ, αὐτῇ ἡ ὥρη[1] συμμαχεῖ τῇ νούσῳ, οἷον καύσῳ θέρος, ὑδρωπικῷ χειμών· ὑπερνικᾷ γὰρ τὸ φυσικόν. φοβερώτερον γάρ ἐστιν ἡ γλῶσσα μελαινομένη καὶ πελίη καὶ αἱματώδης. ὅτι ἂν[2] τούτων ἀπῇ τῶν σημείων καὶ τὸ πάθος
10 ἀσθενέστερον δηλοῖ. περὶ θανάτων σημείων.[3] ταῦτα μὲν ἐν τοῖς πυρετοῖς τοῖς ὀξέσι σημειοῦσθαι χρή, ὁπότε μέλλει ἀποθνήσκειν καὶ ὁπότε σωθήσεται. ὁ ὄρχις ὁ[4] δεξιὸς ψυχόμενός τε καὶ ἀνασπώμενος, θανατῶδες. ὄνυχες μελαινόμενοι καὶ δάκτυλοι ποδῶν ψυχροὶ καὶ μέλανες καὶ σκληροὶ καὶ ἐγκύπτοντες[5] ἐγγὺς τὸν θάνατον δηλοῦσιν.[6] καὶ τὰ ἄκρα τῶν δακτύλων πελιδνὰ[7] καὶ χείλη πέλια ὑπολελυμένα[8] καὶ ἐξεστραμμένα[9] θανατώδεα. καὶ σκοτοδινιῶν καὶ[10] ἀπο-
20 στρεφόμενος, τῇ τε ἠρεμίῃ[11] ἡδόμενος, καὶ ὕπνῳ καὶ κώματι[12] πολλῷ κατεχόμενος, ἀνέλπιστος. καὶ ὑπολυσσέων ἀτρέμα καὶ ἀγνοέων καὶ μὴ[13] ἀκούων μηδὲ συνιεὶς θανατῶδες.[14] καὶ ἐμέων[15] διὰ ῥινῶν ὅταν πίνῃ θανατῶδες.[16] μέλλουσί τε[17] ἀποθνήσκειν ταῦτα σαφέστερα γίνεται. εὐθέως[18] καὶ αἱ κοιλίαι ἐπαίρονται καὶ φυσῶνται. ὅρος

[1] δευτέρων ἐν τῇ ὥρῃ C′ : δεύτερον δὲ ἣν μὲν ἐν τῇ ὥρῃ νουσέῃ αὐτὴ ἡ ὥρη V.
[2] So C′ : V has φοβερώτερον δὲ σπληνί· γλῶσσα μελαινομένη καὶ αἱματώδης· ὅταν.
[3] So C′ as a title. V omits.
[4] V omits ὁ and (lower down) σκληροί.
[5] So C′ : V has ἐκκύπτοντες.
[6] Here V has σημαίνουσι (a gloss).
[7] πελιδνὰ omitted by V, which reads πελιδνὰ ἢ καὶ for πέλια.
[8] ὑπολελυμένα C′ : ἀπολελυμένα V.
[9] Here V adds καὶ ψυχρά.

some are actually mortal. Secondly, if the patient be ill in the ⟨kindred⟩ season, the very season is an ally of the disease; for example, summer of ardent fever,[1] winter of dropsy. For the natural element wins a decisive victory. For a more fearful symptom is the tongue becoming black, dark and blood-stained. Whatever of these symptoms is not present, it shows that the lesion is less violent. The signs of death. These are the symptoms that in acute fevers must foretell the death or recovery of the patient. The right testicle cold and drawn up is a mortal sign. Blackening nails and toes cold, black, hard and bent forward show that death is near. The tips of the fingers livid, and lips dark, pendulous and turned out, are mortal symptoms. The patient who is dizzy and turns away, pleased with quiet and oppressed by deep sleep and coma,[2] is past hope. If he is slightly raving,[3] does not recognise his friends, and cannot hear or understand, it is a mortal symptom. Vomiting through the nostrils when he drinks is a mortal symptom. When patients are about to die these clearer symptoms occur. Immediately the bowels swell and are puffed up. The boundary of death is passed when the heat of the soul has risen above the navel to the part above

[1] *I.e.* Summer heat makes the heat of fever worse, and the wet of winter is bad for the water of dropsy.

[2] Can the MSS. reading (*καύματι*) be correct? Littré apparently adopts it.

[3] *ὑπολυσσάω* is not recognised by the dictionaries.

---

[10] Here C′ has *ἀνθρώποις*.

[11] *ἠρεμία* (*sic*) C′ V.

[12] *καύματι* C′ V.

[13] For *μὴ* V has *μῆδὲ*.

[14] V has *θανατώδης*.

[15] *αἰμέων* C′.

[16] V omits *καὶ . . . θανατῶδες*.

[17] *τε* V: *δὲ* C′.

[18] V omits *εὐθέως*.

δὲ[1] θανάτου· ἐπειδὰν[2] τὸ τῆς ψυχῆς θερμὸν ἐπανέλθῃ ὑπὲρ τοῦ ὀμφαλοῦ ἐς τὸ ἄνω τῶν φρενῶν,[3] καὶ συγκαυθῇ τὸ ὑγρὸν ἅπαν. ἐπειδὰν
30 ὁ πνεύμων καὶ ἡ καρδία τὴν ἰκμάδα ἀποβάλωσιν[4] τοῦ θερμοῦ ἀθροοῦντος ἐν τοῖς θανατώδεσι τόποις, ἀποπνεῖ ἄθροον[5] τὸ πνεῦμα τοῦ θερμοῦ, ὅθενπερ συνέστη τὸ ὅλον, ἐς τὸ ὅλον πάλιν, τὸ μὲν διὰ τῶν σαρκῶν τὸ δὲ διὰ τῶν ἐν τῇ[6] κεφαλῇ ἀναπνοέων, ὅθεν τὸ ζῆν καλέομεν. ἀπολείπουσα δὲ[7] ἡ ψυχὴ τὸ τοῦ σώματος σκῆνος[8] τὸ ψυχρὸν καὶ τὸ θνητὸν εἴδωλον ἅμα καὶ χολῇ καὶ αἵματι καὶ φλέγματι καὶ σαρκὶ
39 παρέδωκεν.[9]

[1] τοῦ θανάτου V.
[2] ἐπὰν V.
[3] τὸν ἄνω τῶν φρενῶν τόπον V.
[4] ἀποβλέπωσι V.
[5] ὠθοῦν C′.
[6] V omits τῇ.
[7] V omits δὲ.
[8] Here V adds καὶ.
[9] C′ has εἴδωλον αἷμα καὶ χολὴν καὶ φλέγμα καὶ σάρκας.

the diaphragm, and all the moisture has been burnt up. When the lungs and the heart have cast out the moisture of the heat that collects in the places of death,[1] there passes away all at once the breath of the heat (wherefrom the whole[2] was constructed) into the whole again, partly through the flesh and partly through the breathing organs in the head, whence we call it the "breath of life."[3] And the soul, leaving the tabernacle of the body, gives up the cold, mortal image to bile, blood, phlegm and flesh.[4]

[1] "The places of death" might mean either (*a*) the vital parts or (*b*) the places fatally attacked by disease.

[2] Is "the whole" the individual organism or the universe? The first instance of τὸ ὅλον seems to refer to the individual, the second to the universe. Perhaps the warm life of the individual is supposed to be re-absorbed into the cosmic warmth. See, however, the next note.

[3] Is ζῆν here supposed to be related to ζέω (boil)? Perhaps, however, both ὅθεν τὸ ζῆν καλέομεν and (above) ὅθενπερ συνέστη τὸ ὅλον are glosses. At any rate their omission improves both the construction and the meaning of the whole sentence.

[4] Notice the poetic language (τὸ τοῦ σώματος σκῆνος, τὸ ψυχρὸν καὶ τὸ θνητὸν εἴδωλον). The words σκῆνος and εἴδωλον suggest Orphic thought.

REGIMEN

ΠΕΡΙ ΔΙΑΙΤΗΣ

ΤΟ ΠΡΩΤΟΝ

I. Εἰ μέν μοί τις ἐδόκει τῶν πρότερον συγγραψάντων περὶ διαίτης ἀνθρωπίνης τῆς πρὸς ὑγείην ὀρθῶς ἐγνωκὼς συγγεγραφέναι πάντα διὰ παντός, ὅσα δυνατὸν ἀνθρωπίνῃ γνώμῃ περιληφθῆναι, ἱκανῶς εἶχεν ἄν μοι, ἄλλων ἐκπονησάντων, γνόντα τὰ ὀρθῶς ἔχοντα, τούτοισι χρῆσθαι, καθότι ἕκαστον αὐτῶν ἐδόκει χρήσιμον εἶναι. νῦν δὲ πολλοὶ μὲν ἤδη συνέγραψαν, οὐδεὶς δέ πω ἔγνω ὀρθῶς καθότι ἦν αὐτοῖς συγγραπτέον· ἄλλοι δὲ
10 ἄλλο ἐπέτυχον· τὸ δὲ ὅλον οὐδείς πω τῶν πρότερον. μεμφθῆναι μὲν οὖν οὐδενὶ αὐτῶν ἄξιόν ἐστιν εἰ μὴ ἐδυνήθησαν ἐξευρεῖν, ἐπαινέσαι δὲ πάντας ὅτι ἐπεχείρησαν γοῦν[1] ζητῆσαι. ἐλέγχειν μὲν οὖν τὰ μὴ ὀρθῶς εἰρημένα οὐ παρεσκεύασμαι· προσομολογεῖν δὲ τοῖς καλῶς[2] ἐγνωσμένοις διανενόημαι· ὅσα μὲν γὰρ ὀρθῶς ὑπὸ τῶν πρότερον εἴρηται, οὐχ οἷόν τε ἄλλως πως ἐμὲ συγγράψαντα ὀρθῶς συγγράψαι· ὅσα δὲ μὴ ὀρθῶς εἰρήκασιν, ἐλέγχων μὲν ταῦτα, διότι οὐχ οὕτως ἔχει, οὐδὲν
20 περανῶ· ἐξηγεύμενος δὲ καθότι δοκεῖ μοι ὀρθῶς ἔχειν ἕκαστον, δηλώσω ὃ βούλομαι. διὰ τοῦτο

[1] ἀλλ' ἐπεχείρησαν γ' οὖν θ: ἀλλ' ἐπεχειρήσαντο M with ἀλλ' and -το erased: ἀλλ' ἐπεχείρησάν γε Diels.

[2] καλῶς θ: ἱκανοῖς M.

REGIMEN

BOOK I

I. If I thought that any one of my predecessors to write on human regimen in its relation to health had throughout written with correct knowledge everything that the human mind can comprehend about the subject, it would have been enough for me to learn what had been correctly worked out by the labours of others, and to make use of these results in so far as they severally appeared to be of use. As a matter of fact, while many have already written on this subject, nobody yet has rightly understood how he ought to treat it. Some indeed have succeeded in one respect and others in another, but nobody among my predecessors has successfully treated the whole subject. Now none of them is blameworthy for being unable to make complete discoveries; but all are praiseworthy for attempting the research. Now I am not prepared to criticise their incorrect statements; nay, I have resolved to accept what they have well thought out. The correct statements of my predecessors it is impossible for me to write correctly by writing them in some other way; as to the incorrect statements, I shall accomplish nothing by exposing their incorrectness. If, however, I explain how far each of their statements appears to me correct I shall set forth my wish. These preliminary remarks are made

δὲ τὸν λόγον τοῦτον προκατατίθεμαι, ὅτι οἱ[1] πολλοὶ τῶν ἀνθρώπων ὁκόταν τινὸς προτέρου ἀκούσωσι περί τινος ἐξηγευμένου, οὐκ ἀποδέχονται τῶν ὕστερον διαλεγομένων περὶ τούτων, οὐ γινώσκοντες ὅτι τῆς αὐτῆς ἐστὶ διανοίης γνῶναι τὰ ὀρθῶς εἰρημένα, ἐξευρεῖν τε τὰ μήπω εἰρημένα. ἐγὼ οὖν, ὥσπερ εἶπον, τοῖσι μὲν ὀρθῶς εἰρημένοισι προσομολογήσω· τὰ δὲ μὴ ὀρθῶς εἰρημένα δηλώσω
30 ποῖα ἐστιν· ὁκόσα δὲ μηδὲ ἐπεχείρησε μηδεὶς τῶν πρότερον δηλῶσαι, ἐγὼ ἐπιδείξω καὶ ταῦτα
32 οἷά ἐστι.

II. Φημὶ δὲ δεῖν τὸν μέλλοντα ὀρθῶς συγγράφειν περὶ διαίτης ἀνθρωπίνης[2] πρῶτον μὲν παντὸς φύσιν ἀνθρώπου γνῶναι καὶ διαγνῶναι· γνῶναι μὲν ἀπὸ τίνων συνέστηκεν ἐξ ἀρχῆς, διαγνῶναι δὲ ὑπὸ τίνων μερῶν κεκράτηται· εἴτε γὰρ τὴν ἐξ ἀρχῆς σύστασιν μὴ γνώσεται, ἀδύνατος ἔσται τὰ ὑπ᾽ ἐκείνων γινόμενα γνῶναι· εἴτε μὴ γνώσεται τὸ ἐπικρατέον ἐν τῷ σώματι, οὐχ ἱκανὸς ἔσται τὰ συμφέροντα προσενεγκεῖν
10 τῷ ἀνθρώπῳ. ταῦτα μὲν οὖν δεῖ[3] γινώσκειν τὸν συγγράφοντα, μετὰ δὲ ταῦτα σίτων καὶ ποτῶν ἁπάντων, οἷσι διαιτώμεθα, δύναμιν ἥντινα ἕκαστα[4] ἔχει καὶ τὴν κατὰ φύσιν καὶ τὴν δι᾽ ἀνάγκην καὶ τέχνην ἀνθρωπίνην.[5] δεῖ γὰρ ἐπίστασθαι τῶν τε ἰσχυρῶν φύσει ὡς χρὴ τὴν δύναμιν ἀφαιρεῖσθαι, τοῖσι τε ἀσθενέσιν ὅκως χρὴ ἰσχὺν προστιθέναι διὰ τέχνης, ὅκου ἂν ὁ καιρὸς ἑκάστῳ[6] παραγένηται. γνοῦσι δὲ τὰ εἰρημένα οὔπω αὐτάρκης ἡ θεραπείη τοῦ ἀνθρώπου, διότι οὐ δύναται

[1] οἱ omitted by M.
[2] ἀνθρωπίης θ.
[3] δεῖ θ M: χρὴ Littré and vulgate.

for the following reasons: most men, when they have already heard one person expounding a subject, refuse to listen to those who discuss it after him, not realising that it requires the same intelligence to learn what statements are correct as to make original discoveries. Accordingly, as I have said, I shall accept correct statements and set forth the truth about those things which have been incorrectly stated. I shall explain also the nature of those things which none of my predecessors has even attempted to set forth.

II. I maintain that he who aspires to treat correctly of human regimen must first acquire knowledge and discernment of the nature of man in general—knowledge of its primary constituents and discernment of the components by which it is controlled. For if he be ignorant of the primary constitution, he will be unable to gain knowledge of their effects; if he be ignorant of the controlling thing in the body he will not be capable of administering to a patient suitable treatment. These things therefore the author must know, and further the power possessed severally by all the foods and drinks of our regimen, both the power each of them possessed by nature and the power given them by the constraint of human art. For it is necessary to know both how one ought to lessen the power of these when they are strong by nature, and when they are weak to add by art strength to them, seizing each opportunity as it occurs. Even when all this is known, the care of a man is not yet complete, because

[4] ἥντινα ἕκαστα ἔχει M: ἥντινα ἔχουσι θ.
[5] ἀνθρωπηίην M. [6] ἑκάστωι θ: ἑκά στων M.

20 ἐσθίων ὁ ἄνθρωπος ὑγιαίνειν, ἢν μὴ καὶ πονῇ. ὑπεναντίας μὲν γὰρ ἀλλήλοισιν ἔχει τὰς δυνάμιας σῖτα καὶ πόνοι, συμφέρονται δὲ πρὸς ἄλληλα πρὸς ὑγείην· πόνοι μὲν γὰρ πεφύκασιν ἀναλῶσαι τὰ ὑπάρχοντα· σῖτα[1] δὲ καὶ ποτὰ ἐκπληρῶσαι τὰ κενωθέντα. δεῖ δέ, ὡς ἔοικε, τῶν πόνων διαγινώσκειν τὴν δύναμιν καὶ τῶν κατὰ φύσιν καὶ τῶν διὰ βίης γινομένων, καὶ τίνες αὐτῶν αὔξησιν παρασκευάζουσιν ἐς σάρκας καὶ τίνες ἔλλειψιν, καὶ οὐ μόνον ταῦτα, ἀλλὰ καὶ τὰς συμ-
30 μετρίας τῶν πόνων πρὸς τὸ πλῆθος τῶν σίτων καὶ τὴν φύσιν τοῦ ἀνθρώπου καὶ τὰς ἡλικίας τῶν σωμάτων, καὶ πρὸς τὰς ὥρας τοῦ ἐνιαυτοῦ καὶ πρὸς τὰς μεταβολὰς τῶν πνευμάτων, πρός τε τὰς θέσεις τῶν χωρίων[2] ἐν οἷσι διαιτέονται, πρός τε τὴν κατάστασιν τοῦ ἐνιαυτοῦ. ἄστρων τε ἐπιτολὰς καὶ δύσιας γινώσκειν δεῖ, ὅκως ἐπίστηται τὰς μεταβολὰς καὶ ὑπερβολὰς φυλάσσειν καὶ σίτων καὶ ποτῶν καὶ πνευμάτων καὶ τοῦ ὅλου κόσμου, ἐξ ὧνπερ τοῖσιν ἀνθρώποισι
40 αἱ νοῦσοι εἰσίν.[3] ταῦτα δὲ πάντα διαγνόντι οὔπω αὔταρκες τὸ εὕρεμά ἐστιν· εἰ μὲν γὰρ ἦν εὑρετὸν ἐπὶ τούτοισι πρὸς ἑκάστου[4] φύσιν σίτου μέτρον καὶ πόνων ἀριθμὸς σύμμετρος μὴ ἔχων ὑπερβολὴν μήτε ἐπὶ τὸ πλέον μήτε ἐπὶ τὸ ἔλασσον, εὕρητο ἂν ὑγείη τοῖσιν ἀνθρώποισιν ἀκριβῶς. νῦν δὲ τὰ μὲν προειρημένα πάντα εὕρηται, ὁκοῖά ἐστι, τοῦτο δὲ ἀδύνατον εὑρεῖν. εἰ μὲν οὖν παρείη τις καὶ ὁρῴη, γινώσκοι ἂν τὸν ἄνθρωπον ἐκδύνοντά τε καὶ ἐν τοῖσι γυμνασίοισι

[1] σῖτα θ: σιτία M. [2] χωρέων Zwinger Diels.

eating alone will not keep a man well; he must also take exercise. For food and exercise, while possessing opposite qualities, yet work together to produce health. For it is the nature of exercise to use up material, but of food and drink to make good deficiencies. And it is necessary, as it appears, to discern the power of the various exercises, both natural exercises and artificial, to know which of them tends to increase flesh and which to lessen it; and not only this, but also to proportion exercise to bulk of food, to the constitution of the patient, to the age of the individual, to the season of the year, to the changes of the winds, to the situation of the region in which the patient resides, and to the constitution of the year. A man must observe the risings and settings of stars, that he may know how to watch for change and excess in food, drink, wind and the whole universe, from which diseases exist among men. But even when all this is discerned, the discovery is not complete. If indeed in addition to these things it were possible to discover for the constitution of each individual a due proportion of food to exercise, with no inaccuracy either of excess or of defect, an exact discovery of health for men would have been made. But as it is, although all the things previously mentioned have been discovered, this last discovery cannot be made. Now if one were present and saw, he would have knowledge[1] of the patient as he stripped and

[1] With the reading of Ermerins and Diels: "saw the patient as he stripped . . . he would know how it is necessary to keep him," etc.

---

[3] φύονται vulgate, Littré. [4] ἑκάστου θ: ἑκάστην M.

50 γυμναζόμενον, ὥστε[1] φυλάσσειν ὑγιαίνοντα, τῶν μὲν ἀφαιρέων, τοῖσι δὲ προστιθείς· μὴ παρεόντι δὲ ἀδύνατον ὑποθέσθαι ἐς ἀκριβείην σῖτα καὶ πόνους· ἐπεὶ ὁκόσον γε δυνατὸν εὑρεῖν ἐμοὶ εἴρηται. ἀλλὰ γὰρ εἰ καὶ πάνυ μικρὸν ἐνδεέστερα[2] τῶν ἑτέρων γίνοιτο, ἀνάγκη κρατηθῆναι ἐν πολλῷ χρόνῳ τὸ σῶμα ὑπὸ τῆς ὑπερβολῆς καὶ ἐς νοῦσον ἀφικέσθαι. τοῖσι μὲν οὖν ἄλλοισι μέχρι τούτου ἐπικεχείρηται ζητηθῆναι· εἴρηται[3] δὲ οὐδὲ ταῦτα· ἐμοὶ δὲ ταῦτα ἐξεύρηται, καὶ πρὸ 60 τοῦ κάμνειν τὸν ἄνθρωπον ἀπὸ τῆς ὑπερβολῆς, ἐφ' ὁκότερον[4] ἂν γένηται, προδιάγνωσις. οὐ γὰρ εὐθέως αἱ νοῦσοι τοῖσιν ἀνθρώποισι γίνονται, ἀλλὰ κατὰ μικρὸν συλλεγόμεναι ἀθρόως[5] ἐκφαίνονται. πρὶν οὖν κρατεῖσθαι ἐν τῷ ἀνθρώπῳ τὸ ὑγιὲς ὑπὸ τοῦ νοσεροῦ, ἃ πάσχουσιν ἐξεύρηταί μοι, καὶ ὅκως χρὴ ταῦτα καθιστάναι ἐς τὴν ὑγείην. τούτου δὲ προσγενομένου πρὸς τοῖσι γεγραμμένοισι, τελευτᾷ[6] τὸ ἐπιχείρημα τῶν 69 διανοημάτων.

III. Συνίσταται μὲν οὖν τὰ ζῷα τά τε ἄλλα πάντα καὶ ὁ ἄνθρωπος ἀπὸ δυοῖν, διαφόροιν μὲν τὴν δύναμιν, συμφόροιν δὲ τὴν χρῆσιν, πυρὸς καὶ ὕδατος. ταῦτα δὲ συναμφότερα αὐτάρκεά ἐστι τοῖσί τε ἄλλοισι πᾶσι καὶ ἀλλήλοισιν, ἑκάτερον δὲ χωρὶς οὔτε αὐτὸ ἑωυτῷ οὔτε ἄλλῳ οὐδενί. τὴν μὲν οὖν δύναμιν αὐτῶν ἑκάτερον

[1] Diels (after Ermerins) puts γινώσκοι ἂν after γυμναζόμενον, reading ὡς δεῖ φυλάσσειν. θ has ὥστε διαφυλάσσειν.

[2] After ἐνδεέστερα Diels (perhaps rightly) adds τὰ ἕτερα.

[3] εἴρηται θ M : εὕρηται has been suggested.

[4] ἀφ' ὁκοτέρων Diels, from the *de qua provenit* of P.

practised his exercises, so as to keep him in health by taking away here and adding there. But without being present it is impossible to prescribe the exact amount of food and exercise, since how far it is possible to make discoveries I have already set forth. In fact, if there occur even a small deficiency of one or the other, in course of time the body must be overpowered by the excess and fall sick. Now the other investigators have attempted to carry their researches to this point, but they have not gone on to set them forth.[1] But I have discovered these things, as well as the forecasting of an illness before the patient falls sick, based upon the direction in which is the excess. For diseases do not arise among men all at once; they gather themselves together gradually before appearing with a sudden spring. So I have discovered the symptoms shown in a patient before health is mastered by disease, and how these are to be replaced by a state of health. When to the things already written this also has been added, the task I have set before myself will be accomplished.

III. Now all animals, including man, are composed of two things, different in power but working together in their use, namely, fire and water. Both together these are sufficient for one another and for everything else, but each by itself suffices neither for itself nor for anything else. Now the power that

[1] Or, "but neither have these things been set forth (discovered)." The conjecture εὕρηται would suggest that the writer had been successful in making a discovery which other authorities had unsuccessfully tried to reach.

---

[5] ἀθρόον θ.
[6] τελευτᾷ θ: τελέεται M: *finem accipit* P.

ἔχει τοιήνδε· τὸ μὲν γὰρ πῦρ δύναται πάντα διὰ παντὸς κινῆσαι, τὸ δὲ ὕδωρ πάντα διὰ παντὸς
10 θρέψαι· ἐν μέρει δὲ ἑκάτερον κρατεῖ καὶ κρατεῖται ἐς τὸ μήκιστον καὶ ἐλάχιστον[1] ὡς ἀνυστόν. οὐδέτερον γὰρ κρατῆσαι παντελῶς δύναται διὰ τόδε· τὸ μὲν πῦρ ἐπεξιὸν ἐπὶ τὸ ἔσχατον τοῦ ὕδατος ἐπιλείπει ἡ τροφή· ἀποτρέπεται οὖν ὁκόθεν μέλλει τρέφεσθαι· τὸ δὲ ὕδωρ ἐπεξιὸν ἐπὶ τὸ ἔσχατον τοῦ πυρός, ἐπιλείπει ἡ κίνησις· ἵσταται οὖν ἐν τούτῳ· ὁκόταν δὲ στῇ, οὐκέτι ἐγκρατές ἐστιν, ἀλλ' ἤδη τῷ ἐμπίπτοντι πυρὶ ἐς τὴν τροφὴν καταναλίσκεται. οὐδέτερον δὲ
20 διὰ ταῦτα δύναται κρατῆσαι παντελῶς· εἰ δέ ποτε κρατηθείη καὶ ὁκότερον πρότερον,[2] οὐδὲν ἂν εἴη τῶν νῦν ἐόντων ὥσπερ ἔχει νῦν· οὕτω δὲ ἐχόντων αἰεὶ ἔσται τὰ αὐτά, καὶ οὐδέτερα καὶ οὐδὲ ἅμα[3] ἐπιλείψει. τὸ μὲν οὖν πῦρ καὶ τὸ ὕδωρ, ὥσπερ εἴρηταί μοι, αὐτάρκεά ἐστι πᾶσι διὰ παντὸς ἐς
26 τὸ μήκιστον καὶ τοὐλάχιστον ὡσαύτως.

IV. Τούτων δὲ προσκεῖται ἑκατέρῳ τάδε· τῷ μὲν πυρὶ τὸ θερμὸν καὶ τὸ ξηρὸν, τῷ δὲ ὕδατι τὸ ψυχρὸν καὶ τὸ ὑγρόν· ἔχει δὲ ἀπ' ἀλλήλων τὸ μὲν πῦρ ἀπὸ τοῦ ὕδατος τὸ ὑγρόν· ἔνι γὰρ ἐν πυρὶ[4] ὑγρότης· τὸ δὲ ὕδωρ ἀπὸ τοῦ πυρὸς τὸ ξηρόν· ἔνι γὰρ ἐν ὕδατι ξηρόν. οὕτω δὲ τούτων ἐχόντων, πολλὰς καὶ παντοδαπὰς ἰδέας ἀποκρίνονται ἀπ' ἀλλήλων καὶ σπερμάτων καὶ ζῴων, οὐδὲν ὁμοίων[5] ἀλλήλοισιν οὔτε τὴν ὄψιν οὔτε

[1] Before ἐλάχιστον Littré adds τὸ.
[2] Several authorities would omit πρότερον.
[3] Some would read καὶ οὐδέτερον οὐδαμὰ, "and neither will fail altogether." This is very likely the correct reading.

each of them possesses is this. Fire can move all things always, while water can nourish all things always; but in turn each masters or is mastered to the greatest maximum or the least minimum possible. Neither of them can gain the complete mastery for the following reason. The fire, as it advances to the limit of the water, lacks nourishment, and so turns to where it is likely to be nourished; the water, as it advances to the limit of the fire, find its motion fail, and so stops at this point. When it stops its force ceases, and hereafter is consumed to nourish the fire which assails it. Neither, however, can become completely master for the following reasons. If ever either were to be mastered first, none of the things that are now would be as it is now. But things being as they are, the same things will always exist, and neither singly nor all together will the elements fail. So fire and water, as I have said, suffice for all things throughout the universe unto their maximum and the minimum alike.

IV. These elements have severally the following attributes. Fire has the hot and the dry, water the cold and the moist. Mutually too fire has the moist from water, for in fire there is moisture, and water has the dry from fire, for there is dryness in water also. These things being so, they separate off from themselves many forms of many kinds, both of seeds and of living creatures, which are like to one another neither in their appearance nor in their power.[1]

[1] Probably δύναμις here means φύσις, "nature," "essence."

---

[4] For ἐν πυρὶ M has ἀπὸ τοῦ ὕδατος.

[5] ὅμοιον θ M: ὁμοίων Zwinger. A. L. Peck has [ἀπ' αὐτῶν] καὶ σπέρματα καὶ ζῷα, καὶ οὐδὲν ὅμοιον ἄλλο ἄλλῳ. Before πολλὰς Fredrich places ἐς.

10 τὴν δύναμιν· ἅτε γὰρ οὔποτε κατὰ τωὐτὸ ἱστάμενα, ἀλλ' αἰεὶ ἀλλοιούμενα ἐπὶ τὰ καὶ ἐπὶ τά,[1] ἀνόμοια ἐξ ἀνάγκης γίνεται καὶ τὰ ἀπὸ τούτων ἀποκρινόμενα. ἀπόλλυται μέν νυν οὐδὲν ἁπάντων χρημάτων, οὐδὲ γίνεται ὅ τι μὴ καὶ πρόσθεν ἦν· συμμισγόμενα δὲ καὶ διακρινόμενα ἀλλοιοῦται· νομίζεται δὲ ὑπὸ τῶν ἀνθρώπων τὸ μὲν ἐξ Ἅιδου ἐς φάος αὐξηθὲν γενέσθαι, τὸ δὲ ἐκ τοῦ φάεος ἐς Ἅιδην μειωθὲν ἀπολέσθαι· ὀφθαλμοῖσι γὰρ πιστεύουσι μᾶλλον ἢ γνώμῃ, οὐχ ἱκανοῖς 20 ἐοῦσιν οὐδὲ περὶ τῶν ὁρεομένων κρῖναι·[2] ἐγὼ δὲ τάδε γνώμῃ ἐξηγέομαι. ζῷα[3] γὰρ κἀκεῖνα καὶ τάδε· καὶ οὔτε, εἰ ζῷον, ἀποθανεῖν οἷόν τε, εἰ μὴ μετὰ πάντων· ποῖ[4] γὰρ ἀποθανεῖται; οὔτε τὸ μὴ ὂν γενέσθαι, πόθεν γὰρ ἔσται;[5] ἀλλ' αὔξεται πάντα καὶ μειοῦται ἐς τὸ μήκιστον καὶ ἐς τὸ ἐλάχιστον, τῶν γε δυνατῶν. ὅ τι δ' ἂν διαλέγωμαι γενέσθαι ἢ[6] ἀπολέσθαι, τῶν πολλῶν εἵνεκεν ἑρμηνεύω· ταῦτα[7] δὲ συμμίσγεσθαι καὶ διακρίνεσθαι δηλῶ· ἔχει δὲ καὶ[8] ὧδε· γενέσθαι καὶ ἀπολέσθαι 30 τωὐτό, συμμιγῆναι καὶ διακριθῆναι τωὐτό, αὐξηθῆναι καὶ μειωθῆναι τωὐτό, γενέσθαι, συμμιγῆναι

[1] ἐπὶ τὰ καὶ ἐπὶ τὰ Corais: ἔπειτα καὶ ἔπειτα MSS.

[2] ὀφθαλμοῖσι δὲ πιστεύεσθαι μᾶλλον, ἢ γνῶμαι· ἐγὼ δὲ τάδε γνώμῃ κ.τ.ε. M.

[3] ζῷα M: ζώει Littré (from θ's ζῶ εἰ γὰρ). For εἰ ζῷον Fredrich and Gomperz read τὸ ἀείζωον.

[4] ποῦ MSS: ποῖ A. L. Peck after H. Rackham.

[5] This is practically the reading of θ M has καὶ οὔτε τὸ ζῶον ἀποθανεῖν οἷόν τε μὴ μετὰ πάντων, καὶ γὰρ ἀποθανεῖται· οὔτε τὸ μὴ ὂν γενέσθαι, κόθεν παραγενήσεται. Both MSS. have ὂν not ἐόν.

[6] ὅτι δ' ἂν διαλέγομαι θ: ὅτι δὴν διαλέγομαι M. ἢ θ: καὶ τὸ M.

[7] ταὐτὰ Bywater after Bernays.

[8] καὶ omitted by M.

For as they never stay in the same condition, but are always changing to this or to that, from these elements too are separated off things which are necessarily unlike. So of all things nothing perishes, and nothing comes into being that did not exist before. Things change merely by mingling and being separated.[1] But the current belief among men is that one thing increases and comes to light from Hades, while another thing diminishes and perishes from the light into Hades. For they trust eyes rather than mind, though these are not competent to judge even things that are seen. But I use mind to expound thus. For there is life in the things of the other world, as well as in those of this. If there be life, there cannot be death, unless all things die with it. For whither will death take place? Nor can what is not come into being. For whence will it come? But all things increase and diminish to the greatest possible maximum or the least possible minimum. Whenever I speak of "becoming" or "perishing" I am merely using popular expressions; what I really mean is "mingling" and "separating." The facts are these. "Becoming" and "perishing" are the same thing; "mixture" and "separation" are the same thing; "increase" and "diminution" are the same thing; "becoming" and "mixture" are the same thing;

[1] The passage, "So of all things . . . and being separated," is almost verbally the same as a fragment of Anaxagoras quoted by Simplicius (*Phys.* 163, 20). It runs: τὸ δὲ γίνεσθαι καὶ ἀπόλλυσθαι οὐκ ὀρθῶς νομίζουσιν οἱ Ἕλληνες· οὐδὲν γὰρ χρῆμα γίνεται οὐδὲ ἀπόλλυται, ἀλλ' ἀπὸ ἐόντων χρημάτων συμμίσγεταί τε, καὶ διακρίνεται καὶ οὕτως ἂν ὀρθῶς καλοῖεν τό τε γίνεσθαι συμμίσγεσθαι καὶ τὸ ἀπόλλυσθαι διακρίνεσθαι.

τωὐτό, ἀπολέσθαι, μειωθῆναι, διακριθῆναι[1] τωὐτὸ, ἕκαστον πρὸς πάντα καὶ πάντα πρὸς ἕκαστον τωὐτὸ, καὶ οὐδὲν πάντων τωὐτό.[2] ὁ
35 νόμος γὰρ τῇ φύσει περὶ τούτων ἐναντίος.

V. Χωρεῖ[3] δὲ πάντα καὶ θεῖα καὶ ἀνθρώπινα ἄνω καὶ κάτω ἀμειβόμενα. ἡμέρη καὶ εὐφρόνη ἐπὶ τὸ μήκιστον καὶ ἐλάχιστον· ὡς καὶ τῇ σελήνῃ τὸ μήκιστον καὶ τὸ ἐλάχιστον,[4] πυρὸς ἔφοδος καὶ ὕδατος, ἥλιος[5] ἐπὶ τὸ μακρότατον καὶ βραχύτατον, πάντα ταὐτὰ καὶ οὐ ταὐτά. φάος Ζηνί, σκότος Ἅιδῃ, φάος Ἅιδῃ, σκότος Ζηνί, φοιτᾷ κεῖνα ὧδε, καὶ τάδε κεῖσε, πᾶσαν ὥρην, πᾶσαν χώρην[6] διαπρησσόμενα κεῖνά τε τὰ
10 τῶνδε, τάδε τ' αὖ τὰ κείνων.[7] καὶ ἃ[9] μὲν πρήσσουσιν οὐκ οἴδασιν, ἃ δὲ οὐ[8] πρήσσουσι δοκέουσιν εἰδέναι· καὶ ἃ[9] μὲν ὁρέουσιν οὐ γινώσκουσιν, ἀλλ' ὅμως αὐτοῖσι πάντα γίνεται δι' ἀνάγκην θείην καὶ ἃ βούλονται καὶ ἃ μὴ βούλονται. φοιτεόντων δ' ἐκείνων ὧδε, τῶν δέ τε κεῖσε,[10] συμμισγομένων πρὸς ἄλληλα, τὴν πεπρωμένην μοίρην ἕκαστον ἐκπληροῖ, καὶ ἐπὶ τὸ μέζον καὶ ἐπὶ τὸ μεῖον.

[1] ἀπολέσθαι ⟨καὶ⟩ διακριθῆναι Diels: Bywater brackets μειωθῆναι.

[2] καὶ οὐδὲν . . . τωὐτό omitted by M: P 7027 has *nihil ex omnibus idem est.*

[3] χωρεῖ Bernays: χωρὶς MSS.

[4] ὡς καὶ . . . ἐλάχιστον omitted by θ. Burnet in his *Early Greek Philosophy* suggests the following reading of the passage. ἡμέρη καὶ εὐφρόνη ἐπὶ τὸ μήκιστον καὶ ἐλάχιστον· ἥλιος, σελήνη ἐπὶ τὸ μήκιστον καὶ ἐλάχιστον· πυρὸς ἔφοδος καὶ ὕδατος. This is very Heracleitean, and may represent the passage of Heracleitus paraphrased by the author of περὶ διαίτης.

[5] οὕτως before ἥλιος Diels.

"perishing," "diminution" and "separation" are the same thing, and so is the relation of the individual to all things, and that of all things to the individual. Yet nothing of all things is the same. For in regard to these things custom is opposed to nature,[1]

V. But all things, both human and divine, are in a state of flux upwards and downwards by exchanges. Day and night, to the maximum and minimum; just as the moon has its maximum and minimum, the ascendancy of fire and of water, so the sun has its longest and its shortest course—all the same things and not the same things. Light for Zeus, darkness for Hades; light for Hades, darkness for Zeus—the things of the other world come to this, those of this world go to that, and during every season throughout every place the things of the other world do the work of this, and those of this world do the work of that. And what men work they know not, and what they work not they think that they know; and what they see they do not understand, but nevertheless all things take place for them through a divine necessity, both what they wish and what they do not wish. And as the things of the other world come to this, and those of this world go to that, they combine with one another, and each fulfils its allotted destiny, both unto the greater and unto the less. And destruction

[1] This and the following chapters contain a mixture of the philosophies of Empedocles, Anaxagoras and Heracleitus. See the Introduction, p. xliii.

---

[6] *πᾶσαν χώρην* omitted by M.
[7] *τάδε τ' αὖ τὰ κείνων* Diels (*ταῦτα* M).
[8] M omits *οὐ*.
[9] *τὰ θ* M: *θ' ἃ* Littré.
[10] *τῶν δε τι κεῖσε θ*: *τῶν δέ τε κεῖοσι* M.

φθορὴ δὲ πᾶσιν ἀπ' ἀλλήλων, τῷ μέζονι ἀπὸ τοῦ μείονος καὶ τῷ μείονι ἀπὸ τοῦ μέζονος, αὐξάνεται
20 τὸ μέζον ἀπὸ τοῦ ἐλάσσονος, καὶ τὸ ἔλασσον ἀπὸ
21 τοῦ μέζονος.[1]

VI. Τὰ δ' ἄλλα πάντα, καὶ ψυχὴ ἀνθρώπου, καὶ σῶμα ὁκοῖον ἡ ψυχή, διακοσμεῖται. ἐσέρπει δὲ ἐς ἄνθρωπον μέρεα μερέων, ὅλα ὅλων, ἔχοντα σύγκρησιν πυρὸς καὶ ὕδατος, τὰ μὲν ληψόμενα, τὰ δὲ δώσοντα· καὶ τὰ μὲν λαμβάνοντα πλεῖον ποιεῖ, τὰ δὲ διδόντα μεῖον. πρίουσιν ἄνθρωποι ξύλον· ὁ μὲν ἕλκει, ὁ δὲ ὠθεῖ. τὸ δ' αὐτὸ τοῦτο ποιέουσι, μεῖον δὲ ποιέοντες πλεῖον ποιέουσι. τοιοῦτον φύσις ἀνθρώπων, τὸ μὲν ὠθεῖ, τὸ δὲ
10 ἕλκει· τὸ μὲν δίδωσι, τὸ δὲ λαμβάνει· καὶ τῷ μὲν δίδωσι, τοῦ δὲ λαμβάνει· καὶ τῷ μὲν δίδωσι[2] τοσούτῳ πλέον, οὗ δὲ[3] λαμβάνει τοσούτῳ μεῖον. χώρην δὲ ἕκαστον φυλάσσει τὴν ἑωυτοῦ, καὶ τὰ μὲν ἐπὶ τὸ μεῖον ἰόντα διακρίνεται ἐς τὴν ἐλάσσονα χώρην· τὰ δὲ ἐπὶ τὸ μέζον πορευόμενα, συμμισγόμενα ἐξαλλάσσει ἐς τὴν μέζω τάξιν· τὰ δὲ ξεῖνα μὴ ὁμότροπα[4] ὠθεῖται[5] ἐκ χώρης ἀλλοτρίης. ἑκάστη δὲ ψυχὴ μέζω καὶ ἐλάσσω ἔχουσα περιφοιτᾷ τὰ μόρια τὰ ἑωυτῆς, οὔτε προσθέσιος
20 οὔτε[6] ἀφαιρέσιος δεομένη τῶν μερέων, κατὰ δὲ αὔξησιν τῶν ὑπαρχόντων καὶ μείωσιν δεομένη χώρης, ἕκαστα διαπρήσσεται ἐς ἥντινα ἂν

[1] καὶ τὸ . . . μέζονος omitted by M. Diels writes αὔξη τε τῶι μέζονι ἀπὸ τοῦ ἐλάσσονος καὶ τῶι ἐλάσσονι ἀπὸ τοῦ μέζονος. So Fredrich and Wil.
[2] τοῦ δὲ . . . δίδωσι omitted by θ.
[3] οὐδὲν θ M: τοῦ δὲ Littré. Bywater reads οὗ δὲ λαμβάνει.
[4] μὴ ὁμοιότροπα bracketed by Bywater: καὶ μὴ ὁμοιότροπα Diels.

comes to all things from one another mutually, to the greater from the less, and to the less from the greater, and the greater increases from the smaller, and the smaller from the greater.

VI. All other things are set in due order, both the soul of man and likewise his body. Into man enter parts of parts and wholes of wholes, containing a mixture of fire and water, some to take and others to give. Those that take give increase, those that give make diminution. Men saw a log; the one pulls and the other pushes, but herein they do the same thing, and while making less they make more. Such is the nature of man. One part pushes, the other pulls; one part gives, the other takes. It gives to this and takes from that, and to one it gives so much the more, while that from which it takes is so much the less.[1] Each keeps its own place; the parts going to the less are sorted out to the smaller place, those advancing to the greater mingle and pass to the greater rank, and the strange parts, being unsuitable, are thrust from a place that is not theirs. Each individual soul, having greater and smaller parts, makes the round of its own members; needing neither to add to, nor to take from, its parts, but needing space to correspond to increase or decrease of what exists already, it fulfils its several duties into whatsoever space it enters, and receives the

[1] Should we read *ᾧ* for *τῷ*?

[5] *ὠθέεται* M : *ἐκχωρέεται* θ.

[6] *οὐ προσθέσιος οὐδὲ ἀφαιρέσιος δεομένης* M. Diels adds *αὐτὴ δ'* before *οὔτε προσθέσιος*.

ἐσέλθῃ, καὶ δέχεται τὰ προσπίπτοντα. οὐ γὰρ δύναται τὸ μὴ ὁμότροπον ἐν τοῖσιν ἀσυμφόροισι χωρίοισιν ἐμμένειν·[1] πλανᾶται μὲν γὰρ ἀγνώμονα· συγγινόμενα[2] δὲ ἀλλήλοισι γινώσκει πρὸς ὃ προσίζει· προσίζει γὰρ τὸ σύμφορον[3] τῷ συμφόρῳ, τὸ δὲ ἀσύμφορον πολεμεῖ καὶ μάχεται καὶ διαλλάσσει ἀπ' ἀλλήλων. διὰ τοῦτο
30 ἀνθρώπου ψυχὴ ἐν ἀνθρώπῳ αὔξανεται, ἐν ἄλλῳ δὲ οὐδενί· καὶ τῶν ἄλλων ζῴων τῶν μεγάλων ὡσαύτως· ὅσα ἄλλως, ἀπ' ἄλλων[4] ὑπὸ βίης
33 ἀποκρίνεται.

VII. Περὶ μὲν[5] τῶν ἄλλων ζῴων ἐάσω, περὶ δὲ ἀνθρώπου δηλώσω. ἐσέρπει δὲ[6] ἐς ἄνθρωπον ψυχὴ πυρὸς καὶ ὕδατος σύγκρησιν ἔχουσα, μοίρην σώματος ἀνθρώπου· ταῦτα δὲ καὶ θήλεα καὶ ἄρσενα πολλὰ καὶ παντοῖα τρέφεταί τε[7] καὶ αὔξεται διαίτῃ τῇ περὶ τὸν ἄνθρωπον[8] ἀνάγκη δὲ τὰ μέρεα ἔχειν πάντα τὰ ἐσιόντα· οὗτινος γὰρ μὴ ἐνείη μοίρη ἐξ ἀρχῆς οὐκ ἂν[9] αὐξηθείη οὔτε πολλῆς τροφῆς ἐπιούσης οὔτε ὀλίγης, οὐ γὰρ ἔχει
10 τὸ προσαυξόμενον· ἔχον δὲ πάντα, αὔξεται ἐν χώρῃ τῇ ἑωυτοῦ ἕκαστον, τροφῆς ἐπιούσης ἀπὸ ὕδατος ξηροῦ καὶ πυρὸς ὑγροῦ,[10] τὰ μὲν εἴσω βιαζόμενα, τὰ δὲ ἔξω. ὥσπερ οἱ τέκτονες τὸ

[1] ἐμμένων M : μὴ ὁμονοεῖν θ.
[2] συνγινόμενα θ : συμμ·σγόμενα M : συγγνώμονα Diels.
[3] συγγινώσκει προσίζει γὰρ τὸ σύμφορον κ τ. ἑ. M.
[4] ὡσαύτως ὅσα διαλλάσσει ἀπ' ἀλλήλων, Diels : ὅσα ἄλλως θ : ὁκόσα δ' ἄλλως M. ὅσα ἀλλοῖα ἀπ' αὐτῶν Wil.
[5] μὲν θ : μὲν οὖν M.
[6] δὲ θ : γὰρ M.
[7] τρέφεται τὲ καὶ αὔξεται θ : τρέφεται· τρέφεται δὲ καὶ αὔξεται M.
[8] τῆι περὶ τὸν ἄνθρωπον M : τηπερ ἄνθρωπος θ Diels.

attacks that are made. For that which is not suitable cannot abide in regions not adapted to it. Now such wander without thought, but combining with one another they realise what they are joining.[1] For the suitable joins the suitable, while the unsuitable wars and fights and separates itself. For this reason a man's soul grows in a man, and in no other creature. It is the same with the other large animals. When it is otherwise, there is forcible separation from others.

VII. I shall say nothing about the other animals, confining my attention to man. Into man there enters a soul, having a blend of fire and water,[2] a portion of a man's body. These, both female and male, many and of many kinds, are nourished and increased by human diet. Now the things that enter must contain all the parts. For that of which no part were present would not grow at all, whether the nutriment that were added were much or little, as having nothing to grow on to it. But having all, each grows in its own place, nutriment being added from dry water and moist fire, some things being forced inside, others outside. As carpenters saw the log, and one pulls and the

[1] Diels' reading would mean: "They wander when at variance, but when they are of one mind they realise," etc.

[2] That soul is a mixture of fire and water, and that the character of soul is relative to that mixture, is doctrine directly derived from Heracleitus. See p. 493. "It is death to souls to become water."

---

[9] ἂν is not in θ. It could easily fall out before αὐξηθείη; on the other hand, potential optatives without ἂν are not infrequent in the Hippocratic Collection. Bywater puts a comma at ἀρχῆς.

[10] Before τὰ θ has καί.

ξύλον πρίζουσι,[1] καὶ ὁ μὲν ἕλκει, ὁ δὲ ὠθεῖ, τωὐτὸ ποιέοντες· κάτω δ' ὁ πιέζων τὸν ἄνω ἕλκει,[2] οὐ γὰρ ἂν παραδέχοιτο[3] κάτω ἰέναι· ἢν δὲ βιάζωνται,[4] παντὸς ἁμαρτήσονται. τοιοῦτον τροφὴ ἀνθρώπου· τὸ μὲν ἕλκει, τὸ δὲ ὠθεῖ· εἴσω δὲ βιαζόμενον[5] ἔξω ἕρπει· ἢν δὲ βιῆται παρὰ καιρόν, παντὸς ἀποτεύξεται.

19 VIII. Χρόνον δὲ τοσοῦτον ἕκαστα τὴν αὐτὴν τάξιν ἔχει,[6] ἄχρι μηκέτι δέχηται ἡ τροφή, μηδὲ χώρην[7] ἱκανὴν ἔχῃ ἐς τὸ μήκιστον τῶν δυνατῶν· ἔπειτ' ἐναμείβει ἐς τὴν μέζονα χώρην, θήλεα καὶ ἄρσενα, τὸν αὐτὸν τρόπον ὑπὸ βίης καὶ ἀνάγκης διωκόμενα· ὁκόσα δ' ἂν πρότερον ἐμπλήσῃ τὴν πεπρωμένην μοίρην, ταῦτα διακρίνεται πρῶτα, ἅμα δὲ καὶ συμμίσγεται· ἕκαστον μὲν γὰρ διακρίνεται πρῶτα, ἅμα δὲ καὶ συμ-

10 μίσγεται· χώρην δὲ ἀμείψαντα καὶ τυχόντα ἁρμονίης ὀρθῆς ἐχούσης συμφωνίας τρεῖς, συλλήβδην διεξιὸν[8] διὰ πασέων, ζώει καὶ αὔξεται τοῖσιν αὐτοῖσιν οἷσι καὶ πρόσθεν· ἢν δὲ μὴ τύχῃ τῆς ἁρμονίης, μηδὲ σύμφωνα τὰ βαρέα τοῖσιν ὀξέσι γένηται ἐν τῇ πρώτῃ συμφωνίῃ, ἢ τῇ δευτέρῃ, ἢ τῇ διὰ παντός,[9] ἑνὸς[10] ἀπογενομένου

[1] τρυπῶσι θ M: πρίζουσι in corrector's hand over τρυπῶσι θ.

[2] δὲ πιεζόντων ἀνέρπει θ: δὲ πιέζοντον ἄνω ἕρπει M: δ' ὁ πιέζων τὸν ἄνω ἕλκει (from several Paris MSS.) Littré: κάτω δὲ πιεζόντων ἄνω ἕρπει Diels: πιεζόμενον ἄνω Fred.

[3] οὐ γὰρ ἂν παρὰ ⟨καιρὸν⟩ δέχοιτο Diels. See Appendix, p. 296.

[4] βιάζωνται . . . ἁμαρτήσονται M: βιάζηται . . . ἁμαρτήσεται θ.

[5] βιαζόμενον M: βιαζομένου θ. Perhaps the readings of θ are a correction due to a scribe or editor who did not realise that besides the deponent βιάζομαι there exists βιάζω.

other pushes, though they do the same thing. The one that presses below pulls the one above, otherwise the saw could not descend. If force be applied they will lose all. Such is the nutriment of a man. One part pulls, the other pushes; what is forced inside comes outside. But if untimely violence be applied there is no success.[1]

VIII. Each keeps the same position until nourishment no longer receives it, and it has not sufficient room for the greatest possible extension; then it passes into larger room, female and male, driven along in the same manner by force and necessity. Such as first fill the allotted portion are the first to be separated, and at the same time they also commingle. For each separates first, and at the same time also commingles. And if, on changing position, they achieve a correct attunement, which has three harmonic proportionals, covering altogether the octave, they live and grow by the same things as they did before. But if they do not achieve the attunement, and the low harmonize not with the high in the interval of the fourth, of the fifth, or in the octave, then the failure of one makes the whole

[1] There is a kind of "one-way traffic" through the body. Interference with the circuit means disease or death.

---

[6] ἕκαστον τὴν αὐτὴν τάξιν ἔχει θ: ἕκαστα τὴν αὐτὴν τάξιν ἔχει M: ἕκαστα τὴν αὐτὴν ἔχει τάξιν Littré.

[7] ἡ χώρη· μηδὲ τροφὴν M: ἡ τροφῆι· μὴδὲ χώρην θ.

[8] διεξιόντα Erm.: διεξιούσας Mack. See Littré VII. liv. for Bernay's conjecture συλλαβὴν δι' ὀξειῶν.

[9] γένηται. ἡ πρώτηι συμφωνίηι· ἡ δὲ δευτέρηι γένεσις. τὸ διαπαντὸς θ: γένηται, ἡ πρώτη συμφωνίη· ἣν δὲ δευτέρη γένεσις ᾖ, τὸ διαπαντὸς M: γένηται, ἣν ἡ πρώτη συμφωνίη, ἣν ἡ δευτέρη γεννηθῇ ἢ τὸ διὰ παντὸς Littré ("mais le passage est désespéré"): γένηται ἐν τῇ πρώτῃ συμφωνίῃ ἢ τῇ δευτέρῃ ἢ τῇ διὰ παντός, Diels.

[10] For ἑνὸς θ has τινός.

πᾶς ὁ τόνος μάταιος· οὐ γὰρ ἂν προσαείσαι· ἀλλ' ἀμείβει ἐκ τοῦ μέζονος ἐς τὸ μεῖον πρὸ
19 μοίρης· διότι οὐ γινώσκουσιν ὅ τι ποιέουσιν.

IX. Ἀρσένων μὲν οὖν καὶ θηλέων διότι ἑκάτερα γίνεται, προϊόντι τῷ λόγῳ δηλώσω. τούτων δὲ ὁκότερον ἂν τύχῃ ἐλθὸν καὶ[1] τύχῃ τῆς ἁρμονίης, ὑγρὸν ἐὸν κινεῖται ὑπὸ τοῦ πυρός· κινεόμενον δὲ ζωπυρεῖται καὶ προσάγεται τὴν τροφὴν ἀπὸ τῶν ἐσιόντων ἐς τὴν γυναῖκα σίτων καὶ πνεύματος, τὰ μὲν πρῶτα πάντῃ ὁμοίως, ἕως ἔτι ἀραιόν ἐστιν, ὑπὸ δὲ τῆς κινήσιος καὶ τοῦ πυρὸς ξηραίνεται καὶ στερεοῦται· στερεούμενον δὲ πυκνοῦται πέριξ,
10 καὶ τὸ πῦρ ἐγκατακλειόμενον οὐκέτι τὴν τροφὴν ἱκανὴν ἔχει ἐπάγεσθαι, οὐδὲ τὸ πνεῦμα ἐξωθεῖ διὰ τὴν πυκνότητα τοῦ περιέχοντος· ἀναλίσκει οὖν τὸ ὑπάρχον ὑγρὸν εἴσω. τὰ μὲν οὖν στερεὰ τὴν φύσιν ἐν τῷ συνεστηκότι καὶ ξηρῷ οὐ καταναλίσκεται τῷ πυρὶ ἐς τὴν τροφήν· ἀλλ' ἐγκρατέα γίνεται καὶ συνίσταται τοῦ ὑγροῦ ἐκλείποντος, ἅπερ ὀστέα καὶ νεῦρα ὀνομάζεται. τὸ δὲ πῦρ ἐκ τοῦ συμμιγέντος κινεόμενον,[2] τοῦ ὑγροῦ, διακοσμεῖται τὸ σῶμα κατὰ φύσιν διὰ τοιήνδε
20 ἀνάγκην· διὰ μὲν τῶν στερεῶν καὶ ξηρῶν οὐ δύναται τὰς διεξόδους χρονίας ποιεῖσθαι, διότι οὐκ ἔχει τροφήν· διὰ δὲ τῶν ὑγρῶν καὶ μαλακῶν δύναται· ταῦτα γάρ ἐστιν αὐτῷ τροφή· ἔνι δὲ καὶ ἐν τούτοισι ξηρότης οὐ καταναλισκομένη ὑπὸ

[1] ἐλθὸν καὶ θ: omitted by M.
[2] κινεόμενον Diels: the MSS. have the genitive.

scale of no value, as there can be no consonance, but they change from the greater to the less before their destiny. The reason is they know not what they do.

IX. As for males and females, later on in my discourse I shall explain why each severally come to be. But whichever of the two happens to come and achieves the attunement, it is moist and is kept in movement by the fire. Being in movement it gets inflamed, and draws to itself its nourishment from the food and breath that enter the woman. At first, while it is still rare, this occurs equally throughout; but owing to the movement and the fire it dries and solidifies; as it solidifies it hardens all round, and the fire being imprisoned can no longer draw to itself its nourishment in sufficient quantity, while it does not expel the breath owing to the hardness of its envelope. So it consumes the available moisture inside. Now the parts in the compacted, dry mass that are solid in substance are not consumed by the fire for its nourishment, but they prove powerful, and as the moisture fails they become compact, and are called bones and sinews. The fire, meanwhile, being moved[1] out of the moisture which was mixed with it, arranges the body according to nature through the following necessity. Through the hard and dry parts it cannot make itself lasting passages, because it has no nourishment; but it can through the moist and soft, for these are its nourishment. Yet in these too there is dryness not consumed by the fire,

[1] The MSS. reading (*κινουμένου* or *κινευμένου*) will give the rendering: "out of the moisture mixed with it, and put in motion by it, arranges," etc.

τοῦ πυρός· ταῦτα δὲ συνίσταται πρὸς ἄλληλα. τὸ μὲν οὖν ἐσωτάτω καταφραχθὲν πῦρ καὶ πλεῖστόν ἐστι καὶ μεγίστην τὴν διέξοδον ἐποιήσατο· πλεῖστον γὰρ τὸ ὑγρὸν ἐνταῦθα ἐνῆν, ὅπερ κοιλίη καλεῖται· καὶ ἐξέπεσεν ἐντεῦθεν,
30 ἐπεὶ οὐκ εἶχε τροφήν, ἔξω, καὶ ἐποιήσατο τοῦ πνεύματος διεξόδους καὶ τροφῆς ἐπαγωγὴν καὶ διάπεμψιν· τὸ δὲ ἀποκλεισθὲν ἐς[1] ἄλλο σῶμα περιόδους ἐποιήσατο τρισσάς,[2] ὅπερ ἦν ὑγρότατον τοῦ πυρός, ἐν τούτοισι τοῖσι χωρίοισιν, αἵτινες φλέβες καλέονται κοῖλαι· ἐς δὲ τὰ μέσα τούτων τὸ ὑπολειπόμενον τοῦ ὕδατος συνιστάμενον
37 πήγνυται, ὅπερ καλεῖται σάρκες.

X. Ἑνὶ δὲ λόγῳ πάντα διεκοσμήσατο κατὰ τρόπον αὐτὸ ἑωυτῷ τὰ ἐν τῷ σώματι τὸ πῦρ, ἀπομίμησιν τοῦ ὅλου, μικρὰ πρὸς μεγάλα καὶ μεγάλα πρὸς μικρά· κοιλίην μὲν τὴν μεγίστην, ὕδατι ξηρῷ καὶ ὑγρῷ ταμεῖον, δοῦναι πᾶσι καὶ λαβεῖν παρὰ πάντων, θαλάσσης δύναμιν, ζῴων συμφόρων[3] τροφόν, ἀσυμφόρων δὲ φθορόν· περὶ δὲ ταύτην ὕδατος ψυχροῦ καὶ ὑγροῦ σύστασιν, διέξοδον πνεύματος ψυχροῦ καὶ θερμοῦ· ἀπομίμησιν
10 γῆς, τὰ ἐπεισπίπτοντα πάντα ἀλλοιούσης. καταναλίσκον δὲ καὶ αὖξον[4] σκέδασιν ὕδατος λεπτοῦ καὶ πυρὸς ἐποιήσατο ἠερίου,[5] ἀφανέος καὶ φανεροῦ, ἀπὸ τοῦ συνεστηκότος ἀπόκρισιν, ἐν ᾧ φερόμενα ἐς τὸ φανερὸν ἀφικνεῖται ἕκαστον μοίρῃ

[1] Between ἐς and ἄλλο Diels inserts τό.

[2] περιόδους ἐποιήσατο τρισσάς is placed by Diels after χωρίοισιν. Fredrich marks an hiatus after κοῖλαι.

[3] συμφόρων Wilamowitz: συντρόφων θ: ἐντρόφων M.

and these dry parts become compacted one with another. So the fire shut up in the innermost part both is most abundant and made for itself the greatest passage. For there the moisture was most abundant, and it is called the belly. Therefrom the fire burst forth, since it had no nourishment, and made passages for the breath and to supply and distribute nourishment. The fire shut up in the rest of the body made itself three passages, the moistest part of the fire being in those places called the hollow veins. And in the middle of these that which remains of the water becomes compacted and congeals. It is called flesh.

X. In a word, all things were arranged in the body, in a fashion conformable to itself, by fire, a copy of the whole, the small after the manner of the great and the great after the manner of the small. The belly is made the greatest, a steward for dry water and moist, to give to all and to take from all, the power of the sea, nurse of creatures suited to it, destroyer of those not suited. And around it a concretion of cold water and moist, a passage for cold breath and warm, a copy of the earth, which alters all things that fall into it. Consuming and increasing,[1] it made a dispersion of fine water and of ethereal fire, the invisible and the visible, a secretion from the compacted substance, in which things are carried and come to light, each

[1] With the reading of Diels: "Consuming some and increasing other."

---

[4] καταναλίσκοντα δὲ αὖξον (αὔξον θ) θM: καταναλίσκον δὲ καὶ αὖξον Zwinger, Littré: καὶ τὰ μὲν καταναλίσκον, τὰ δὲ αὖξον Diels.

[5] Should we read ἀραιοῦ?

πεπρωμένῃ. ἐν δὲ τούτῳ ἐποιήσατο τὸ πῦρ[1] περιόδους τρισσάς, περαινούσας πρὸς ἀλλήλας καὶ εἴσω καὶ ἔξω· αἱ μὲν πρὸς τὰ κοῖλα τῶν ὑγρῶν, σελήνης δύναμιν, αἱ δὲ πρὸς τὴν ἔξω περιφορήν,[2] πρὸς τὸν περιέχοντα πάγον, ἄστρων
20 δύναμιν, αἱ δὲ μέσαι καὶ εἴσω καὶ ἔξω περαίνουσαι.[3] τὸ θερμότατον καὶ ἰσχυρότατον πῦρ, ὅπερ πάντων ἐπικρατεῖται, διέπον ἅπαντα κατὰ φύσιν, ἄϊκτον[4] καὶ ὄψει καὶ ψαύσει, ἐν τούτῳ ψυχή, νόος, φρόνησις, αὔξησις, κίνησις, μείωσις, διάλλαξις,[5] ὕπνος, ἔγερσις· τοῦτο πάντα διὰ παντὸς κυβερνᾷ,
26 καὶ τάδε καὶ ἐκεῖνα, οὐδέποτε ἀτρεμίζον.

XI. Οἱ δὲ ἄνθρωποι ἐκ τῶν φανερῶν τὰ ἀφανέα σκέπτεσθαι οὐκ ἐπίστανται· τέχνῃσι γὰρ χρεόμενοι ὁμοίῃσιν ἀνθρωπίνῃ φύσει οὐ γινώσκουσιν· θεῶν γὰρ νόος ἐδίδαξε μιμεῖσθαι τὰ ἑωυτῶν, γινώσκοντας ἃ ποιέουσι, καὶ οὐ γινώσκοντας ἃ μιμέονται. πάντα γὰρ ὅμοια, ἀνόμοια ἐόντα· καὶ σύμφορα πάντα, διάφορα ἐόντα· διαλεγόμενα, οὐ διαλεγόμενα· γνώμην ἔχοντα, ἀγνώμονα· ὑπεναντίος ὁ τρόπος ἑκάστων, ὁμολογεόμενος.
10 νόμος γὰρ καὶ φύσις, οἷσι πάντα διαπρησσόμεθα, οὐχ ὁμολογεῖται ὁμολογεόμενα· νόμον γὰρ ἄνθρωποι ἔθεσαν αὐτοὶ ἑωυτοῖσιν, οὐ γινώσκοντες περὶ ὧν ἔθεσαν, φύσιν δὲ πάντων θεοὶ διεκόσμησαν.[6]

[1] τὸ πῦρ Diels : πυρὸς MSS.
[2] αἱ δὲ ⟨ὡς⟩ πρὸς τὴν ἔξω περιφορὴν Diels. Perhaps a gloss.
[3] After περαίνουσαι Diels has a comma, followed by ⟨πρὸς τὰς ἑτέρας, ἡλίου δύναμιν,⟩ τὸ θερμότατον.
[4] ἀοικτον θ : ἄψοφον M : ἄϊκτον Littré : ἄθικτον Bernays, Diels. Bernays himself preferred ἄψαυστον.
[5] θ omits κίνησις, μείωσις, διάλλαξις

according to its allotted portion. And in this fire made for itself three groups of circuits, within and without each bounded by the others: those towards the hollows of the moist, the power of the moon; those towards the outer circumference, towards the solid enclosure, the power of the stars; the middle circuits, bounded both within and without. The hottest and strongest fire, which controls all things, ordering all things according to nature, imperceptible to sight or touch, wherein are soul, mind, thought, growth, motion, decrease, mutation, sleep, waking. This governs all things always, both here and there, and is never at rest.

XI. But men do not understand how to observe the invisible through the visible. For though the arts they employ are like the nature of man, yet they know it not. For the mind of the gods taught them to copy their[1] own functions, and though they know what they are doing yet they know not what they are copying. For all things are like, though unlike, all compatible though incompatible, conversing though not conversing, intelligent without intelligence. The fashion of each is contrary, though in agreement. For custom and nature, by means of which we accomplish all things, do not agree though they do agree. For custom was settled by men for themselves without their knowing those things about which they settled the custom; but the nature of all things was

[1] Probably "the operations of their own bodies," but Littré translates: "les opérations divines."

[6] *θεὸς διεκόσμησεν θ*, altered to *θεοὶ διεκόσμησαν* (or the reverse). Is the latter a Christian correction?

τὰ μὲν οὖν ἄνθρωποι διέθεσαν οὐδέποτε κατὰ τωὐτὸ ἔχει οὔτε ὀρθῶς οὔτε μὴ ὀρθῶς· ὁκόσα δὲ θεοὶ διέθεσαν ἀεὶ ὀρθῶς ἔχει· καὶ τὰ ὀρθὰ καὶ
17 τὰ μὴ ὀρθὰ τοσοῦτον διαφέρει.

XII. Ἐγὼ δὲ δηλώσω τέχνας φανερὰς ἀνθρώπου παθήμασιν ὁμοίας ἐούσας καὶ φανεροῖσι καὶ ἀφανέσι. μαντικὴ τοιόνδε· τοῖσι μὲν φανεροῖσι τὰ ἀφανέα γινώσκει, καὶ τοῖσιν ἀφανέσι τὰ φανερά, καὶ τοῖσιν ἐοῦσι τὰ μέλλοντα, καὶ τοῖσιν ἀποθανοῦσι τὰ ζῶντα, καὶ τῷ ἀσυνέτῳ[1] συνίασιν, ὁ μὲν εἰδὼς ἀεὶ ὀρθῶς, ὁ δὲ μὴ εἰδὼς ἄλλοτε ἄλλως. φύσιν ἀνθρώπου καὶ βίον ταῦτα μιμεῖται· ἀνὴρ γυναικὶ συγγενόμενος παιδίον
10 ἐποίησε· τῷ φανερῷ τὸ ἄδηλον γινώσκει ὅτι οὕτως ἔσται. γνώμη[2] ἀνθρώπου ἀφανὴς γινώσκουσα τὰ φανερὰ ἐκ παιδὸς ἐς ἄνδρα μεθίσταται· τῷ ἐόντι τὸ μέλλον γινώσκει. οὐχ ὅμοιον ἀποθανὼν ζώοντι· τῷ τεθνηκότι οἶδεν τὸ ζῷον.[3] ἀσύνετον γαστήρ· ταύτῃ συνίεμεν ὅτι διψῇ ἢ πεινῇ. ταῦτα[4] μαντικῆς τέχνης καὶ φύσιος ἀνθρωπίνης πάθεα, τοῖσι μὲν γινώσκουσιν ἀεὶ ὀρθῶς, τοῖσι δὲ μὴ γινώσκουσιν ἀεὶ ἄλλοτε
19 ἄλλως.

[1] τῶν ἀσυνέτων (θ omits τῶν) MSS. : τῷ ἀσυνέτῳ Bywater.

[2] ὅτι ⟨συνέλαβεν⟩· οὕτως ἔσται γνώμη Diels.

[3] οὐχ ὅμοιον ἀποθανὼν ζώοντι· τῷ τεθνηκότι οἶδεν τὸ ζῷιον Diels : οὐχ ὅμοιον ἀποθανὼν ζῶντι τῶ τεθνηκότι οἶδεν τὸ ζωιον θ : οὐχ ὅμοιον ἀπὸ θανάτου ζώοντι. τῶι τεθνηκότι τὸ ζῶον οἶδε M : οὐχ ὁ μὴ ὢν ἀπὸ θανάτου, ζῶον δὲ, κ.τ.ε. Littré : διότι οὐχ ὅμοιον τὸ ἀποθανὸν τῷ ζώοντι Ermerins.

arranged by the gods. Now that which men arrayed never remains constant, whether right or wrong; but whatsoever things were arranged by the gods always remain right. So great the difference between the right and the wrong.

XII. But I will show that arts are visibly like to the affections of man, both visible and invisible. Seercraft is after this fashion. By the visible it gets knowledge of the invisible, by the invisible knowledge of the visible, by the present knowledge of the future, by the dead knowledge of the living, and by means of that which understands not men have understanding—he who knows, right understanding always, he who knows not, sometimes right understanding, sometimes wrong. Seercraft herein copies the nature and life of man. A man by union with a woman begets a child; by the visible he gets knowledge of the invisible that so it will be. The invisible human intelligence, getting knowledge of the visible, changes from childhood to manhood; by the present it gets knowledge of the future. A corpse is not like a living creature; by the dead he knows the living. The belly is without consciousness, yet by it we are conscious of hunger and thirst. The characteristics of seercraft and of human nature are these:[1] for those who know, always rightly interpreted;[2] for those who know not, sometimes rightly and sometimes not.

[1] Or (with ταὐτά) "the same."

[2] The grammar is curious. With ὀρθῶς some participle (or verb) must be understood, perhaps γινωσκόμενα or γινώσκεται. The ἀεὶ before ἄλλοτε is suspicious.

---

[4] For ταῦτα perhaps we should read ταὐτά.

XIII. Σιδήρου ὄργανα· † τέχνησι †[1] τὸν σίδηρον πυρὶ[2] τήκουσι, πνεύματι ἀναγκάζοντες τὸ πῦρ, τὴν ὑπάρχουσαν τροφὴν ἀφαιρέονται,[3] ἀραιὸν δὲ ποιήσαντες παίουσι καὶ συνελαύνουσιν, ὕδατος δὲ ἄλλου τροφῇ ἰσχυρὸν γίνεται. ταῦτα[4] πάσχει ἄνθρωπος ὑπὸ παιδοτρίβου· τὴν ὑπάρχουσαν τροφὴν πυρὶ ἀφαιρεῖται, ὑπὸ πνεύματος ἀναγκαζόμενος· ἀραιούμενος[5] κόπτεται, τρίβεται, καθαίρεται, ὑδάτων δὲ ἐπαγωγῇ[6] ἄλλοθεν ἰσχυρὸς
10 γίνεται.

XIV. Καὶ οἱ γναφέες τοῦτο[7] διαπρήσσονται· λακτίζουσι, κόπτουσιν, ἕλκουσι, λυμαινόμενοι ἰσχυρότερα ποιέουσι,[8] κείροντες τὰ ὑπερέχοντα καὶ παραπλέκοντες καλλίω ποιέουσι· ταὐτὰ
5 πάσχει ἄνθρωπος.

XV. Σκυτέες τὰ ὅλα κατὰ[9] μέρεα διαιρέουσι, καὶ τὰ μέρεα ὅλα ποιέουσι, τάμνοντες δὲ καὶ κεντέοντες τὰ σαθρὰ ὑγιέα ποιέουσιν. καὶ ἄνθρωπος δὲ ταὐτὰ[10] πάσχει· ἐκ τῶν ὅλων μέρεα διαιρεῖται, καὶ ἐκ τῶν μερέων συντιθεμένων ὅλα γίνεται· κεντεόμενοί τε καὶ τεμνόμενοι τὰ σαθρὰ ὑπὸ τῶν ἰητρῶν ὑγιάζονται·[11] καὶ τόδε ἰητρικῆς· τὸ λυπέον ἀπαλλάσσειν, καὶ ὑφ' οὗ πονεῖ ἀφαιρέοντα ὑγιέα ποιεῖν. ἡ φύσις αὐτομάτη ταῦτα[12]
10 ἐπίσταται· καθήμενος πονεῖ ἀναστῆναι, κινεό-

[1] τέχνησι (θ) is corrupt, and so is the τέχνης of M. Perhaps we should read τεχνῖται.
[2] πυρὶ τήκουσι Bywater: περιτήκουσι MSS.
[3] ἀφαιρέονται θ: ἀφαιρέοντες M.
[4] ταῦτα MSS.: ταὐτὰ Ermerins and Diels.
[5] ἀραιούμενα θ: ὑπὸ τῶν θ: ἀραιούμενος δὲ M.
[6] ἐπαγωγῆι θ: ὑπαγωγὴ M: ὑπὸ τῶν for ὑδάτων θ.
[7] τοῦτο θ: τὠυτὸ M.

XIII. Iron tools. Craftsmen melt the iron with fire, constraining the fire with breath; they take away the nourishment it has already; when they have made it rare, they beat it and weld it; and with the nourishment of other water it grows strong. Such is the treatment of a man by his trainer. By fire the nourishment he has already is taken away, breath constraining him. As he is made rare, he is struck, rubbed and purged. On the application of water from elsewhere he becomes strong.

XIV. This do also the fullers. They trample, strike and pull; by maltreating they make stronger; by cutting off the threads that project, or by weaving them in, they beautify. The same happens to a man.

XV. Cobblers divide wholes into parts and make the parts wholes; cutting and stitching they make sound what is rotten. Man too has the same experience. Wholes are divided into parts, and from union of the parts wholes are formed. By stitching and cutting, that which is rotten in men is healed by physicians. This too is part of the physician's art: to do away with that which causes pain, and by taking away the cause of his suffering to make him sound. Nature of herself knows how to do these things. When a man is sitting it is a labour to rise; when he is moving it is a labour to come

---

[8] κόπτουσιν· ἕλκουσι· λυμαινόμενοι· ἰσχυρότερα ποιέουσι· θ: παίουσι λυμαινόμενοι κόπτουσι ἕλκουσι λυμαινόμενοι ἰσχυρότερα ποιέουσι· M.

[9] κατα M: καὶ τὰ θ.

[10] ταῦτα θ: τῶιϋτὸ M: ταὐτὰ Bywater.

[11] ὑγιάζονται Ermerins: ὑγιαίνονται MSS.

[12] Bywater has ταὐτά.

μενος πονεῖ ἀναπαύσασθαι, καὶ ἄλλα τὰ αὐτὰ
12 ἔχει ἡ φύσις ἰητρική.[1]

XVI. Τέκτονες πρίοντες ὁ μὲν ὠθεῖ, ὁ δὲ ἕλκει· τὸ αὐτὸ ποιέοντες ἀμφοτέρως·[2] † τρυπῶσιν, ὁ μὲν ἕλκει, ὁ δὲ ὠθεῖ· † πιεζόντων ἄνω ἕρπει, τὸ δὲ κάτω· μείω ποιέοντες[3] πλείω ποιέουσι.[4] φύσιν ἀνθρώπου μιμέονται. πνεῦμα τὸ μὲν ἕλκει, τὸ δὲ ὠθεῖ· τὸ αὐτὸ ποιεῖ ἀμφοτέρως·[5] τὰ μὲν κάτω πιέζεται, τὰ δὲ ἄνω ἕρπει. ἀπὸ μιῆς ψυχῆς διαιρεομένης πλείους καὶ μείους καὶ
9 μέζονες καὶ ἐλάσσονες.

XVII. Οἰκοδόμοι ἐκ διαφόρων σύμφορον[6] ἐργάζονται, τὰ μὲν ξηρὰ ὑγραίνοντες, τὰ δὲ ὑγρὰ ξηραίνοντες, τὰ μὲν ὅλα διαιρέοντες, τὰ δὲ διῃρημένα συντιθέντες· μὴ οὕτω δὲ ἐχόντων οὐκ ἂν ἔχοι ᾗ δεῖ.[7] δίαιταν ἀνθρωπίνην μιμεῖται· τὰ μὲν ξηρὰ ὑγραίνοντες, τὰ δὲ ὑγρὰ ξηραίνοντες, τὰ μὲν ὅλα διαιρέουσι, τὰ δὲ διῃρημένα συντι-
8 θέασι· ταῦτα πάντα διάφορα ἐόντα συμφέρει.[8]

XVIII. [Μουσικῆς ὄργανον ὑπάρξαι δεῖ πρῶτον, ἐν ᾧ δηλώσει ἃ βούλεται] ἁρμονίης συντάξιες[9] ἐκ τῶν αὐτῶν οὐχ αἱ αὐταί, ἐκ τοῦ ὀξέος, ἐκ τοῦ βαρέος, ὀνόματι μὲν ὁμοίων, φθόγγῳ δὲ οὐχ

[1] Bywater's emendation. The MSS. have τοιαῦτα or τὰ τοιαῦτα and ἰητρικῆς.

[2] τωὕτὸ ποιέει. ἀμφοτέρως φέρει M: τῶιυτο ποιέοντες ἀμφότεροι θ. See Appendix, p. 296.

[3] μιον οἷοι ἐόντες M: μιω ποιέοντες θ.

[4] After ποιέουσι Littré adds καὶ πλείω ποιέοντες μείω ποιέουσι because the Latin MS. 7027 has *et maius facientes minuunt.*

[5] ἀμφοτέρως φέρει, σίτων M: καὶ ἀμφοτέρως· θ. M also has ποιέειν before ἀμφοτέρως.

[6] θ omits σύμφορον.

[7] ηδει θ: ἰδίως M.

to rest. In other respects too nature is the same as the physician's art.

XVI. When carpenters saw, one pushes and the other pulls, in both cases doing the same thing. [When boring, one pulls and the other pushes.] When they press the tool, this goes up, that goes down.[1] When they diminish they increase. They are copying the nature of man. This draws breath in, that expels it; in both cases the same thing is done. Some parts ⟨of the food⟩ are pressed down, some come up. From one soul when divided come more and less, greater and smaller.

XVII. Builders out of diverse materials fashion a harmony, moistening what is dry, drying what is moist, dividing wholes and putting together what is divided. Were this not so, the result would not be what it should. It is a copy of the diet of man; moistening the dry, drying the moist, they divide wholes and put together what is divided. All these being diverse are harmonious.

XVIII. [First there must be an instrument of music, whereby to set forth what is intended.] From the same notes come musical compositions that are not the same, from the high and from the low, which are alike in name[2] but not alike in

[1] Probably this means that as the saw goes down the log appears to come up and *vice versa*. Perhaps, however, τὸ δὲ applies to a different action of the saw, "When they press, it first goes up, then down."

[2] *I.e.* they are all called "notes."

---

[8] After συμφέρει M adds τῆι φύσει.

[9] ἁρμονίη συντάξιες M: ἁρμονίης σύνταξις θ. The words Μουσικῆς . . . βούλεται should probably be deleted as a marginal note which has been incorporated into the text.

ὁμοίων· τὰ πλεῖστον[1] διάφορα μάλιστα συμφέρει, τὰ δὲ ἐλάχιστον διάφορα ἥκιστα συμφέρει· εἰ δὲ ὅμοια πάντα ποιήσει τις, οὐκέτι[2] τέρψις· αἱ πλεῖσται μεταβολαὶ καὶ πολυειδέσταται μάλιστα τέρπουσιν.

10 Μάγειροι ὄψα σκευάζουσιν ἀνθρώποισι διαφόρων, συμφόρων, παντοδαπὰ συγκρίνοντες, ἐκ τῶν αὐτῶν οὐ τὰ αὐτά, βρῶσιν καὶ πόσιν ἀνθρώπῳ·[3] ἢν δὲ πάντα ὅμοια ποιήσῃ, οὐκ ἔχει τέρψιν· οὐδ' εἰ ἐν τῷ αὐτῷ πάντα συντάξειεν, οὐκ ἂν ἔχοι ὀρθῶς. κρούεται τὰ κρούματα ἐν μουσικῇ τὰ μὲν ἄνω, τὰ δὲ κάτω. γλῶσσα μουσικὴν μιμεῖται διαγινώσκουσα μὲν τὸ γλυκὺ καὶ τὸ ὀξὺ τῶν προσπιπτόντων, καὶ διάφωνα καὶ σύμφωνα· κρούεται δὲ τοὺς φθόγγους[4] ἄνω 20 καὶ κάτω, καὶ οὔτε τὰ ἄνω κάτω κρουόμενα ὀρθῶς ἔχει οὔτε τὰ κάτω ἄνω· καλῶς δὲ ἡρμοσμένης γλώσσης, τῇ συμφωνίῃ[5] τέρψις, ἀναρμόστου δὲ 23 λύπη.

XIX. Νακοδέψαι[6] τείνουσι, τρίβουσι, κτενίζουσι, πλύνουσι· ταὐτὰ[7] παιδίων θεραπείη. πλοκέες ἄγοντες κύκλῳ πλέκουσιν, ἀπὸ τῆς ἀρχῆς ἐς τὴν ἀρχὴν τελευτῶσι· τὸ αὐτὸ[8] περίοδος 5 ἐν τῷ σώματι, ὁκόθεν ἄρχεται, ἐπὶ τοῦτο τελευτᾷ.

XX. Χρυσίον ἐργάζονται, κόπτουσι, πλύνουσι, τήκουσι· πυρὶ μαλακῷ, ἰσχυρῷ δὲ οὔ, συνίσταται· ἀπειργασμένοι πρὸς πάντα χρῶνται· ἄνθρωπος

[1] πλεῖστον and ἐλάχιστον Wilamowitz: πλεῖστα and ἐλάχιστα MSS.

[2] οὐκέτι θ: οὐκ ἔνι M. [3] ἀνθρώπωι θ: ἀνθρώπων M.

[4] τοὺς φθόγγους, bracketed by Diels after Bywater.

[5] τῇ συμφωνίηι θ: τῆς συμφωνίης M.

sound. Those that are most diverse make the best harmony; those that are least diverse make the worst. If a musician composed a piece all on one note, it would fail to please. It is the greatest changes and the most varied that please the most.

Cooks prepare for men dishes of ingredients that disagree while agreeing, mixing together things of all sorts, from things that are the same, things that are not the same, to be food and drink for a man. If the cook make all alike there is no pleasure in them; and it would not be right either if he were to compound all things in one dish. The notes struck while playing music are some high, some low. The tongue copies music in distinguishing, of the things that touch it, the sweet and the acid, the discordant from the concordant. Its notes are struck high and low, and it is well neither when the high notes are struck low nor when the low are struck high. When the tongue is well in tune the concord pleases, but there is pain when the tongue is out of tune.

XIX. Curriers stretch, rub, comb and wash. Children are tended in the same way. Basket-makers turn the baskets round as they plait them; they end at the place from which they begin. The circuit in the body is the same; it ends where it begins.

XX. Men work on gold, beat it, wash it and melt it. With gentle, not strong, fire it is compacted. When they have wrought it they use it for all purposes. So a man beats corn, washes it,

[6] ιακοδέψαι θ: σκυτοδεψαι M. [7] τοὺτὰ Ermerins: ταῦτα MSS.
[8] τὸ αὐτὸ Diels: τοῦτο MSS. M omits συμφόρων (l. 11) and καὶ σύμφωνα (l. 19); and θ omits χρῶνται (l. 8) and reads ἀπεργασάμενοι.

σῖτον κόπτει, πλύνει, ἀλήθει, πυρώσας χρῆται· ἰσχυρῷ μὲν πυρὶ ἐν τῷ σώματι οὐ συνίσταται, 6 μαλακῷ δέ.

XXI. Ἀνδριαντοποιοὶ μίμησιν σώματος ποιέουσιν πλὴν ψυχῆς,[1] γνώμην δὲ ἔχοντα οὐ ποιέουσιν, ἐξ ὕδατος καὶ γῆς, τὰ ὑγρὰ ξηραίνοντες καὶ τὰ ξηρὰ ὑγραίνοντες·[2] ἀφαιρέονται ἀπὸ τῶν ὑπερεχόντων, προστιθέασι πρὸς τὰ ἐλλείποντα, ἐκ τοῦ ἐλαχίστου ἐς τὸ μήκιστον αὔξοντες. ταῦτα[3] πάσχει καὶ ἄνθρωπος· αὔξεται ἐκ τοῦ ἐλαχίστου ἐς τὸ μέγιστον, ἐκ τῶν ὑπερεχόντων ἀφαιρεόμενος, τοῖσιν ἐλλείπουσι προστιθείς, τὰ 10 ξηρὰ ὑγραίνων καὶ τὰ ὑγρὰ ξηραίνων.

XXII. Κεραμέες τροχὸν δινέουσι, καὶ οὔτε πρόσω οὔτε ὀπίσω προχωρεῖ, † ἀμφοτέρως ἅμα τοῦ ὅλου ἀπομιμα τῆς περιφορῆς·†[4] ἐν δὲ τῷ αὐτῷ ἐργάζονται περιφερομένῳ παντοδαπά, οὐδὲν ὅμοιον τὸ ἕτερον τῷ ἑτέρῳ ἐκ τῶν αὐτῶν τοῖσιν αὐτοῖσιν ὀργάνοισιν.[5] ἄνθρωποι ταῦτα[6] πάσχουσι καὶ τἄλλα ζῷα· ἐν τῇ αὐτῇ περιφορῇ πάντα ἐργάζονται, ἐκ τῶν αὐτῶν οὐδὲν ὅμοιον τοῖσιν αὐτοῖσιν ὀργάνοισιν, ἐξ ὑγρῶν ξηρὰ 10 ποιέοντες καὶ ἐκ τῶν ξηρῶν ὑγρά.

XXIII. Γραμματικὴ τοιόνδε· σχημάτων σύνθεσις, σημήϊα ἀνθρωπίνης φωνῆς, δύναμις τὰ παροιχόμενα μνημονεῦσαι, τὰ ποιητέα δηλῶσαι· δι' ἑπτὰ σχημάτων ἡ γνῶσις· ταῦτα πάντα

[1] πλὴν ψυχῆς is bracketed by Diels. It has the appearance of a note that has crept into the text.
[2] καὶ τὰ ξηρὰ ὑγραίνοντες omitted by M.
[3] Perhaps ταὐτά.

grinds it, applies fire and then uses it. With strong fire it is not compacted in the body, but with gentle fire.

XXI. Statue-makers copy the body without the soul, as they do not make intelligent things, using water and earth, drying the moist and moistening the dry. They take from that which is in excess and add to that which is deficient, making their creations grow from the smallest to the tallest. Such is the case of man. He grows from his smallest to his greatest, taking away from that which is in excess, adding to that which is deficient, moistening the dry and drying the moist.

XXII. Potters spin a wheel, which shifts neither forwards nor backwards, yet moves both ways at once, therein copying the revolution of the universe. On this wheel as it revolves they make pottery of every shape, and no two pieces are alike, though they are made from the same materials and with the same tools. Men and the animals too are in the same case. In one and the same revolution they make all things, without two being alike, from the same materials and with the same tools, making dry from moist and moist from dry.

XXIII. The art of writing is of this sort: the putting together of figures, symbols of human voice, a power to recall past events, to set forth what must be done. Through seven figures[1] comes

[1] *I.e.* the seven vowels α, ε, η, ι, ο, υ, ω.

---

[4] ἀμφοτέρως ἅμα τοῦ ὅλου ἀπομιμα (*sic*) τῆς περιφορῆς (θ) is corrupt. M has καὶ ἀμφοτέρωσε, ἅμα τοῦ ὅλου μιμητὴς περιφερῆς. Diels would read ἄγει for ἅμα.

[5] τοῖσιν αὐτοῖσιν ὀργάνοισιν omitted by M.

[6] Perhaps ταὐτά.

ἄνθρωπος διαπρήσσεται, καὶ ὁ ἐπιστάμενος γράμματα καὶ ὁ μὴ ἐπιστάμενος. δι' ἑπτὰ σχημάτων καὶ αἱ αἰσθήσεις ἀνθρώπῳ,[1] ἀκοὴ ψόφου, ὄψις φανερῶν, ῥὶν ὀδμῆς, γλῶσσα ἡδονῆς καὶ ἀηδίης, στόμα διαλέκτου, σῶμα ψαύσιος, θερμοῦ ἢ
10 ψυχροῦ πνεύματος διέξοδοι ἔξω καὶ ἔσω· διὰ
11 τούτων ἀνθρώποισιν γνῶσις, ἀγνωσίη.[2]

XXIV. Παιδοτριβίη τοιόνδε· διδάσκουσι παρανομεῖν κατὰ νόμον, ἀδικεῖν δικαίως, ἐξαπατᾶν, κλέπτειν, ἁρπάζειν, βιάζεσθαι τὰ αἴσχιστα καὶ κάλλιστα·[3] ὁ μὴ ταῦτα ποιέων κακός, ὁ δὲ ταῦτα ποιέων ἀγαθός· ἐπίδειξις τῆς τῶν πολλῶν ἀφροσύνης· θεῶνται ταῦτα καὶ κρίνουσιν ἕνα ἐξ ἁπάντων ἀγαθόν, τοὺς δὲ ἄλλους κακούς· πολλοὶ θαυμάζουσιν, ὀλίγοι γινώσκουσιν. ἐς ἀγορὴν ἐλθόντες ἄνθρωποι ταὐτὰ διαπρήσσονται· ἐξα-
10 πατῶσι ἄνθρωποι πωλέοντες καὶ ὠνεόμενοι· ὁ πλεῖστα ἐξαπατήσας, οὗτος θαυμάζεται. πίνοντες καὶ μαινόμενοι ταὐτὰ διαπρήσσονται. τρέχουσι, παλαίουσι, μάχονται, κλέπτουσιν, ἐξαπατῶσιν· εἷς ἐκ πάντων κρίνεται. ὑποκριτικὴ ἐξαπατᾷ εἰδότας· λέγουσιν ἄλλα καὶ φρονέουσιν ἕτερα,[4] οἱ αὐτοὶ ἐσέρπουσι καὶ ἐξέρπουσιν οὐχ οἱ αὐτοί· ἔνι[5] δὲ ἀνθρώπῳ ἄλλα μὲν λέγειν, ἄλλα δὲ ποιεῖν,[6] καὶ τὸν αὐτὸν μὴ εἶναι τὸν αὐτόν, καὶ τότε μὲν

[1] καὶ αἱ αἰσθήσεις ἀνθρώπων θ: καὶ ἡ αἴσθησις ἡ ἀνθρώπων M: ἀνθρώπῳ Bywater.

[2] So θ. M has γνῶσις ἀνθρώποισι· ἀγωνίη.

[3] τὰ κάλλιστα καὶ αἴσχιστα M: τὰ αἴσχιστα καὶ κάλλιστα θ. We ought perhaps to delete καί, which might easily be a repetition of the first syllable of κάλλιστα.

[4] εἰδότας ἃ λέγουσιν ἀλλὰ καὶ φρονέουσιν· θ: εἰδότας λέγουσι ἄλλα. καὶ φρονέουσιν ἕτερα· M. Diels suggests ἄλλα λέγουσιν

knowledge. All these things a man performs, both he who knows letters and he who knows them not. Through seven figures come sensations for a man; there is hearing for sounds, sight for the visible, nostril for smell, tongue for pleasant or unpleasant tastes, mouth for speech, body for touch, passages outwards and inwards for hot or cold breath. Through these comes knowledge or lack of it.

XXIV. The trainer's art is of this sort: they teach how to transgress the law according to law, to be unjust justly, to deceive, to trick, to rob, to do the foulest violence most fairly. He who does not these things is bad; he who does them is good. It is a display[1] of the folly of the many. They behold these things and judge one man out of all to be good and the others to be bad. Many admire, few know. Men come to the market-place and do the same things; men deceive when they buy and sell. He who has deceived most is admired. When drinking and raving they do the same things. They run, they wrestle, they fight, they trick, they deceive. One out of them all is judged. The actor's art deceives those who know. They say one thing and think another; they come on and go off, the same persons yet not the same. A man too can say one thing and do another; the same man can be not the same; he may be now of one mind, now of

[1] Bernays suggested ἀπόδειξις, "proof."

καὶ ἄλλα φρονέουσιν, which is probably the correct reading. Peck suggests ἰδόντας for εἰδότας. So too Fredrich.

[5] ἑνὶ MSS. I thought of ἔνι before I knew that the suggestion had already been made by Bernays.

[6] ποιεῖν M: ἀκούειν θ.

ἄλλην τότε δὲ ἄλλην ἔχειν γνώμην.[1] οὕτω μὲν
20 αἱ τέχναι πᾶσαι τῇ ἀνθρωπίνῃ φύσει ἐπικοινω-
21 νέουσιν.

XXV. Ἡ δὲ ψυχὴ τοῦ ἀνθρώπου, ὥσπερ μοι καὶ προείρηται, σύγκρησιν ἔχουσα πυρὸς καὶ ὕδατος, μέρεα δὲ ἀνθρώπου, ἐσέρπει ἐς ἅπαν ζῷον, ὅ τι περ ἀναπνεῖ, καὶ δὴ καὶ ἐς ἄνθρωπον πάντα καὶ νεώτερον καὶ πρεσβύτερον. αὔξεται δὲ οὐκ[2] ἐν πᾶσιν ὁμοίως, ἀλλ' ἐν μὲν τοῖσι νέοισι τῶν σωμάτων, ἅτε ταχείης ἐούσης τῆς περιφορῆς καὶ τοῦ σώματος αὐξίμου, ἐκπυρουμένη καὶ λεπτυνομένη καταναλίσκεται ἐς τὴν αὔξησιν
10 τοῦ σώματος· ἐν δὲ τοῖσι πρεσβυτέροισιν, ἅτε βραδέης ἐούσης τῆς κινήσιος καὶ ψυχροῦ τοῦ σώματος, καταναλίσκεται ἐς τὴν μείωσιν τοῦ ἀνθρώπου. ὅσα δὲ τῶν σωμάτων ἀκμάζοντά ἐστι καὶ ἐν[3] τῇσιν ἡλικίῃσι τῇσι γονίμῃσι, δύναται τρέφειν καὶ αὔξειν· δυνάστης δε ἄνθρωπος, ὅστις δύναται πλείστους ἀνθρώπους τρέφειν, οὗτος[4] ἰσχυρός· ἀπολειπόντων δέ, ἀσθενέστερος. τοιοῦτον καὶ ἕκαστα τῶν σωμάτων· ὁκοῖα πλείστας δύναται ψυχὰς τρέφειν, ταῦτα ἰσχυρότατα,[5]
20 ἀπελθόντων δὲ τούτων ἀσθενέστερα.

XXVI. Ὅ τι μὲν ἂν ἐς ἄλλο ἐσέλθῃ, οὐκ αὔξεται· ὅ τι δὲ ἐς τὴν γυναῖκα, αὔξεται, ἢν τύχῃ τῶν προσηκόντων. καὶ διακρίνεται τὰ μέλεα πάντα ἅμα καὶ αὔξεται, καὶ πρότερον οὐδὲν ἕτερον ἑτέρου οὐδ' ὕστερον.[6] τὰ δὲ μέζω

[1] τότε μὲν ἄλλην· τότε δὲ ἄλλην μὴ ἔχειν γνώμην· θ: τότε μὲν ἄλλην ἔχειν γνώμην M.

[2] οὐκ is omitted by θ. [3] ἐν omitted by θ.

[4] οὗτος is omitted by θ, perhaps rightly. M has δυνάσται δὲ ἄνθρωποι. ὅστι κ.τ.ἑ., but reads οὗτος.

another. So all the arts have something in common with the nature of man.

XXV. The soul of man, as I have already said, being a blend of fire and water, and the parts of man, enter into every animal that breathes, and in particular into every man, whether young or old. But it does not grow equally in all; but in young bodies, as the revolution is fast and the body growing, it catches fire, becomes thin and is consumed for the growth of the body; whereas in older bodies, the motion being slow and the body cold, it is consumed for the lessening of the man. Such bodies as are in their prime and at the procreative age can nourish it and make it grow. Just as a potentate is strong who can nourish[1] very many men, but is weaker when they desert him, even so those bodies are severally strongest that can nourish very many souls, but are weaker when this faculty has departed.[2]

XXVI. Whatever enters into something else does not grow; but whatever enters a woman grows if it meets with the things that suit it. And all the limbs are separated and grow simultaneously, none before or after another; although those by nature

[1] *τρέφειν* can mean "to keep" pets or servants, as well as "to nourish" the body.

[2] *ἀπελθόντων δὲ τούτων* is strange. To what does it refer? And how can a body nourish many souls?

---

[5] *τοιούτων καὶ ἕκαστα τῶν σωμάτων πλεῖστα· ὁκοῖα πλεῖστα δύναται ψυχὰς τρέφειν, ταῦτα ἰσχυρότερον·* M: *τοιοῦτον ὁκοι ἕκαστα τῶν σωμάτων πλείστας δύναται τρέφειν ταῦτα ἰσχυρότατα* θ.

[6] *οὔτε πρότερον οὐδὲν ἕτερον ἑτέρου, οὐδ' ὕστερον* M: *οὔτε πρότερον . . . οὔθ' ὕστερον* Littré, following a later correction in H.

φύσει πρότερα φαίνεται τῶν ἐλασσόνων, οὐδὲν πρότερα γινόμενα. οὐκ ἐν ἴσῳ δὲ χρόνῳ πάντα διακοσμεῖται, ἀλλὰ τὰ μὲν θᾶσσον, τὰ δὲ βραδύτερον, ὅκως ἂν καὶ τοῦ πυρὸς τύχῃ ἕκαστα καὶ
10 τῆς τροφῆς· τὰ μὲν οὖν ἐν τεσσαράκοντα ἡμέρῃσιν ἴσχει πάντα φανερά, τὰ δ' ἐν δύο μησί, τὰ δ' ἐν[1] τρισί, τὰ δ' ἐν τετραμήνῳ. ὡς δ' αὔτως καὶ γόνιμα γίνεται τὰ μὲν θᾶσσον ἑπτάμηνα τελέως, τὰ δὲ βραδύτερον ἐννέα μησὶ τελέως· ἐς φάος ἀναδείκνυται ἔχοντα τὴν σύγκρησιν ἥνπερ καὶ
16 διὰ παντὸς ἕξει.[2]

XXVII. Ἄρρενα μὲν οὖν καὶ θήλεα ἐν τῷδε τῷ τρόπῳ γίνοιτ' ἂν ὡς ἀνυστόν· τὰ δὲ θήλεα πρὸς ὕδατος μᾶλλον ἀπὸ τῶν ψυχρῶν καὶ ὑγρῶν καὶ μαλακῶν αὔξεται καὶ σίτων καὶ ποτῶν καὶ ἐπιτηδευμάτων· τὰ δὲ ἄρσενα πρὸς πυρὸς μᾶλλον, ἀπὸ τῶν ξηρῶν καὶ θερμῶν καὶ σίτων καὶ διαίτης. εἰ οὖν θῆλυ τεκεῖν βούλοιτο, τῇ πρὸς ὕδατος διαίτῃ[3] χρηστέον· εἰ δὲ ἄρσεν,[4] τῇ πρὸς πυρὸς ἐπιτηδεύσει διακτέον· καὶ οὐ μόνον[5] τὸν ἄνδρα
10 δεῖ τοῦτο διαπρήσσεσθαι, ἀλλὰ καὶ τὴν γυναῖκα. οὐ γὰρ ἀπὸ τοῦ ἀνδρὸς μόνον ἀποκριθὲν αὔξιμόν ἐστιν, ἀλλὰ καὶ ἀπὸ τῆς γυναικός, διὰ τάδε·[6] ἑκάτερον μὲν τὸ μέρος οὐκ ἔχει ἱκανὴν τὴν κίνησιν τῷ πλήθει τοῦ ὑγροῦ, ὥστε καταναλίσκειν τὸ ἐπιρρέον καὶ συνιστάναι δι' ἀσθενείην τοῦ πυρός· ὁκόταν δὲ κατὰ τωὐτὸ ἀμφότερα συνεκπεσόντα[7] τύχῃ, συμπίπτει[8] πρὸς ἄλληλα, τὸ πῦρ τε πρὸς

[1] πάντα φανερά, . . . τὰ δ' ἐν omitted by M.
[2] ἕξει M: ἔχει θ.
[3] διαίτῃ θ: διαιτήσει M.
[4] ἄρσεν θ: ἄρσενα M.
[5] καὶ οὐ μόνον θ: οὐ μόνον δὲ M.

larger become visible before the smaller, yet they are formed none the earlier. Not all take the same time to form; some take less time, some longer, according as they severally meet with fire and nourishment. Some have everything visible in forty days, some in two months, some in three months and others in four. Similarly also some are formed before others; those that grew quicker are fully formed in seven months, those that grew more slowly in nine months; and they appear in the light with the same blend as they will have always.[1]

XXVII. Males and females would be formed, so far as possible, in the following manner. Females, inclining more to water, grow from foods, drinks and pursuits that are cold, moist and gentle. Males, inclining to fire, grow from foods and regimen that are dry and warm. So if a man would beget a girl, he must use a regimen inclining to water. If he wants a boy, he must live according to a regimen inclining to fire. And not only the man must do this, but also the woman. For growth belongs, not only to the man's secretion, but also to that of the woman, for the following reason. Either part alone has not motion enough, owing to the bulk of its moisture and the weakness of its fire, to consume and to solidify the oncoming water. But when it happens that both are emitted together to one place, they conjoin, the fire to the fire and the

[1] It might perhaps be well to punctuate with a colon at *γίνεται*, no colon at *τελέως* and a comma at *ἀποδείκνυται*.

---

[6] So θ. M reads *οὐ γὰρ τὸ ἀπὸ τοῦ ἀνδρὸς μοῦνον αὔξιμόν ἐστιν ἀποκριθὲν, ἀλλὰ καὶ τὸ ἀπὸ τῆς γυναικὸς διὰ τόδε·*

[7] *συνεκπεσόντα* θ: *συνεμπεσόντα* M.

[8] *συμπίπτει* θ: *περιπίπτει* M.

τὸ πῦρ καὶ τὸ ὕδωρ ὡσαύτως. ἢν μὲν οὖν ἐν ξηρῇ τῇ χώρῃ πέσῃ, κινεῖται, εἰ καὶ κρατεῖ τοῦ
20 συνεκπεσόντος[1] ὕδατος, καὶ ἀπὸ τούτου αὔξεται τὸ πῦρ, ὥστε μὴ κατασβέννυσθαι ὑπὸ τοῦ ἐπιπίπτοντος[2] κλύδωνος, ἀλλὰ τό τε ἐπιὸν δέχεσθαι καὶ συνιστάναι πρὸς τὸ ὑπάρχον· ἢν δὲ ἐς ὑγρὸν πέσῃ, εὐθέως ἀπ᾽ ἀρχῆς κατασβέννυταί τε[3] καὶ διαλύεται ἐς τὴν μείω τάξιν. ἐν μιῇ δὲ ἡμέρῃ τοῦ μηνὸς ἑκάστου δύναται συστῆναι καὶ κρατῆσαι τῶν ἐπιόντων, καὶ ταῦτ᾽ ἢν τύχῃ συνεκπεσόντα[4]
28 παρ᾽ ἀμφοτέρων κατὰ τόπον.

XXVIII. Συνίστασθαι δὲ δύναται καὶ τὸ θῆλυ καὶ τὸ ἄρσεν πρὸς ἄλληλα, διότι καὶ ἐν ἀμφοτέροις ἀμφότερα τρέφεται, καὶ διότι ἡ μὲν ψυχὴ τωὐτὸ πᾶσι τοῖσιν ἐμψύχοισι, τὸ δὲ σῶμα διαφέρει ἑκάστου. ψυχὴ μὲν οὖν αἰεὶ ὁμοίη καὶ ἐν μέζονι καὶ ἐν ἐλάσσονι· οὐ γὰρ ἀλλοιοῦται οὔτε διὰ φύσιν οὔτε δι᾽ ἀνάγκην· σῶμα δὲ οὐδέποτε τωὐτὸ οὐδενὸς οὔτε κατὰ φύσιν οὔθ᾽ ὑπ᾽ ἀνάγκης, τὸ μὲν γὰρ διακρίνεται ἐς πάντα, τὸ δὲ συμ-
10 μίσγεται πρὸς ἅπαντα. ἢν μὲν οὖν ἐς[5] ἄρσενα τὰ σώματα ἀποκριθέντα ἀμφοτέρων[6] τύχῃ, αὔξεται κατὰ τὸ ὑπάρχον, καὶ γίνονται οὗτοι ἄνδρες λαμπροὶ τὰς ψυχὰς καὶ τὸ σῶμα[7] ἰσχυροί, ἢν μὴ ὑπὸ τῆς διαίτης βλαβῶσι τῆς ἔπειτα. ἢν δὲ τὸ μὲν ἀπὸ τοῦ ἀνδρὸς ἄρσεν ἀποκριθῇ, τὸ δὲ ἀπὸ τῆς γυναικὸς θῆλυ, καὶ

[1] εἰ μὲν οὖν ἐν ξηρῆι τῆι χωρῆι πέσηι. κεινέεται· εἰ καὶ κρατέει τοῦ συνεκπεσόντος ὕδατος· θ: ἢν μὲν νῦν ἐν ξηρῆι τῆι χώρηι περικενέεται, κρατέει τοῦ συνεμπεσόντος ὕδατος M.

[2] ἐπιπίπτοντος θ: ἐμπίπτοντος M.

[3] After τε M adds ὑπὸ τοῦ ἐμπίπτοντος κλύδωνος.

[4] συνεκπεσόντα θ: ξυνεμπεσόντα M.

water likewise. Now if the fire fall in a dry place, it is set in motion, if it also master the water emitted with it, and therefrom it grows, so that it is not quenched by the onrushing flood, but receives the advancing water and solidifies it on to what is there already. But if it fall into a moist place, immediately from the first it is quenched and dissolves into the lesser rank.[1] On one day in each month it can solidify, and master the advancing parts, and that only if it happen that parts are emitted from both parents together in one place.

XXVIII. Male and female have the power to fuse into one solid, both because both are nourished in both and also because soul is the same thing in all living creatures, although the body of each is different. Now soul is always alike, in a larger creature as in a smaller, for it changes neither through nature nor through force. But the body of no creature is ever the same, either by nature or by force, for it both dissolves into all things and also combines with all things. Now if the bodies secreted from both happen to be male, they grow up to the limit of the available matter,[2] and the babies become men brilliant in soul and strong in body, unless they be harmed by their subsequent diet. If the secretion from the man be male and that of the woman female, should

[1] Littré translates: "passant au rang de décroissance." Does it refer to "lifeless" matter, *i.e.* matter that cannot form a living embryo?

[2] Littré says: "ils croissent sur le fonds existant."

---

[5] ἐς is omitted by M.

[6] ἀμφοτέρων is omitted by *θ*.

[7] τὸ σῶμα M: *τὰ σώματα θ*.

ἐπικρατήσῃ τὸ ἄρσεν, ἡ μὲν ψυχὴ προσμίσγεται πρὸς τὴν ἰσχυροτέρην ἡ ἀσθενεστέρη, οὐ γὰρ ἔχει πρὸς ὅ τι ὁμοτροπώτερον[1] ἀποχωρήσει τῶν
20 παρεόντων· προσέρχεται γὰρ καὶ ἡ μικρὴ πρὸς τὴν μέζω καὶ ἡ μέζων πρὸς τὴν ἐλάσσονα·[2] κοινῇ δὲ τῶν ὑπαρχόντων κρατέουσι· τὸ δὲ σῶμα τὸ μὲν ἄρσεν αὔξεται, τὸ δὲ θῆλυ μειοῦται καὶ διακρίνεται ἐς ἄλλην μοίρην. καὶ οὗτοι ἧσσον μὲν τῶν προτέρων λαμπροί, ὅμως δέ, διότι ἀπὸ τοῦ ἀνδρὸς τὸ ἄρσεν ἐκράτησεν, ἀνδρεῖοι γίνονται, καὶ τοὔνομα τοῦτο δικαίως ἔχουσιν. ἢν δὲ ἀπὸ μὲν τῆς γυναικὸς ἄρσεν ἀποκριθῇ, ἀπὸ δὲ τοῦ ἀνδρὸς θῆλυ, κρατήσῃ δὲ τὸ ἄρσεν, αὔξεται τὸν
30 αὐτὸν τρόπον τῷ προτέρῳ· τὸ δὲ μειοῦται· γίνονται δὲ οὗτοι ἀνδρόγυνοι καὶ καλέονται τοῦτο ὀρθῶς. τρεῖς μὲν οὖν[3] αὗται γενέσιες τῶν ἀνδρῶν, διάφοροι δὲ πρὸς τὸ μᾶλλον καὶ ἧσσον τὸ τοιοῦτον[4] εἶναι διὰ τὴν σύγκρησιν τοῦ ὕδατος τῶν μερέων καὶ τροφὰς καὶ παιδεύσιας καὶ συνηθείας. δηλώσω δὲ προϊόντι τῷ λόγῳ καὶ
37 περὶ τούτων.

XXIX. Τὸ δὲ θῆλυ γίνεται κατὰ τὸν αὐτὸν τρόπον· ἢν μὲν ἀπ' ἀμφοτέρων θῆλυ ἀποκριθῇ, θηλυκώτατα καὶ εὐφυέστατα γίνεται· ἢν δὲ τὸ μὲν ἀπὸ τῆς γυναικὸς θῆλυ, τὸ δὲ ἀπὸ τοῦ ἀνδρὸς ἄρσεν, κρατήσῃ δὲ τὸ θῆλυ,[5] θρασύτεραι μὲν τῶν πρόσθεν, ὅμως δὲ κόσμιαι καὶ αὗται.[6] ἢν δὲ τὸ μὲν ἀπὸ τοῦ ἀνδρὸς θῆλυ, τὸ δ' ἀπὸ τῆς γυναικὸς

[1] ὁμοτροπωτερον θ: ὁμοτροφώτερον M.

[2] προσδέχεται γὰρ ἡ μικρὴ τὴν μέζω· καὶ ἡ μέζω, τὴν ἐλάσσονα· M: προσέρχεται γὰρ καὶ ἡ μικρῆι πρὸς τὴν μέζωι· καὶ ἡ μέζω πρὸς τὴν ἐλάσσονα· θ.

the male gain the mastery, the weaker soul combines with the stronger, since there is nothing more congenial present to which it can go. For the small goes to the greater and the greater to the less, and united they master the available matter. The male body grows, but the female body decreases into another part.[1] And these, while less brilliant than the former, nevertheless, as the male from the man won the mastery, they turn out brave, and have rightly this name. But if male be secreted from the woman but female from the man, and the male get the mastery, it grows just as in the former case, while the female diminishes. These turn out hermaphrodites ("men-women") and are correctly so called. These three kinds of men are born, but the degree of manliness depends upon the blending of the parts of water, upon nourishment, education and habits. In the sequel I shall discuss these matters also.

XXIX. In like manner the female also is generated. If the secretion of both parents be female, the offspring prove female and fair, both to the highest degree. But if the woman's secretion be female and the man's male, and the female gain the mastery, the girls are bolder than the preceding, but nevertheless they too are modest. But if the man's secretion be female, and the woman's male,

[1] Or, "destiny."

---

[3] *οὖν* is omitted by *θ*.

[4] For *τὸ τοιοῦτον θ* has *τοιοῦτοι*.

[5] A few MSS. (but neither M nor *θ*) have after *θῆλυ* the words *αὔξεται τὸν αὐτὸν τρόπον καὶ*. Littré prints them.

[6] *αὗται θ* : *αὐταί* M.

ἄρσεν, κρατήσῃ δὲ τὸ θῆλυ, αὔξεται τὸν αὐτὸν τρόπον, γίνονται δὲ τολμηρότεραι τῶν προτέρων
10 καὶ ἀνδρεῖαι ὀνομάζονται. εἰ δέ τις ἀπιστεῖ ψυχὴν μὴ προσμίσγεσθαι ψυχῇ,[1] ἀφορῶν ἐς ἄνθρακας, κεκαυμένους πρὸς κεκαυμένους[2] προσβάλλων, ἰσχυροὺς πρὸς ἀσθενέας, τροφὴν αὐτοῖσι διδούς, ὅμοιον τὸ σῶμα πάντες παρασχήσονται καὶ οὐ διάδηλος ἕτερος τοῦ ἑτέρου,[3] ἀλλ' ἐν ὁποίῳ σώματι ζωπυρέονται, τοιοῦτον δὴ τὸ πᾶν[4] ἔσται· ὁκόταν δ' ἀναλώσωσι τὴν ὑπάρχουσαν τροφήν, διακρίνονται ἐς τὸ ἄδηλον· τοῦτο καὶ
19 ἀνθρωπίνη ψυχὴ πάσχει.

XXX. Περὶ δὲ τῶν διδύμων γινομένων ὧδε ὁ λόγος δηλώσει. τὸ μὲν πλεῖστον τῆς γυναικὸς ἡ φύσις αἰτίη τῶν μητρέων· ἢν γὰρ ὁμοίως ἀμφοτέρωσε πεφύκωσι κατὰ τὸ στόμα, καὶ ἀναχάσκωσιν ὁμοίως, καὶ ξηραίνωνται ἀπὸ τῆς καθάρσιος, δύνανται τρέφειν, ἢν τὰ τοῦ ἀνδρὸς συλλαμβάνῃ ὥστε εὐθὺς ἀποσχίζεσθαι[5] ἐς ἀμφοτέρας τὰς μήτρας ὁμοίως. ἢν μὲν οὖν πολὺ ἀπ' ἀμφοτέρων τὸ σπέρμα καὶ ἰσχυρὸν ἀποκριθῇ, δύναται ἐν
10 ἀμφοτέρῃσι τῇσι χώρῃσιν αὔξεσθαι· κρατεῖ γὰρ τῆς τροφῆς τῆς ἐπιούσης. ἢν δέ πως ἄλλως γένηται, οὐ γίνεται δίδυμα. ὁκόταν μὲν οὖν ἄρσενα ἀπ' ἀμφοτέρων ἀποκριθῇ, ἐξ ἀνάγκης

[1] ψυχῆι M: φύσει θ.
[2] Between πρὸς and κεκαυμένους Littré without MS. authority inserts μή. For ἀφορῶν ἐς M has ἀφρῶν ἐστίν.

and the female gain the mastery, growth takes place after the same fashion, but the girls prove more daring than the preceding, and are named "mannish." If anyone doubts that soul combines with soul, let him consider coals. Let him place lighted coals on lighted[1] coals, strong on weak, giving them nourishment. They will all present a like substance, and one will not be distinguished from another, but the whole will be like the body in which they are kindled. And when they have consumed the available nourishment, they dissolve into invisibility. So too it is with the soul of man.

XXX. How twins are born my discourse will explain thus. The cause is chiefly the nature of the womb in woman. For if it has grown equally on either side of its mouth, and if it opens equally, and also dries equally after menstruation, it can give nourishment, if it conceive the secretion of the man so that it immediately divides into both parts of the womb equally. Now if the seed secreted from both parents be abundant and strong, it can grow in both places, as it masters the nourishment that reaches it. In all other cases twins are not formed. Now when the secretion from both parents

[1] Or, with Littré's reading, "unlighted." But it seems more natural, if the male soul be the strong burning coal, for the female soul to be represented by a weak burning coal. When combined, the two coals burn with one flame, with that flame, in fact, appropriate "to the substance in which they are kindled."

---

[3] So M. θ has ἅπαν παρεσχηκότος· καὶ οὐ διάδηλον ἕτερον τοῦ στεροῦ· Peck reads: πᾶν παρασχήσεται καὶ οὐ διάδηλον τὸ ἕτερον τοῦ ἑτέρου—perhaps rightly.

[4] So θ. M has τοιοῦτον ἀπὸ πάντων.

[5] After ἀποσχίζεσθαι M has οὕτω γὰρ ἀνάγκη σκίδνασθαι.

ἐν ἀμφοτέροις ἄρσενα γεννᾶσθαι·[1] ὁκόταν δὲ θήλεα ἀπ' ἀμφοτέρων, θήλεα γίνεται· ὅταν δὲ τὸ μὲν θῆλυ, τὸ δὲ ἄρσεν, ὁκότερον ἂν ἑκατέρου κρατήσῃ, τοιοῦτον ἐπαύξεται. ὅμοια δὲ ἀλλήλοισι τὰ δίδυμα διὰ τάδε[2] γίνεται, ὅτι πρῶτον μὲν τὰ χωρία ὅμοια[3] ἐν οἷσιν αὔξεται, ἔπειτα
20 ἅμα ἀπεκρίθη, ἔπειτα τῇσιν αὐτῇσι τροφῇσιν
21 αὔξεται, γόνιμά τε γίνεται ἅμα[4] ἐς φάος.

XXXI. Ἐπίγονα δὲ τῷδε τῷ τρόπῳ γίνεται· ὅταν αἵ τε μῆτραι θερμαί τε καὶ ξηραὶ φύσει ἔωσιν, ἥ τε γυνὴ τοιαύτη, τό τε σπέρμα ξηρὸν καὶ θερμὸν ἐμπέσῃ, οὐκ ἐπιγίνεται[5] ἐν τῇσι μήτρῃσιν ὑγρασίη οὐδεμίη, ἥτις τὸ εἰσπίπτον[6] σπέρμα κρατήσει· διὰ τοῦτο συνίσταται ἐξ ἀρχῆς καὶ ζώει, διατελεῖν δὲ οὐ δύναται, ἀλλὰ τὸ ὑπάρχον προσδιαφθείρει,[7] διότι οὐ ταὐτὰ συμ-
9 φέρει ἀμφοῖν.[8]

XXXII. Ὕδατος δὲ τὸ λεπτότατον καὶ πυρὸς τὸ ἀραιότατον σύγκρησιν λαβόντα ἐν ἀνθρώπου σώματι ὑγιεινοτάτην ἕξιν ἀποδεικνύει διὰ τάδε, ὅτι ἐν τῇσι μεταβολῇσι τοῦ ἐνιαυτοῦ τῶν ὡρέων τῇσι μεγίστῃσιν οὐκ ἐπιπληροῦται τὸ ἔσχατον οὐδέτερον,[9] οὔτε τὸ ὕδωρ ἐς τὸ πυκνότατον ἐν[10] τῇσι τοῦ ὕδατος ἐφόδοισιν, οὔτε τὸ πῦρ ἐν τῇσι τοῦ πυρὸς, οὔτε τῶν ἡλικιέων[11] ἐν τῇσι μετα-

[1] γεννᾶσθαι θM: γεννᾶται Littré.

[2] τόδε M: τάδε θ.

[3] τὰ χορια ὅμοια θ: τὰ χωρία M.

[4] γόνιμά τε γίνεται ἅμα θ: γονημά τε ἀναγίνεται M: γόνιμά τε ἀνάγεται ἅμα Littré.

[5] οὐκ ἐπιγίνεται θ: οὐκέτι γίνεται M (perhaps rightly).

[6] εἴσπιπτον θ: ἐπεισπίπτον M.

is male, of necessity boys are begotten in both places;[1] but when from both it is female, girls are begotten. But when one secretion is female and the other male, whichever masters the other gives the embryo its sex. Twins are like one another for the following reasons. First, the places are alike in which they grow; then they were secreted together; then they grow by the same nourishment, and at birth they reach together the light of day.

XXXI. Superfetation occurs in the following way. When the womb is naturally hot and dry, and the woman is also such, and the seed that enters it is dry and hot, there is no superfluous moisture in the womb to master the seed that enters. Therefore, though it congeals at first and lives, yet it cannot last, but destroys as well the embryo already there, as the same things do not suit both.

XXXII. (1) The finest water and the rarest fire, on being blended together in the human body, produce the most healthy condition for the following reasons. At the greatest changes in the seasons of the year neither is fulfilled to the extreme limit; the water is not fulfilled to the densest limit at the onsets of the water, nor is the fire at the onsets of the fire, whether these be due to alterations in age or to

[1] If we accept the reading of θ M the grammar is peculiar; we have *γεννᾶσθαι* as though *ἀνάγκη* (and not *ἐξ ἀνάγκης*) had preceded.

---

[7] *προσδιαφθείρει* M: *διαφορει* θ. M has *καὶ* (perhaps rightly) after *ἀλλὰ*.

[8] *ἀμφοῖν* M: *αὐτοῖν* θ, which has *ταῦτα* for *ταὐτὰ*.

[9] M omits *οὐκ . . . οὐδέτερον*. [10] *ἐν* θ: *ἐστι* M.

[11] *ἡλικιῶν* θ: *μικρέων* M.

στάσεσιν, οὔτε τῶν σίτων καὶ ποτῶν ἐν τοῖσι
10 διαιτήμασι. δύνανται γὰρ γένεσίν τε πλείστην δέξασθαι ἀμφότερα καὶ πλησμονήν· χαλκὸς ὁ μαλακώτατός τε[1] καὶ αραιότατος πλείστην κρῆσιν δέχεται καὶ γίνεται κάλλιστος· καὶ ὕδατος τὸ λεπτότατον καὶ πυρὸς τὸ ἀραιότατον σύγκρησιν λαμβάνοντα[2] ὡσαύτως. οἱ μὲν οὖν ταύτην ἔχοντες τὴν φύσιν ὑγιαίνοντες διατελέουσι τὸν πάντα χρόνον, μέχρι τεσσαράκοντα ἐτέων, οἱ δὲ καὶ μέχρι γήρως τοῦ ἐσχάτου· ὁκόσοι δ' ἂν ληφθῶσιν ὑπό τινος νοσήματος
20 ὑπὲρ τεσσαράκοντα ἔτεα, οὐ μάλα ἀποθνήσκουσιν.[3] ὁκόσα δὲ τῶν σωμάτων σύγκρησιν λαμβάνει πυρὸς τοῦ ἰσχυροτάτου καὶ ὕδατος τοῦ πυκνοτάτου, ἰσχυρὰ μὲν καὶ ἐρρωμένα τὰ σώματα γίνεται, φυλακῆς δὲ πολλῆς δεόμενα· μεγάλας γὰρ τὰς μεταβολὰς ἔχει ἐπ' ἀμφότερα, καὶ ἐν τῇσι τοῦ ὕδατος ἐφόδοισιν ἐς νοσήματα πίπτουσι, ἔν τε τῇσι τοῦ πυρὸς ὡσαύτως. τοῖσιν οὖν διαιτήμασι συμφέρει χρῆσθαι τὸν τοιοῦτον πρὸς τὰς ὥρας τοῦ ἔτεος ἐναντιούμενον, ὕδατος μὲν
30 ἐφόδου γινομένης, τοῖσι πρὸς πυρός, πυρὸς δὲ ἐφόδου γενομένης, τοῖσι πρὸς ὕδατος χρῆσθαι, κατὰ μικρὸν μεθιστάντα μετὰ τῆς ὥρης. ὕδατος δὲ τοῦ παχυτάτου καὶ πυρὸς τοῦ λεπτοτάτου συγκρηθέντων ἐν τῷ σώματι, τοιαῦτα συμβαίνει ἐξ ὧν διαγινώσκειν χρὴ ψυχρὴν φύσιν καὶ

[1] M omits τε after μαλακώτατος, as it also does after θερμαί at the beginning of Chapter XXXI.

[2] σύγκρησιν λαμβάνοντα omitted by θ.

[3] ἀποθνήσκουσιν θ : διαφυγγάνουσι M.

the foods and drinks that comprise diet.[1] For both ⟨finest water and rarest fire⟩ can admit the amplest generation and fullness. It is the softest and rarest bronze that admits of the most thorough blending and becomes most beautiful; even so is it with the finest water and the rarest fire, when they are blended together. Now those who have this nature continue in good health all the time till they are forty years of age, some of them till extreme old age, while such of them as are attacked by some illness after the age of forty do not generally die of it.[2] (2) Such bodies as are blended of the strongest fire and the densest water turn out strong and robust physically, but need great caution. For they are subject to great changes in either direction, and fall into illnesses at the onsets of the water and likewise at those of the fire. Accordingly it is beneficial for a man of this type to counteract the seasons of the year in the diet he follows, employing one inclined to fire when the onset of water occurs, and one inclined to water when the onset of fire occurs, changing it gradually as the season itself changes. (3) When the thickest water and the finest fire have been blended in the body, the results are such that we must discern a nature cold and moist. These

[1] Referring apparently to the ἔφοδοι of water or of fire that may take place at the change from one period of life to another, or through peculiarities of diet.

[2] The διαφυγγάνουσι of M ("rarely escape death") is an obvious correction based upon a misunderstanding of the argument. The verb applies not to the average man over forty, who may be "a bad life," but to the man whose elements are a happy blend of "the finest water and the rarest fire." Such a man, the writer asserts, is "a good life" all his days. This passage is a clear proof of the general superiority of θ over our other MSS.

ὑγρήν· ταῦτα τὰ σώματα ἐν τῷ χειμῶνι νοσερώτερα ἢ ἐν τῷ θέρει, καὶ[1] ἐν τῷ ἦρι ἢ ἐν τῷ φθινοπώρῳ. τῶν ἡλικιέων, ὑγιηρότατοι τῶν τοιούτων οἱ παῖδες, δεύτερον νεηνίσκοι, νοσε-
40 ρώτατοι δὲ οἱ πρεσβύτατοι καὶ οἱ ἔγγιστα, καὶ ταχέως γηράσκουσιν αἱ φύσιες αὗται. διαιτᾶσθαι[2] δὲ συμφέρει τοῖσι τοιούτοισιν ὁκόσα θερμαίνει καὶ ξηραίνει καὶ πόνοισι καὶ σίτοισι, καὶ πρὸς τὰ ἔξω τοῦ σώματος μᾶλλον τοὺς πόνους[3] ποεῖσθαι ἢ πρὸς τὰ εἴσω. εἰ δὲ λάβοι[4] πυρός τε τὸ ὑγρότατον καὶ ὕδατος τὸ πυκνότατον σύγκρησιν ἐν τῷ σώματι, τοισίδε γινώσκειν ὑγρὴν καὶ θερμὴν φύσιν· κάμνουσι μὲν μάλιστα οἱ τοιοῦτοι ἐν τῷ ἦρι, ἥκιστα δὲ ἐν τῷ φθινοπώρῳ,
50 ὅτι ἐν μὲν τῷ ἦρι ὑπερβολὴ τῆς ὑγρασίης, ἐν δὲ τῷ φθινοπώρῳ συμμετρίη τῆς ξηρασίης· τῶν δὲ ἡλικιέων νοσερώταται ὅσαι νεώταταί εἰσιν· αὔξεται δὲ τὰ σώματα ταχέως, καταρροώδεις δὲ οἱ τοιοῦτοι γίνονται. διαιτῆσθαι δὲ συμφέρει ὅσα ξηραίνοντα ψύχει καὶ σίτων καὶ ποτῶν καὶ πόνων,[5] τοὺς δὲ πόνους τούτοις εἴσω τοῦ σώματος ποιεῖσθαι συμφέρει μᾶλλον. εἰ δὲ σύγκρησιν λάβοι πυρός τε τὸ ἰσχυρότατον καὶ ὕδατος τὸ λεπτότατον, ξηρὴ φύσις καὶ θερμή. νοῦσος μὲν
60 τοῖσι τοιούτοισιν ἐν τῇσι τοῦ πυρὸς ἐφόδοισιν, ὑγείη δὲ ἐν τῇσι τοῦ ὕδατος· ἡλικίῃσιν ἀκμαζούσῃσι πρὸς σαρκὸς εὐεξίην νοσερώτατοι, ὑγιηρότατοι δὲ οἱ πρεσβύτεροι καὶ τὰ ἔγγιστα ἑκατέρων. δίαιται ὅσαι[6] ψύχουσι καὶ ὑγραί-

[1] καὶ M : ἢ θ.
[2] διαιτᾶσθαι θ : διαιτεῖσθαι M, with η written over the εῖ-,

bodies are more unhealthy in winter than in summer, and in spring more than in autumn. As for age, such persons are most healthy in childhood, next come youths, while the least healthy are the very old and the elderly; such constitutions age rapidly. Such persons ought to use a regimen that warms and dries, whether it be exercise or food, and their exercise should be directed more to the outside of the body than to the inner parts. (4) If the moistest fire and the densest water be blended in the body, from the following signs discern a nature moist and warm. Such persons are sick most in spring and least in autumn, because in spring there is excess of moisture, but in autumn a moderate amount of dryness. As for age, the youngest are the most unhealthy. Their bodies grow quickly, but such persons prove to be subject to catarrhs. Their regimen should consist of such things as dry and cool, both food, drink and exercise, and these persons profit more if their exercise be directed to the internal parts of the body. (5) If there be blended the strongest fire and the finest water, the constitution is dry and warm. Such persons fall sick at the onsets of fire, and are healthy at the onsets of water. It is at the prime of life, and when the body is stoutest, that these fall sick most; the most healthy are the aged. It is the same with those nearest these ages. Regimen should be such

[3] τοὺς πόνους θ: τοῦ πόνου M.
[4] εἰ δὲ λάβοι θ: ἢν δὲ λάβηι M.
[5] θ omits καὶ πόνων and M omits καὶ ποτῶν.
[6] ὁκόσαι M. ὅσαι θ. Littré wrongly says that θ omits the word.

νουσι, καὶ τῶν πόνων ὅσοι ἥκιστα ἐκθερμαίνουσι καὶ συντήκουσι καὶ[1] πλείστην ψύξιν παρασχήσουσιν· αἱ τοιαῦται φύσιες μακρόβιοι καὶ εὔγηροι γίνονται. ἢν δὲ σύγκρησιν λαβῃ πυρὸς τοῦ ἀραιοτάτου καὶ ὕδατος τοῦ ξηροτάτου, ξηρὴ
70 καὶ ψυχρὴ ἡ τοιαύτη φύσις, νοσερὴ μὲν ἐν τῷ φθινοπώρῳ, ὑγιηρὴ δὲ ἐν τῷ ἦρι καὶ τοῖσιν ἔγγιστα ὡσαύτως· ἡλικίαι πρὸς ἔτεα τεσσαράκοντα νοσεραί· παῖδες δὲ ὑγιηρότατοι καὶ τὰ προσέχοντα ἑκατέροισιν. δίαιται ὅσαι θερμαὶ ἐοῦσαι ὑγραίνουσι· καὶ πόνοι ἐξ ὀλίγου προσαγόμενοι, ἡσυχῇ διαθερμαίνοντες, μὴ πολὺ ἀπὸ τῶν ὑπαρχόντων φέροντες. περὶ μὲν οὖν φύσιος διαγνώσιος οὕτω χρὴ διαγινώσκειν τῆς ἐξ ἀρχῆς
79 συστάσιος.

XXXIII. Αἱ δὲ ἡλικίαι αὗται πρὸς ἑωυτὰς ὧδε ἔχουσι· παῖς μὲν οὖν κέκρηται ὑγροῖσι καὶ θερμοῖσι, διότι ἐκ τούτων συνέστηκε καὶ ἐν τούτοισιν ηὐξήθη· ὑγρότατα μὲν οὖν καὶ θερμότατα ὅσα ἔγγιστα γενέσιος, καὶ αὔξεται ὡς πλεῖστον, καὶ τὰ ἐχόμενα ὡσαύτως. νεηνίσκος δὲ θερμοῖσι καὶ ξηροῖσι, θερμοῖσι μέν, ὅτι τοῦ πυρὸς ἐπικρατεῖ ἡ ἔφοδος τοῦ ὕδατος· ξηροῖσι δέ, ὅτι[2] τὸ ὑγρὸν ἤδη κατανάλωται τὸ ἐκ τοῦ παιδός, τὸ
10 μὲν ἐς τὴν αὔξησιν τοῦ σώματος, τὸ δὲ ἐς τὴν κίνησιν τοῦ πυρός, τὸ δὲ ὑπὸ τῶν πόνων. ἀνήρ, ὁκόταν στῇ[3] τὸ σῶμα, ξηρὸς καὶ ψυχρός, διότι τοῦ μὲν θερμοῦ ἡ ἔφοδος οὐκ ἔτι ἐπικρατεῖ, ἀλλ' ἕστηκεν, ἀτρεμίζον δὲ τὸ σῶμα τῆς αὐξήσιος

[1] M has ὁκόσοι ἥκιστα ἐκθερμαίνοντες καὶ συντήκοντες πλείστην ψύξιν παρασχήσουσι.

as cools and moistens, with such exercises as warm and dissolve least and produce the most thorough cooling. Such natures have long life and a healthy old age. (6) Should there be a blending of the rarest fire and the driest water, such a nature is dry and cold, unhealthy in autumn and healthy in spring, "autumn" and "spring" denoting approximate periods. At the age of forty (more or less) they are unhealthy; childhood (and the periods just before and after) is the most healthy time. Regimen should be such as is warm and at the same time moistens. Exercise should be mild at first, gradually increasing, gently warming and not taking too much from the available strength. In this way then ought one to judge of the nature of the original constitution of a man.

XXXIII. The various ages stand thus in relation to each other. A child is blended of moist, warm elements, because of them he is composed and in them he grew. Now the moistest and warmest are those nearest to birth, and likewise those next to it, and these grow the most. A young man is composed of warm and dry elements; warm because the onset of the fire masters the water, dry because the moisture from the child is already spent, partly for the growth of the body, partly for the motion of the fire, and partly through exercise. A man, when his growth is over, is dry and cold, because the onset of the warm no longer has the mastery, but stands, and the body, now that

[2] M has θερμὸς μέν, ὅτι τοῦ πυρὸς ἐπικρατέει ἡ ἔφοδος, τοῦ ὕδατος. ξηρὸς δέ, ὅτι κ.τ.ἕ.
[3] For ἀνήρ, ὁκόταν στῇ θ has ὥστειη.

ἔψυκται· ἐκ δὲ τῆς νεωτέρης ἡλικίης τὸ ξηρὸν ἔνι· ἀπὸ δὲ τῆς ἐπιούσης ἡλικίης καὶ τοῦ ὕδατος τῆς ἐφόδου οὔκω ἔχων τὴν ὑγρασίην, διὰ ταῦτα[1] τοῖσι ξηροῖσι[2] κρατεῖται. οἱ δὲ πρεσβῦται ψυχροὶ καὶ ὑγροί,[3] διότι πυρὸς μὲν ἀποχώρησις,
20 ὕδατος δὲ ἔφοδος· καὶ ξηρῶν μὲν ἀπάλλαξις,
21 ὑγρῶν δὲ κατάστασις.

XXXIV. Τῶν δὲ πάντων τὰ μὲν ἄρσενα θερμότερα καὶ ξηρότερα, τὰ δὲ θήλεα ὑγρότερα καὶ ψυχρότερα διὰ τάδε, ὅτι τε ἀπ' ἀρχῆς ἐν τοιούτοισιν ἑκάτερα ἐγένετο καὶ ὑπὸ τοιούτων αὔξεται, γενόμενα τε τὰ μὲν ἄρσενα[4] τῇσι διάτῃσιν ἐπιπονωτέρῃσι χρῆται, ὥστε ἐκθερμαίνεσθαι καὶ ἀποξηραίνεσθαι, τὰ δὲ θήλεα ὑγροτέρῃσι καὶ ῥᾳθυμοτέρῃσι τῇσι διαίτῃσι χρέονται, καὶ κάθαρσιν[5] τοῦ θερμοῦ ἐκ τοῦ
10 σώματος ἑκάστου μηνὸς ποιέονται.

XXXV. Περὶ δὲ φρονήσιος ψυχῆς ὀνομαζομένης καὶ ἀφροσύνης ὧδε ἔχει· πυρὸς τὸ ὑγρότατον καὶ ὕδατος τὸ ξηρότατον κρῆσιν λαβόντα ἐν σώματι φρονιμώτατον, διότι τὸ μὲν πῦρ ἔχει ἀπὸ τοῦ ὕδατος τὸ ὑγρόν, τὸ δὲ ὕδωρ ἀπὸ τοῦ πυρὸς τὸ ξηρόν· ἑκάτερον δὲ οὕτως αὐταρκέστατον·[6] οὔτε

[1] θ omits from τὸ ξηρὸν to διὰ ταῦτα. οὔκω is an emendation of Littré; M has οὐκό.

[2] τοῖσι ξηροῖσι θ: ξηροῖσι καὶ ὑγρυγοῖσι (sic) M.

[3] πρεσβῦται ψυχροὶ καὶ ὑγροὶ M: πρεσβύτεροι ψυχροισι θ.

[4] τὰ μὲν ἄρσενα is omitted by θ and M. Littré gives this reading on the authority of his MS. K'.

[5] For καὶ κάθαρσιν θ has καὶ θράυσι (an interesting haplography from καὶ κάθαρσιν).

[6] There is a large gap in θ here. Fol. 191 v. ends ἀπὸ τοῦ ὕδατος τὸ ὑ, while fol. 192 r. begins τὴν τροφήν· συμφέρει δὲ καὶ ἐμέτοισι χρέεσθαι. See p. 284. οὔπω for οὕτως Fred.

it has rest from growth, is cooled. But the dryness from the younger age is still in him, and he is mastered by the dry elements because he has not yet got the moisture which advancing years and the onset of the water will bring. Old men are cold and moist,[1] because fire retreats and there is an onset of water; the dry elements have gone and the moist have established themselves.

XXXIV. The males of all species are warmer and drier, and the females moister and colder, for the following reasons: originally each sex was born in such things and grows thereby, while after birth males use a more rigorous regimen, so that they are well warmed and dried, but females use a regimen that is moister and less strenuous, besides purging the heat out of their bodies every month.

XXXV. The facts are as follow with regard to what are called the intelligence[2] of the soul and the want of it. The moistest fire and the driest water, when blended in a body, result in the most intelligence, because the fire has the moisture from the water, and the water the dryness from the fire. Each is thus most self-sufficing. The fire is not in

[1] The reading of θ, "older men are mastered by cold elements," is less adapted to the context than that of M, but may possibly be right.

[2] *φρόνησις* seems to mean the power of the soul to perceive things, whether by the mind or by the senses. "Intelligence" is not a satisfactory rendering, nor yet is "sensitiveness," which has been suggested as an equivalent. Perhaps "quick at the uptake." Peck thinks that sensations only, not mind, are referred to, and would translate everywhere "sensitive" and "sensitiveness."

τὸ πῦρ τῆς τροφῆς ἐνδεέστερον ἐπὶ πολὺ φοιτᾷ, οὔτε τὸ[1] ὕδωρ τῆς κινήσιος δεόμενον κωφοῦται· αὐτό τε οὖν ἑκάτερον οὕτως αὐταρκέστατόν ἐστι 10 πρὸς ἄλληλά τε κρηθέντα. ὅ τι γὰρ ἐλάχιστα τῶν πέλας δεῖται, τοῦτο μάλιστα τοῖσι παρεοῦσι προσέχει, πυρός τε τὸ ἥκιστα κινεόμενον μὴ ὑπ' ἀνάγκης, καὶ ὕδατος τὸ μάλιστα μὴ ὑπὸ βίης. ἐκ τούτων δὲ ἡ ψυχὴ συγκρηθεῖσα φρονιμωτάτη καὶ μνημονικωτάτη· εἰ δέ τινι ἐπαγωγῇ χρεομένη τούτων ὁκοτερονοῦν αὐξηθείη ἢ[2] μαραίνοι, ἀφρονέστατον ἂν γένοιτο, διότι οὕτως ἔχοντα αὐταρκέστατα. εἰ δὲ πυρὸς τοῦ εἰλικρινεστάτου καὶ ὕδατος[3] σύγκρησιν λάβοι, ἐνδεέστερον δὲ τὸ πῦρ εἴη τοῦ 20 ὕδατος ὀλίγον, φρόνιμοι μὲν καὶ οὗτοι, ἐνδεέστεροι δὲ τῆς προτέρης, διότι κρατεόμενον τὸ πῦρ ὑπὸ τοῦ ὕδατος καὶ[4] βραδείην τὴν κίνησιν ποιεόμενον, νωθρότερον προσπίπτει πρὸς τὰς αἰσθήσιας· παραμόνιμοι δ' εἰσὶν ἐπιεικέως αἱ τοιαῦται ψυχαὶ πρὸς ὅ τι ἂν προσέχωσιν· εἰ δὲ ὀρθῶς διαιτῷτο,[5] καὶ φρονιμώτερος καὶ ὀξύτερος γένοιτο[6] παρὰ τὴν φύσιν. συμφέρει δὲ τῷ τοιούτῳ τοῖσι πρὸς πυρὸς διαιτήμασι μᾶλλον χρῆσθαι καὶ μὴ πλησμονῇσι μήτε σίτων μήτε πομάτων. δρόμοισιν οὖν 30 χρῆσθαι ὀξέσιν, ὅκως τοῦ τε ὑγροῦ κενῶται τὸ σῶμα καὶ τὸ ὑγρὸν ἐφιστῆται[7] θᾶσσον· πάλῃσι

[1] οὔτε τὸ Littré (with apparently the authority of some Paris MSS.): τό τε M.

[2] ἢ omitted by M.

[3] An adjective agreeing with ὕδατος seems to have fallen out here.

[4] καὶ omitted by M.

[5] διαιτῷντο M (and Littré, with plurals to follow).

want of nourishment so as to wander far, nor is the water in such need of motion as to be dulled. So each is thus most self-sufficing by itself, as are both when blended with one another. For that which has least need of its neighbours attends most closely to the things at hand, as is the case with such fire as moves the least and not by necessity, and by such water as moves the most and not by force. The soul blended of these is most intelligent and has the best memory. But if by the influence of some addition one or the other of these grow or diminish, there will result something most unintelligent, because things blended in the original way are most self-sufficing.[1] If there be a blend of the purest fire and water, and the fire fall a little short of the water, such persons too are intelligent, but fall short of the former blend, because the fire, mastered by the water and so making slow motion, falls rather dully on the senses. But such souls are fairly constant in their attention, and this kind of man under right regimen may become more intelligent and sharper than natural endowment warrants. Such a one is benefited by using a regimen inclining rather towards fire, with no surfeit either of foods or of drinks. So he should take sharp runs, so that the body may be emptied of moisture and the moisture may be stayed

[1] It is difficult to be satisfied with this sentence, although the MS. tradition shows no important variants. Can τινι ἐπαγωγῇ χρεομένῃ, "by an addition which uses them," with no expressed object, be right? Littré's "par l'usage de quelque addition," seems very strange. One might conjecture (without much confidence) χρεόμενον or ἐρχομένῃ.

---

[6] Before παρὰ Zwinger adds ἂν, but in the *Corpus* the plain optative is often equivalent to the optative with ἄν.

[7] ἐφιστῆται Littré: ἐφίσταται Mack: ἐπίσταται M.

δὲ καὶ τρίψεσι καὶ τοῖσι τοιούτοισι γυμνασίοισιν οὐ συμφέρει χρῆσθαι, ὅκως μὴ κοιλοτέρων τῶν πόρων γινομένων πλησμονῆς[1] πληρῶνται, βαρύνεσθαι γὰρ ἀνάγκη τῆς ψυχῆς τὴν κίνησιν ὑπὸ τῶν τοιούτων· τοῖσι περιπάτοισι συμφέρει χρῆσθαι καὶ ἀπὸ δείπνου καὶ ὀρθρίοισι καὶ ἀπὸ τῶν δρόμων, ἀπὸ δείπνου μέν, ὅκως τροφὴν ξηροτέρην ἡ ψυχὴ δέχηται ἀπὸ τῶν ἐσιόντων, 40 ὄρθρου δέ, ὅκως αἱ διέξοδοι κενῶνται τοῦ ὑγροῦ καὶ μὴ φράσσωνται οἱ πόροι τῆς ψυχῆς, ἀπὸ δὲ τῶν γυμνασίων, ὅκως μὴ ἐγκαταλείπηται ἐν τῷ σώματι τὸ ἀποκριθὲν ἀπὸ τοῦ δρόμου, μηδὲ συμμίσγηται τῇ ψυχῇ, μηδὲ ἐμφράσσῃ τὰς διεξόδους, μηδὲ συνταράσσῃ τὴν τροφήν. συμφέρει δὲ καὶ ἐμέτοισι χρῆσθαι, ὅκως ἀποκαθαίρηται τὸ σῶμα, εἴ τι ἐνδεέστερον οἱ πόνο διαπρήσσονται, προσάγειν δὲ ἀπὸ τῶν ἐμέτων, κατὰ μικρὸν προστιθέντα τούτοισι πλείονας 50 ἡμέρας ἢ[2] τέσσαρας τὰς ἐλαχίστας. χρίεσθαι δὲ συμφορώτερον ἢ λούεσθαι, λαγνεύειν δὲ ὕδατος ἐφόδων γινομένων,[3] ἐν δὲ τῇσι τοῦ πυρὸς[4] μεῖον. εἰ δέ τινι[5] ἐνδεεστέρην τὴν δύναμιν τὸ πῦρ λάβοι τοῦ ὕδατος, βραδυτέρην[6] ἀνάγκη ταύτην εἶναι, καὶ καλέονται οἱ τοιοῦτοι ἠλίθιοι· ἅτε γὰρ βραδείης ἐούσης τῆς περιόδου, κατὰ βραχύ τι προσπίπτουσιν αἱ αἰσθήσιες, ὀξεῖαι ἐοῦσαι, καὶ ἐπ' ὀλίγον συμμίσγονται διὰ βραδυτῆτα τῆς περιόδου· αἱ γὰρ αἰσθήσιες τῆς ψυχῆς ὁκόσαι μὲν δι' ὄψιος

[1] πλησμονὴν M (perhaps rightly).
[2] τούτοισι πλείονας ἡμέρας ἢ θ: τὸ σιτίον, ἐς ἡμέρας M.
[3] After γινομένων M adds πλείονα, with μείονα at the end of the sentence.
[4] After πυρὸς some authorities add ἐπιφορῇσι.

sooner. But it is not beneficial for such to use wrestling, massage or like exercises, for fear lest, the pores[1] becoming too hollow, they be filled with surfeit. For the motion of the soul is of necessity weighed down by such things. Walks, however, are beneficial, after dinner, in the early morning and after running; after dinner, that the soul may receive drier nourishment from the things that enter; in the early morning, that the passages may be emptied of moisture and the pores of the soul may not be obstructed; after exercise, in order that the secretion from running may not be left behind in the body to contaminate the soul, obstruct the passages and trouble the nourishment. It is beneficial also to use vomiting, so that the body may be cleansed of impurities left behind owing to any failure of exercise to purify, and after the vomiting gradually to increase the amount of food for more than four days at least. Unction is more beneficial to such persons than baths, and sexual intercourse should take place when the onsets of water occur, less, however, at the onsets of fire. If in any case fire receive a power inferior to that of water, such a soul is of necessity slower, and persons of this type are called silly. For as the circuit is slow, the senses, being quick, meet their objects spasmodically, and their combination is very partial owing to the slowness of the circuit. For the senses of the soul that act through sight or hearing are quick; while

[1] The word πόρος means any "passage" or "way" in the body, and is not limited to the pores of the skin.

---

[5] τινι M : τι θ.
[6] βραδυτέρην Zwinger and others: βραχυτέρην θM.

60 ἢ ἀκοῆς εἰσίν, ὀξεῖαι, ὁκόσαι δὲ διὰ ψαύσιος, βραδύτεραι καὶ εὐαισθητότεραι. τούτων μὲν οὖν αἰσθάνονται καὶ οἱ τοιοῦτοι οὐδὲν ἧσσον, τῶν ψυχρῶν καὶ τῶν θερμῶν καὶ τῶν τοιούτων· ὅσα δὲ δι' ὄψιος ἢ δι' ἀκοῆς αἰσθέσθαι[1] δεῖ, ἃ μὴ πρότερον ἐπίστανται, οὐ δύνανται αἰσθάνεσθαι· ἢν γὰρ μὴ σεισθῇ ἡ ψυχὴ ὑπὸ τοῦ πυρὸς πεσόντος, οὐκ ἂν αἴσθοιτο ὁκοῖόν ἐστιν. αἱ οὖν τοιαῦται ψυχαὶ οὐ πάσχουσι τοῦτο διὰ παχύτητα· εἰ δὲ ὀρθῶς διαιτῷντο, βελτίους γίνοιντο 70 ἂν καὶ οὗτοι. συμφέρει δὲ τὰ διαιτήματα ἅπερ τῷ προτέρῳ, ξηροτέροισι καὶ ἐλάσσοσι,[2] τοῖσι δὲ πόνοισι πλείοσι καὶ ὀξυτέροισι· συμφέρει δὲ καὶ πυριῆσθαι, καὶ ἐμέτοισι χρῆσθαι ἐκ τῶν πυριησίων, καὶ προσαγωγῇσι τῇσιν[3] ἐκ τῶν ἐμέτων ἐκ πλείονος χρόνου ἢ τὸ πρότερον, καὶ ταῦτα ποιέων ὑγιεινότερος ἂν καὶ φρονιμώτερος εἴη. εἰ δὲ κρατηθείη ἐπὶ πλεῖον τὸ πῦρ ὑπὸ τοῦ ἐόντος ὕδατος, τούτους ἤδη οἱ μὲν ἄφρονας ὀνομάζουσιν, οἱ δὲ ἐμβροντήτους. ἔστι δ' ἡ μανίη τοιούτων[4] ἐπὶ 80 τὸ βραδύτερον·[5] οὗτοι κλαίουσί τε οὐδενὸς ἕνεκα,[6] δεδίασί τε τὰ μὴ φοβερά, λυπέονταί τε ἐπὶ τοῖσι μὴ προσήκουσι, αἰσθάνονταί τε ἐτεῇ οὐδενὸς ὡς[7] προσήκει τοὺς φρονέοντας. συμφέρει δὲ τούτοισι πυριῆσθαι καὶ ἐλλεβόροισιν καθαίρεσθαι[8] ἐκ τῶν πυριησίων, καὶ τῇ διαίτῃ χρῆσθαι ᾗπερ πρότερον· ἰσχνασίης[9] δεῖται καὶ ξηρασίης. εἰ δὲ τὸ ὕδωρ ἐνδεεστέρην τὴν δύναμιν λάβοι, τοῦ πυρὸς εἰλι-

[1] So M. θ has διόψιος αἰσθάνονται δεῖ.
[2] M has ξηροτέροισι δὲ καὶ ἐλάσσοσι τοῖσι σιτίοισι.
[3] For προσαγωγῇσι τῇσιν M. has προσαγωσι.
[4] τοιοῦτο M : τοῦτο θ : τοιούτων Littré.

those that act through touch are slower, and produce a deeper impression. Accordingly, persons of this kind perceive as well as others the sensations of cold, hot and so on, but they cannot perceive sensations of sight or hearing unless they are already familiar with them. For unless the soul be shaken by the fire that strikes it, it cannot perceive its character. Souls of such a kind have this defect because of their coarseness. But if their regimen be rightly regulated, even these may improve. The regimen that benefits is the same as in the former case, with food drier and less, and with exercise more in amount and more vigorous. Vapour baths too are beneficial, as is the use of vomiting after them, and the food after the vomiting should be increased at longer intervals than in the former case; following such a regimen will make such men more healthy and more intelligent. But if the fire should be mastered to a greater extent by the water in the soul, we have then cases of what are called by some "senseless" people, and by others "grossly stupid." Now the imbecility of such inclines to slowness; they weep for no reason, fear what is not dreadful, are pained at what does not affect them, and their sensations are really not at all those that sensible persons should feel. These persons are benefited by vapour baths followed by purging with hellebore, the diet to be the same as before. Reduction of flesh and drying are called for. But if the power of the water prove insufficient, and the

[5] βραδύτερον M: βραχύτερον θ.
[6] For ἕνεκα some MSS. have λυπέοντος ἢ τύπτοντος.
[7] ητιη οὐδενως θ: αἰτίη οὐδὲν οὖν M.
[8] ἐκκαθαίρεσθαι M.
[9] After ἰσχνασίης M has τε πλεύμονος οὗτος.

κρινῇ τὴν σύγκρησιν ἔχοντος, ἐν ὑγιαίνουσι σώμασι φρόνιμος ἡ τοιαύτη ψυχὴ καὶ ταχέως
90 αἰσθανομένη τῶν προσπιπτόντων καὶ οὐ μεταπίπτουσα πολλάκις. φύσις μὲν οὖν ἡ τοιαύτη ψυχῆς ἀγαθῆς· βελτίων δὲ καὶ οὗτος ὀρθῶς διαιτεόμενος γίνοιτο ἄν, καὶ κακίων μὴ ὀρθῶς. συμφέρει δὲ τῷ τοιούτῳ τῇ διαίτῃ χρῆσθαι τῇ πρὸς ὕδατος μᾶλλον, ὑπερβολὰς φυλασσόμενον καὶ σίτων καὶ ποτῶν καὶ πόνων, καὶ δρόμοισι καμπτοῖσι καὶ διαύλοισι καὶ πάλῃ καὶ τοῖσιν ἄλλοισι γυμνασίοισιν πᾶσιν ὑπερβολὴν οὐδενὸς ποιεόμενον. ἢν γὰρ ἔχῃ ὑγιηρῶς τὸ σῶμα[1] καὶ
100 μὴ ὑπ' ἄλλου τινὸς συνταράσσηται,[2] τῆς ψυχῆς φρόνιμος ἡ σύγκρησις. εἰ δ' ἐπὶ πλεῖον κρατηθείη ἡ τοῦ ὕδατος δύναμις ὑπὸ τοῦ πυρός, ὀξυτέρην μὲν τοσούτῳ ἀνάγκη εἶναι τὴν ψυχὴν ὅσῳ θᾶσσον κινεῖται, καὶ πρὸς τὰς αἰσθήσιας θᾶσσον προσπίπτειν, ἧσσον δὲ μόνιμον[3] τῶν πρότερον,[4] διότι θᾶσσον ἐκκρίνεται[5] τὰ παραγινόμενα καὶ ἐπὶ πλείονα ὁρμᾶται διὰ ταχυτῆτα. συμφέρει δὲ τῷ τοιούτῳ διαιτῆσθαι τῇ πρὸς ὕδατος[6] διαίτῃ μᾶλλον ἢ τῇ προτέρῃ· καὶ μάζῃ μᾶλλον ἢ τῷ
110 ἄρτῳ, καὶ ἰχθῦσιν ἢ κρέασι· τῷ ποτῷ ὑδαρεστέρῳ· καὶ λαγνείῃσιν ἐλασσόσι χρῆσθαι· καὶ τῶν πόνων τοῖσι κατὰ φύσιν μάλιστα καὶ πλείστοισι· τοῖσι δ' ὑπὸ βίης χρῆσθαι μὲν ἀνάγκῃ, ἐλάττοσι δέ· καὶ ἐμέτοισιν ἐκ[7] τῶν πλησμονῶν, ὅκως κενῶται

[1] ἢν γὰρ ὑπογήρως ἔχηται τὸ σῶμα· θ: the text is that of M.
[2] After συνταράσσηται θ adds τὸ σῶμα.
[3] μονίμους θ.
[4] προτέρων θ.
[5] διὸ θάσσον ἐκκρίνεται θ: διότι κρίνεται M.

fire have a pure blend, the body is healthy, and such a soul is intelligent, quickly perceiving without frequent variations the objects that strike it. Such a nature implies a good soul; correct regimen, however, will make it too better, and bad regimen will make it worse. Such a person is benefited by following a regimen inclining to water, and by avoiding excess, whether of food, drink or exercise, with exercises on the circular and double[1] tracks, wrestling and all other forms of athletics, but he must in no case fall into excess. For if his body be in a healthy state and be not troubled from any source, the blend of his soul is intelligent. But if the power of the water be further mastered by the fire, the soul must be quicker, in proportion to its more rapid motion, and strike its sensations more rapidly, but be less constant than the souls discussed above, because it more rapidly passes judgment on the things presented to it, and on account of its speed rushes on to too many objects. Such a person is benefited by a regimen inclining more to water than the preceding; he must eat barley bread rather than wheaten, and fish rather than meat; his drink should be well diluted and his sexual intercourse less frequent; exercises should be as far as possible natural and there should be plenty of them; violent exercise should be sparingly used, and only when necessary; vomiting should be employed after surfeits, in such

[1] The δίαυλος was a race to the end of the 200 yards track and then back again.

---

[6] After ὕδατος M has a long passage, which is practically a repetition of the preceding lines, beginning *μᾶλλον ὑπερβολῆς φυλασσόμενον*.

[7] Before ἐκ θ adds *καὶ*.

μὲν τὸ σῶμα, θερμαίνηται δὲ ὡς ἥκιστα. συμφέρει δὲ καὶ ἀσαρκεῖν τοῖσι τοιούτοισι πρὸς τὸ φρονίμους εἶναι· πρὸς γὰρ σαρκὸς εὐεξίην καὶ αἵματος[1] φλεγμονὴν ἀνάγκη γίνεσθαι· ὁκόταν δὲ τοῦτο πάθῃ ἡ τοιαύτη ψυχή, ἐς μανίην καθίσ-
120 ταται, κρατηθέντος τοῦ ὕδατος, ἐπισπασθέντος[2] τοῦ πυρός. συμφέρει δὲ τοῖσι τοιούτοισι καὶ τὰς πρήξιας πρήσσειν βεβρωκόσι μᾶλλον ἢ ἀσίτοισι· στασιμωτέρη γὰρ ἡ ψυχὴ τῇ τροφῇ καταμισγομένη τῇ συμφόρῳ μᾶλλον ἢ ἐνδεὴς ἐοῦσα τροφῆς. εἰ δέ τινι[3] πλέον ἐπικρατηθείη τὸ ὕδωρ ὑπὸ[4] τοῦ πυρός, ὀξείη[5] ἡ τοιαύτη ψυχὴ ἄγαν, καὶ τούτους ὀνειρώσσειν ἀνάγκη·[6] καλέουσι δὲ αὐτοὺς ὑπομαινομένους·[7] ἔστι δὲ ἔγγιστα μανίης τὸ τοιοῦτον· καὶ γὰρ ἀπὸ βραχείης φλεγμονῆς ἀσυμφόρου
130 μαίνονται, καὶ ἐν τῇσι μέθῃσι καὶ ἐν τῇσιν εὐεξίῃσι τῆς σαρκὸς καὶ ὑπὸ τῶν κρεηφαγιῶν. ἀλλὰ χρὴ τὸν τοιοῦτον τούτων μὲν πάντων ἀπέχεσθαι καὶ τῆς ἄλλης πλησμονῆς, καὶ γυμνασίων τῶν ἀπὸ βίης γινομένων, μάζῃ δὲ ἀτρίπτῳ διαιτῆσθαι, καὶ λαχάνοισιν ἑφθοῖσι πλὴν τῶν καθαρτικῶν, καὶ ἰχθυδίοισιν ἐν ἅλμῃ, καὶ ὑδροποτεῖν βέλτιστον, εἰ δύναιτο· εἰ δὲ μή, ὅ τι ἐγγιστότατα τούτου, ἀπὸ μαλακοῦ οἴνου καὶ λευκοῦ· καὶ τοῖσι περιπάτοισι τοῖσιν ὀρθρίοισι
140 πολλοῖσιν, ἀπὸ δείπνου δὲ ὁκόσον ἐξαναστῆναι,

υ

[1] For καὶ αἵματος M has κάματος.

[2] M has ἐπισπασθεῖσα without τοῦ πυρός.

[3] τινι θ: τι M. [4] ὑπὸ M: ἀπὸ θ.

[5] Folio 193 v. of θ ends with the ὀ- of ὀξείη and 194 r. begins -τασπᾶσαι ἀφανεα οὐχοιονται· ὡς αὔτως δὲ καὶ τῆς φωνῆς. See p. 292.

a way as to empty the body with a minimum of heat. To reduce the flesh of such persons conduces to their intelligence; for abundance of flesh cannot fail to result in inflammation of the blood, and when this happens to a soul of this sort it turns to madness, as the water has been mastered and the fire attracted. Such persons are also benefited if they eat a meal before they go about their duties, instead of doing them without food, as their soul is more stable when it is mixed with its appropriate nourishment than when it lacks nourishment. But if in any case the water be yet more mastered by the fire, such a soul is too quick, and men of this type inevitably suffer from dreams. They are called "half-mad"; their condition, in fact, is next door to madness, as even a slight untoward inflammation results in madness, whether arising from intoxication, or from over-abundance of flesh, or from eating too much meat. Such persons ought to abstain from all these things and from surfeit of every kind, as well as from violent forms of exercise; their diet should consist of unkneaded barley bread, boiled vegetables (except those that purge), and sardines, while to drink water only is best, should that be possible, otherwise the next best thing is a soft white wine. There should be plenty of walking in the morning, but after dinner

---

[6] *ἀνάγκη* K′: M omits.

[7] *οἱ δέ, ὑπομαίνεσθαι·* M: *αὐτοὺς ὑπομαινομένους* Littré, from second hand in H. The first hand reads *τὸ ὑπομαίνεσθαι*, and Littré's E has *ὑπομαίνεσθαι* after *ὑπομαινομένους*. The reading of M ("others <give> *ὑπομαίνεσθαι*") is a note which has crept into the text.

ὅκως τὰ μὲν σῖτα μὴ ξηραίνωνται ἀπὸ τῶν ἀπὸ δείπνου περιπάτων, τὸ δὲ σῶμα κενῶται ὑπὸ τοῦ ὀρθρίου· λούεσθαι δὲ χλιερῷ ὕδατι περικλύδην μᾶλλον ἢ χρίεσθαι· συμφέρει δὲ καὶ ἐν τῷ θέρει τῆς ἡμέρας ὕπνοισι χρῆσθαι βραχέσι μὴ[1] πολλοῖσιν, ὅκως μὴ ἀποξηραίνηται τὸ σῶμα ὑπὸ τῆς ὥρης· ἐπιτήδειον δὲ τοῦ ἦρος καὶ ἐλλεβόροισι καθαίρειν προπυριηθέντας, εἶτα ἐπάγειν πρὸς τὴν διαίτην ἡσυχῇ, καὶ μὴ ἄσιτον τὰς πρήξιας
150 μηδὲ τοῦτον ποιεῖσθαι· ἐκ ταύτης τῆς ἐπιμελείης
151 ἡ τοιαύτη ψυχὴ φρονιμωτάτη ἂν εἴη.

XXXVI. Περὶ μὲν οὖν φρονίμου καὶ ἄφρονος ψυχῆς ἡ σύγκρησις αὕτη αἰτίη ἐστίν, ὥσπερ μοι καὶ γέγραπται· καὶ δύναται ἐκ τῆς διαίτης καὶ βελτίων καὶ χείρων γίνεσθαι. δρόμοισι δὲ πυρὸς ἐπικρατέοντος, τῷ ὕδατι προστιθέναι δυνατὸν[2] ἂν ἴσως, καὶ τοῦ ὕδατος ἐπικρατέοντος ἐν τῇ συγκρήσει τὸ πῦρ αὐξῆσαι· ἐκ τούτων δὲ φρονιμώτεραι καὶ ἀφρονέστεραι γίνονται. τῶν δὲ τοιούτων οὐκ ἐστὶν ἡ σύγκρησις αἰτίη· οἷον
10 ὀξύθυμος, ῥάθυμος, δόλιος, ἁπλοῦς, δυσμενής, εὔνους· τῶν τοιούτων ἁπάντων ἡ φύσις τῶν πόρων δι' ὧν ἡ ψυχὴ πορεύεται, αἰτίη ἐστί· δι' ὁκοίων γὰρ ἀγγείων ἀποχωρεῖ καὶ πρὸς ὁκοῖά τινα προσπίπτει καὶ ὁκοίοις τισὶ καταμίσγεται, τοιαῦτα φρονέουσι· διὰ τοῦτο οὐ[3] δυνατὸν τὰ τοιαῦτα ἐκ διαίτης μεθιστάναι· φύσιν γὰρ μεταπλάσαι ἀφανέα οὐχ οἷόν τε. ὡσαύτως δὲ καὶ

[1] βραχέσιν ἢ M : βραχέσι μὴ Littré after van der Linden.
[2] προστιθέντα ἀδύνατον M : προστιθέναι ἀδύνατον Mack: προστιθέναι δυνατὸν Littré.

only just enough to unbend the limbs; the object is to empty the body by the morning walk, but not to dry the food as the result of walking after dinner. Preferable to unction is a tepid shower-bath. It is also beneficial to have in summer a short, occasional siesta, to prevent the body being dried up by the season. In spring it is a good thing to purge with hellebore after a vapour bath; then the usual diet should be restored gradually, as this type of man, like the preceding, must not go about duties fasting. With this treatment such a soul may be highly intellectual.

XXXVI. It is this blending, then, that is, as I have now explained, the cause of the soul's intelligence or want of it; regimen can make this blending either better or worse. When the fire prevails in his courses, it is doubtless possible to add to the water, and, when the water prevails in the blend, to increase the fire. These things are the source of greater or less intelligence in souls. But in the following cases the blend is not the cause of the characteristic:—irascibility, indolence, craftiness, simplicity, quarrelsomeness and benevolence. In all these cases the cause is the nature of the passages through which the soul passes. For such dispositions of the soul depend upon the nature of the vessels through which it passes, upon that of the objects it encounters and upon that of the things with which it mixes. It is accordingly impossible to change the above dispositions through regimen, for invisible nature cannot be moulded differently. Similarly, the nature of voice too depends

[3] *οὐ* A. L. Peck: *οὖν* M: *γοῦν* Littré: *ἀδύνατον* (for *οὖν δυνατὸν*) Linden, Mack.

τῆς φωνῆς ὁκοίη τις ἂν ᾖ, οἱ πόροι αἴτιοι[1] τοῦ πνεύματος· δι' ὁκοίων γὰρ ἄν[2] τινων κινῆται ὁ
20 ἠὴρ καὶ πρὸς ὁκοίους τινὰς προσπίπτῃ,[3] τοιαύτην ἀνάγκη τὴν φωνὴν εἶναι. †καὶ ταύτην[4] μὲν δυνατὸν καὶ βελτίω καὶ χείρω ποιεῖν, διότι λειοτέρους καὶ τραχυτέρους[5] τοὺς πόρους τῷ πνεύματι δυνατὸν ποιῆσαι, κεῖνο[6] δὲ ἀδύνατον
25 ἐκ διαίτης ἀλλοιῶσαι.†

[1] For αἴτιοι θ has αὔξονται.
[2] γὰρ ἂν omitted by θ.
[3] προσπίπτει θ: προσπίπτειν M.
[4] ταύτην K′: ταῦτα θM.
[5] πλειοτέρους καὶ βραδυτέρους θ: λειοτέρους καὶ βραχυτέρους M: λειοτέρους καὶ τραχυτέρους Littré. θ omits τοὺς πόρους and M reads τοὺς πόνους. A. L. Peck would read βελτίω ποιεῖν, διότι ⟨δὲ⟩ λειοτέρους καὶ . . . ἀδύνατον ποιῆσαι, κεῖνο ἀδύνατον ἐκ διαίτης.
[6] Query: κεῖνα.

upon the passages of the breath. The character of voice inevitably depends upon the nature of the passages through which the air moves, and upon the nature of those it encounters. In the case of voice, indeed, it is possible to make it better or worse, because it is possible to render the passages smoother or rougher for the breath, but the aforesaid characteristics cannot be altered by regimen.[1]

[1] I am satisfied with no restoration of this sentence. Peck's reading makes good sense logically, but requires us to take *ταῦτα* = characteristics that can be changed and *κεῖνο* = characteristics that cannot—a strange use of the words to say the least. If with K′ (and Mack) we read *ταύτην*, and alter *κεῖνο* to *κεῖνα* the sense is: we *can* change the *πόροι* (throat, nose) that give characteristics to voice, but we cannot get at the internal *πόροι* along which *ψυχή* travels. *κεῖνα* = the characteristics (or vessels) mentioned above. But no MS. gives *κεῖνα*, and it is hard to see why it should have been changed to *κεῖνο*.

APPENDIX

In Chapter VII both θ and M give τρυπῶσι, though a second hand in θ has written over it πρίζουσι. In Chapter XVI occurs the sentence τρυπῶσιν, ὁ μὲν ἕλκει, ὁ δὲ ὠθεῖ, which some editors bracket. Boring with an auger seems an impossible action to represent by ἕλκει and ὠθεῖ, and so these editors regard the words as a stupid note which has crept into the text. But the MS. authority is very strong, and causes misgivings. Dr. Peck thinks that a horizontal auger could be worked up and down by a leather thong. But though you can pull a thong you cannot push it. Perhaps there is a reference to the working of an auger by means of a bow, the string of which was twisted round the top of the auger, and then the bow was worked just as a saw. See the *Dictionary of Antiquities, s. v. terebrum.* I do not, however, feel confident enough to adopt the reading τρυπῶσι, though it is quite possible that it is right. Diels' reading of the next sentence in VII will mean: "As they press below, up it comes, for it could not admit of going down at a wrong time," with reference to παρὰ καιρὸν lower down. Timely force works well, untimely force spoils everything. βιαζόμενα, βιάζωνται and βιαζόμενον are probably passives, although a meaning might be wrung out of the middle voice.

REGIMEN II

XXXVII. Χωρέων δὲ θέσιν καὶ φύσιν ἑκάστων χρὴ ὧδε διαγινώσκειν. κατὰ παντὸς μὲν εἰπεῖν ὧδε ἔχει· ἡ πρὸς μεσημβρίην κειμένη θερμοτέρη καὶ ξηροτέρη τῆς πρὸς τὰς ἄρκτους κειμένης, διότι ἐγγυτάτω[1] τοῦ ἡλίου ἐστίν. ἐν δὲ ταύτῃσι τῇσι χώρῃσιν ἀνάγκη καὶ τὰ ἔθνεα τῶν ἀνθρώπων καὶ τὰ φυόμενα ἐκ τῆς γῆς ξηρότερα καὶ θερμότερα καὶ ἰσχυρότερα εἶναι ἢ ἐν τῇσιν ἐναντίῃσιν· οἷον τὸ Λιβυκὸν ἔθνος πρὸς τὸ Ποντικὸν καὶ τὰ
10 ἔγγιστα ἑκατέρων. αὐταὶ[2] δὲ καθ' ἑωυτὰς αἱ χῶραι ὧδε ἔχουσι· τὰ ὑψηλὰ καὶ αὐχμηρὰ καὶ πρὸς μεσημβρίην κείμενα ξηρότερα τῶν πεδίων τῶν ὁμοίως κειμένων, διότι ἐλάσσους[3] ἰκμάδας ἔχει· τὰ μὲν γὰρ οὐκ ἔχει στάσιν τῷ ὀμβρίῳ ὕδατι, τὰ δὲ ἔχει. τὰ δὲ λιμναῖα καὶ ἑλώδεα ὑγραίνει καὶ θερμαίνει· θερμαίνει μέν, διότι κοῖλα καὶ περιέχεται[4] καὶ οὐ διαπνεῖται· ὑγραίνει δέ, διότι τὰ φυόμενα ἐκ τῆς γῆς ὑγρότερα, οἷσι τρέφονται οἱ ἄνθρωποι, τό τε πνεῦμα ὃ ἀναπνέο-
20 μεν[5] παχύτερον διὰ τὸ ὕδωρ ἀπὸ τῆς ἀκινησίης. τὰ δὲ κοῖλα καὶ μὴ ἔνυδρα ξηραίνει καὶ θερμαίνει· θερμαίνει μέν, ὅτι κοῖλα καὶ περιέχεται, ξηραίνει δὲ διά τε τῆς τροφῆς τὴν ξηρότητα, καὶ διότι τὸ πνεῦμα, ὃ ἀναπνέομεν, ξηρὸν ἐόν, ἕλκει ἐκ τῶν

[1] ἐγγυτάτω θ : ἐγγυτέρωι M.
[2] αὐταὶ my emendation : αὗται θ M.

REGIMEN II

XXXVII The way to discern the situation and nature of various districts is, broadly speaking, as follows: The southern countries are hotter and drier than the northern; because they are very near the sun. The races of men and plants in these countries must of necessity be drier, hotter and stronger than those which are in the opposite countries. For example, compare the Libyan race with the Pontic, and also the races nearest to each. Countries considered by themselves have the following characters. Places which are high and scorched and are situated to the south are drier than plains though so situated, because they have less moisture; for they do not retain the rain that falls, but the others do. Marshy and boggy places moisten and heat. They heat because they are hollow and encompassed about, and there is no current of air. They moisten, because the things that grow there, on which the inhabitants feed, are more moist, while the air which is breathed is thicker, because the water there stagnates. Hollows that are without water dry and heat. They heat because they are hollow and encompassed; they dry both by reason of the dryness of the food, and by reason that the air which is breathed, being dry, attracts the moisture from our bodies for

[3] ἐλάσσους Littré: ἐλάσσωι θ: ἐλάσσω M.
[4] περιέχεται θ: περιεχόμενα M.
[5] ἀναπνέομεν θ: ἀναφέρομεν M.

σωμάτων τὸ ὑγρὸν ἐς τροφὴν ἑωυτῷ, οὐκ ἔχον πρὸς ὅ τι ἂν ὑγρότερον προσπῖπτον τρέφηται. ὅκου δὲ τοῖσι χωρίοισιν ὄρεα προσκεῖται πρὸς νότου,[1] ἐν τούτοισιν αὐχμώδεες[2] οἱ νότοι καὶ νοσεροὶ προσπνέουσιν. ὅκου δὲ βόραθεν ὄρη
30 πρόσκειται, ἐν τούτοισιν οἱ βορέαι ταράσσουσι καὶ νούσους ποιέουσιν. ὅκου δὲ βόραθεν κοῖλα χωρία τοῖσιν ἄστεσι προσκεῖται, ἢ καὶ ἐκ θαλάσσης νῆσος ἀντίκειται,[3] πρὸς[4] τῶν θερινῶν πνευμάτων θερμὸν καὶ νοσερὸν τοῦτο τὸ χωρίον, διότι οὔτε βορέης διαπνέων καθαρὴν τὴν ἐπαγωγὴν τοῦ πνεύματος παρέχει, οὔτε ὑπὸ τῶν θερινῶν πνευμάτων διαψύχεται. τῶν δὲ νήσων αἱ μὲν ἐγγὺς τῶν ἠπείρων δυσχειμερώτεραί εἰσιν, αἱ δὲ πόντιαι ἀλεεινότεραι τὸν χειμῶνα, διότι αἱ χιόνες
40 καὶ πάγοι ἐν μὲν τῇσιν ἠπείροισιν ἔχουσι στάσιν καὶ τὰ πνεύματα ψυχρὰ πέμπουσιν ἐς τὰς ἐγγὺς νήσους, τὰ δὲ πελάγια οὐκ ἔχει στάσιν ἐν
43 χειμῶνι.

XXXVIII. Περὶ δὲ πνευμάτων ἥντινα φύσιν ἔχει καὶ δύναμιν ἕκαστα, ὧδε χρὴ διαγινώσκειν. φύσιν μὲν ἔχει τὰ πνεύματα πάντα ὑγραίνειν καὶ ψύχειν τά τε σώματα τῶν ζῴων καὶ τὰ φυόμενα ἐκ τῆς γῆς διὰ τάδε· ἀνάγκη τὰ πνεύματα ταῦτα πνεῖν ἀπὸ χιόνος καὶ κρυστάλλου καὶ πάγων ἰσχυρῶν καὶ ποταμῶν καὶ λιμνέων καὶ γῆς ὑγρανθείσης καὶ ψυχθείσης. καὶ τὰ μὲν ἰσχυρότερα τῶν πνευμάτων ἀπὸ μεζόνων καὶ ἰσχυρο-
10 τέρων, τὰ δὲ ἀσθενέστερα ἀπὸ μειόνων καὶ ἀσθενεστέρων· ὥσπερ γὰρ καὶ τοῖσι ζῴοισι πνεῦμα ἔνεστιν, οὕτω καὶ τοῖσιν ἄλλοισι πᾶσι

[1] θ omits πρὸς νότου.

its own nourishment, having nothing moister to assail in order to nourish itself therefrom. In places where mountains are situated to the south, the south winds that blow are parching and unhealthy; where the mountains are situated to the north, there northern winds occasion disorders and sickness. Where there are hollows on the north side of a town, or where it is faced by an island to the north, such a district becomes hot and sickly with the summer winds, because no north wind blows across to bring a pure current of air, nor is the land cooled by the summer winds. Islands which are near the mainland have very severe winters; but those which are further out to sea are milder in winter. The reason is because the snow and ice on the mainland remain, and send cold winds to the neighbouring islands; but islands situated in mid-ocean have no snow remaining in the winter.

XXXVIII. You may distinguish the nature and power of every particular wind in the following way. All winds have a power of moistening and cooling both animal and vegetable bodies for this reason; because all these winds must come either from snow or ice or places severely frozen, or from rivers or lakes, or from moist and cold land. The stronger winds come from these conditions when widely extended and strongly intensified, weaker winds from these conditions less widely extended and less intensified. As there is breath in the animals, so there is in

---

[2] θ omits from αὐχμώδεες to τούτοισιν.

[3] M omits ἢ καὶ . . . ἀντίκειται.

[4] θ has πρὸ τῶν θερι at the end of fol. 194v; 194r begins ρης. οὐκέτι ὅμοιος παραγίνεται. See Chapter XXXVIII, p. 302.

τοῖσι μὲν ἔλασσον, τοῖσι δὲ[1] κατὰ μέγεθος. φύσιν μὲν οὖν ἔχει ψύχειν καὶ ὑγραίνειν τὰ πνεύματα πάντα. διὰ θέσιν δὲ χωρίων καὶ τόπους, δι' ὧν παραγίνεται τὰ πνεύματα ἐς τὰς χώρας ἑκάστας, διάφορα γίνεται ἀλλήλων, ψυχρότερα, θερμότερα, ὑγρότερα, ξηρότερα, νοσερώτερα, ὑγιεινότερα. τὴν δὲ αἰτίην ἑκάστων
20 ὧδε χρὴ γινώσκειν· ὁ μὲν βορέας ψυχρὸς καὶ ὑγρὸς πνεῖ, ὅτι ὁρμᾶται ἀπὸ τοιούτων χωρίων, πορεύεταί τε διὰ τοιούτων τόπων, οὕστινας ὁ ἥλιος οὐκ ἐφέρπει, οὐδ' ἀποξηραίνων τὸν ἠέρα ἐκπίνει τὴν ἰκμάδα, ὥστε παραγίνεται ἐπὶ τὴν οἰκεομένην, τὴν ἑωυτοῦ δύναμιν ἔχων, ὅκου μὴ διὰ τὴν θέσιν τῆς χώρης διαφθείρεται· καὶ τοῖσι μὲν οἰκέουσιν ἔγγιστα ψυχρότατος, τοῖσι δὲ προσωτάτω ἥκιστα. ὁ δὲ νότος πνεῖ μὲν ἀπὸ τῶν ὁμοίων τὴν φύσιν τῷ βορέᾳ· ἀπὸ γὰρ τοῦ
30 νοτίου πόλου πνέων, ἀπὸ χιόνος πολλῆς καὶ κρυστάλλου καὶ πάγων ἰσχυρῶν ὁρμώμενος, τοῖσι μὲν ἐκεῖσε πλησίον αὐτοῦ οἰκέουσιν ἀνάγκη τοῖον πνεῖν ὁκοῖόν περ ἡμῖν ὁ βορέας. ἐπὶ δὲ πᾶσαν χώρην οὐκ ἔτι ὅμοιος παραγίνεται· διὰ γὰρ τῶν ἐφόδων τοῦ ἡλίου καὶ ὑπὸ τὴν μεσημβρίην πνέων, ἐκπίνεται τὸ ὑγρὸν ὑπὸ τοῦ ἡλίου· ἀποξηραινόμενος δὲ ἀραιοῦται· διὸ ἀνάγκη θερμὸν αὐτὸν καὶ ξηρὸν ἐνθάδε παραγίνεσθαι. ἐν μὲν οὖν τοῖσιν ἔγγιστα χωρίοισιν ἀνάγκη τοιαύτην
40 δύναμιν ἀποδιδόναι θερμὴν καὶ ξηρήν, καὶ ποιεῖ τοῦτο ἐν τῇ Λιβύῃ· τά τε γὰρ φυόμενα ἐξαυαίνει,[2] καὶ τοὺς ἀνθρώπους λανθάνει ἀποξηραίνων· ἅτε γὰρ οὐκ ἔχων οὔτε ἐκ θαλάσσης ἰκμάδα λαβεῖν οὔτε ἐκ ποταμοῦ, ἐκ τῶν ζῴων καὶ ἐκ τῶν φυομένων

everything else; some have less, some more according to size. Now all winds have a cooling and moistening nature. But winds differ from one another according to the situation of the countries and places through which they come to the various regions, being colder, hotter, moister, drier, sicklier or healthier. You may know the cause of each in the following way. The north wind blows cold and moist, because it blows from such countries, and passes through places which the sun does not approach to dry the air and consume the moisture, so that it comes to the habitable earth with its own power, unless this be destroyed by the situation of the place. It is most cold to those who dwell nearest to these places and least to those who are farthest from them. The south blows sometimes from places that are of the same nature as the north; for when it blows from the south pole and starts from much snow, ice and severe frosts, it must of necessity blow to those who dwell there near it after the same manner as the north does to us. But it does not come the same to every country; for instance, when it blows through the approaches of the sun under the south, the moisture is absorbed by the sun. As it dries it becomes rare, and therefore of necessity it must reach here hot and dry. Therefore in the most adjacent countries it must impart such a hot and dry quality, as it does in Libya, where it parches the plants, and insensibly dries up the inhabitants. For as it cannot get any moisture either from sea or river, it drinks up the moisture of animals and

[1] Before κατὰ μέγεθος M has πν̄ᾱ.

[2] ἐξαναίνεται M.

ἐκπίνει τὸ ὑγρόν. ὅταν δὲ τὸ πέλαγος περαιώσῃ, ἅτε θερμὸς ἐὼν καὶ ἀραιός, πολλῆς ὑγρασίης ἐμπίμπλησι τὴν χώρην ἐμπίπτων·[1] ἀνάγκη δὲ τὸν νότον θερμόν τε καὶ ὑγρὸν εἶναι, ὅπου μὴ τῶν χωρίων αἱ θέσιες αἴτιαί εἰσιν. ὡσαύτως
50 δὲ καὶ αἱ τῶν ἄλλων πνευμάτων δυνάμιες ἔχουσιν. κατὰ δὲ τὰς χώρας ἑκάστας τὰ πνεύματα ἔχει ὧδε· τὰ μὲν ἐκ θαλάσσης πνεύματα ἐς τὰς χώρας[2] ἐσπίπτοντα, ἢ ἀπὸ χιόνος ἢ πάγων ἢ λιμνέων ἢ ποταμῶν, ἅπαντα ὑγραίνει καὶ ψύχει καὶ τὰ φυτὰ καὶ τὰ ζῷα, καὶ ὑγείην τοῖσι σώμασι παρέχει ὅσα μὴ ὑπερβάλλει ψυχρότητι· καὶ ταῦτα δὲ βλάπτει, διότι μεγάλας τὰς μεταβολὰς ἐν τοῖσι σώμασιν ἐμποιεῖ τοῦ θερμοῦ καὶ τοῦ ψυχροῦ· ταῦτα δὲ πάσχουσιν ὅσοι ἐν χωρίοις
60 οἰκέουσιν ἑλώδεσι καὶ θερμοῖσιν ἐγγὺς ποταμῶν ἰσχυρῶν. τὰ δ' ἄλλα τῶν πνευμάτων ὅσα πνεῖ ἀπὸ τῶν προειρημένων, ὠφελεῖ, τόν τε ἠέρα καθαρὸν[3] καὶ εἰλικρινέα παρέχοντα καὶ τῷ τῆς ψυχῆς θερμῷ ἰκμάδα διδόντα. ὅσα δὲ τῶν πνευμάτων κατὰ γῆν παραγίνεται, ξηρότερα ἀνάγκη εἶναι, ἀπό τε τοῦ ἡλίου ἀποξηραινόμενα καὶ ἀπὸ τῆς γῆς· οὐκ ἔχοντα δὲ τροφὴν ὁκόθεν ἐπαγάγηται,[4] τὰ πνεύματα, ἐκ τῶν ζῴων ἕλκοντα τὸ ὑγρόν, βλάπτει καὶ τὰ φυτὰ καὶ τὰ ζῷα. καὶ
70 ὅσα ὑπὲρ τὰ ὄρεα ὑπερπίπτοντα παραγίνεται ἐς τὰς πόλιας, οὐ μόνον ξηραίνει, ἀλλὰ καὶ ταράσσει τὸ πνεῦμα ὃ ἀναπνέομεν, καὶ τὰ σώματα τῶν ἀνθρώπων, ὥστε νούσους ἐμποιεῖν. φύσιν μὲν οὖν καὶ δύναμιν ἑκάστων οὕτω χρὴ γινώσκειν· ὅπως δὲ χρὴ πρὸς ἕκαστα παρεσκευάσθαι, προϊόντι τῷ
76 λόγῳ δηλώσω.

plants. But when the wind, being hot and rare, has passed the ocean, it fills the country where it strikes with much moisture. The south wind must necessarily be hot and moist, where the situation of the countries does not cause it to be otherwise. The powers of other winds too are similarly conditioned. The properties of winds due to varieties of region are as follow. The winds which strike regions from off the sea, or from snow, frost, lakes or rivers, all moisten and cool both plants and animals, and are healthy unless they be cold to an excess, when they are hurtful by reason of the great changes of cold and heat which they make in bodies. Those are subject to these changes who inhabit marshy and hot places near great rivers. All other winds which blow from the foresaid places are beneficial, as they afford a pure and serene air, and a moisture to temper the heat of the soul. The winds which come by land must necessarily be drier, being dried both by the sun and the earth. These winds, not having a place whence to draw nourishment, and attracting moisture from living creatures, hurt both plants and animals. The winds which pass over mountains to reach cities do not only dry, but also disturb the air which we breathe, and the bodies of men, so as to engender diseases. This is the way to judge of the nature and power of various winds. I will show in the subsequent discourse how we must provide against each.

---

[1] ἐμπίπτων θ : ἐκπίπτων M.
[2] M has ἐσπίπτοντα. ξηρότερα πῶς ἐστι· τὰ δὲ ἀπο χιόνος.
[3] θ has καθαίροντα.
[4] ἐπαγάγηται θ : σπάσηται M : σπάσεται Littré.

XXXIX. Σίτων δὲ καὶ ποτῶν[1] δύναμιν ἑκάστων καὶ τὴν κατὰ φύσιν καὶ τὴν διὰ τέχνης ὧδε χρὴ γινώσκειν. ὅσοι μὲν κατὰ παντὸς ἐπεχείρησαν εἰπεῖν περὶ τῶν γλυκέων ἢ λιπαρῶν ἢ ἁλμυρῶν ἢ περὶ ἄλλου τινὸς τῶν τοιούτων τῆς δυνάμιος, οὐκ ὀρθῶς γινώσκουσιν· οὐ γὰρ τὴν αὐτὴν δύναμιν ἔχουσιν οὔτε τὰ γλυκέα ἀλλήλοισιν οὔτε τὰ λιπαρὰ οὔτε τῶν ἄλλων τῶν τοιούτων οὐδέν· πολλὰ γὰρ τῶν γλυκέων διαχωρεῖ, τὰ δ' 10 ἵστησι, τὰ δὲ ξηραίνει, τὰ δὲ ὑγραίνει. ὡσαύτως δὲ καὶ τῶν ἄλλων ἁπάντων· ἔστι δὲ ὅσα στύφει καὶ διαχωρεῖται, τὰ δὲ οὐρεῖται, τὰ δὲ οὐδέτερα τούτων. ὡσαύτως δὲ καὶ τῶν θερμαντικῶν καὶ τῶν ἄλλων ἁπάντων, ἄλλην ἄλλα δύναμιν ἔχει. περὶ μὲν οὖν ἁπάντων οὐχ οἷόν τε δηλωθῆναι ὁποῖά τινά ἐστι· καθ' ἕκαστα δὲ ἥντινα δύναμιν 17 ἔχει διδάξω.

XL. Κριθαὶ φύσει μὲν ψυχρὸν καὶ ὑγρὸν καὶ ξηραίνει· ἔνι δὲ καὶ καθαρτικόν τι[2] ἀπὸ τοῦ χυλοῦ τοῦ ἀχύρου· τεκμήριον δέ· εἰ μὲν ἐθέλοις[3] κριθὰς ἀπτίστους ἑψῆσαι, καθαίρει ὁ χυλὸς ἰσχυρῶς· εἰ δὲ πτίσας,[4] ψύχει μᾶλλον καὶ ἵστησιν· ὅταν δὲ πυρωθῶσι, τὸ μὲν ὑγρὸν καὶ καθαρτικὸν ὑπὸ τοῦ πυρὸς παύεται,[5] τὸ δὲ καταλειπόμενον ψυχρὸν καὶ ξηρόν. ὁκόσα δὲ δεῖ ψῦξαι καὶ ξηρῆναι, ἄλφιτον διαπρήσσεται ὧδε 10 χρεομένῳ[6] μάζῃ παντοδαπῇ· δύναμιν δὲ ἔχει ἡ μᾶζα τοιήνδε. τὰ συγκομιστὰ ἄλευρα τροφὴν μὲν ἔχει ἐλάσσω, διαχωρεῖ δὲ μᾶλλον· τὰ δὲ

[1] σιτῶν δὲ καὶ ποτῶν θ : σιτίων δὲ καὶ πομάτων M.
[2] τι omitted by θ.
[3] τεκμήριον μέν· εἰ μὲν θέλεις M : τεκμήριον δὲ εἰ μὲν ἐθέλοις θ.

XXXIX. The power of various foods and drinks, both what they are by nature and what by art, you should judge of thus. Those who have undertaken to treat in general either of sweet, or fat, or salt things, or about the power of any other such thing, are mistaken. The same power does not belong to all sweet things, nor to all fat things, nor to all particulars of any other class. For many sweet things are laxative, many binding, many drying, many moistening. It is the same with all other kinds; some are astringent or laxative, some diuretic; there are some that are neither. It is the same with things which are heating and with all other things, one has one power, another, another. Since therefore it is impossible to set forth these things in general, I will show what power each one has in particular.

XL. Barley in its own nature is cold, moist and drying, but it has something purgative from the juice of the husks. This is proved by boiling unwinnowed barley, the decoction of which is very purgative; but if it be winnowed, it is more cooling and astringent. When it is parched, the moist and purgative quality is removed by the fire, and that which is left is cool and dry. When, therefore, it is necessary to cool and dry, barley meal thus used will do it, no matter how the cake is prepared; such, in fact, is the power of the barley cake.[1] The meal together with the bran has less nourishment, but passes better by stool. That which is cleaned from

[1] The words μάζῃ . . . . . τοιήνδε seem out of place. Should the words μᾶζα παντοδαπή· δύναμιν κ.τ.ἑ. be transposed and placed after ἧσσον δὲ διαχωρεῖ?

[4] πτίσας θ: πτίσαι M: ἐπτισμένας K Mack Littré.
[5] παύεται θ: οἴχεται M. [6] χρώμεθα M.

καθαρὰ τροφιμώτερα, ἧσσον δὲ διαχωρεῖ. μᾶζα προφυρηθεῖσα, ῥαντή, ἄτριπτος, κούφη, καὶ διαχωρεῖ, καὶ ψύχει· ψύχει μὲν διότι[1] ψυχρῷ ὕδατι ὑγρὴ ἐγένετο, διαχωρεῖ δὲ διότι ταχέως πέσσεται, κούφη δὲ διότι πολλὴ τῆς τροφῆς μετὰ τοῦ πνεύματος ἔξω ἀποκρίνεται. στενότεραι[2] γὰρ αἱ διέξοδοι τῇ τροφῇ[3] ἐοῦσαι ἄλλην ἐπιοῦσαν
20 οὐκ ἐπιδέχονται· καὶ τὸ μὲν σὺν τῷ πνεύματι λεπτυνόμενον ἀποκρίνεται ἔξω, τὸ δ' αὐτοῦ μένον[4] φῦσαν ἐμποιεῖ· καὶ τὸ μὲν ἄνω ἐρυγγάνεται, τὸ δὲ κάτω ὑποχωρεῖ· πολλὴ οὖν τῆς τροφῆς ἀπὸ τοῦ σώματος ἀπογίνεται.[5] εἰ δὲ ἐθέλοις[6] εὐθέως συμφυρήσας τὴν μᾶζαν[7] διδόναι, ἡ τοιαύτη ξηραντική· ἅτε γὰρ τὸ ἄλφιτον ξηρὸν ἐὸν καὶ ἀπὸ τοῦ ὕδατος διάβροχον οὕτω[8] γεγενημένον, ἐμπεσὸν ἐς τὴν κοιλίην, ἕλκει ἐξ αὐτῆς τὸ ὑγρὸν θερμὸν ἐόν· πέφυκε γὰρ τὸ μὲν θερμὸν ψυχρὸν
30 ἕλκειν, τὸ δὲ ψυχρὸν τὸ θερμόν· καταναλισκομένου δὲ τοῦ ὑγροῦ ἐκ τῆς κοιλίης ἀνάγκη ξηραίνεσθαι, τοῦ δὲ ὕδατος τοῦ σὺν τῇ μάζῃ ἐσελθόντος † ψύχει ψύχεσθαι ἐπαγόμενον.†[9] ὅσα

[1] M has ὅτι (three times).
[2] M has στενοτοποροι.
[3] τῆς τροφῆς M.
[4] μένον θ : ἐμμένον M.
[5] ἀπογίνεται θ : ἀποπνέεται M.
[6] εἰ δὲ ἐθέλοις θ : εἰ δὲ θέλεις M : ἢν δὲ θέλῃς Littré.
[7] τὴν μάζαν εὐθέως φυρήσας M.
[8] οὕπω ἰδιάβροχον θ : διάβροχον οὕτω M.
[9] ἐπαγόμενον ὂν θ : ἐπαγόμενον M. θ has ψύχεσθαι without ψύχει, M ψύχει without ψύχεσθαι. I give Littré's reading within daggers.

---

[1] προφυρηθεῖσα seems to mean "mixed some time before it is cooked (or required)."
[2] This is a very perplexing sentence. Whether we take the reading of θ or that of M the grammar is abnormal.

the bran is more nourishing, but does not pass so well by stool. Barley cake made into a paste betimes,[1] sprinkled with water but not well kneaded, is light, passes easily by stool, and cools. It cools because it is moistened with cold water; it passes by stool because that it is soon digested, and it is light because that a great part of the nourishment is secreted outside with the breath. For the passages, being too narrow for the nourishment, will not receive a new addition, and part of it is attenuated and secreted outside with the breath, while a part remains and causes flatulence; of this some is belched upwards, and some passes out downwards. A great part, therefore, of the nourishment passes out of the body. If you will give the barley cake as soon as it is mixed, it is drying, for the barley meal, being dry, and moist only by the water which is mixed with it, coming into the belly attracts its moisture as being hot; for it is natural for the hot to attract the cold, and the cold the hot. The moisture of the belly being consumed it must necessarily grow dry, and when the water mixed with the barley cake has entered the belly it must grow cool.[2] So when

Littré, combining the two readings, translates: "Le liquide qui est dans le ventre se consume et se dessèche nécessairement, et celui qui y est appelé se refroidit par le froid de l'eau introduite avec la polenta." He takes τὸ ὑγρὸν as the subject of both infinitives and ψύχει as a noun. But we should certainly require τῷ ψύχει and τὸ ἐπαγόμενον, and it is also hard to distinguish (as Littré does) the ὑγρὸν ἐπαγόμενον from the ὑγρὸν ἐσελθόν. I am tempted to think that ψύχει ἐπαγόμενον ("cools when introduced") is a note that has crept into an original text which read τοῦ . . . ἐσελθόντος ψύχεσθαι, and that the subject of both infinitives is τὴν κοιλίην.

οὖν δεῖ ψῦξαι ἢ ξηρῆναι ἢ διαρροίῃ ἐχόμενον[1] ἢ ἄλλῃ τινὶ θερμασίῃ, ἡ τοιαύτη μᾶζα διαπρήσσεται. ἡ δὲ ξηρὴ τριπτὴ ξηραίνει μὲν οὐχ ὁμοίως διὰ τὸ πεπιλῆσθαι ἰσχυρῶς, τροφὴν δὲ τῷ σώματι πλείστην δίδωσιν, ἅτε γὰρ ἡσυχῇ τηκομένης δέχονται τὴν τροφὴν αἱ δίοδοι·[2] διαχωρεῖ μὲν οὖν
40 βραδέως, φῦσαν δὲ οὐκ ἐμποιεῖ οὐδὲ ἐρυγγάνεται, ἡ δὲ προφυρηθεῖσα τριπτὴ τρέφει μὲν ἧσσον,
42 διαχωρεῖ δὲ καὶ φῦσαν ἐμποιεῖ μᾶλλον.

XLI. Κυκεὼν δὲ σὺν ἀλφίτοισι[3] μοῦνον ἐφ' ὕδατι μὲν ψύχει καὶ τρέφει, ἐπ' οἴνῳ δὲ θερμαίνει καὶ τρέφει καὶ ἵστησιν· ἐπὶ μέλιτι δὲ θερμαίνει μὲν ἧσσον καὶ τρέφει, διαχωρεῖ δὲ μᾶλλον, ἢν μὴ ἄκρητον[4] ᾖ τὸ μέλι· εἰ δὲ μή, ἵστησιν. ἐπὶ δὲ γάλακτι τρόφιμοι μὲν πάντες, ἀτὰρ τὸ μὲν ὄϊον[5] ἵστησι, τὸ δὲ αἴγειον μᾶλλον διαχωρεῖ, τὸ δὲ βόειον[6] ἧσσον, τὸ δὲ ἵππειον καὶ τὸ ὄνειον
9 μᾶλλον διαχωρεῖ.

XLII. Πυροὶ ἰσχυρότεροι κριθῶν καὶ τροφιμώτεροι, διαχωρέουσι δὲ ἧσσον καὶ αὐτοὶ καὶ ὁ χυλός. ἄρτος δὲ ὁ μὲν συγκομιστὸς ξηραίνει καὶ διαχωρεῖ, ὁ δὲ καθαρὸς τρέφει μὲν μᾶλλον, διαχωρεῖ δὲ ἧσσον. αὐτῶν δὲ τῶν ἄρτων ὁ μὲν ζυμίτης κοῦφος διαχωρεῖ· κοῦφος μέν, ὅτι ἀπὸ τῆς ζύμης τοῦ ὀξέος τὸ ὑγρὸν προανάλωται,[7] ὅπερ ἐστὶν ἡ τροφή· διαχωρεῖ δὲ ὅτι

[1] χεόμενον M.
[2] ὅδοι θ: δίοδοι M.
[3] M omits σὺν ἀλφίτοισι.
[4] μὴ ἄκρητον M: ατηκτον θ.
[5] βόϊον μὲν θ: μὲν ὄϊον M.
[6] ὄϊον (ὄϊον?) θ: βόειον M.
[7] προσανάλωται M.

---

[1] The base of cyceon was barley meal, mixed with water, wine or milk. To this was added honey, or salt or herbs.

it is necessary to cool or to dry a sufferer from diarrhœa or from any sort of inflammation, barley cake of this sort serves well. Barley cake that is dry and well kneaded does not dry so much, by reason that it is more tightly compressed, but it is very nourishing, because as it gently dissolves the passages admit the nourishment; so it passes slowly without occasioning wind either downwards or upwards. That which has been mixed beforehand and well kneaded nourishes less, but passes by stool and causes more wind.

XLI. Cyceon made with barley only[1] added to water cools and nourishes, with wine it heats, nourishes and is astringent. With honey it heats and nourishes less, but is more laxative unless the honey be unmixed;[2] with unmixed honey it is astringent. With milk all cyceons are nourishing; made with sheep's milk they are astringent, with goats' milk they are more laxative, with cows' milk less, but with mares' or asses' milk they are more laxative.

XLII. Wheat is stronger and more nourishing than barley, but both it and its gruel are less laxative. Bread made of it without separating the bran dries and passes; when cleaned[3] from the bran it nourishes more, but is less laxative. Of the various breads themselves the fermented is light and passes. It is light because the moisture is quickly used up owing to the acid of the leaven, and this is the nourishment.[4] It passes, because it is

[2] With ἄτηκτον: "if the honey be unmelted."

[3] *I.e.* "white" bread, as opposed to "brown" (συγκομιστός).

[4] *I.e.* the consumption of moisture is nourishment.

ταχέως πέσσεται. ὁ δὲ ἄζυμος διαχωρεῖται[1]
10 μὲν ἧσσον, τρέφει δὲ μᾶλλον. ὁ δὲ τῷ χυλῷ
πεφυρημένος κουφότατος,[2] καὶ τρέφει ἱκανῶς, καὶ
διαχωρεῖ· τρέφει μὲν ὅτι καθαρός, κοῦφος δέ, ὅτι
τῷ κουφοτάτῳ πεφύρηται καὶ ἐζύμωται ὑπὸ
τούτου καὶ πεπύρωται· διαχωρεῖ δὲ ὅτι τὸ γλυκὺ
καὶ διαχωρητικὸν τοῦ πυροῦ[3] συμμέμικται. καὶ
αὐτῶν δὲ τῶν ἄρτων οἱ μέγιστοι τροφιμώτατοι,
ὅτι ἥκιστα ἐκκαίονται ὑπὸ τοῦ πυρὸς τὸ ὑγρόν·
καὶ οἱ ἰπνῖται τροφιμώτεροι τῶν ἐσχαριτῶν καὶ
ὀβελιέων, διότι[4] ἧσσον ἐκκαίονται ὑπὸ τοῦ
20 πυρός. οἱ δὲ κλιβανῖται καὶ οἱ ἐγκρυφίαι
ξηρότατοι, οἱ μὲν διὰ τὴν σποδόν, οἱ δὲ διὰ τὸ
ὄστρακον ἐκπινόνται τὸ ὑγρόν. οἱ δὲ σεμιδα-
λῖται ἰσχυρότατοι τούτων πάντων, ἔτι δὲ μᾶλλον
οἱ ἐκ τοῦ χόνδρου καὶ τρόφιμοι σφόδρα, οὐ
μέντοι διαχωρέουσιν ὁμοίως. ἄλητον καθαρὸν καὶ
πινόμενον ἐφ' ὕδατι ψύχει, καὶ πλῦμα σταιτὸς
ἐπὶ πυρί. πιτύρων χυμὸς ἑφθὸς[5] κοῦφος καὶ
διαχωρεῖ. τὰ δὲ ἐν γάλακτι ἑψόμενα[6] ἄλητα
διαχωρεῖ μᾶλλον ἢ τὰ ἐν τῷ ὕδατι, διὰ τοὺς
30 ὀρρούς, καὶ μάλιστα ἐν τοῖσι διαχωρητικοῖσιν.
ὁκόσα δὲ σὺν μέλιτι καὶ ἐλαίῳ ἕψεται ἢ
ὀπτᾶται ἐξ ἀλήτων, πάντα καυσώδεα καὶ
ἐρευγματώδεα·[7] ἐρευγματώδεα μὲν διότι τρόφιμα
ἐόντα οὐ διαχωρητικά ἐστι, καυσώδεα δὲ διότι
λιπαρὰ καὶ γλυκέα καὶ ἀσύμφορα ἀλλήλοισιν
ἐόντα,[8] οὐ τῆς αὐτῆς καθεψήσιος δεόμενα, ἐν τῷ

[1] διαχωρέει M. [2] κουφότερος M. [3] πυρὸς M.
[4] After διότι θ has περιπλάσσεται τος ταις ὀβολίσκοις. This looks like a marginal note; τος perhaps represents ἄρτος.
[5] ὀπτὸς θ: ἑφθὸς M. [6] διδόμενα M.
[7] ἐρευγμώδεα M.

soon digested; but that which is not fermented does not pass so well, but nourishes more. That which is mixed with wheat gruel is lightest, affords good nourishment, and passes. It nourishes because it is made of pure wheat. It is light because it is tempered with what is most light, and is fermented by it and baked. It passes because it is mixed with the sweet and laxative part of the wheat. Of loaves themselves the largest are the most nourishing, because the moisture of these is least consumed by the fire. Those which are baked in an oven are more nourishing than those which are baked on the hearth or on a spit, because that they are less burnt by the fire. Those which are baked in a pan or under the ashes are the most dry; the latter by reason of the ashes, the former by reason of the earthen pan which imbibes their moisture. The bread made of finest flour called similago is the most strengthening of all, except that which is made of groats, which is very nourishing, but does not pass so well by stool. Fine flour mixed with water and drunk is refreshing, and so is the water wherein flour of spelt has been washed over a fire. A decoction of bran when boiled is light and passes well by stool. Meal boiled in milk passes better by stool than that boiled in water by reason of the whey, and especially if it is mixed with laxatives. All foods from meals boiled or fried with honey and oil are heating and windy; windy because they are very nourishing and do not pass by stool, heating because in one place are fat, sweet and ill-assorted ingredients, which should not be

---

[8] καὶ ἀσύμφορα δὲ ἀλλήλοισιν ἐόντα· θ: ξύμφορα ἀλλήλοις ὄντα M.

αὐτῷ ἐστί. σεμίδαλις καὶ χόνδρος ἑφθά,[1] ἰσχυρὰ
38 καὶ τρόφιμα, οὐ μέντοι διαχωρεῖ.

XLIII. Τίφη, ζειὰ[2] κουφότερα πυρῶν, καὶ τὰ ἐξ αὐτῶν γινόμενα ὁμοίως ὥσπερ ἐκ τῶν πυρῶν, καὶ διαχωρεῖ δὲ μᾶλλον. βρόμος ὑγραίνει καὶ
4 ψύχει ἐσθιόμενος καὶ ῥόφημα πινόμενος.[3]

XLIV. Τὰ πρόσφατα ἄλφιτα καὶ ἄλητα ξηρότερα τῶν παλαιῶν, διότι ἔγγιον τοῦ πυρὸς καὶ τῆς ἐργασίης εἰσί· παλαιούμενα δέ, τὸ μὲν θερμὸν ἐκπνεῖ, τὸ δὲ ψυχρὸν ἐπάγεται. ἄρτοι θερμοὶ μὲν ξηραίνουσι, ψυχροὶ δὲ ἧσσον, ἕωλοι
6 δέ τι ἧσσον,[4] ἰσχνασίην δέ τινα παρέχουσιν.

XLV. Κύαμοι, τρόφιμον καὶ στατικὸν καὶ φυσῶδες· φυσῶδες μὲν ὅτι οὐ δέχονται οἱ πόροι τὴν τροφὴν ἀλέα ἐπιοῦσαν· στάσιμον δὲ ὅτι ὀλίγην[5] ἔχει τὴν ὑποστάθμην τῆς τροφῆς. οἱ δὲ πισοὶ φυσῶσι μὲν ἧσσον, διαχωρέουσι δὲ μᾶλλον. ὠχροὶ καὶ δόλιχοι διαχωρητικώτεροι[6] τούτων, ἧσσον δὲ φυσώδεες, τρόφιμοι δέ. ἐρέβινθοι λευκοὶ διαχωρέουσι καὶ οὐρέονται καὶ τρέφουσι· τρέφει μὲν τὸ σαρκῶδες· οὐρεῖται δὲ
10 τὸ γλυκύ· διαχωρεῖται δὲ τὸ ἁλμυρόν. κέγχρων χόνδροι καὶ κυρήβια,[7] ξηρὸν καὶ στάσιμον,[8] μετὰ σύκων ἰσχυρὸν τοῖσι πονέουσιν·[9] αὐτοὶ δὲ οἱ

[1] ἑφθὸς M.

[2] τιφηζεια θ: στρύγις (and ἐξ αὐτῆς) M.

[3] πινόμενος θ: γενόμενος M.

[4] ἕωλοι δέ τι ἧσσον omitted by M. I suggest δ' ἔτι "yesterday's still less."

[5] ὅλην M.

[6] διαχωρητικοὶ θ: διαχωρητικώτερα M. Also φυσώδεα and τρόφιμα.

[7] χονδρια κυρηβαξια θ: χόνδροι· κυβηρια M.

[8] ξηρὰ καὶ στάσιμα M.

cooked in the same way. Similago and groats boiled are strengthening and very nourishing, but do not pass by stool.

XLIII. The spelts[1] are lighter than wheat, and preparations therefrom are as light as[2] those from wheat, and more laxative. Oats, whether eaten or drunk as a decoction, moisten and cool.

XLIV. Freshly cooked meal and flour are drier than those which are stale, because they are nearer the fire with which they were prepared; for as they grow stale the heat exhales and the cold succeeds. Hot bread dries, cold dries less, yesterday's bread somewhat less, and causes a certain amount of leanness.

XLV. Beans afford an astringent and flatulent nourishment; flatulent because that the passages do not admit the abundant nourishment which is brought, astringent because that it has only a small residue from its nourishment. Peas are less windy and pass better by stool. The chick-pea, called *ochrus*, and the bean called *dolichus* pass better by stool than these, and are less windy but nourishing. The white chick-pea passes by stool and urine, and nourishes. The substantial part nourishes, the sweet passes by urine, and the saline passes by stool. Millet groats and husks are dry and binding; with figs they are strong nourishment for hard workers. Whole millet by itself boiled is

[1] *Triticum monococcum* and *triticum spelta*.

[2] I am not satisfied with θ's reading (in the text), nor with Littré's τοῖς for ὥσπερ. An old emendation, τῶν, has more to be said for it: "preparations therefrom are similarly lighter than those from wheat."

[a] ἰσχυρῶν τοῖσι πόνοισι M.

κέγχροι ἑφθοὶ τρόφιμοι, οὐ μέντοι διαχωρέουσιν. φακοὶ καυσώδεες καὶ ταρακτικοί,[1] οὔτε διαχωρέουσιν οὔτε ἵστασιν. ὄροβοι στάσιμον καὶ ἰσχυρὸν καὶ παχύνει καὶ πληροῖ καὶ εὔχρουν ποιεῖ τὸν ἄνθρωπον. λίνου καρπὸς τρόφιμον καὶ[2] στάσιμον· ἔχει δέ τι καὶ ψυκτικόν. ὁρμίνου καρπὸς παραπλήσια διαπρήσσεται.
20 θέρμοι φύσει μὲν ἰσχυρόν καὶ θερμόν, διὰ δὲ τὴν ἐργασίην κουφότερον καὶ ψυκτικώτερον καὶ διαχωρεῖ. ἐρύσιμον ὑγραίνει καὶ διαχωρεῖ. σικύου σπέρμα διουρεῖται μᾶλλον ἢ διαχωρεῖ. σήσαμα ἄπλυτα διαχωρεῖται,[3] πληροῖ δὲ καὶ παχύνει· διαχωρεῖ μὲν διὰ τὸ ἄχυρον τὸ ἔξω, παχύνει δὲ διὰ τὴν σάρκα· πεπλυμένα δὲ διαχωρεῖ μὲν ἧσσον,[4] παχύνει δὲ καὶ πληροῖ μᾶλλον, αὐαίνει[5] δὲ καὶ καίει διὰ τὸ λιπαρὸν καὶ πῖον. κνίκος διαχωρεῖ.[6] μήκων στάσιμον, μᾶλλον ἡ
30 μέλαινα, ἀτὰρ καὶ ἡ λευκή· τρόφιμον μέντοι καὶ ἰσχυρόν. τούτων δὲ οἱ χυλοὶ[7] διαχωρητικώτεροι τῆς σαρκός· δεῖ οὖν τῇ ἐργασίῃ φυλάσσειν, ὁκόσα μὲν βούλει ξηραίνειν, τοὺς χυλοὺς[8] ἀφαιρέοντα τῇ σαρκὶ χρῆσθαι· ὁκόσα δὲ διαχωρῆσαι,[9] τῷ μὲν χυλῷ πλέονι, τῇ δὲ σαρκὶ
36 ἐλάσσονι καὶ εὐχυλοτέρῃ.[10]

XLVI. Περὶ δὲ τῶν ζῴων τῶν ἐσθιομένων ὧδε χρὴ γινώσκειν. βοὸς[11] κρέα ἰσχυρὰ καὶ στάσιμα

[1] καταρρηκτικόν M. [2] τρόφιμον καὶ omitted by M.
[3] σήσαμα ἄπλυτα διαχωρεῖται omitted by M.
[4] After ἧσσον M adds δέ. [5] αὐαίνει θ: ὑγραίνει M.

nourishing, but it does not pass by stool. Lentils are heating and trouble the bowels; they are neither laxative nor astringent. Bitter vetches are binding, strengthening, fattening, filling, and give a person a good colour. Linseed is nourishing, astringent, and somewhat refreshing. Clary seed is much of the same nature as linseed. Lupins are in their nature strengthening and heating, but by preparation they become more light and cooling than they are naturally, and pass by stool. Hedge-mustard seed moistens and passes by stool. Cucumber seeds pass better by urine than by stool. Unwashed sesame seeds pass by stool, fill and fatten; they pass by stool by reason of their outward skins, they are fattening by reason of their substance; when washed they pass less by stool, but they fatten and fill more; they dry and heat because they are fat and oily. Wild saffron passes by stool. Poppy is binding, the black more than the white, but the white also. It is nourishing, however, and strengthening. Of all these seeds the juices are more laxative than their substance. When, therefore, you have a mind to dry, you must take care in preparation to remove their juices, and to make use of their substance; when you have a mind to loosen, to make use of more of their juices, less of their substance, and only of those that are very succulent.

XLVI. As to animals which are eatable, you must know that beef is strong and binding, and hard of

[6] θ omits κνίκος διαχωρεῖ. [7] χυλοὶ θ: χυμοὶ M.
[8] χυλοὺς θ: χυμοὺς M.
[9] διαχωρῆσαι θ: διαχωρεέι M.
[10] ἐνχυλοτέρηι θ: ἐνχυμοτερα M, which also has χυμῶι
[11] βοὸς θ: βόεια M.

καὶ δύσπεπτα τῇσι κοιλίῃσι, διότι παχύαιμον καὶ πολύαιμόν ἐστι τοῦτο τὸ ζῷον· καὶ τὰ κρέα βαρέα ἐς τὸ σῶμα,[1] καὶ αὐταὶ αἱ σάρκες καὶ τὸ γάλα καὶ τὸ αἷμα. ὁκόσων δὲ τὸ γάλα λεπτὸν καὶ τὸ αἷμα ὅμοιον, καὶ αἱ σάρκες παραπλήσιοι. τὰ δὲ αἴγεια κουφότερα τούτων καὶ διαχωρεῖ μᾶλλον. τὰ δὲ ὕεια ἰσχὺν μὲν τῷ σώματι 10 ἐμποιεῖ μᾶλλον τούτων, διαχωρεῖ δὲ ἱκανῶς διότι λεπτὰς τὰς φλέβας ἔχει καὶ ὀλιγαίμους, σάρκα δὲ πολλήν. ἄρνεια δὲ κουφότερα οἴων, καὶ ἐρίφεια αἰγείων, καὶ διότι ἀναιμότερα καὶ ὑγρότερα. ξηρὰ γὰρ καὶ ἰσχυρὰ φύσει καὶ τὰ ζῷα, ὁκόταν μὲν ἁπαλὰ ᾖ, διαχωρεῖ, ὁκόταν δὲ αὐξηθῇ, οὐχ ὁμοίως· καὶ τὰ μόσχεια τῶν βοείων ὡσαύτως. τὰ δὲ χοίρεια τῶν συείων βαρύτερα· φύσει γὰρ εὔσαρκον ὂν τὸ ζῷον καὶ ἄναιμον ὑπερβολὴν ὑγρασίης ἔχει τέως ἂν νέον ᾖ· ὁκόταν οὖν οἱ 20 πόροι μὴ δέχωνται τὴν τροφὴν ἐπιοῦσαν, ἐμμένον θερμαίνει καὶ ταράσσει τὴν κοιλίην. τὰ δὲ ὄνεια διαχωρεῖ, καὶ τῶν πώλων ἔτι μᾶλλον, καὶ τὰ ἵππεια δ' ἔτι κουφότερα. κύνεια ξηραίνει καὶ θερμαίνει καὶ ἰσχὺν ἐμποιεῖ, οὐ μέντοι διαχωρεῖ· σκυλάκεια δὲ ὑγραίνει καὶ διαχωρεῖ, οὐρεῖται δὲ μᾶλλον. ὑὸς ἀγρίου ξηραίνει καὶ ἰσχὺν παρέχει καὶ διαχωρεῖ. ἐλάφου δὲ ξηραίνει μέν, ἧσσον δὲ διαχωρεῖ, οὐρεῖται δὲ μᾶλλον. λαγῷα ξηρὰ καὶ στάσιμα, οὔρησιν δέ τινα παρέχει. ἀλω- 30 πέκων ὑγρότερα, καὶ οὐρεῖται δέ· καὶ ἐχίνων χερσαίων οὐρητικά, ὑγραίνει δέ.

XLVII. Ὀρνίθων δὲ πέρι ὧδε ἔχει· σχεδόν τι πάντα ξηρότερα ἢ τὰ τετράποδα· ὁκόσα γὰρ

[1] σῶμα θ: στόμα M: ἐν τῷ σταθμῷ Zwinger.

digestion, because this animal abounds with a gross thick blood. The meat is heavy to the body, the flesh itself, the milk and the blood. Those animals which have a thin milk, and the blood the same, have flesh too of the like nature. Goats' flesh is lighter than these, and passes better by stool. Swine's flesh affords more strength to the body than these and passes well by stool, because this animal has small anaemic veins, but much flesh. Lambs' flesh is lighter than sheep's, and kids' than goats', because they do not abound with so much blood, and are more moist. For animals too which are naturally dry and strong, when tender, pass by stool; but when they are grown up, not so much; it is just the same with veal compared to beef. But young pigs' flesh is heavier than pork; for this animal, abounding naturally in flesh and not in blood, has excess of moisture whilst young; so when the passages refuse the entering nourishment, it remains, grows hot, and deranges the belly. The flesh of asses passes by stool, and that of their foals still better, though horseflesh is somewhat lighter. Dogs' flesh dries, heats, and affords strength, but does not pass by stool. The flesh of puppies moistens and passes by stool, still more by urine. Wild boars' flesh is drying and strengthening, and passes by stool. Deer's flesh is drying and passes not so well by stool, but better by urine. Hares' flesh is dry and constipating, but is somewhat diuretic. Foxes' flesh is moister, and passes by urine. Hedgehogs' is diuretic and moistens.

XLVII. With birds it is as follows. All birds almost are drier than beasts, for those creatures

κύστιν οὐκ ἔχει οὔτε οὐρεῖ οὔτε σιαλοχοεῖ[1] διὰ θερμότητα τῆς κοιλίης· ἀναλίσκεται γὰρ τὸ ὑγρὸν ἐκ τοῦ σώματος ἐς τὴν τροφὴν τῷ θερμῷ, ὥστε οὔτε οὐρεῖται οὔτε σιαλοχοεῖ· ἐν οἵῳ δὲ μὴ ἔνι τοιαύτη ὑγρασίη,[2] ξηρὰ εἶναι ἀνάγκη· ξηρότατον μὲν οὖν φαίνεται φάσσης, δεύτερον πέρδικος, τρίτον περιστερῆς καὶ ἀλεκτρυόνος καὶ
10 τρυγόνος· ὑγρότατον δὲ χηνός. ὅσα δὲ σπερμολογεῖ ξηρότατα τῶν ἑτέρων. νήσσης δὲ καὶ τῶν ἄλλων ὁκόσα ἐν ἕλεσι[3] διαιτῆται ἢ ἐν ὕδασι,
13 πάντα ὑγρά.

XLVIII. Τῶν δὲ ἰχθύων ξηρότατοι μὲν οἵδε, σκορπίος, δράκων, καλλιώνυμος, κόκκυξ, γλαῦκος, πέρκη, θρίσσα· κοῦφοι δὲ οἱ πετραῖοι σχεδόν τι πάντες, οἷον κίχλη, φυκίς, κωβιός, ἐλεφιτίς·[4] οἱ τοιοῦτοι τῶν ἰχθύων κουφότεροι τῶν πλανήτων· ἅτε γὰρ ἀτρεμίζοντες ἀραιὴν τὴν σάρκα ἔχουσιν καὶ κούφην. οἱ δὲ πλανῆται καὶ κυματοπλῆγες τεθρυμμένοι τῷ πόνῳ στερεωτέρην καὶ βαθυτέρην τὴν σάρκα ἔχουσιν. νάρκαι δὲ καὶ ῥῖναι καὶ
10 ψῆσσαι καὶ τὰ τοιαῦτα κοῦφα. ὁκόσοι δὲ ἐν τοῖσι πηλώδεσι καὶ ὑγροῖσι[5] χωρίοισι τὰς τροφὰς ἔχουσιν, οἷον κέφαλοι, κεστραῖοι, ἐγχέλυες, οἱ τοιοῦτοι τῶν ἰχθύων βαρύτεροί εἰσι, διότι ἀπὸ τοῦ ὕδατος καὶ τοῦ πηλοῦ καὶ τῶν ἐν τούτοις φυομένων τὰς τροφὰς ἔχουσιν, ἀφ' ὧν καὶ τὸ πνεῦμα ἐσιὸν ἐς τὸν ἄνθρωπον βλάπτει καὶ βαρύνει. οἱ δὲ ποτάμιοι καὶ λιμναῖοι ἔτι βαρύ-

[1] I have adopted here the readings of θ. M has: σιαλοχοέει διὰ γὰρ θερμότητα τῆς κοιλίης ἀναλίσκεται τὸ ὑγρὸν κ.τ.ἕ.

[2] So θ. M has ὅτωι δὲ μὴ ἔνι τοιαῦται ὑγρασίαι ξηραίνειν ἀναγκη.

which have no bladder neither make urine nor have spittle, by reason of the heat of the belly. For the moisture of the body is consumed to nourish the heat; wherefore they neither urinate nor spit. Therefore that which wants such moisture must necessarily be dry. The flesh of ringdoves is the driest, secondly partridges, thirdly pigeons, cocks and turtles. The flesh of geese is the most moist. Those which feed on seed are drier than the others. Ducks and other fowls that feed on marshes or waters are all moist.

XLVIII. As to the flesh of fish, these are the driest. The scorpion fish, dragon fish,[1] the fish called callionymos, the piper, the grey fish, the perch, the fish called thrissa. The fish that frequent stony places are almost all light, as the thrush fish, the hake, the gudgeon and elephitis. These are lighter than those which move from place to place, for these remaining quiet have a rare and light flesh, but those which wander and are wave-tossed have a more solid and deeper flesh, being much battered by the toil. The torpedo, skate, turbot and such-like are light. All those fish that feed in muddy and marshy places, as mullet, cestreus, eels and the like are heavier (of digestion), because they feed upon muddy water and other things which grow therein. The air of which also, entering a person, hurts and oppresses him. The fish of rivers and ponds are heavier than these. The

[1] The great weever.

[3] θ has εαεσι—an interesting survival of a mistake made when the manuscripts were in uncials; ΕΛΕΣΙ and ΕΑΕΣΙ.

[4] Said to be corrupt. Corrected by Coraes to ἀλφηστής.

[5] ὑγροῖσι θ: ἱδρηλοῖσι M.

τεροι τούτων. πολύποδες δὲ καὶ σηπίαι καὶ τὰ τοιαῦτα οὔτε κοῦφα, ὡς δοκεῖ, ἐστὶν οὔτε
20 διαχωρητικά, τοὺς δ' ὀφθαλμοὺς ἀπαμβλύνουσιν·[1] οἱ μέντοι χυμοὶ τούτων διαχωρέουσιν. τὰ δὲ κογχύλια, οἷον πίνναι, λεπάδες, πορφύραι, κήρυκες, ὄστρεα, αὐτὴ μὲν ἡ σὰρξ ξηραίνει,[2] οἱ δὲ χυλοὶ διαχωρητικοί· μύες δὲ καὶ κτένες καὶ τελλῖναι μᾶλλον τούτων διαχωρέουσιν· αἱ δὲ κνίδαι μάλιστα· καὶ τὰ σελάχεα ὑγραίνει καὶ διαχωρεῖ. ἐχίνων ᾠὰ καὶ τὸ ὑγρὸν καράβων διαχωρεῖ, καὶ ἄρκοι,[3] καὶ καρκίνοι, μᾶλλον μὲν οἱ ποτάμιοι, ἀτὰρ καὶ οἱ θαλάσσιοι, καὶ οὐρεῖται.[4]
30 οἱ τάριχοι ξηραίνουσι καὶ ἰσχναίνουσι· τὰ δὲ πίονα [5] διαχωρεῖ ἐπιεικέως· ξηρότατοι μὲν τῶν ταρίχων οἱ θαλάσσιοι, δεύτερον δὲ οἱ ποτάμιοι, ὑγρότατοι δὲ οἱ λιμναῖοι. αὐτῶν δὲ τῶν ταρίχων,
34 οἵπερ καὶ ἰχθύες ξηρότατοι, οὗτοι καὶ τάριχοι.[6]

XLIX. Τῶν δὲ ζῴων τῶν τιθασσῶν,[7] τὰ ὑλόνομα καὶ ἀγρόνομα[8] τῶν ἔνδον τρεφομένων ξηρότερα, ὅτι πονοῦντα ξηραίνεται καὶ ὑπὸ τοῦ ἡλίου καὶ ὑπὸ τοῦ ψύχεος, καὶ τῷ πνεύματι ξηροτέρῳ χρῆται.[9] τὰ δὲ ἄγρια τῶν ἡμέρων ξηρότερα,[10] καὶ τὰ ὀλιγοφάγα τῶν πολυφάγων, καὶ τὰ †χλωροφάγα†[11] τῶν ποηφάγων, καὶ τὰ καρ-

[1] ἀπαμβλύνουσιν θ: βαρύνουσιν M.

[2] αὐτὴ μὲν ἡ σὰρξ ξηραίνει θ: αὐτὰ μὲν ξηρὰ M.

[3] καὶ ἄρκοι omitted by θ, which also reads οἱ for καὶ (before καρκίνοι). M has καράβου μύες καὶ ἄρκοι καὶ καρκινοι.

[4] M has καὶ διαχωρεει καὶ οὐρέεται.

[5] πίονα θ: πλείονα M.

[6] αὐτῶν δὲ τῶν ταρίχων οἱ περ καὶ ἰχθύες M: αὐτῶν δὲ τῶν ταρίχων οἳ περκαὶ ἰχθῦες θ: αὐτέων δὲ τῶν θαλασσίων οἱ λεγόμενοι πέρκαι ἰχθύες Littré (from Paris MSS.), suggesting αἱ λεγόμεναι πηλαμύδες.

polypus, cuttle and the like are neither light, as they are thought to be, nor do they pass by stool, but they dull the eyes. The broth of them, however, passes by stool. Shell-fish, as the pinna, limpet, purple fish, trumpet and oysters, have a flesh that dries, but their broths pass by stool. Mussels, cockles and tellines pass better than these by stool; sea-nettles do so especially; fish that are cartilaginous moisten and pass by stool. The spawn of urchins and the juice of spiny lobsters pass by stool; arcos too and crabs, the river variety more than others, but also sea-crabs; they are also diuretic. Pickled fish are drying and attenuating; oily ones are gently laxative. The driest of pickled fish are those of the sea, the next those of the rivers, while the moistest are those of the lakes. Of pickled fish considered by themselves those are driest which are made from the driest fish.

XLIX. As to animals which are tamed, those which feed in the woods and fields are drier than those fed within doors, because their labours in the sun and the cold dry them, while they breathe an air that is drier. Wild beasts are drier than tame; small eaters than great eaters; hay eaters than grass eaters;[1] fruit eaters than non-fruit eaters; small drinkers than

[1] This is evidently the sense of the sentence, but neither the χλωροφάγα of θ nor the καρποφάγα of M can bear the meaning "hay eating." Perhaps we should adopt the conjecture of Zwinger.

---

[7] For τιθασσῶν θ has πόλεων.
[8] ἀγρόνομα M : ὑγρόνομα θ.
[9] χρῆται θ : τρέφεται M.
[10] M has καὶ τὰ ὠμοφάγα· καὶ τὰ ὑλοφάγα. after ξηρότερα.
[11] χλωροφάγα θ: καρποφάγα M : χορτοφάγα Zwinger.

ποφάγα τῶν μὴ καρποφάγων, καὶ τὰ ὀλιγόποτα τῶν πολυπότων, καὶ τὰ πολύαιμα τῶν ἀναίμων
10 καὶ ὀλιγαίμων, καὶ τὰ ἀκμάζοντα μᾶλλον ἢ τὰ λίην παλαιὰ καὶ τὰ νέα, καὶ τὰ ἄρσενα τῶν θηλείων, καὶ τὰ ἔνορχα τῶν ἀνόρχων, καὶ μέλανα λευκῶν, καὶ τὰ δασέα ψιλῶν· τὰ δ' ἐναντία ὑγρότερα. αὐτῶν δὲ τῶν ζῴων ἰσχυρόταται μὲν αἱ σάρκες αἱ μάλιστα πονέουσαι καὶ ἐναιμόταται καὶ ἐν ᾗσι κατακλίνεται, κουφόταται δὲ τῶν σαρκῶν αἱ ἥκιστα πονέουσαι καὶ ὀλιγαιμόταται,[1] καὶ ἐκ τῆς σκιῆς, καὶ ὅσαι ἐσώταται τοῦ ζῴου. τῶν δὲ ἀναίμων ἐγκέφαλος καὶ μυελὸς ἰσχυρό-
20 τατα· κουφότατα δὲ κεφαλαί, πόδες, κτένες, καὶ μύες. τῶν δὲ ἰχθύων ξηρότατά ἐστι τὰ ἄνω, κουφότατα δὲ τὰ ὑπογάστρια, καὶ κεφαλαὶ
23 ὑγρότεραι διὰ τὴν πιμελὴν καὶ τὸν ἐγκέφαλον.

L. Ὠιὰ δὲ ὀρνίθων ἰσχυρὸν καὶ τρόφιμον καὶ φυσῶδες· ἰσχυρὸν μέν, ὅτι γένεσίς ἐστι ζῴου, τρόφιμον δέ, ὅτι γάλα ἐστὶ τοῦ ζῴου, φυσῶδες
4 δέ, ὅτι ἐκ μικροῦ ὄγκου ἐς πολὺ διαχεῖται.

LI. Τυρὸς δὲ ἰσχυρὸν καὶ καυσῶδες καὶ τρόφιμον καὶ στάσιμον·[2] ἰσχυρὸν μέν, ὅτι ἔγγιστα γενέσιος, τρόφιμον δέ, ὅτι τοῦ γάλακτος τὸ σαρκῶδές ἐστιν ὑπόλοιπον, καυσῶδες δέ, ὅτι λιπαρόν, στάσιμον δέ, ὅτι ὀπῷ καὶ πυτίῃ
6 συνέστηκεν.

LII. Ὕδωρ ψυχρὸν καὶ ὑγρόν·[3] οἶνος θερμὸν καὶ ξηρόν· ἔχει δέ τι καὶ καθαρτικὸν ἀπὸ τῆς ὕλης. τῶν δὲ οἴνων οἱ μέλανες καὶ αὐστηροὶ

[1] καὶ ὀλιγαιμόταται is omitted by M.
[2] καὶ στάσιμον omitted by M.
[3] ψυκτικόν M : ψυχρὸν καὶ ὑγρόν θ.

great drinkers; those which abound in blood than those which have little or no blood; those which are in their vigour than those which are very old or young; males than females; entire than gelded; the black than the white; the hairy than those which have little or no hair. The opposite to these are more moist. As to the flesh of animals as a class, that is the strongest which labours most, abounds most in blood, and on which they lie. Those are lightest which have laboured least, have least blood, are most in the shade, and are placed most inwardly in the animal. Of the bloodless parts the brain and the marrow are the strongest; the lightest parts are the head, the feet, the region of the genitals and those that are tendinous.[1] Of fish, the driest parts are the upper, the lightest those below the stomach; the head is more moist by reason of the fat and brain.

L. Birds' eggs are strong, nourishing and windy. An egg is strong because it is the origin of an animal; nourishing because it is the milk of the animal; windy, because from small bulk it expands to a great one.

LI. Cheese is strong, heating, nourishing and binding; it is strong because it is nearest to a creature's origin; it is nourishing because the fleshy part of the milk remains in it; it is heating because it is fat; binding, because it is coagulated by fig juice or rennet.

LII. Water is cooling and moist. Wine is hot and dry, and it has something purgative from its original substance. Dark and harsh wines are more dry, and

[1] For the meaning of μύες see Littré's note.

ξηρότεροι καὶ οὔτε διαχωρέονται οὔτε οὐρέονται[1] οὔτε πτύονται·[2] ξηραίνουσι δὲ τῇ θερμασίῃ,[3] τὸ ὑγρὸν ἐκ τοῦ σώματος καταναλίσκοντες. οἱ δὲ μαλακοὶ μέλανες ὑγρότεροι, καὶ φυσῶσι καὶ διαχωρέουσι μᾶλλον. οἱ δὲ γλυκέες μέλανες ὑγρότεροι καὶ ἀσθενέστεροι,[4] καὶ φυσῶσιν ὑγρα-
10 σίην ἐμποιέοντες. οἱ δὲ λευκοὶ[5] αὐστηροὶ θερμαίνουσι μέν, οὐ μὴν ξηραίνουσιν, οὐρέονται δὲ μᾶλλον ἢ διαχωρέουσιν. οἱ νέοι μᾶλλον τῶν οἴνων διαχωρέουσι, διότι ἐγγυτέρω τοῦ γλεύκεός εἰσι καὶ τροφιμώτεροι, καὶ οἱ ὄζοντες τῶν ἀνόδμων τῆς αὐτῆς ἡλικίης, διότι πεπειρότεροί εἰσι, καὶ οἱ παχέες τῶν λεπτῶν. οἱ δὲ λεπτοὶ οὐρέονται μᾶλλον· καὶ οἱ λευκοὶ καὶ οἱ λεπτοὶ γλυκέες οὐρέονται μᾶλλον ἢ διαχωρέουσι, καὶ ψύχουσι μὲν καὶ ἰσχναίνουσι καὶ ὑγραίνουσι τὸ
20 σῶμα,[6] καὶ τὸ αἷμα ἀσθενὲς ποιέουσιν, αὔξοντες τὸ ἀντίπαλον τῷ αἵματι ἐν τῷ σώματι.[7] γλεῦκος φυσᾷ καὶ ἐκταράσσει καὶ τὴν κοιλίην ὑπάγει·[8] φυσᾷ μέν, ὅτι θερμαίνει, ὑπάγει δὲ ἐκ τοῦ σώματος ὅτι καθαίρει,[9] ταράσσει δὲ ζέον ἐν τῇ κοιλίῃ καὶ διαχωρεῖ. οἱ ὀξίναι οἶνοι ψύχουσι καὶ ὑγραίνουσι καὶ ἰσχναίνουσι, ψύχουσι μὲν καὶ ἰσχναίνουσι κένωσιν[10] τοῦ ὑγροῦ ἐκ τοῦ σώματος ποιεόμενοι, ὑγραίνουσι δὲ ἀπὸ τοῦ ἐσιόντος ὕδατος σὺν τῷ οἴνῳ. ὄξος ψυκτικόν,

[1] οὔτε οὐρέονται omitted by θ.
[2] πτύονται θ: πτύουσι M.
[3] τῆι θερμασίηι θ: τὴν θερμασίην M.
[4] καὶ ἀσθενέστεροι θ: θερμαίνουσι M.
[5] After λευκοὶ M adds καὶ.
[6] οἱ δὲ λεπτοὶ γλυκέες· οὐρέονται μᾶλλον καὶ διαχωρέουσι καὶ ὑγραίνουσι τὸ σῶμα M: οἱ δὲ λεπτοὶ οὐραίονται μᾶλλον· καὶ οἱ

they pass well neither by stool nor by urine, nor by spittle. They dry by reason of their heat, consuming the moisture out of the body. Soft dark wines are moister; they are flatulent and pass better by stool. The sweet dark wines are moister and weaker; they cause flatulence because they produce moisture. Harsh white wines heat without drying, and they pass better by urine than by stool. New wines pass by stool better than other wines because they are nearer the must, and more nourishing; of wines of the same age, those with bouquet pass better by stool than those without, because they are riper, and the thicker wines better than the thin. Thin wines pass better by urine. White wines and thin sweet wines pass better by urine than by stool; they cool, attenuate and moisten the body, but make the blood weak, increasing in the body that which is opposed to the blood. Must causes wind, disturbs the bowels and empties them. It causes wind because it heats; it empties the body because it purges; it disturbs by fermenting in the bowels and passing by stool. Acid wines cool, moisten and attenuate; they cool and attenuate by emptying the body of its moisture; they moisten from the water that enters with the wine. Vinegar is refreshing,

---

λευκοὶ καὶ οἱ λεπτοὶ γλυκέες οὐραίονται μᾶλλον. ἢ διαχωρέουσι· καὶ ψύχουσι μὲν καὶ ἰσχναίνουσι· καὶ ὑγραίνουσι τὸ σῶμα θ.

[7] αὔξονταί τε ἐς τὸ ἀντίπαλον τῶ αἵματι ἐν τῶι σώματι θ: αὔξοντές τε τὸ ἀντίπαλον τοῦ σώματος τὸ αἷμα ἐν τῶι σώματι M.

[8] φυσᾷ καὶ ὑπάγει καὶ ἐκταράσσεται ζέον ἐν τῆι κοιλίηι καὶ διαχωρέει M: φυσᾶ καὶ ἐκταράσσει· καὶ τὴν κοιλίην ὑπάγει· θ.

[9] ὅτι καθαίρει Littré: κάθαρσιν θ M.

[10] ψύχουσι μὲν καὶ ἰσχναίνουσι omitted by θ. M has κενώσει, θ κένωσι. Perhaps some ancient texts had κένωσιν ποιεόμενοι and others κενώσει (sc. τοῦ ὑγροῦ).

30 διότι τῆκον τὸ ὑγρὸν τὸ ἐν τῷ σώματι καταναλίσκει, ἵστησι δὲ μᾶλλον ἢ διαχωρεῖ διότι οὐ τρόφιμον καὶ δριμύ. ἕψημα θερμαίνει καὶ ὑγραίνει καὶ ὑπάγει, θερμαίνει μὲν ὅτι οἰνῶδες, ὑγραίνει δὲ ὅτι τρόφιμον, ὑπάγει δὲ ὅτι γλυκὺ καὶ πρός, καθηψημένον[1] ἐστίν. τρύγες στεμφυλίτιδες ὑγραίνουσι καὶ ὑπάγουσι καὶ φυσῶσι,
37 διότι[2] καὶ τὸ γλεῦκος τὸ αὐτὸ ποιεῖ.

LIII. Μέλι θερμαίνει καὶ ξηραίνει ἄκρητον, σὺν ὕδατι δὲ ὑγραίνει καὶ διαχωρεῖ τοῖσι χολώδεσι, τοῖσι δὲ φλεγματώδεσιν ἵστησιν. ὁ δὲ γλυκὺς οἶνος διαχωρεῖ μᾶλλον τοῖσι φλεγ-
5 ματίῃσι.

LIV. Περὶ δὲ λαχάνων ὧδε ἔχει. σκόροδον θερμὸν καὶ διαχωρητικὸν καὶ οὐρεῖται, ἀγαθὸν τοῖσι σώμασι, τοῖσι δ' ὀφθαλμοῖσι φλαῦρον· κάθαρσιν γὰρ ἐκ τοῦ σώματος πολλὴν ποιεόμενον, τὴν ὄψιν ἀπαμβλύνει· διαχωρεῖ δὲ καὶ οὐρεῖται, διὰ τὸ καθαρτικόν· ἑφθὸν ἀσθενέστερον ἢ ὠμόν· φῦσαν δὲ ἐμποιεῖ διὰ τοῦ πνεύματος τὴν ἐπίστασιν.[3] κρόμμυον τῇ μὲν ὄψει ἀγαθόν, τῷ δὲ σώματι κακόν, διότι θερμὸν καὶ καυσῶδές ἐστι
10 καὶ οὐ διαχωρεῖ· τροφὴν μὲν γὰρ οὐ δίδωσι τῷ σώματι οὐδὲ ὠφελείην· θερμαῖνον δὲ ξηραίνει διὰ τὸν ὀπόν. πράσον θερμαίνει μὲν ἧσσον, οὐρεῖται δὲ καὶ διαχωρεῖ· ἔχει δέ τι καὶ καθαρτικόν· ὑγραίνει δὲ καὶ ὀξυρεγμίην παύει· ὕστατον δὲ ἐσθίειν. ῥαφανὶς ὑγραίνει διαχέουσα τὸ φλέγμα τῇ δριμύτητι, τὰ δὲ φύλλα ἧσσον. πρὸς τὰ ἀρθριτικὰ μοχθηρὸν ἡ ῥίζη, ἐπιπολάζον δὲ καὶ

[1] καθάπερ ἡψημένον M: προσκαθήμενον θ: προσκαθεψημένον (*sic*) Littré.

because it dissolves and consumes the moisture in the body; it is binding rather than laxative because it affords no nourishment and is sharp. Boiled-down wine warms, moistens and sends to stool. It warms because it is vinous, moistens because it is nutritious, and sends to stool because it is sweet and moreover boiled-down. Wine from grape-husks moistens, sends to stool and fills with wind, because must also does the same.

LIII. Honey unmixed warms and dries; mixed with water it moistens, sends to stool those of bilious temperament, but binds those who are phlegmatic. But sweet wine tends to send the phlegmatic to stool.

LIV. The qualities of vegetables are as follow. Garlic warms, passes well by stool and by urine, and is good for the body though bad for the eyes. For making a considerable purgation of the body it dulls the sight. It promotes stools and urine because of the purgative qualities it possesses. When boiled it is weaker than when raw. It causes flatulence because it causes stoppage of wind. The onion is good for sight, but bad for the body, because it is hot and burning, and does not lead to stool; for without giving nourishment or help to the body it warms and dries on account of its juice. The leek warms less, but passes well by urine and by stool; it has also a certain purgative quality. It moistens and it stops heartburn, but you must eat it last. The radish moistens through melting the phlegm by its sharpness, but the leaves do so less. The root is bad for arthritis, and it repeats and is hard to digest. Cress

[2] διότι M: ὅπερ θ.
[3] ἐπίσταισιν M: ἐπίσπασι θ.

δύσπεπτον. κάρδαμον θερμαντικὸν καὶ τὴν σάρκα τῆκον· συνίστησι φλέγμα λευκόν, ὥστε 20 στραγγουρίην ἐμποιεῖν. νᾶπυ θερμόν· διαχωρεῖ, δυσουρεῖται δὲ καὶ τοῦτο· καὶ εὔζωμον παραπλήσια τούτοισι διαπρήσσεται. κορίανον θερμὸν καὶ στατικόν, καὶ ὀξυρεγμίην παύει, ὕστατον δ' ἐπεσθιόμενον καὶ ὑπνοποιεῖ. θρίδαξ ψυχρότερον πρὶν τὸν ὀπὸν ἔχειν· ἀσθενείην δ' ἐνίοτε[1] ἐμποιεῖ τῷ σώματι. ἄνηθον[2] θερμὸν καὶ στατικόν, καὶ πταρμὸν παύει ὀσφραινόμενον. σέλινον οὐρεῖται μᾶλλον ἢ διαχωρεῖ, καὶ αἱ ῥίζαι μᾶλλον ἢ αὐτὸ διαχωρέουσιν. ὤκιμον ξηρὸν καὶ 30 θερμὸν[3] καὶ στάσιμον. πήγανον οὐρεῖται μᾶλλον ἢ διαχωρεῖ, καὶ συστρεπτικόν τι ἔχει, καὶ πρὸς τὰ φάρμακα τὰ βλαβερὰ ὠφελεῖ προπινόμενον. ἀσπάραγος ξηρὸν καὶ στάσιμον. ἐλελίσφακον ξηρὸν καὶ στατικόν. στρύχνος ψύχει καὶ ἐξονειρώσσειν οὐκ ἐᾷ. ἀνδράχνη ψύχει ἡ ποταινίη,[4] τεταριχευμένη δὲ θερμαίνει. κνίδη[5] καθαίρει. καλαμίνθη θερμαίνει καὶ καθαίρει.[6] μίνθη θερμαίνει καὶ οὐρεῖται καὶ ἐμέτους ἵστησι, καὶ ἢν πολλάκις ἐσθίῃ τις, τὴν γονὴν τήκει ὥστε 40 ῥεῖν, καὶ ἐντείνειν κωλύει, καὶ τὸ σῶμα ἀσθενὲς ποιεῖ. λάπαθον θερμαῖνον διαχωρεῖ. ἀνδράφαξις ὑγρόν, οὐ μέντοι διαχωρεῖ. βλίτον θερμόν, οὐ διαχωρητικόν.[7] κράμβη θερμαίνει καὶ διαχωρεῖ· χολώδεα δὲ ἄγει. σεύτλου ὁ μὲν χυλὸς διαχωρεῖ, αὐτὸ δὲ ἵστησιν, αἱ δὲ ῥίζαι τῶν σεύτλων διαχωρητικώτεραι. κολοκύντη θερμαίνει[8] καὶ ὑγραίνει

[1] δ' ἐνίοτε θ: δὲ τινὰ M.
[2] Before θερμὸν θ adds ἧσσον.
[3] καὶ θερμὸν is omitted by θ.

is heating and melts the flesh; it congeals white phlegm, so as to produce strangury. Mustard is hot and passes well by stool; it too passes hardly by urine. Rocket also has effects like those of mustard. Coriander is hot and astringent; it stops heartburn, and when eaten last also causes sleep. Lettuce is rather cooling before it has its juice, but sometimes it produces weakness in the body. Anise is hot and astringent, and the smell of it stops sneezing. Celery passes better by urine than by stool, and the root passes by stool better than does the stalk. Basil is dry, hot and astringent. Rue passes better by urine than by stool, and it has a certain congealing quality, while if drunk beforehand it is a prophylactic against poisons. Asparagus is dry and astringent. Sage is dry and astringent. Night-shade cools and prevents nightly pollutions. Purslane when fresh cools, when preserved it warms. Nettles purge. Catmint warms and purges. Mint warms, passes easily by urine, and stops vomiting; if eaten often it melts the seed and makes it run, preventing erections and weakening the body. Sorrel warms and passes well by stool. Orach is moist without passing well by stool. Blite is warm without passing well by stool. Cabbage warms, passes well by stool and evacuates bilious matters. Beet juice passes well by stool, though the vegetable itself is astringent; the roots of beet are rather more aperient. The pumpkin

[4] *ποταινίῃ* Foes (in note), Mack, Littré: *ποταμιηι* θ: *ποταμίῃ* M.

[5] For *κνίδη* M has *καὶ*.

[6] θ omits *καθαίρει*. *μίνθη θερμαίνει καὶ*.

[7] *βλίτον· θερμόν, οὐδιαχωρητικόν* M. Omitted by θ, while Littré has *οὐ θερμόν, διαχωρητικόν*.

[8] *ψύχει* Littré: *θερμαίνει* θ M.

καὶ διαχωρεῖ, οὐκ οὐρεῖται δέ. γογγυλὶς καυσῶδες, ὑγραίνει δὲ καὶ ταράσσει τὸ σῶμα, οὐ μέντοι διαχωρεῖ, δυσουρεῖται[1] δέ. γλήχων θερ-
50 μαίνει καὶ διαχωρεῖ. ὀρίγανον θερμαίνει, ὑπάγει δὲ χολώδεα. θύμβρη παραπλήσια διαπρήσσεται. θύμον θερμόν, διαχωρεῖ καὶ οὐρεῖται, ἄγει δὲ φλεγματώδεα. ὕσσωπος θερμαίνει καὶ ὑπάγει φλεγματώδεα. τῶν δὲ ἀγρίων λαχάνων ὅσα ἐν τῷ στόματι θερμαντικὰ καὶ εὐώδεα, ταῦτα θερμαίνει καὶ οὐρεῖται μᾶλλον ἢ διαχωρεῖ· ὁκόσα δὲ ὑγρὴν φύσιν ἔχει καὶ ψυχρὴν καὶ μωρὴν ἢ ὀσμὰς βαρείας, ὑποχωρεῖται μᾶλλον ἢ οὐρεῖται· ὁκόσα δέ ἐστι στρυφνὰ ἢ αὐστηρά, στάσιμα·
60 ὅσα δὲ δριμέα καὶ εὐώδεα, διουρεῖται· ὁκόσα δὲ δριμέα καὶ ξηρὰ ἐν τῷ στόματι, ταῦτα ξηραίνει· ὁκόσα δὲ ὀξέα,[2] ψυκτικά. οἱ δὲ χυμοὶ διουρητικοί, κρήθμου, σελίνου, σκορόδου ἀποβρέγματα, κυτίσου, μαράθρου, πράσου,[3] ἀδιάντου, στρύχνου· ψύχει σκολοπένδριον,[4] μίνθη, σέσελι, σέρις, καυκαλίδες, ὑπερικόν, κνίδαι· διαχωρητικοὶ δὲ καὶ καθαρτικοί, ἐρεβίνθων, φακῆς, κριθῆς, σεύτλων, κράμβης, λινοζώστιος, ἀκτῆς, κνήκου·
69 ταῦτα μᾶλλον ὑποχωρεῖται ἢ διουρεῖται.

LV. Περὶ δὲ ὀπώρης ὧδε ἔχει. τὰ μὲν

[1] For δυσουρεῖται θ has οὐραίεται.
[2] Before ψυκτικά M has καί.
[3] θ has μαράθου πράσων, and M μαράθων· πράσον·

warms,[1] moistens, and passes easily by stool though not by urine. The turnip is heating, moistening, and disturbing to the body; but it does not pass easily, either by stool or by urine.[2] Pennyroyal warms and passes easily by stool. Marjoram warms, and also evacuates bilious matters. Savory acts in a similar way. Thyme is hot, passes easily by stool and urine, and evacuates phlegmatic humours. Hyssop is warming and expels phlegmatic humours. Of wild vegetables, those that are warming in the mouth, and of a sweet smell, warm and pass more readily by urine than by stool; those that have a moist, cold and sluggish nature, or a strong smell, pass more easily by stool than by urine; those that are rough or harsh, are binding; those that are sharp and of a sweet smell pass easily by urine; those that are sharp and dry in the mouth are drying; those that are acid are cooling. Diuretic juices are those of samphire, celery, garlic (in infusions), clover, fennel, leek, maiden-hair, nightshade. Cooling are hart's tongue, mint, *seseli*, endive, bur-parsley, hypericum, nettles. Juices that send to stool or purge are those of chick-pea, lentils, barley, beet, cabbage, mercury, elder, carthamus. These help stools rather than urine.

LV. The following are the qualities of fruits.

[1] It is difficult to accept this reading, although the authority for it is very strong. Littré's reading (*ψύχει*, but he does not give his authority) may be correct, but it is difficult to see why it should have been changed to *θερμαίνει*.

[2] With the reading of *θ*: "does not pass easily by stool, though it does by urine."

---

[4] *θ* has *ἀδιάντου καὶ ψύχει στρυχνόν. καὶ τοῦτο ψύχει· καὶ σκολοπένδριον.*

ἐγκάρπια[1] διαχωρητικώτερα, τὰ δὲ χλωρὰ τῶν ξηρῶν. ἡ δὲ δύναμις εἰρήσεται[2] αὐτῶν. μόρα θερμαίνει καὶ ὑγραίνει καὶ διαχωρεῖ. ἄπιοι πέπειροι θερμαίνουσι καὶ ὑγραίνουσι καὶ διαχωρέουσιν· αἱ δὲ σκληραὶ στάσιμον· ἀχράδες δὲ χειμέριοι πέπειροι διαχωρέουσι καὶ τὴν κοιλίην καθαίρουσιν·[3] αἱ δὲ ὠμαὶ στάσιμον. μῆλα γλυκέα δύσπεπτα, ὀξέα δὲ πέπονα ἧσσον· κυδώνια
10 στυπτικὰ καὶ οὐ διαχωρέουσιν·[4] οἱ δὲ χυλοὶ τῶν μήλων πρὸς τοὺς ἐμέτους στατικοὶ καὶ οὐρητικοί· καὶ ὀδμαὶ πρὸς τοὺς ἐμέτους· τὰ δὲ ἄγρια μῆλα στατικά, ἑφθὰ δὲ μᾶλλον διαχωρεῖ· πρὸς δὲ τὴν ὀρθοπνοίην οἵ τε χυλοὶ αὐτῶν καὶ αὐτὰ πινόμενα ὠφελεῖ. οὖα[5] δὲ καὶ μέσπιλα καὶ κράνια καὶ ἡ τοιαύτη ὀπώρη στατικὴ καὶ στρυφνή. ῥοιῆς γλυκείης χυλὸς διαχωρεῖ, καυσῶδες δέ τι ἔχει· αἱ οἰνώδεες φυσώδεες· αἱ δὲ ὀξεῖαι ψυκτικώτεραι·[6] οἱ δὲ πυρῆνες πασέων στάσιμον. σίκυοι ὠμοὶ
20 δύσπεπτον·[7] πέπονες δὲ οὐρέονται καὶ διαχωρέουσι,[8] φυσώδεες δέ. βότρυες θερμὸν καὶ ὑγρὸν

[1] For ἐγκάρπια θ has κάρπιμα.

[2] εἰρήσεται θ: εἴρηται M.

[3] καθαίρουσιν θ: καθαίρει M.

[4] θ has μῆλα κυδώνια δύσπεπτα ὀξέα πέπονα ἧσσον· ἔχει δὲ τι στυπτικόν.

[5] For οὖα θ has a blank space.

[6] θ has ὁ οἰνώδης φυσώδης· ἡ δὲ ὀξία ψυκτικωτέρη· οἱ δὲ πύρινες πάντων στάσιμον. M. has αἱ οἰνώδεες. ἧσσον καυσώδεες· αἱ δὲ ὀξεῖαι ψυκτικώτεραι· οἱ δὲ πύρινες πάντων στάσιμοι. Littré reads αἱ οἰνώδεες τῶν ῥοιῶν φυσώδεες· αἱ δὲ ὀξεῖαι ψυκτικώτεραι· οἱ δὲ πυρῆνες πασέων στάσιμοι.

[7] The text is that of θ. The reading of M is σικυοὶ ὠμοὶ ψυχροὶ καὶ δύσπεπτοι. Littré has the reading of M, and continues: οἱ δὲ πέπονες οὐρέονται.

[8] θ has διαχωρέουσι δὲ, M διαχωρεῦνται.

Fruit generally[1] is rather relaxing, more so when fresh than when dry. The properties of fruits shall now be given. Mulberries warm, moisten and pass easily by stool. Pears when ripe warm, moisten and pass easily by stool, but when hard they are binding. Wild winter pears when ripe pass easily by stool and purge the bowels; when unripe they are binding. Sweet apples are indigestible, but acid apples when ripe are less so. Quinces are astringent, and do not pass easily by stool. Apple juice stops vomiting and promotes urine. The smell too of apples is good for vomiting. Wild apples are astringent, but when cooked they pass more easily by stool. For orthopnœa their juice, and the apples themselves when a draught is made of them, are beneficial. Service berries, medlars, cornel berries and such fruit generally are binding and astringent. The juice of the sweet pomegranate is laxative, but has a certain burning quality. Vinous pomegranates are flatulent.[2] The acid are more cooling. The seeds of all[3] are astringent. Unripe gourds[4] are indigestible; ripe gourds[5] pass easily by urine and stool, but are flatulent. Grapes are warming and moist, passing easily by

[1] *ἐγκάρπιος* means literally, "containing seed within it." It may therefore mean here "with the seed formed," *i.e.* "ripe," as Littré takes it. I prefer, however, to make *ἐγκάρπια* = fruit generally, those things "whose seed is in themselves." The reading of θ (*κάρπιμα*) can scarcely be right, as *κάρπιμος* means "fruitful" or "fruit-bearing." It is possible that *ἐγκάρπια* refers to fruit as distinguished from nuts. *ὀπώρα* includes both.

[2] With the reading of M, "less burning."

[3] The reading *πάντων* has overwhelming authority. Can it mean "of all fruits" (pomegranates included)?

[4] Apparently the cucumber.

[5] Apparently the melon.

καὶ διαχωρεῖ, μάλιστα μὲν οἱ λευκοί· οἱ μὲν οὖν γλυκέες θερμαίνουσιν ἰσχυρῶς, διότι πολὺ ἤδη τοῦ θερμοῦ ἔχουσιν· οἱ δὲ ὀμφακώδεες ἧσσον θερμαίνουσι, καθαίρουσι δὲ πινόμενοι· ἀσταφίδες δὲ καυσῶδες, διαχωρεῖ δέ. σῦκον χλωρὸν ὑγραίνει καὶ διαχωρεῖ καὶ θερμαίνει· ὑγραίνει μὲν διὰ τὸ ἔγχυλον εἶναι,[1] θερμαίνει δὲ διὰ τὸν γλυκὺν ὀπὸν καὶ διαχωρεῖ· τὰ πρῶτα τῶν σύκων κά-
30 κιστα, ὅτι ὀπωδέστατα, βέλτιστα δὲ τὰ ὕστατα· ξηρὰ σῦκα καυσώδεα μέν, διαχωρεῖ δέ. αἱ ἀμυγδάλαι καυσῶδες, τρόφιμον δέ· καυσῶδες μὲν διὰ τὸ λιπαρόν, τρόφιμον δὲ διὰ τὸ σαρκῶδες. κάρυα στρογγύλα παραπλήσια· τὰ δὲ πλατέα τρόφιμα πέπονα, καὶ διαχωρεῖ[2] καθαρὰ ἐόντα, καὶ φῦσαν ἐμποιεῖ· οἱ δὲ χιτῶνες αὐτῶν στάσιμον. ἄκυλοι δὲ καὶ βάλανοι δρύϊνοι[3] στατικὰ ὠμά·[4]
38 ἑφθὰ ἧσσον.

LVI. Τὰ πίονα τῶν κρεῶν καυσώδεα, διαχωρεῖ δέ. κρέα ταριχηρὰ ἐν οἴνῳ μὲν ξηραίνει καὶ τρέφει, ξηραίνει μὲν διὰ τὸν οἶνον, τρέφει δὲ διὰ τὴν σάρκα· ἐν ὄξει δὲ τεταριχευμένα θερμαίνει μὲν ἧσσον διὰ τὸ ὄξος, τρέφει δὲ ἱκανῶς· ἐν ἁλὶ δὲ κρέα ταριχηρὰ τρόφιμα μὲν ἧσσον, διὰ τὸ ἅλας[5] τοῦ ὑγροῦ ἀπεστερημένα, ἰσχναίνει δὲ καὶ ξηραίνει καὶ διαχωρεῖ ἱκανῶς. τὰς δὲ δυνάμιας ἑκάστων ἀφαιρεῖν καὶ προστιθέναι ὧδε χρή,
10 εἰδότα ὅτι[6] πυρὶ καὶ ὕδατι πάντα συνίσταται

[1] So θ: M has διότι ἔγχυλόν ἐστι.

[2] τὰ δὲ πλατεα πεπονα. τρόφιμον καὶ διαχωρέει M: τὰ δὲ πλατέα κάρεα. τρόφιμα· πέπονα· καὶ διαχωρέει θ.

[3] δρυΐνοι θ: καὶ φηγηι M: καὶ φηγοὶ Littré.

[4] After ὠμά M adds καὶ ὀπτά.

stool; white grapes are especially so. Sweet grapes are very heating, because by the time they are sweet they have absorbed much heat. Unripe grapes are less warming, but a draught made from them is purgative. Raisins are burning, but pass well by stool. The green fig moistens, passes well by stool and warms; it moistens because it is juicy, warms and passes well because of its sweet juice. The first crop of figs is the worst, because such figs have most juice; the latest are the best. Dry figs are burning, but pass well by stool. Almonds are burning but nutritious; burning because they are oily, and nutritious because they are fleshy. Round nuts[1] are similar. Flat nuts[2] are nutritious when ripe, pass easily by stool when peeled, and cause flatulence. Their skins, however, are binding. Ilex nuts and acorns are binding when raw, but less so when boiled.

LVI. Rich meats are burning, but pass well by stool. Meats preserved in wine are drying and nutritious; drying because of the wine, and nourishing because of the flesh. When preserved in vinegar they are less warming because of the vinegar, but they are quite nutritious. Meats preserved in salt are less nutritious, because the brine has deprived them of their moisture, but they attenuate, dry, and pass by stool quite well. The powers of foods severally ought to be diminished or increased in the following way, as it is known that out of fire and water are composed all things, both animal and

[1] Ordinary nuts. [2] Chestnuts.

---

[5] θ has αλι and τὸ ἅλα, M αλσὶ and τὸ ἅλες. Two MSS. have the late form τὸ ἅλας (so Mack and Littré).
[6] εἰδότα ὅτι omitted by M.

καὶ ζῷα καὶ φυτά, καὶ ὑπὸ τούτων αὔξεται καὶ ἐς ταῦτα διακρίνεται. τῶν μὲν οὖν ἰσχυρῶν σιτίων ἑψῶντα πολλάκις καὶ διαψύχοντα τὴν δύναμιν ἀφαιρεῖν, τῶν δὲ ὑγρῶν πυροῦντα καὶ φώζοντα τὴν ὑγρασίην ἐξαιρεῖν, τῶν δὲ ξηρῶν βρέχοντα καὶ νοτίζοντα, τῶν δὲ ἁλμυρῶν βρέχοντα καὶ ἑψῶντα, τῶν δὲ πικρῶν καὶ δριμέων τοῖσι γλυκέσι διακιρνῶντα, τῶν δὲ στρυφνῶν τοῖσι λιπαροῖσι· καὶ τῶν ἄλλων
20 πάντων ἐκ τῶν προειρημένων χρὴ γινώσκειν. ὁκόσα πυρούμενα ἢ φωζόμενα στάσιμά ἐστι[1] μᾶλλον τῶν ὠμῶν, διότι τὸ ὑγρὸν ὑπὸ τοῦ πυρὸς ἀφῄρηται καὶ τὸ ὀπῶδες καὶ τὸ λιπαρόν· ὅταν οὖν ἐς τὴν κοιλίην ἐμπέσῃ, ἕλκει τὸ ὑγρὸν ἐκ τῆς κοιλίης ἐφ' ἑωυτά, καὶ συγκαίει[2] τὰ στόματα τῶν φλεβῶν, ξηραίνοντα καὶ θερμαίνοντα, ὥστε ἵστησι τὰς διεξόδους τῶν ὑγρῶν.[3] τὰ δὲ ἐκ τῶν ἀνύδρων καὶ ξηρῶν καὶ πνιγηρῶν χωρίων ἅπαντα ξηρότερα καὶ θερμότερα καὶ ἰσχὺν πλείω
30 παρέχεται ἐς τὸ σῶμα, διότι ἐκ τοῦ ἴσου ὄγκου βαρύτερα καὶ πυκνότερα καὶ πολύνοστά[4] ἐστιν ἢ τὰ ἐκ τῶν ὑγρῶν τε καὶ ἀρδομένων καὶ ψυχρῶν· ταῦτα δὲ ὑγρότερα καὶ κουφότερα καὶ ψυχρότερα. οὔκουν[5] δεῖ τὴν δύναμιν αὐτοῦ μόνον γνῶναι τοῦ τε σίτου καὶ τοῦ πόματος καὶ τῶν ζῴων, ἀλλὰ καὶ τῆς πατρίδος[6] ὁκόθεν εἰσίν. ὅταν μὲν οὖν βούλωνται τροφὴν ἰσχυροτέρην τῷ σώματι προσενεγκεῖν ἀπὸ τῶν αὐτῶν σίτων, τοῖσιν ἐκ τῶν ἀνύδρων χωρίων χρηστέον καὶ σιτίοισι καὶ
40 πόμασι καὶ ζῴοισιν· ὁκόταν δὲ κουφοτέρῃ τροφῇ

[1] ὅκως ἀπυρούμενα ἢ φωζόμενα στασιμά ἐστι M : ὅσα πυροῦται η φωζομενα στατικά ἐστι θ.

vegetable, and that through them all things grow, and into them they are dissolved. Take away their power from strong foods by boiling and cooling many times; remove moisture from moist things by grilling and roasting them; soak and moisten dry things, soak and boil salt things, bitter and sharp things mix with sweet, and astringent things mix with oily. All other cases judge in accordance with what has been already said. Foods grilled or roasted are more binding than raw, because the fire has taken away the moisture, the juice and the fat. So when they fall into the belly they drag to themselves the moisture from the belly, burning up the mouths of the veins, drying and heating them so as to shut up the passages for liquids. Things coming from waterless, dry and torrid regions are all drier and warmer, and provide the body with more strength, because, bulk for bulk, they are heavier, more compact and more nutritious[1] than those from moist regions that are well-watered and cold, the latter foods being moister, lighter and colder. Accordingly, it is necessary to know the property, not only of foods themselves, whether of corn, drink or meat, but also of the country from which they come. So those who wish to give the body a stronger nourishment, without increasing the bulk of the food, must

[1] Or (reading *πολύναστα*) "more compressed."

---

[2] *ἐπ' αὐτὰ συγκλείων* M : *ἐφεωυτο· καὶ συγκαίων* θ. Perhaps we should read *ἐφ' ἑωυτό, συγκαῖον κ.τ.ἕ.* with singular participles following. The subject then would be "such food as this."

[3] *τῶν ὑγρῶν* θ : *τοῦ ὑγροῦ* M.

[4] *πολύνοστά* M : *πολυναστα* θ.

[5] *οὐκοῦν* θ M.

[6] *τὰς πατρίδας* θ : *τῆς πατρίδος* M.

καὶ ὑγροτέρῃ, τοῖς ἐκ τῶν ἀρδομένων χρηστέον. τὰ γλυκέα καὶ τὰ δριμέα καὶ τὰ ἁλυκὰ καὶ τὰ πικρὰ καὶ τὰ αὐστηρὰ καὶ τὰ σαρκώδεα θερμαίνειν πέφυκε, καὶ ὅσα ξηρά ἐστι καὶ ὅσα ὑγρά.[1] ὁκόσα μὲν οὖν ξηροῦ μέρος πλέον ἐν αὐτοῖσι ἔχει, ταῦτα μὲν θερμαίνει καὶ ξηραίνει· ὁκόσα δὲ ὑγροῦ μέρος ἔχει πλέον, ταῦτα πάντα θερμαίνοντα ὑγραίνει καὶ διαχωρεῖ μᾶλλον ἢ τὰ ξηρά· τροφὴν γὰρ μᾶλλον ἐς τὸ σῶμα διδόντα, ἀντί-
50 σπασιν ποιεῖται ἐς τὴν κοιλίην,[2] καὶ ὑγραίνοντα διαχωρεῖ. ὅσα θερμαίνοντα ξηραίνει ἢ σῖτα ἢ ποτά, οὔτε πτύσιν οὔτε διούρησιν οὔτε διαχώρησιν ποιέοντα ξηραίνει τὸ σῶμα διὰ τάδε· θερμαινόμενον τὸ σῶμα κενοῦται τοῦ ὑγροῦ, τὸ μὲν ὑπ' αὐτῶν τῶν σιτίων, τὸ δὲ ἐς[3] τὴν τροφὴν τῷ τῆς ψυχῆς θερμῷ καταναλίσκεται, τὸ δὲ διὰ τοῦ χρωτὸς ἐξωθεῖται θερμαινόμενον καὶ λεπτυνόμενον. τὰ γλυκέα καὶ τὰ πίονα καὶ τὰ λιπαρὰ πληρωτικά ἐστι, διότι ἐξ ὀλίγου ὄγκου πολύχοά
60 ἐστι· θερμαινόμενα δὲ καὶ διαχεόμενα πληροῖ τὸ θερμὸν ἐν τῷ σώματι καὶ γαληνίζειν[4] ποιεῖ. τὰ δὲ ὀξέα καὶ δριμέα καὶ αὐστηρὰ καὶ στρυφνὰ καὶ[5] συγκομιστὰ καὶ ξηρὰ οὐ πληροῖ, διότι τὰ στόματα τῶν φλεβῶν ἀνέωξέ τε καὶ διεκάθηρε· καὶ τὰ μὲν ξηραίνοντα, τὰ δὲ δάκνοντα, τὰ δὲ στύφοντα φρῖξαι καὶ συστῆναι ἐς ὀλίγον ὄγκον ἐποίησεν τὸ ὑγρὸν τὸ ἐν τῇ σαρκί· καὶ τὸ κενὸν πολὺ ἐγένετο ἐν τῷ σώματι. ὅταν οὖν βούλῃ ἀπ' ὀλίγων πληρῶσαι ἢ ἀπὸ πλειόνων κενῶσαι,

[1] πικρά θ: ὑγρὰ M.
[2] So θ M. The vulgate has ἀντίστασιν and ἐν τῇ κοιλίῃ.
[3] δ' εἰς θ: δὲ M.

use corn, drink and meat from waterless regions. When they need lighter and moister nourishment, they must use things from well-watered regions. Things sweet, or sharp, or salt, or bitter, or harsh, or fleshy are naturally heating, whether they are dry or moist. Things that have in themselves a greater portion of the dry, these warm and dry; those that have a greater portion of the moist in all cases warm, moisten and pass by stool better than things that are dry; for being more nourishing to the body they cause a revulsion to the belly, and, moistening, pass readily by stool. Such foods or drinks as warm and dry, producing neither spittle nor urine nor stools, dry the body for the following reasons. The body growing warm is emptied of its moisture, partly by the foods themselves, while part is consumed in giving nourishment to the warmth of the soul, while yet another part, growing warm and thin, forces its way through the skin. Things sweet, or fat, or oily are filling, because though of small bulk they are capable of wide diffusion. Growing warm and melting they fill up the warmth in the body and make it calm. Things acid, sharp, harsh, astringent, †—† and dry are not filling, seeing that they open and thoroughly cleanse the mouths of the veins; and some by drying, others by stinging, others by contracting, make the moisture in the flesh shiver and compress itself into a small bulk, and so the void in the body becomes great. So when you wish to fill with little food, or empty with more, use foods of

[4] *γαληνίζειν* θ : *γαληνιάζειν* M.
[5] *συγκομιστὰ* M : *δυσκόμιστα* θ. The true reading has been lost, as we need a word meaning harsh or dry. *συγκομιστὸς* means "assorted" and *δυσκόμιστος* "intolerable."

70 τοιούτοισι χρῆσθαι. τὰ πρόσφατα πάντα ἰσχὺν παρέχεται πλείω τῶν ἄλλων διὰ τόδε, ὅτι ἔγγιον τοῦ ζῶντός ἐστι· τὰ δὲ ἕωλα καὶ σαπρὰ διαχωρεῖ μᾶλλον τῶν προσφάτων, διότι ἔγγιον τῆς σηπεδόνος ἐστί. τὰ δὲ ἔνωμα στροφώδεα καὶ ἐρευγμώδεα, διότι ἃ δεῖ τῷ πυρὶ κατεργάζεσθαι, ταῦτα ἡ κοιλίη διαπρήσσεται ἀσθενεστέρη ἐοῦσα τῶν ἐσιόντων. τὰ δὲ ἐν τοῖσιν ὑποτρίμμασιν ὄψα σκευαζόμενα καυσώδεα καὶ ὑγρά, ὅτι λιπαρὰ καὶ πυρώδεα καὶ θερμὰ καὶ ἀνομοίους τὰς 80 δυνάμιας ἀλλήλοισιν ἔχοντα ἐν τῷ αὐτῷ ἵζει.[1] τὰ δὲ ἐν ἅλμῃ ἢ ὄξει βελτίω καὶ οὐ καυσώδεα.

LVII. Περὶ δὲ λουτρῶν ὧδε ἔχει· ὕδωρ πότιμον ὑγραίνει καὶ ψύχει, δίδωσι γὰρ τῷ σώματι ὑγρασίην· τὸ δὲ ἁλμυρὸν λουτρὸν θερμαίνει καὶ ξηραίνει, φύσει γὰρ θερμὸν ἕλκει ἀπὸ τοῦ σώματος τὸ ὑγρόν. τὰ δὲ θερμὰ λουτρὰ νῆστιν μὲν ἰσχναίνει καὶ ψύχει· φέρει γὰρ ἀπὸ τοῦ σώματος τὸ ὑγρὸν τῇ θερμασίῃ· κενουμένης δὲ τῆς σαρκὸς τοῦ ὑγροῦ, ψύχεται τὸ σῶμα· βεβρωκότα δὲ θερμαίνει καὶ ὑγραίνει, διαχέοντα τὰ ὑπάρχοντα 10 ἐν τῷ σώματι ὑγρὰ ἐς πλείονα ὄγκον. ψυχρὰ δὲ λουτρὰ τοὐναντίον· κενῷ μὲν τῷ σώματι δίδωσι θερμόν τι †ψυχρὸν ἐόν· βεβρωκότος δὲ ἀφαιρεῖ ὑγροῦ ἐόντος ξηρὸν ἐόν, καὶ πληροῖ τοῦ ὑπάρχον-

this kind. Fresh foods in all cases give more strength than others, just because they are nearer to the living creature. But stale and putrid things pass more readily by stool than do fresh because they are nearer to corruption. Raw things cause colic and belching, because what ought to be digested by the fire is dealt with by the belly, which is too weak for the substances that enter it. Meats prepared in sauces[1] are burning and moist, because there are united in one place things oily, fiery, warm, and with mutually opposite properties. Preparations in brine or vinegar are better and are not burning.

LVII. As to baths, their properties are these. Drinkable[2] water moistens and cools, as it gives moisture to the body. A salt bath warms and dries, as having a natural heat it draws the moisture from the body. Hot baths, when taken fasting, reduce and cool, for they carry the moisture from the body owing to their warmth, while as the flesh is emptied of its moisture the body is cooled. Taken after a meal they warm and moisten, as they expand to a greater bulk the moisture already existing in the body. Cold baths have an opposite effect. To an empty body they give a certain amount of heat; after a meal they take away moisture and fill with

[1] The *ὑπότριμμα* (like the Latin *moretum*) was a piquant dish of various ingredients grated together.
[2] *I.e.* what we call "fresh" water.

---

[1] *καὶ ἀνόμοια ἐς τὰς δυνάμιας· ἀλλήλοισιν ἀῦτις ἔχοντα ἐν τω αὐτῷι ἴζει. θ*: *καὶ ἀνομοίας τὰς δυνάμιας ἀλλήλοισι ἔχοντα ἐν τῶι αὐτέωι ἴζει* M.

τος ξηροῦ.†[1] ἀλουσίη ξηραίνει καταναλισκομένου
15 τοῦ ὑγροῦ, καὶ ἀνηλειψίη ὡσαύτως.[2]

LVIII. Λίπος δὲ θερμαίνει καὶ ὑγραίνει καὶ μαλάσσει. ἥλιος δὲ καὶ πῦρ ξηραίνει διὰ τάδε· θερμὰ ἐόντα καὶ ξηρὰ ἕλκει ἐκ τοῦ σώματος τὸ ὑγρόν. σκιὴ δὲ καὶ ψύχεα τὰ μέτρια ὑγραίνει· δίδωσι γὰρ μᾶλλον ἢ λαμβάνει. ἱδρῶτες πάντες ἀπιόντες καὶ ξηραίνουσι καὶ ἰσχναίνουσιν, ἐκλείποντος τοῦ ὑγροῦ ἐκ τοῦ σώματος. λαγνείη ἰσχναίνει καὶ ὑγραίνει καὶ θερμαίνει· θερμαίνει μὲν διὰ τὸν πόνον καὶ τὴν ἀπόκρισιν τοῦ ὑγροῦ,
10 ἰσχναίνει δὲ διὰ τὴν κένωσιν, ὑγραίνει δὲ διὰ τὸ ὑπολειπόμενον ἐν τῷ σώματι τῆς συντήξιος τῆς
12 ὑπὸ τοῦ πόνου.

LIX. Ἔμετοι ἰσχναίνουσι διὰ τὴν κένωσιν τῆς τροφῆς, οὐ μέντοι ξηραίνουσιν, ἢν μή τις τῇ ὑστεραίῃ θεραπεύῃ ὀρθῶς, ἀλλ' ὑγραίνουσι μᾶλλον διὰ τὴν πλήρωσιν[3] καὶ διὰ τὴν σύντηξιν τῆς σαρκὸς τὴν ὑπὸ τοῦ πόνου· ἢν δέ τις ἐάσῃ

[1] κένωσι μὲν τῶι σώματι δίδωσι· θερμῶι ἐόντι ψυχρὸν ἐόν· βεβρωκότος δὲ ἀφαιρέει θερμοῦ ἐόντος· καὶ πληροῖ ψυχροῦ ἐόντος τοῦ ὑπάρχοντος ὑγροῦ θ: κενῶι μὲν τῶι σώματι δίδωσι θερμόν τι ψυχρόν· βεβρωκότι δὲ ἀφαιρέεται ὑγροῦ ἐόντος· καὶ πληροῖ ψυχρὸν ἐὸν τοῦ ὑπάρχοντος ξηροῦ M.

The text within daggers is Littré's, but does not claim to be the original, which probably will never be recovered. Littré, however, is right when he says: "le sens est déterminé par opposition." Perhaps the reading originally was something like this; κενῷ μὲν τῷ σώματι δίδωσι θερμόν τι· βεβρωκότος δὲ ἀφαιρεῖ ὑγροῦ ἐόντος καὶ πληροῖ ψυχροῦ ἐόντος τοῦ ὑπάρχοντος ξηροῦ. We should certainly expect, from the sentence ψυχρὰ δὲ λουτρὰ τοὐναντίον, a passage of which the correct summary is:—

(1) θερμὰ λουτρὰ
  (a) νῆστιν ἰσχναίνει καὶ ψύχει.
  (b) βεβρωκότα θερμαίνει καὶ ὑγραίνει.

their dryness, which is cold.[1] To refrain from baths dries, as the moisture is used up, and so does to refrain from oiling.

LVIII. Oiling warms, moistens and softens. The sun and fire dry for the following reason. Being warm and dry, they draw the moisture from the body. Shade and moderate cold moisten, for they give more than they receive. All sweats on their departure both dry and reduce, as the moisture of the body leaves it. Sexual intercourse reduces, moistens and warms. It warms owing to the fatigue and the excretion of moisture; it reduces owing to the evacuation; it moistens because of the remnant in the body of the matters melted by the fatigue.

LIX. Vomitings reduce through the evacuation of the nourishment. They do not, however, dry, unless appropriate treatment be applied on the following day; they tend rather to moisten through the repletion[2] and through the melting of flesh caused by the fatigue. But if on the morrow one

[1] See critical note on this passage.

[2] The "repletion" must mean fulness caused by the added emetic. This does not give a very good sense, and one is tempted to think that the *πικρωσι* (*i.e.* *πίκρωσιν*) of *θ* is either the correct reading or at least a near corruption of it. Perhaps the sharp taste of certain emetics is referred to. which tends to extract moisture from glands. See p. 51.

---

(2) *ψυχρὰ λουτρὰ*
  (*a*) *νῆστιν πληροῖ καὶ θερμαίνει.*
  (*b*) *βεβρωκότα ψύχει καὶ ξηραίνει.*

I have in my translation given the general sense of the passage as I conceive it to have been originally written.

[2] *καὶ ἀναλιψιηι ὡσαύτως θ* : *καὶ ἀναληφίη ὡσαύτως* M : *ὡσαύτως δὲ καὶ ἡ ἀνηλειψίη* Littré.

[3] *πλήρωσιν* M : *πικρωσι θ*.

ταῦτα καταναλωθῆναι τῇ ὑστεραίῃ ἐς τὴν τροφὴν τῷ θερμῷ, καὶ τῇ διαίτῃ ἡσύχως προσαγάγῃ, ξηραίνουσιν. κοιλίην δὲ συνεστηκυῖαν διαλύει ἔμετος, καὶ διαχωροῦσαν μᾶλλον τοῦ καιροῦ
10 ἵστησι, τὴν μὲν διυγραίνων, τὴν δὲ ξηραίνων·[1] ὁκόταν μὲν οὖν στῆσαι βούλῃ, τὴν ταχίστην φαγόντα χρὴ ἐμεῖν, πρὶν ἂν ὑγρὸν ἐὸν τὸ σιτίον καταβιβασθῇ κάτω,[2] καὶ τοῖσι στρυφνοῖσι καὶ τοῖσιν αὐστηροῖσι σιτίοισι μᾶλλον χρῆσθαι· ὁκόταν δὲ λῦσαι τὴν κοιλίην βούλῃ, ἐνδιατρίβειν ἐν τοῖσι σιτίοισιν ὡς πλεῖστον χρόνον συμφέρει, καὶ τοῖσι δριμέσι καὶ ἁλμυροῖσι καὶ λιπαροῖσι
18 καὶ γλυκέσι σιτίοισι καὶ πόμασι χρῆσθαι.

LX. Ὕπνοι δὲ νῆστιν μὲν ἰσχναίνουσι καὶ ψύχουσιν, ἢν μὴ μακροὶ ἔωσι, κενοῦντες τοῦ ὑπάρχοντος ὑγροῦ· ἢν δὲ[3] μᾶλλον, ἐκθερμαίνοντες συντήκουσι τὴν σάρκα, καὶ διαλύουσι τὸ σῶμα, καὶ ἀσθενὲς ποιέουσι· βεβρωκότα δὲ θερμαίνοντες ὑγραίνουσι, τὴν τροφὴν ἐς τὸ σῶμα διαχέοντες· ἀπὸ δὲ τῶν ὀρθρίων περιπάτων ὕπνος μάλιστα ξηραίνει. ἀγρυπνίη δὲ ἐν μὲν τοῖσι σιτίοισι βλάπτει, οὐκ ἐῶσα τὸ σιτίον
10 τήκεσθαι· ἀσίτῳ δὲ ἰσχνασίην μέν τινα δίδωσι, βλάπτει δὲ ἧσσον. ῥᾳθυμίη ὑγραίνει καὶ ἀσθενὲς τὸ σῶμα ποιεῖ. ἀτρεμίζουσα γὰρ ἡ ψυχὴ οὐκ ἀναλίσκει τὸ ὑγρὸν ἐκ τοῦ σώματος· πόνος δὲ ξηραίνει καὶ τὸ σῶμα ἰσχυρὸν ποιεῖ. μονοσιτίη ἰσχναίνει καὶ ξηραίνει καὶ τὴν κοιλίην ἵστησι, διότι τῷ τῆς ψυχῆς θερμῷ τὸ ὑγρὸν ἐκ τῆς

[1] ξηραίνων Littré: ἀντισπῶν θ: M omits τὴν δὲ ξηραίνων. The ἀντισπῶν of θ is possibly correct, and ξηραίνων a gloss.

lets the moisture be consumed by the warmth for its nourishment, and increase nourishment gradually, vomitings dry. Constipated bowels are relaxed by vomiting, and too relaxed bowels are bound thereby; it moistens the former and dries the latter. When, therefore, you wish to bind the bowels, take a meal and administer an emetic as quickly as possible, before the food can be moistened and drawn downwards; the food used should by preference be astringent and dry. But when you wish to loosen the bowels, it is beneficial to keep the food as long as possible, and to take food and drink that are sharp, salt, greasy and sweet.

LX. Sleep when fasting reduces and cools, if it be not prolonged, as it empties the body of the existing moisture; if, however, it be prolonged, it heats and melts the flesh, dissolves the body and enfeebles it. After a meal sleep warms and moistens, spreading the nourishment over the body. It is especially after early-morning walks that sleep is drying. Want of sleep, after a meal, is injurious, as it prevents the food from dissolving; to a fasting person it is less injurious, while it tends to reduce flesh. Inaction moistens and weakens the body; for the soul, being at rest, does not consume the moisture out of the body. But labour dries and strengthens the body. Taking one meal[1] a day reduces, dries and binds the bowels, becausc, through the warmth of the soul the moisture

[1] The *μονόσιτοι* took the *δεῖπνον* only; others took the *ἄριστον* as well.

---

[2] *πρὶν διυγρηνθῆναι τὸν σῖτον· καὶ κατασπασθῆναι κάτω* M: *πρὶν ἂν ὑγρὸν ἐὸν τὸ σιτίον καταβιβασθῆι κάτωι θ.*

[3] After *δὲ θ* adds *μακροὶ ἐῶσι.*

κοιλίης καὶ τῆς σαρκὸς καταναλίσκεται·[1] ἄριστον δὲ τἀναντία διαπρήσσεται τῇ μονοσιτίῃ. ὕδωρ πόμα[2] θερμὸν ἰσχναίνει πάντα, καὶ ψυχρόν
20 ὡσαύτως. τὸ δὲ ὑπερβάλλον ψυχρὸν καὶ πνεῦμα καὶ σιτίον καὶ ποτὸν πήγνυσι τὸ ὑγρὸν τὸ ἐν τῷ σώματι καὶ τὰς κοιλίας συνίστησι τῇ πήξει καὶ ψύξει· κρατεῖ γὰρ τοῦ τῆς ψυχῆς ὑγροῦ. καὶ τοῦ θερμοῦ δὲ πάλιν αἱ ὑπερβολαὶ πηγνύουσι, καὶ οὕτως ὥστε μὴ διάχυσιν ἔχειν. ὅσα δὲ θερμαίνοντα τὸ σῶμα, τροφὴν μὴ διδόντα, κενοῖ τοῦ ὑγροῦ τὴν σάρκα μὴ ὑπερβολὴν ποιέοντα, πάντα ψύξιν τῷ ἀνθρώπῳ παραδίδωσι· κενουμένου γὰρ τοῦ ὑπάρχοντος ὑγροῦ, πνεύματος
30 ἐπακτοῦ πληρεύμενον ψύχεται.

LXI. Περὶ δὲ τῶν πόνων ἥντινα ἔχουσι δύναμιν διηγήσομαι. εἰσὶ γὰρ οἱ μὲν κατὰ φύσιν, οἱ δὲ διὰ βίης· οἱ μὲν οὖν κατὰ φύσιν αὐτῶν εἰσιν[3] ὄψιος πόνος, ἀκοῆς, φωνῆς, μερίμνης. ὄψιος μὲν οὖν δύναμις τοιήδε·[4] προσέχουσα ἡ ψυχὴ τῷ ὁρατῷ[5] κινεῖται καὶ θερμαίνεται· θερμαινομένη δὲ ξηραίνεται, κεκενωμένου τοῦ ὑγροῦ. διὰ δὲ τῆς ἀκοῆς ἐσπίπτοντος τοῦ ψόφου σείεται ἡ ψυχὴ καὶ πονεῖ, πονέουσα δὲ θερμαίνεται καὶ ξηραί-
10 νεται. ὅσα μεριμνᾷ ἄνθρωπος, κινεῖται ἡ ψυχὴ

[1] καταναλίσκει θ: καταναλίσκεται M.
[2] πόμα M: πολὺ θ.
[3] Before ὄψιος θ has οἱ δὲ.
[4] τοιῆδε θ: τοιαύτη M.
[5] M has ὁρεομένῳ, perhaps rightly.

[1] The word *πόνος* cannot always be represented by the same English equivalent. It may mean "toil" generally, voluntary toil (or "exercise"), or even the "pain" caused by toil (usually *κόπος*). The division of *πόνοι* into natural

is consumed from out the belly and the flesh. To take lunch has effects opposite to those of taking one meal only. Hot water as a drink is a general reducer of flesh, and cold water likewise. But excessive cold, whether of breath, food or drink, congeals the moisture in the body, and binds the bowels by the congealing and the cold; for it overpowers the moisture of the soul. Then again excess of heat too causes congealing, to such an extent as to prevent diffusion. Such things as warm the body without affording nourishment, and empty the flesh of its moisture, even when there is no excess, in all cases cause chill in a man; for, the existing moisture being emptied out, the body is filled with breath from outside and grows cold.

LXI. I will now discuss the properties of exercises.[1] Some exercises are natural and some violent. Natural exercises are those of sight, hearing, voice and thought. The nature[2] of sight is as follows. The soul, applying itself to what it can see, is moved and warmed. As it warms it dries, the moisture having been emptied out. Through hearing, when noise strikes the soul, the latter is shaken and exercised, and as it is exercised it is warmed and dried. By all the thoughts that come to a man the

and violent corresponds to no modern division, as is proved by the enumeration of "natural" exercises, while by "violent" exercise we mean "excessive" exercise, but *οἱ διὰ βίης πόνοι* means rather exercises that are artificial, the result of conscious and forced effort. Apparently all muscular exercises are "violent."

[2] The word *δύναμις* means much the same thing as *δύναμιν* in the first sentence. The essential qualities are referred to in both cases, but it seems preferable to use different equivalents in the translation, as *δύναμιν* refers mostly to the *qualities* and *δύναμις* to the *essence* of exercises.

ὑπὸ τούτων καὶ θερμαίνεται καὶ ξηραίνεται, καὶ τὸ ὑγρὸν καταναλίσκουσα πονεῖ, καὶ κενοῖ[1] τὰς σάρκας, καὶ λεπτύνει τὸν ἄνθρωπον. ὁκόσοι δὲ πόνοι φωνῆς, ἢ λέξιες ἢ ἀναγνώσιες ἢ ᾠδαί,[2] πάντες οὗτοι κινέουσι τὴν ψυχήν· κινεομένη δὲ θερμαίνεται καὶ ξηραίνεται, καὶ τὸ ὑγρὸν κατανα-
17 λίσκει.

LXII. Οἱ δὲ περίπατοι κατὰ φύσιν μὲν εἰσί, καὶ οὗτοι μάλιστα τῶν λοιπῶν, ἔχουσι δέ τι βίαιον. δύναμις δὲ αὐτῶν ἑκάστων[3] τοιήδε· ὁ ἀπὸ δείπνου περίπατος ξηραίνει τήν τε κοιλίην καὶ τὸ σῶμα, καὶ τὴν γαστέρα οὐκ ἐᾷ πίειραν γίνεσθαι[4] διὰ τάδε· κινευμένου τοῦ ἀνθρώπου, θερμαίνεται τὰ σιτία καὶ τὸ σῶμα· ἕλκει οὖν τὴν ἰκμάδα ἡ σάρξ, καὶ οὐκ ἐᾷ περὶ τὴν κοιλίην συνίστασθαι· τὸ μὲν οὖν σῶμα πληροῦται, ἡ δὲ
10 κοιλίη λεπτύνεται. ξηραίνεται δὲ διὰ τάδε· κινευμένου τοῦ σώματος καὶ θερμαινομένου, τὸ λεπτότατον τῆς τροφῆς καταναλίσκεται, τὸ μὲν ὑπὸ τοῦ συμφύτου θερμοῦ, τὸ δὲ σὺν τῷ πνεύματι ἀποκρίνεται ἔξω, τὸ δὲ καὶ διουρεῖται· ὑπολείπεται δὲ τὸ ξηρότατον ἀπὸ τῶν σιτίων ἐν τῷ σώματι,[5] ὥστε τὴν κοιλίην ἀποξηραίνεσθαι καὶ τὴν σάρκα. καὶ οἱ ὄρθριοι περίπατοι ἰσχναίνουσι, καὶ τὰ περὶ τὴν κεφαλὴν κοῦφά τε καὶ εὐαγέα[6] καὶ εὐήκοα παρασκευάζουσι, καὶ τὴν
20 κοιλίην λύουσιν· ἰσχναίνουσι μὲν ὅτι κινεύμενον τὸ σῶμα θερμαίνεται, καὶ τὸ ὑγρὸν λεπτύνεται καὶ καθαίρεται, τὸ μὲν ὑπὸ τοῦ πνεύματος, τὸ δὲ μύσσεται καὶ χρέμπτεται, τὸ δὲ ἐς τὴν τροφὴν

[1] M transposes κενοῖ and λεπτύνει.
[2] So θ. M has λέξις· ἢ ἀνάγνωσις· ἢ ᾠδή·

soul is warmed and dried; consuming the moisture it is exercised, it empties the flesh and it makes a man thin. Exercises of the voice, whether speech, reading or singing, all these move the soul. And as it moves it grows warm and dry, and consumes the moisture.

LXII. Walking is a natural exercise, much more so than the other exercises, but there is something violent about it. The properties of the several kinds of walking are as follow. A walk after dinner dries the belly and body; it prevents the stomach becoming fat for the following reasons. As the man moves, the food and his body grow warm. So the flesh draws the moisture, and prevents it accumulating about the belly. So the body is filled while the belly grows thin. The drying is caused thus. As the body moves and grows warm, the finest part of the nourishment is either consumed by the innate heat, or secreted out with the breath or by the urine. What is left behind in the body is the driest part from the food, so that the belly and the flesh dry up. Early-morning walks too reduce [the body], and render the parts about the head light, bright and of good hearing, while they relax the bowels. They reduce because the body as it moves grows hot, and the moisture is thinned and purged, partly by the breath, partly when the nose is blown and the throat cleared, partly being consumed by

---

[3] *ἑκάστων* is omitted by M, which reads in its place *ἐστὶ*.

[4] *γίνεσθαι θ*: *γενέσθαι* M.

[5] *ἐν τῷ σώματι* is omitted by M, perhaps rightly.

[6] *εὐαγέα* Littré (after Foës, Zwinger and Mack): *εὐπαγῆ θ*: *εὐπαγέα* M.

τῷ τῆς ψυχῆς θερμῷ καταναλίσκεται· τὴν δὲ κοιλίην λύουσι, διότι θερμῇ ἐούσῃ τοῦ ψυχροῦ πνεύματος ἐπεισπίπτοντος[1] ἄνωθεν, ὑποχωρεῖ τὸ θερμὸν τῷ ψυχρῷ. κοῦφα δὲ τὰ περὶ τὴν κεφαλὴν ποιεῖ διὰ τάδε· ὅταν κενωθῇ ἡ κοιλίη, ἕλκει ἐς ἑωυτὴν ἔκ τε τοῦ ἄλλου σώματος καὶ
30 ἐκ τῆς κεφαλῆς τὸ ὑγρὸν θερμὴ ἐοῦσα. κενουμένης δὲ τῆς κεφαλῆς, ἀποκαθαίρεται ἥ τε ὄψις καὶ ἡ ἀκοή· καὶ γίνεται εὐαγής.[2] οἱ δὲ ἀπὸ τῶν γυμνασίων περίπατοι καθαρὰ τὰ σώματα παρασκευάζουσι καὶ ἰσχνά,[3] οὐκ ἐῶντες τὴν σύντηξιν τῆς σαρκὸς τὴν ὑπὸ τοῦ πόνου συν-
36 ίστασθαι, ἀλλ' ἀποκαθαίρουσιν.

LXIII. Τῶν δὲ δρόμων δύνανται οἱ μὴ καμπτοὶ καὶ μακροί,[4] ἐξ ὀλίγου προσαγόμενοι, θερμαίνοντες τὴν σάρκα συνεψεῖν καὶ διαχεῖν, καὶ τῶν σίτων τὴν δύναμιν τὴν ἐν τῇ σαρκὶ καταπέσσουσι,[5] βραδύτερά τε καὶ παχύτερα τὰ σώματα παρασκευάζουσι τῶν τρόχων· τοῖσι δὲ πολλὰ ἐσθίουσι συμφορώτεροι, καὶ χειμῶνος μᾶλλον ἢ θέρεος. οἱ δὲ ἐν τῷ ἱματίῳ δρόμοι τὴν μὲν δύναμιν τὴν αὐτὴν ἔχουσι, θᾶσσον δὲ διαθερμαί-
10 νοντες[6] ὑγρότερα τὰ σώματα ποιέουσιν, ἀχρο-

[1] θερμηι ἐοῦσα τοῦ ψυχροῦ . . . ἐπισπίπτοντος θ: θερμὴ ἐοῦσα . . . ἐσπίπτοντος M. Some MSS. have θερμοῦ for ψυχροῦ.

[2] γίνεται εὐαγής θ: γίνονται εὐαγέες M.

[3] ἴσχνα θ (which also reads καθαρώτατα): ἰσχναίνουσι M.

[4] τῶν δὲ δρόμων δύνανται· οἱ μὲν καμπτοὶ καὶ μακροὶ. θ: τῶν δὲ δρόμων γίνονται· οἱ μὲν μακροὶ καὶ καμπτοὶ M, with δύνανται after διαχέειν. ἄκαμπτοι and μὴ καμπτοὶ have been suggested by early editors.

[5] For καταπέσσουσι θ has καταπέσσει, and βαθύτερα for παχύτερα.

the heat of the soul for the nourishment thereof. They relax the bowels because, cold breath rushing into them from above while they are hot, the heat gives way before the cold. It makes light the parts about the head for the following reasons. When the bowels have been emptied, being hot they draw to themselves the moisture from the body generally, and especially from the head; when the head is emptied sight and hearing are purged, and the man becomes bright.[1] Walks after gymnastics render the body pure and thin, prevent the flesh melted by exercise from collecting together, and purge it away.

LXIII. Of running exercises, such as are not double[2] and long, if increased gradually, have the power to heat, concoct and dissolve the flesh; they digest the power of the foods that is in the flesh, making the body slower and more gross than do circular runnings, but they are more beneficial to big eaters, and in winter rather than in summer. Running in a cloak has the same power, but heating more rapidly it makes the body more moist but less

[1] It is tempting to give εὐαγής here and above an active sense, "with clear vision." It is not possible, however, to find a parallel, except perhaps Euripides, *Supp.* 652: ἔστην θεατὴς πύργον εὐαγῆ λαβών, where εὐαγῆ may mean, not "clearly seen," but "affording a clear view." "Affording a clear view," however, is not the same thing as "having good eyesight." So one has to fall back upon the general sense of "bright" or "clear." Perhaps "alert."

[2] The "double" exercise consisted in running along a double track to a goal and back again to a starting-point. It was of a fixed length and could not be "increased gradually" as readily as could distances along a single track.

---

[6] M has μᾶλλον δὲ διαθερμαίνουσι καὶ.

τερα δέ, διότι οὐκ ἀποκαθαίρει προσπῖπτον τὸ πνεῦμα τὸ εἰλικρινές, ἀλλ' ἐν τῷ αὐτῷ ἐγγυμνάζεται πνεύματι· συμφέρει οὖν τοῖσι ξηροῖσι καὶ τοῖσι πολυσάρκοισιν, ὅστις καθελεῖν τὴν σάρκα βούλεται, καὶ τοῖσι πρεσβυτέροισι διὰ ψύξιν[1] τοῦ σώματος. οἱ δὲ δίαυλοι καὶ ὑπηέριοι[2] τὴν μὲν σάρκα ἧσσον διαχέουσιν, ἰσχναίνουσι δὲ μᾶλλον, διότι τοῖς εἴσω[3] τῆς ψυχῆς μέρεσιν οἱ πόνοι ὄντες ἀντισπῶσιν[4] ἐκ τῆς σαρκὸς τὸ ὑγρὸν
20 καὶ τὸ σῶμα λεπτύνουσι καὶ ξηραίνουσιν. οἱ δὲ τρόχοι τὴν μὲν σάρκα ἥκιστα διαχέουσιν, ἰσχναίνουσι δὲ καὶ προσστέλλουσι[5] τήν τε σάρκα καὶ τὴν κοιλίην μάλιστα, διότι ὀξυτάτῳ τῷ πνεύματι χρώμενοι τάχιστα τὸ ὑγρὸν ἕλκουσιν
25 ἐφ' ἑωυτούς.

LXIV. Τὰ δὲ παρασείσματα ξηροῖσι μὲν καὶ ἐξαπίνης,[6] ἀσύμφορα· σπάσματα γὰρ ἐμποιεῖ διὰ τόδε. τεθερμασμένον τὸ σῶμα, τὸ μὲν δέρμα

[1] διὰ ψῦξιν θ : διαψύχειν M.

[2] οἱ δὲ δίαυλοι καὶ ὑπηέριοι· θ : οἱ δε δίαυλοι καὶ ἤπειροι ἵπποι. M : οἱ δὲ δίαυλοι καὶ ὑπηέριοι ἵπποι Littré. The ἵπποι of M is probably a corruption of ὑπηέριοι.

[3] εἴσω θ : ἔσω M : ἔξω Littré, with inferior MS. authority.

[4] ἀντισπῶσιν M : ἀντισπῶντες θ : ἀνασπῶσιν Zwinger.

[5] πρὸς στέλλουσι θ : διαστέλλουσι M.

[6] After ἐξαπίνης M adds οὐκ ἐπιτήδεια καὶ. It also has διὰ τεθερμασμένον, while θ has τεθερμασμένον only. Littré reads διατεθερμασμένον.

---

[1] This means that the body becomes thinner but less flabby. The δίαυλος was a καμπτὸς δρόμος of roughly 200 yards each way, *i.e.* of 400 yards in all.

[2] Both the reading and the interpretation of this sentence are uncertain. Probably the mental strain of the "quarter-mile" is referred to; it is the most strenuous of the foot races and may well be said to be concerned with the "inner

tanned, because this is not cleansed by meeting the rush of pure air, but remains in the same air while it is exercised. So this kind of running is beneficial to those who have a dry body, to those who have excess of flesh which they wish to reduce, and, because of the coldness of their bodies, to those who are getting on in years. The double course, with the body exposed to the air, dissolves the flesh less, but reduces the body more,[1] because the exercises, being concerned with the inner parts[2] of the soul, draw by revulsion the moisture out of the flesh, and render the body thin and dry. Running in a circle dissolves the flesh least, but reduces and contracts the flesh and the belly most, because, as it causes the most rapid respiration, it is the quickest to draw the moisture to itself.

LXIV. Swinging the arms, for persons of dry flesh, and when jerky, is inexpedient, as it causes sprains, in the following way. The body having been warmed,[3] this swinging makes the skin consider-

parts of soul." Probably the reading ἔξω is an attempt to connect psychologically this mental strain with the profuse perspiration caused by the δίαυλος. I believe that ἵπποι is a mere corruption of ὑπηέριοι, but its adoption may have been encouraged by a desire to explain the introduction of "mental exercises"; the comparative inaction of riding suggests an active mental factor.

[3] This sentence appears to contain such an undoubted instance of a *nominativus pendens* that it renders less likely my substitution (in Chapter LXII) of θερμῇ ἐούσῃ for θερμὴ ἐοῦσα in order to avoid such an anacoluthon. One way out of the grammatical difficulty would be to take as the subject τεθερμασμένον τὸ σῶμα, in the sense of "bodily heat," but it seems too violent to say σῶμα λεπτύνει. Another way would be to read διὰ τεθερμασμένον (with M). The chief objection to this is that local διὰ with the accusative appears to be confined to the poets. Fortunately the general sense is clear, that the flesh becomes hot, dry and brittle.

ἰσχυρῶς λεπτύνει, τὴν δὲ σάρκα ἧσσον συνίστησι τῶν τρόχων, κενοῖ δὲ τὴν σάρκα τοῦ ὑγροῦ. τὰ δὲ ἀνακινήματα καὶ ἀνακουφίσματα τὴν μὲν σάρκα ἥκιστα διαθερμαίνει,[1] παροξύνει δὲ καὶ τὸ σῶμα καὶ τὴν ψυχήν, καὶ τοῦ πνεύματος κενοῖ. πάλη δὲ καὶ τρῖψις τοῖσι μὲν ἔξω τοῦ 10 σώματος παρέχει τὸν πόνον μᾶλλον, θερμαίνει δὲ τὴν σάρκα καὶ στερεοῖ καὶ αὔξεσθαι ποιεῖ διὰ τόδε· τὰ μὲν στερεὰ φύσει τριβόμενα† συνίστησι†, τὰ δὲ κοῖλα αὔξεται,[2] ὅσαι φλέβες εἰσί· θερμαινόμεναι δὲ αἱ σάρκες καὶ ξηραινόμεναι ἕλκουσιν ἐφ᾽ ἑωυτὰς τὴν τροφὴν διὰ τῶν πόρων,[3] εἶτα αὔξονται. ἀλίνδησις παραπλήσια πάλῃ διαπρήσσεται, ξηραίνει δὲ μᾶλλον διὰ τὴν κόνιν καὶ σαρκοῖ ἧσσον. ἀκροχειρισμὸς[4] ἰσχναίνει καὶ τὰς σάρκας ἕλκει ἄνω, καὶ κωρυ-20 κομαχίη καὶ χειρονομίη παραπλήσια διαπρήσσεται. πνεύματος δὲ κατάσχεσις τοὺς πόρους διαναγκάσαι καὶ τὸ δέρμα λεπτῦναι καὶ τὸ 23 ὑγρὸν ἐκ τοῦ δέρματος ἐξῶσαι δύναται.

LXV. Τὰ ἐν κόνει καὶ τὰ ἐν ἐλαίῳ[5] γυμνάσια διαφέρει τοσόνδε· κόνις μὲν ψυχρόν, ἔλαιον δὲ θερμόν· ἐν μὲν τῷ χειμῶνι τὸ ἔλαιον αὐξιμώτερον, διότι τὸ ψῦχος κωλύει φέρειν ἀπὸ τοῦ σώματος· ἐν δὲ τῷ θέρει τὸ ἔλαιον ὑπερβολὴν θερμασίης ποιεῦν[6] τήκει τὴν σάρκα, ὅταν ὑπὸ τῆς ὥρης ἐκθερμαίνηται καὶ τοῦ ἐλαίου καὶ τοῦ

[1] M has διαθερμαίνουσιν and later κενοῦσι.

[2] After αὔξεται M reads τῆς γοῦν σαρκὸς τὸ μὲν πυκνὸν τριβόμενον ξυνίσταται· τὰ δὲ κοῖλα αὔξεται καὶ ὁκόσαι φλέβες εἰσί κ.τ.ἕ. This appears to be an attempt to mend the grammar of the corrupt sentence preceding. It has probably crept into the text from the margin.

ably thinner, but contracts the flesh less than running in a circle, and empties the flesh of its moisture. Sparring and raising the body[1] heat the flesh least, but they stimulate both body and soul, while they empty the body of breath. Wrestling and rubbing give exercise more to the exterior parts of the body, but they warm the flesh, harden it and make it grow, for the following reason. Parts that are naturally hard are compressed by rubbing, while hollow parts grow, such as are veins. For the flesh, growing warm and dry, draws to itself the nourishment through the passages, and then it grows. Wrestling in the dust has effects like to those of ordinary wrestling, but it dries more because of the dust, and it increases flesh less. Wrestling with the fingers reduces and draws the flesh upwards; the punch-ball and arm exercises have like effects. Holding the breath has the property of forcing open the passages, of thinning the skin, and of expelling therefrom the moisture.

LXV. Exercises in dust differ from those in oil thus. Dust is cold, oil is warm. In winter oil promotes growth more, because it prevents the cold from being carried from the body. In summer, oil, producing excess of heat, melts the flesh, when the latter is heated by the season, by the oil and by the

[1] Or, "the arms." The *lexica* neglect this word. I take it to refer either to raising the body from a prone position or to arm exercises.

---

[3] πόρων θ: φλεβῶν M.
[4] ἀκροχειρισμὸς θ: ἀκροχεῖριξ δ' M.
[5] ἐν κονίηι καὶ ἐλαίωι M.
[6] ποιεῦν θ: ποιεύμενον M.

πόνου. ἡ δὲ κόνις ἐγγυμνάζεσθαι ἐν τῷ θέρει αὐξιμώτερον·[1] ψύχουσα γὰρ τὸ σῶμα οὐκ ἐᾷ
10 ἐκθερμαίνεσθαι ἐς ὑπερβολήν· ἐν δὲ τῷ χειμῶνι διαψυκτικὸν καὶ κρυμνῶδες·[2] ἐνδιατρίβειν δὲ ἐν τῇ κόνει μετὰ τοὺς πόνους ἐν τῷ θέρει, ὀλίγον μὲν χρόνον ὠφελεῖ ψύχουσα, πολὺν δὲ ὑπερξηραίνει καὶ τὰ σώματα σκληρὰ καὶ ξυλώδεα ἀποδεικνύει. τρῖψις ἐλαίου σὺν ὕδατι μαλάσσει
16 καὶ οὐ ἐᾷ πολλὰ[3] διαθερμαίνεσθαι.

LXVI. Περὶ δὲ κόπων τῶν ἐν τοῖσι σώμασιν γινομένων ὧδε ἔχει· οἱ μὲν ἀγύμναστοι τῶν ἀνθρώπων ἀπὸ παντὸς πόνου κοπιῶσι· οὐδὲν γὰρ τοῦ σώματος διαπεπόνηται πρὸς οὐδένα πόνον· τὰ δὲ γεγυμνασμένα τῶν σωμάτων ὑπὸ[4] τῶν ἀηθῶν[5] πόνων κοπιᾷ· τὰ δὲ καὶ ὑπὸ τῶν συνήθων γυμνασίων κοπιᾷ, ὑπερβολῇ χρησάμενα. τὰ μὲν οὖν εἴδεα τῶν κόπων ταῦτά[6] ἐστιν· ἡ δὲ δύναμις αὐτῶν ὧδε ἔχει· οἱ μὲν οὖν ἀγύμνα-
10 στοι ὑγρὴν τὴν σάρκα ἔχοντες, ὅταν πονήσωσι, θερμαινομένου τοῦ σώματος, σύντηξιν πολλὴν ἀφιᾶσιν· ὅ τι μὲν οὖν ἂν ἐξιδρώσῃ ἢ καὶ σὺν πνεύματι ἀποκαθαρθῇ, οὐ παρέχει πόνον ἄλλον ἢ τῷ κενωθέντι τοῦ σώματος παρὰ τὸ ἔθος· ὅ τι δ' ἂν ἐμμείνῃ τῆς συντήξιος, οὐ μόνον τῷ κενωθέντι τοῦ σώματος παρὰ τὸ ἔθος παρέχει πόνον,[7] ἀλλὰ καὶ τῷ δεξαμένῳ τὸ ὑγρόν· οὐ γάρ ἐστι σύντροφον[8] τῷ σώματι, ἀλλὰ πολέμιον. ἐς μὲν δὴ τὰ ἄσαρκα τῶν σωμάτων οὐ συνίσταται

[1] M has ἐγγυμνάζεται, δὲ after ἐν and αὐξιμώτερος.

[2] κρυμνῶδες θ : κρυμῶδες M.

[3] οὐκ εᾶ πολλὰ θ : οὐ δεινῶς ἐᾶ M.

[4] For ὑπὸ Linden and Mack would read ἀπὸ (probably rightly).

exercise. In summer it is exercise in dust that promotes growth more, for by cooling the body it prevents its being heated to excess. But in winter dust is chilling, or even freezing. To remain in the dust after exercise in summer benefits by its cooling property, if it be for a short time; if it be for long, it dries the body to excess and renders it hard as wood. Rubbing with oil and water softens the body, and prevents its becoming over-heated.

LXVI. The fatigue pains that arise in the body are as follow. Men out of training suffer these pains after the slightest exercise, as no part of their body has been inured to any exercise; but trained bodies feel fatigue pains after unusual exercises, some even after usual exercises if they be excessive. These are the various kinds of fatigue pains; their properties are as follow. Untrained people, whose flesh is moist, after exercise undergo a considerable melting, as the body grows warm. Now whatever of this melted substance passes out as sweat, or is purged away with the breath, causes pain only to the part of the body that has been emptied contrary to custom; but such part of it as remains behind causes pain not only to the part of the body emptied contrary to custom, but also to the part that has received the moisture, as it is not congenial to the body but hostile to it. It tends to gather, not at the fleshless, but at the fleshy parts of the body, in

---

[5] *ἀήθων* θ: *ἀνεθίστων* M.

[6] *ταῦτά* θ: *τοιαυτά* M.

[7] M has *ὅτι δ' ἣν ἐμμένηι τῆς ἀποκρήσιος οὐ παρέχει τὸν πόνον κ.τ.ἕ.*

[8] *σύντροφον* θ: *σύμφορον* M.

20 ὁμοίως, ἐς δὲ τὰ σαρκώδεα, ὥστε τούτοισι πόνον παρέχειν ἕως ἂν ἐξέλθῃ. ἅτε δὴ οὐκ ἔχον περίοδον,[1] ἀτρεμίζον ἐκθερμαίνεται αὐτό τε καὶ τὰ προσπίπτοντα· ἢν μὲν οὖν πολὺ γένηται τὸ ἀποκριθέν, ἐκράτησε καὶ τοῦ ὑγιαίνοντος, ὥστε συνεκθερμανθῆναι τὸ πᾶν[2] σῶμα, καὶ ἐνεποίησε πυρετὸν ἰσχυρόν.[3] θερμανθέντος γὰρ τοῦ αἵματος καὶ ἐπισπασθέντος, ταχείην ἐποίησε[4] τὴν περίοδον τὰ ἐν τῷ σώματι, καὶ τό τε ἄλλο σῶμα καθαίρεται ὑπὸ τοῦ πνεύματος, καὶ τὸ συν-
30 εστηκὸς θερμαινόμενον λεπτύνεταί τε καὶ ἐξωθεῖται[5] ἐκ τῆς σαρκὸς ἔξω ὑπὸ τὸ δέρμα, ὅπερ ἱδρὼς καλεῖται θερμός. τούτου δ' ἀποκριθέντος, τό τε αἷμα καθίσταται ἐς τὴν κατὰ φύσιν κίνησιν,[6] καὶ ὁ πυρετὸς ἀνίησι, καὶ ὁ κόπος παύεται μάλιστα τριταῖος. χρὴ δὲ τοὺς τοιούτους κόπους[7] ὧδε θεραπεύειν· πυρίῃσι καὶ λουτροῖσι θερμοῖσι διαλύοντα τὸ συνεστηκός, περιπάτοισί τε μὴ[8] βιαίοισιν, ὡς ἀποκαθαίρωνται, καὶ ὀλιγοσιτίῃσι καὶ ἰσχνασίῃσι συνι-
40 στάναι τῆς σαρκὸς τὴν κένωσιν, καὶ ἀλείφεσθαι[9] τῷ ἐλαίῳ ἡσυχῇ πολὺν χρόνον, ὅκως μὴ βιαίως διαθερμαίνωνται,[10] καὶ τοῖσι χρίσμασι τοῖσιν ἱδρωτικοῖσι[11] χρίεσθαι καὶ μαλακευνεῖν[12] συμφέρει. τοῖσι δὲ γυμναζομένοισιν ἀπὸ τῶν ἀνεθίστων πόνων διὰ τάδε γίνεται ὁ κόπος· ὅ τι ἂν μὴ

[1] πάροδον θ : περίοδον M. [2] τὸ πᾶν θ : ὅλον τὸ M.
[3] ἰσχυρόν is omitted by θ.
[4] ἐποιήσατο θ : ἐποίησε M. θ has τὴν ἐν and M τὰ ἐν.
[5] M has συνεξωθέεται. [6] κείνησι θ : σύστασιν M.
[7] τὸν τοιοῦτον κόπον M. [8] M omits τε μή.
[9] ἀλίφεσθαι θ : τρίβεσθαι M
[10] ἀναθερμαίνεται M.

such a way as to cause them pain until it has passed out. Now as it has no circulation, it remains still and grows hot, as do also the things that touch it. Now if the secretion prove abundant it overpowers even that which is healthy, so that the whole body is heated and a high fever follows. For when the blood has been attracted and heated, the things in the body set up a rapid circulation, and the body generally is cleansed by the breath, while the collected moisture, becoming warm, is thinned and forced outwards from the flesh to the skin, and is called "hot sweat." When the secretion of this is over, the blood is restored to its natural motion,[1] the fever subsides, and the fatigue pains cease about the third day. Pains of this sort should be treated thus. Break up the collected humour by vapour baths, and by hot baths, and make firm the reduced flesh[2] by gentle walks, in order to effect purgation, by restricted diet and by practices that cause leanness; it is beneficial to apply oil gently to the body for a long time, that the heating be not violent, to use sudorific unguents, and to lie on a soft bed. Those in training suffer fatigue pains from unaccustomed exercises for the following reasons. Any

[1] I retain θ's κείνησι (*i.e.* κίνησιν) If correct it throws light upon the early history of the circulation of the blood. But M's σύστασιν is quite probably correct.

[2] Littré says "on soutient la réduction de la chair," but this can scarcely represent συνιστάναι. The whole of this sentence is grammatically loose; it is difficult, for instance, to decide how far συμφέρει extends its influence, and what infinitives (if any) are imperatival.

---

[11] τοῖσιν ἰδιωτικοῖσι θ: τοῖσι ἰδρωτικοῖσι· καὶ τοῖσι μαλακτικοῖσι M.

[12] μαλακευνεῖν Littré: μαλακυνεῖν θ: μαλακύνειν M.

πεπονήκῃ τὸ σῶμα, ὑγρὴν ἀνάγκη τὴν σάρκα εἶναι πρὸς τοῦτον τὸν τόπον,[1] πρὸς ὃν μὴ εἴθισται πονεῖν, ὥσπερ ἀγυμνάστων πρὸς ἕκαστα· τὴν μὲν οὖν[2] σάρκα συντήκεσθαι ἀνάγκη καὶ
50 ἀποκρίνεσθαι καὶ συνίστασθαι ὥσπερ τῷ προτέρῳ. συμφέρει δὲ θεραπεύεσθαι ὧδε· τοῖσι μὲν γυμνασίοισι χρῆσθαι τοῖσι συνήθεσιν, ὅκως τὸ συνεστηκὸς θερμαινόμενον λεπτύνηται καὶ ἀποκαθαίρηται, καὶ τὸ ἄλλο σῶμα μὴ ὑγραίνηται, μηδ' ἀγύμναστον γίνηται. τοῖσι δὲ λουτροῖσι θερμοῖσι καὶ τούτοισι συμφέρει[3] χρῆσθαι, καὶ τῇ τρίψει ὁμοίως ὡς καὶ τοῖς ἔμπροσθεν.[4] τῆς δὲ πυριήσιος οὐδὲν δεῖται· οἱ πόνοι γὰρ ἱκανοὶ θερμαίνοντες λεπτύνειν καὶ ἀποκαθαίρειν τὸ
60 συστάν. οἱ δὲ ἀπὸ τῶν συνήθων γυμνασίων κόποι τῷδε τῷ τρόπῳ γίνονται· ἀπὸ μὲν συμμέτρου πόνου κόπος οὐ γίνεται·[5] ὅταν δὲ πλεῖον τοῦ καιροῦ πονήσῃ,[6] ὑπερεξήρηνε τὴν σάρκα· κενωθεῖσα δὲ τοῦ ὑγροῦ, θερμαίνεταί τε καὶ ἀλγεῖ καὶ φρίσσει καὶ ἐς πυρετὸν μακρότερον καθίσταται, ἢν μή τις ἐκθεραπεύσῃ ὀρθῶς. χρὴ δὲ πρῶτον μὲν αὐτὸν τῷ λουτρῷ μὴ σφόδρα πολλῷ μηδὲ θερμῷ ἄγαν λοῦσαι, εἶτα πίσαι[7] αὐτὸν ἐκ τοῦ λουτροῦ μαλθακὸν οἶνον, καὶ
70 δειπνεῖν ὡς πλεῖστα καὶ παντοδαπὰ σιτία, καὶ

[1] τόπον M : πόνον θ. [2] μὲν οὖν M : γοῦν θ.

[3] τοῖσι θερμοῖσι ξυμφέρει καὶ τοῦτον χρέεσθαι M.

[4] ὅμοια. ὡς καὶ τοὺς ἔμπροσθεν θ : ὁμοίως ὡς καὶ τὸν πρόσθεν M.

[5] γίγνεται θ : κινέεται M.

[6] ὅτ' ἂν δὲ πλεῖον τοῦ καιροῦ πονήσηι θ : ὁκόταν δὲ πλεῖον τοῦ καιροῦ πόνος ἦι M : ὁκόταν δὲ πλείων τοῦ καιροῦ πόνος ᾖ Littré.

unexercised part of the body must of necessity have its flesh moist, just as persons out of training are moist generally throughout.[1] So the flesh must of necessity melt, secrete itself and collect itself, as in the former case. Beneficial treatment of such cases is as follows. Accustomed exercises should be practised, so that the collected humour may grow warm, become thin, and purge itself away, while the body generally may become neither moist nor yet unexercised. It is beneficial to employ hot baths in these cases also, with rubbing as before. But there is no need of vapour baths, as the exercises, being warming, are sufficient to thin and purge away the humour that has collected. Fatigue pains from accustomed exercises arise in the following way. Moderate toil is not followed by pain; but when immoderate it dries the flesh overmuch, and this flesh, being emptied of its moisture, grows hot, painful and shivery, and falls into a longish fever, unless proper treatment be applied. First the patient should be washed in a bath not too copious nor yet over-hot; then after the bath give him to drink a soft wine; he should eat as heartily as possible of a many-coursed dinner, and drink copiously of a soft

[1] This is the general sense of the passage, with the reading *τόπον*. It must be confessed, however, that the accusative is strange to express (with *πρὸς*) "place where," and grammatically the reading *πόνον* is superior. But how can flesh be "moist in relation to one particular exercise"? If for *εἶναι* the MSS. had *γενέσθαι* one would without hesitation read *πόνον*, and translate: "whatever be the unusual exercise, the flesh must become moist with this exercise, just as persons out of training become moist with any exercise."

[7] λοῦσαι· εἶτα πεῖσαι θ: λούεσθαι· εἶτα πισας M.

ποτῷ ὑδαρεῖ, μαλθακῷ δ' οἴνῳ χρῆσθαι καὶ πολλῷ,[1] εἶτ' ἐνδιατρῖψαι πλέω χρόνον μέχρι ἂν αἱ φλέβες πληρωθεῖσαι ἀρθῶσιν· εἶτα ἐξεμείτω, καὶ ἐξαναστάντα[2] ὀλίγον καθεύδειν μαλθακῶς· εἶτα προσάγειν ἡσυχῇ τοῖσι σιτίοισι καὶ τοῖσι πόνοισι τοῖσι συνήθεσιν ἐς ἡμέρας ἕξ, ἐν ταύτῃσι δὲ καταστῆσαι ἐς τὸ σύνηθες καὶ σίτου καὶ ποτοῦ. δύναμιν δὲ ἔχει ἡ θεραπείη τοιήνδε· ἀνεξηρασμένον τὸ σῶμα ἐς ὑπερβολὴν
80 ἐξυγρῆναι δύναται ἄτερ ὑπερβολῆς· εἰ μὲν οὖν δυνατὸν ἦν, τὴν ὑπερβολὴν τοῦ πόνου[3] γνόντα ὁκόση τίς ἐστι, τοῦ σίτου τῇ συμμετρίῃ ἀκέσασθαι, εὖ ἂν εἶχεν οὕτω· νῦν δὲ τὸ μὲν ἀδύνατον, τὸ δὲ ῥᾴδιον· ἐξηρασμένον γὰρ τὸ σῶμα, σίτων ἐμπεσόντων παντοδαπῶν, ἕλκει τὸ σύμφορον αὐτὸ ἑωυτῷ[4] ἕκαστον τοῦ σώματος ἑκάστου σίτου, πληρωθὲν δὲ καὶ ὑγρανθέν, κενωθείσης τῆς κοιλίης ὑπὸ τοῦ ἐμέτου, ἀφίησι πάλιν τὴν ὑπερβολήν· ἡ δὲ κοιλίη κενὴ ἐοῦσα ἀντισπᾷ.
90 τὸ μὲν οὖν ὑπερβάλλον ὑγρὸν ἐξερεύγεται ἡ σάρξ, τὸ δὲ σύμμετρον οὐκ ἀφίησιν, ἢν μὴ διὰ βίης ἢ φαρμάκων ἢ πόνων ἢ ἄλλης τινὸς ἀντισπάσιος. τῇ δὲ προσαγωγῇ χρησάμενος καταστήσεις τὸ σῶμα ἐς τὴν ἀρχαίην δίαιταν[5]
95 ἡσυχῇ.

[1] οἴνωι δὲ μαλακῶι χρέεσθαι πολλῶι M.
[2] ἐξαναστάντα M: ἀναστὰς θ.
[3] τὴν ὑπερβολὴν τοῦ πόνου omitted by θ.

wine, well diluted; then he should let a longish interval pass, until the veins become filled and inflated. Then let him vomit, and, having gone a short stroll, sleep on a soft bed. Then increase gradually his food and usual[1] exercises for six days, in which you must restore him to his usual food and drink. The treatment has the property of moistening without excess the body which has been dried to excess. Now if it were possible to discover the amount of the excess and cure it by an appropriate amount of food, all would be well thus. But as it is, this is impossible, but the other course is easy. For the body, in a state of dryness, after the entrance of all sorts of food, draws to itself what is beneficial from the several foods for the several parts of the body; on being filled and moistened, the belly having been emptied by the emetic, it casts away the excess, while the belly, being empty, exercises a revulsion. So the flesh rejects the excessive moisture, but it does not cast away that which is of an appropriate amount, unless it be under the constraint of drugs, of exercises,[2] or of some revulsion. By employing gradation, you will restore the body gently to its old regimen.

[1] Or, "usual food and exercises.'
[2] Or, "fatigue."

---

[4] *σιτῶν ἐμπεσόντων παντοδαπῶν. ἕλκει τὸ σύμφορον αὐτὸ ἑωυτωι θ: τῶν ἐμπεσόντων παντοδαπῶν· ἀφ' ὧν λαμβάνει τὸ ξυμφερον αὐτὸ ἑωϋτῶι* M.
[5] *κατέστησε τὸ σῶμα τὴν δίαιταν* M.

ΠΕΡΙ ΔΙΑΙΤΗΣ

LXVII. Περὶ διαίτης ἀνθρωπίνης, ὥσπερ μοι καὶ πρόσθεν εἴρηται, συγγράψαι μὲν οὐχ οἷόν τε ἐς ἀκριβείην, ὥστε πρὸς τὸ πλῆθος τοῦ σίτου τὴν συμμετρίην τῶν πόνων ποιεῖσθαι· πολλὰ γὰρ τὰ κωλύοντα. πρῶτον μὲν αἱ φύσιες τῶν ἀνθρώπων διάφοροι ἐοῦσαι· καὶ γὰρ [1] ξηραὶ αὐταὶ [2] ἑωυτῶν πρὸς αὐτὰς [3] καὶ πρὸς ἀλλήλας [4] μᾶλλον καὶ ἧσσον ξηραί, καὶ ὑγραὶ ὡσαύτως, καὶ αἱ ἄλλαι πᾶσαι· ἔπειτα αἱ ἡλικίαι οὐ τῶν
10 αὐτῶν δεόμεναι· ἔτι δὲ καὶ τῶν χωρίων αἱ θέσιες, καὶ τῶν πνευμάτων αἱ μεταβολαί, τῶν τε ὡρέων αἱ μεταστάσιες, καὶ τοῦ ἐνιαυτοῦ αἱ καταστάσιες. αὐτῶν τε τῶν σίτων πολλαὶ αἱ διαφοραί· [5] πυροί τε γὰρ πυρῶν καὶ οἶνος οἴνου καὶ τἄλλα οἷς [6] διαιτεόμεθα, πάντα διάφορα ἐόντα ἀποκωλύει μὴ [7] δυνατὸν εἶναι ἐς ἀκριβείην συγγραφῆναι. ἀλλὰ γὰρ αἱ διαγνώσιες [8] ἔμοιγε ἐξευρημέναι εἰσὶ τῶν ἐπικρατεόντων ἐν τῷ σώματι, ἤν τε οἱ πόνοι ἐπικρατέωσι τῶν σίτων, ἤν τε τὰ
20 σῖτα τῶν πόνων, καὶ ὡς χρὴ ἕκαστα ἐξακεῖσθαι, προκαταλαμβάνειν τε ὑγείην, ὥστε τὰς νούσους μὴ προσπελάζειν, [9] εἰ μή τις πάνυ μεγάλα ἐξαμαρτάνοι καὶ πολλάκις· ταῦτα δὲ φαρμάκων

[1] After γὰρ M adds αἱ. [2] αὐταὶ M: αὗται θ.
[3] ἑωϋτὰς M: αὐτὰς θ. [4] ἀλλήλας θ: ἄλλας M.
[5] πολλαὶ αἱ διαφοραί· θ: πολλὴ διαφορα M.

REGIMEN III

LXVII. As I have said above, it is impossible to treat of the regimen of men with such a nicety as to make the exercises exactly proportionate to the amount of food. There are many things to prevent this. First, the constitutions of men differ; dry constitutions, for instance, are more or less dry as compared with themselves or as compared with one another. Similarly with moist constitutions, or with those of any other kind. Then the various ages have different needs. Moreover, there are the situations of districts, the shiftings of the winds, the changes of the seasons, and the constitution of the year. Foods themselves exhibit many differences; the differences between wheat and wheat, wine and wine, and those of the various other articles of diet, all prevent its being possible to lay down rigidly exact rules in writing. But the discovery that I have made is how to diagnose what is the overpowering element in the body, whether exercises overpower food or food overpowers exercises; how to cure each excess, and to insure good health so as to prevent the approach of disease, unless very serious and many blunders be made. In such cases there is

---

[6] οἷς θ: ὅσα M. [7] μὴ omitted by M.

[8] διαγνώσιες θ: προγνώσιες M.

[9] προκαταλαμβάνειν τὲ ὑγιείην· ὥστε τὰς νούσους προσπελάζειν θ: προκαταμανθάνειν τε ὑγιέας τὰς φύσεις· μὴ προσπελάζειν τε τὰς νούσους M.

δεῖται ἤδη, ἔστι δ' ἅσσα οὐδ' ὑπὸ τῶν φαρμάκων δύναται ὑγιάζεσθαι. ὡς μὲν οὖν δυνατὸν εὑρεθῆναι, ἔγγιστα τοῦ ὅρου[1] ἐμοὶ εὕρηται, τὸ δὲ
27 ἀκριβὲς οὐδενί.

LXVIII. Πρῶτον μὲν οὖν τοῖσι πολλοῖσι τῶν ἀνθρώπων συγγράψω ἐξ ὧν μάλιστα ἂν ὠφελοῖντο οἵτινες σίτοισί τε καὶ πόμασι τοῖσι προστυχοῦσι χρῶνται, πόνοισί τε τοῖσιν ἀναγκαίοισιν, ὁδοιπορίῃσί τε τῇσι πρὸς ἀνάγκας, θαλασσουργίῃσί τε τῇσι πρὸς[2] συλλογὴν τοῦ βίου, θαλπόμενοί τε παρὰ τὸ σύμφορον,[3] ψυχόμενοί τε παρὰ τὸ ὠφέλιμον, τῇ τε ἄλλῃ διαίτῃ ἀκαταστάτῳ χρεόμενοι. τούτοισι δὴ συμφέρει
10 ἐκ τῶν ὑπαρχόντων ὧδε διαιτῆσθαι· τὸν μὲν οὖν[4] ἐνιαυτὸν ἐς τέσσαρα μέρεα διαιρέω,[5] ἅπερ μάλιστα γινώσκουσιν οἱ πολλοί, χειμῶνα, ἦρ, θέρος, φθινόπωρον· χειμῶνα μὲν ἀπὸ πλειάδων δύσιος ἄχρι ἰσημερίης ἠαρινῆς, ἦρ δὲ ἀπὸ ἰσημερίης μέχρι πλειάδων ἐπιτολῆς,[6] θέρος δὲ ἀπὸ πλειάδων μέχρι ἀρκτούρου ἐπιτολῆς, φθινόπωρον δὲ ἀπὸ ἀρκτούρου μέχρι πλειάδων δύσιος. ἐν μὲν οὖν τῷ χειμῶνι συμφέρει πρὸς τὴν ὥρην, ψυχρήν τε καὶ συνεστηκυίην, ὑπεναντιούμενον
20 τοῖσι διαιτήμασιν ὧδε χρῆσθαι. πρῶτον μὲν μονοσιτίῃσι χρὴ διάγειν, ἢν μὴ πάνυ ξηρὴν τις τὴν κοιλίην ἔχῃ· ἢν[7] δὲ μή, μικρὸν ἀριστῆν· τοῖσι δὲ διαιτήμασι χρῆσθαι τοῖσι ξηραντι-

[1] ἔγγιστα τοῦ ὅρου M : εἴ τις τὰ τοῦ ὀρθοῦ θ.
[2] After πρὸς M adds τὴν.

need of drugs, while some there are that not even drugs can cure. So as far as it is possible to make discoveries, to the utmost limit my discoveries have been made, but absolute accuracy has been attained by nobody.

LXVIII. Now first of all I shall write, for the great majority of men, the means of helping such as use any ordinary food and drink, the exercises that are absolutely necessary, the walking that is necessary, and the sea-voyages required to collect the wherewithal to live—the persons who suffer heat contrary to what is beneficial and cold contrary to what is useful, making use of a regimen generally irregular. These are benefited by living as follows, so far as their circumstances allow. I divide the year into the four parts most generally recognised—winter, spring, summer, autumn. Winter lasts from the setting of the Pleiads to the spring equinox, spring from the equinox to the rising of the Pleiads, summer from the Pleiads to the rising of Arcturus, autumn from Arcturus to the setting of the Pleiads. Now in winter it is beneficial to counteract the cold and congealed season by living according to the following regimen. First a man should have one meal a day only, unless he have a very dry belly; in that case let him take a light luncheon. The articles of diet to be used are such as are of a drying

---

[3] So M, but θ has ἡλιουμένοι τὲ παρὰ τὸ συμφέρον, a reading so attractive that it is difficult to choose between it and that of M.

[4] οὖν is omitted by M.

[5] διτιρέωι ἐς θ: διαιρέωσιν. M: διαιρέουσιν Littré. The reading in the text is that of Mack.

[6] ἐπιτολῆς M: ὑπερβολῆς θ. [7] ἦν θ: εἰ M.

κοῖσι[1] καὶ θερμαντικοῖσι καὶ συγκομιστοῖσι καὶ ἀκρήτοισιν, ἀρτοσιτίῃ τε[2] μᾶλλον, καὶ τοῖσιν ὀπτοῖσι τῶν ὄψων μᾶλλον ἢ ἑφθοῖσι, καὶ τοῖσι πόμασι μέλασιν ἀκρητεστέροισι καὶ ἐλάσσοσι· λαχάνοισιν ὡς ἥκιστα χρή, πλὴν τοῖσι θερμαντικοῖσι καὶ ξηροῖσι, καὶ χυλοῖσι καὶ ῥοφήμασιν
30 ὡς ἥκιστα· τοῖσι δὲ πόνοισι πολλοῖσιν ἅπασι, τοῖσί τε δρόμοισι καμπτοῖσιν ἐξ ὀλίγου προσάγοντα, καὶ τῇ πάλῃ ἐν ἐλαίῳ μακρῇ, ἀπὸ κούφων προσαναγκάζοντα· τοῖσί τε[3] περιπάτοισιν ἀπὸ τῶν γυμνασίων ὀξέσιν, ἀπὸ δὲ τοῦ δείπνου βραδέσιν ἐν ἀλέῃ, ὀρθρίοισί τε πολλοῖσιν ἐξ ὀλίγου ἀρχόμενον, προσάγοντα[4] ἐς τὸ σφοδρόν, ἀποπαύοντά τε ἡσυχῇ· καὶ σκληροκοιτίῃσι[5] καὶ νυκτοβατίῃσι[6] καὶ νυκτοδρομίῃσι[7] χρῆσθαι συμφέρει· πάντα γὰρ ταῦτα ἰσχναίνει καὶ
40 θερμαίνει· χρίεσθαί τε πλείω.[8] ὁκόταν δὲ ἐθέλῃ λούεσθαι,[9] ἢν μὲν ἐκπονήσῃ ἐν παλαίστρῃ, ψυχρῷ λουέσθω· ἢν δὲ ἄλλῳ τινὶ πόνῳ χρήσηται, τὸ θερμὸν συμφορώτερον. χρῆσθαι δὲ καὶ λαγνείῃ πλέον ἐς ταύτην τὴν ὥρην, καὶ τοὺς πρεσβυτέρους μᾶλλον ἢ τοὺς νεωτέρους. χρῆσθαι δὲ καὶ τοῖσιν ἐμέτοισι, τοὺς μὲν ὑγροτέρους τρὶς τοῦ μηνός, τοὺς δὲ ξηροτέρους δὶς ἀπὸ σίτων παντοδαπῶν, ἐκ δὲ τῶν ἐμέτων προσάγειν ἡσυχῇ πρὸς τὸ εἰθισμένον σιτίον ἐς ἡμέρας τρεῖς, καὶ

[1] τοῖσι δὲ ξηραντικοῖσι θ: τοῖσι ξηροῖσι καὶ αὐστηροῖσι M.
[2] ἀρτοσιτίῃι τε θ: ἀρτοσιτέειν δὲ M. [3] τε omitted by M.
[4] προσάγοντα (without ἐς τὸ) M: πρὸς ἅπαντα ἐς τὸ θ.
[5] σκληρευνίῃισι θ: σκληροκοιτίῃισι M.
[6] νυκτοβαδίῃισι θ.
[7] νυκτοδρομίῃισι θ: κοινοβατίῃισι καὶ κυνοδρομίῃισι M.
[8] χρίεσθαι τὲ τὰ πλείωι θ: χρέεσθαί τε πλείω M.

nature, of a warming character, assorted[1] and undiluted; wheaten bread is to be preferred to barley cake, and roasted to boiled meats; drink should be dark, slightly diluted wine, limited in quantity: vegetables should be reduced to a minimum, except such as are warming and dry, and so should barley water and barley gruel. Exercises should be many and of all kinds; running on the double track increased gradually; wrestling after being oiled, begun with light exercises and gradually made long; sharp walks after exercises, short walks in the sun after dinner; many walks in the early morning, quiet to begin with, increasing until they are violent, and then gently finishing. It is beneficial to sleep on a hard bed and to take night walks and night runs, for all these things reduce and warm; unctions should be copious. When a bath is desired, let it be cold after exercise in the palaestra; after any other exercise, a hot bath is more beneficial. Sexual intercourse should be more frequent at this season, and for older men more than for the younger. Emetics are to be used three times a month by moist constitutions, twice a month by dry constitutions, after a meal of all sorts of food; after the emetic three days should pass in slowly increasing the food to the

[1] *συγκομιστὸς* is rendered here by Littré "de substances grossières," by Liddle and Scott (after Foës) "mixed." I suppose that the objection to the latter is its apparent inconsistency with *ἀκρήτοισιν*. But *συγκομιστὸς* applies to foods and *ἄκρητος* to wine. My own objection to translating *συγκομιστὸς* (with Littré) "coarse," "of unbolted meal" (as in *Ancient Medicine*), is that it limits too much the foods to which it applies. I think (with Foës) that "a mixed diet," as we term it, is referred to.

[a] λούεσθαι θ: λούσασθαι M.

50 τοῖσι πόνοισι κουφοτέροισι καὶ ἐλάσσοσι τοῦτον τὸν χρόνον· ἀπὸ δὲ βοείων καὶ χοιρείων κρεῶν ἢ τῶν ἄλλων ὅ τι ἂν ὑπερβάλλῃ πλησμονῇ,[1] ἐμεῖν συμφέρει,[2] καὶ ἀπὸ τυρωδέων καὶ γλυκέων καὶ λιπαρῶν ἀνεθίστων πλησμονῆς ἐμεῖν συμφέρει· καὶ ἀπὸ μέθης καὶ σίτων μεταβολῆς καὶ χωρίων μεταλλαγῆς ἐμεῖν βέλτιον. διδόναι δὲ καὶ τῷ ψύχει ἑωυτὸν θαρσέων,[3] πλὴν ἀπὸ σίτων[4] καὶ γυμνασίων, ἀλλ' ἔν τε τοῖσιν ὀρθρίοισι περιπάτοισιν, ὅταν ἄρξηται τὸ σῶμα διαθερμαί-
60 νεσθαι, καὶ ἐν[5] τοῖσι δρόμοισι καὶ ἐν τῷ ἄλλῳ χρόνῳ, ὑπερβολὴν φυλασσόμενος· οὐκ ἀγαθὸν γὰρ τῷ σώματι μὴ χειμάζεσθαι ἐν τῇ ὥρῃ·[6] οὐδὲ γὰρ τὰ δένδρεα μὴ χειμασθέντα ἐν τῇ ὥρῃ δύναται καρποφορεῖν,[7] οὐδ' αὐτὰ ἐρρῶσθαι. χρῆσθαι δὲ καὶ τοῖσι πόνοισι πολλοῖσι ταύτην τὴν ὥρην ἅπασιν· ὑπερβολὴν γὰρ οὐκ ἔχει, ἢν μὴ οἱ κόποι ἐγγίνωνται· τοῦτο τὸ[8] τεκμήριον διδάσκω τοὺς ἰδιώτας. διότι δὲ οὕτως ἔχει φράσω· τῆς ὥρης ἐούσης ψυχρῆς καὶ συνε-
70 στηκυίης, παραπλήσια πέπονθε καὶ τὰ ζῷα· βραδέως οὖν διαθερμαίνεσθαι ἀνάγκη τὰ σώματα ὑπὸ τοῦ πόνου, καὶ τοῦ ὑγροῦ μικρόν τι μέρος ἀποκρίνεσθαι τοῦ ὑπάρχοντος· εἶτα τοῦ χρόνου ὅντινα μὲν πονεῖν ἀποδέδοται, ὀλίγος· ὅντινα δὲ ἀναπαύεσθαι, πολύς·[9] ἡ μὲν γὰρ ἡμέρη βραχείη,

[1] πλησμονῆι θ: πλησσομέν–ν M.
[2] συμφέρει omitted by M.
[3] εωυτον θαρσέων θ: καθαρων ἑωϋτὸν M.
[4] ἀπὸ σιτῶν θ: ἀπὸ τῶν σιτίων M. [5] ἐν omitted by M.

usual amount, and exercises should be lighter and fewer during this time. Emetics are beneficial after beef, pork, or any food causing excessive surfeit; also after excess of unaccustomed foods, cheesy, sweet or fat. Further, it is better to take an emetic after drunkenness, change of food or change of residence. One may expose oneself confidently to cold, except after food and exercise, but exposure is wise in early-morning walks, when the body has begun to warm up, in running, and during the other times, though excess should be avoided. For it is not good for the body not to be exposed to the cold of winter, just as trees that have not felt winter's cold can neither bear fruit nor themselves be vigorous. During this season, take also plenty of all sorts of exercise. For there is no risk of excess, unless fatigue-pains follow; this is the sign that I teach laymen, and the reason I will now proceed to explain. As the season is cold and congealed, animals too have the qualities of the season. So the body perforce warms up slowly under exercise, and only a small part of the available moisture is excreted. Then the time devoted to exercise is little, and that devoted to rest is much, as in winter days are short and nights are long. For these reasons neither the length nor the character of the exercise can be excessive. So in this way should this season be

---

[6] οὐκ ἀγαθὸν γὰρ τῶι σώματι μὴ χειμάζεσθαι ἐν τῇ ὥρηι· θ: ἀγαθὸν γὰρ τῶι σώματι χειμάζεσθαι ἢ γυμνάζεσθαι ἐν τῆι ὥρηι M.

[7] καρποφορεῖν θ: καρπὸν φέρειν M.

[8] τὸ omitted by M.

[9] ὅντινα μὲν πονέει. ἀποδέδοται ὀλίγος. ὅντινα δὲ ἀναπαύεται. πολύς. M: ὅντινα μὲν πονέειν ἀποδέδοται ὁ λόγος· ὅντινα δὲ ἀναπαύεσθαι πολὺς θ.

ἡ δὲ εὐφρόνη[1] μακρή· διὰ ταῦτα οὐκ ἔχει ὑπερβολὴν ὁ χρόνος καὶ ὁ πόνος. χρὴ οὖν ταύτην τὴν ὥρην οὕτω διαιτῆσθαι, ἀπὸ πλειάδων δύσιος μέχρις ἡλίου τροπῶν ἡμέρας τεσσα-
80 ράκοντα τέσσαρας· περὶ δὲ τὴν τροπὴν ἐν φυλακῇ ὡς μάλιστα εἶναι, καὶ ἀπὸ τροπῆς ἡλίου ἄλλας τοσαύτας ἡμέρας τῇ αὐτῇ διαίτῃ χρῆσθαι. μετὰ δὲ ταῦτα ὥρη ἤδη ζέφυρον πνεῖν, καὶ μαλακωτέρη ἡ ὥρη· χρὴ δὴ καὶ τῇ διαίτῃ μετὰ τῆς ὥρης ἐφέπεσθαι ἡμέρας πεντεκαίδεκα. εἶτα δὲ ἀρκτούρου ἐπιτολή, καὶ χελιδόνα ὥρη ἤδη φαίνεσθαι,[2] τὸν ἐχόμενον δὲ χρόνον ποικιλώτερον ἤδη ἄγειν[3] μέχρις ἰσημερίης ἡμέρας τριήκοντα δύο. δεῖ οὖν καὶ τοῖσι διαιτή-
90 μασιν ἕπεσθαι[4] τῇ ὥρῃ διαποικίλλοντα μαλακωτέροισι[5] καὶ κουφοτέροισι, τοῖσί τε σιτίοισι καὶ τοῖσι[6] πόνοισι, προσάγοντα ἡσυχῇ πρὸς τὸ ἦρ. ὁκόταν δὲ ἰσημερίη γένηται, ἤδη μαλακώτεραι αἱ ἡμέραι καὶ μακρότεραι, αἱ νύκτες δὲ βραχύτεραι, καὶ ἡ ὥρη ἡ ἐπιοῦσα[7] θερμή τε καὶ ξηρή, ἡ δὲ παρεοῦσα τρόφιμός τε καὶ εὔκρητος. δεῖ οὖν, ὥσπερ καὶ τὰ δένδρεα παρασκευάζεται ἐν ταύτῃ τῇ ὥρῃ αὐτὰ αὐτοῖς[8] ὠφελείην ἐς τὸ θέρος, οὐκ ἔχοντα γνώμην, αὔξησίν τε καὶ σκιήν,
100 οὕτω καὶ τὸν ἄνθρωπον· ἐπεὶ γὰρ γνώμην ἔχει, τῆς σαρκὸς τὴν αὔξησιν δεῖ ὑγιηρὴν παρασκευάζειν. χρὴ οὖν, ὡς μὴ ἐξαπίνης τὴν δίαιταν μεταβάλῃ, διελεῖν τὸν χρόνον ἐς μέρεα ἓξ κατὰ

[1] συφρονηι θ: νὺξ M.

[2] ζέφυρον πνέειν καὶ μαλακωτέρη ἡ ὥρηι· χρὴ δὴ· καὶ τῆ διαίτη μετὰ τῆς ὥρης ἐφέπεσθαι. ἡμέρας πεντεκαίδεκα· εἶτα δὲ ἀρκτούρου ἐπιτοληι· καὶ χελιδόνα ὥρηι ἤδη φαίνεσθαι· θ: ζέφυρον

passed, for forty-four days, from the setting of the Pleiads to the solstice. Near the solstice itself the greatest possible caution is required, and for the same number of days after the solstice the same regimen should be adopted. After this interval it is now time for the west wind to blow, and the season is milder; so for fifteen days regimen should be assimilated to the season. Then Arcturus rises, and it is now the season for the swallow to appear; from this time onwards live a more varied life for thirty-two days until the equinox. It is accordingly right to assimilate regimen to the season, varying it with the milder and lighter foods and exercises, with a gentle gradation until spring comes. When the equinox has come, the days are now milder and longer, the nights shorter; the coming season is hot and dry, the actual season is nourishing and temperate. Accordingly, just as trees, which have no intelligence, prepare for themselves growth and shade to help them in summer, even so man, seeing that he does possess intelligence, ought to prepare an increase of flesh that is healthy. It is accordingly necessary, in order that regimen may not be changed suddenly, to divide the time into six parts of eight

---

καὶ μαλακωτέρη ἡ ὥρη ἤδη· δεῖ οὖν καὶ τῆι διαίτηι μετὰ τῆς ὥρη ἕπεσθαι ἡμέρας πεντεκαίδεκα εἶτεδ' ἀρκτούρου ἐπιτολῆ καὶ χελιδόνα ἤδη φέρεσθαι M.

[3] *ἄγειν* θ M : *διάγειν* Littré, who says: "*διάγειν* om., restit. al. manu H."

[4] *ἕπεσθαι* θ: *χρέεσθαι* M.

[5] *μαλακωτέροισι* θ M : *φαυλοτέροισι* Littré (without giving authority).

[6] Before *πόνοισι* Littré has **ποτοῖσι καὶ** without giving authority. θ M omit.

[7] M omits ἡ before **ἐπιοῦσα.** [8] *αὐτὰ αὐτοῖς* θ : *αὐτοῖσι* M

ὀκτὼ ἡμέρας. ἐν μὲν οὖν[1] τῇ πρώτῃ μοίρῃ χρὴ τῶν τε πόνων ἀφαιρεῖν καὶ τοῖσι λοιποῖσιν ἠπιωτέροισι[1] χρῆσθαι, τοῖσί τε σιτίοισι μαλακωτέροισι καὶ καθαρωτέροισι, τοῖσί τε πόμασιν ὑδαρεστέροισι καὶ λευκοτέροισι, καὶ τῇ πάλῃ σὺν τῷ ἐλαίῳ ἐν τῷ ἡλίῳ χρῆσθαι· ἐν ἑκάστῃ
110 δὲ ὥρῃ ἕκαστα τῶν διαιτημάτων μεθιστάναι κατὰ μικρόν· καὶ τῶν περιπάτων ἀφαιρεῖν, τῶν ἀπὸ μὲν τοῦ δείπνου[2] πλέους, τῶν δὲ ὀρθρίων ἐλάσσους· καὶ τῆς μάζης ἀντὶ τῶν ἄρτων προστίθεσθαι, καὶ τῶν λαχάνων τῶν ἑψανῶν προσάγειν, καὶ τὰ ὄψα ἀνισάζειν τὰ ἑφθὰ τοῖσιν ὀπτοῖσι, λουτροῖσί τε χρῆσθαι, καί τι καὶ ἐναριστῆν μικρόν, ἀφροδισίοισι δὲ ἐλάσσοσι, καὶ τοῖσιν ἐμέτοισι, τὸ μὲν πρῶτον ἐκ τῶν τριῶν δύο ποιεῖσθαι,[3] εἶτα διὰ πλείονος χρόνου, ὅπως
120 ἂν καταστήσῃ τὸ σῶμα σεσαρκωμένον καθαρῇ σαρκί, καὶ τὴν δίαιταν μαλθακὴν ἐν τούτῳ τῷ χρόνῳ μέχρι πλειάδων ἐπιτολῆς. ἐν τούτῳ θέρος, καὶ τὴν δίαιταν ἤδη δεῖ[4] πρὸς τοῦτο ποιεῖσθαι· χρὴ οὖν, ἐπειδὰν πλειὰς ἐπιτείλῃ, τοῖσί τε σίτοισι μαλακωτέροισι καὶ καθαρωτέροισι καὶ ἐλάσσοσι χρῆσθαι, εἶτα τῇ μάζῃ πλεῖον ἢ τῷ ἄρτῳ, ταύτῃ δὲ προφυρητῇ[5] ἀτριπτοτέρῃ, τοῖσι δὲ πόμασι μαλακοῖσι, λευκοῖσιν, ὑδαρέσιν, ἀρίστῳ δὲ ὀλίγῳ, καὶ ὕπνοισιν ἀπὸ
130 τοῦ ἀρίστου βραχέσι, καὶ πλησμονῇσιν ὡς

[1] For ἠπιωτέροισι M has ὀξυτέροισι.

[2] M has τοὺς πλείους and Littré has τοὺς before ἐλάσσους (θ M omit).

[3] καὶ τοῖσιν ἐμέτοισι· τὸ μὲν πρῶτον ἐκ τῶν τριῶν. δύο ποιέεσθαι θ: καὶ τοῖσι ἐμέτοισι· τὸ μὲν πρῶτον ἐκ τῶν δύο

days apiece. So in the first portion one ought to lessen the exercises, and such as one adopts should be of a milder type, with foods softer and purer, and drinks more diluted and whiter, with wrestling in the sun, the body oiled. In each season the various items of regimen should be changed gradually. Walks should be lessened, those after dinner more, early-morning walks less. Take barley cake instead of wheaten bread, and eat boiled vegetables; make boiled meats equal to roast; use baths; have a little luncheon; use sexual intercourse less, and also your emetics. At first vomit twice instead of thrice, in the same period,[1] then at longer intervals, so as to furnish the body with permanent pure flesh, while regimen should be mild during this period until the rising of the Pleiads. Then it is summer, and hereafter regimen should be adapted to that season. So when that constellation has risen, eat softer, purer and less food, more barley-cake than wheaten bread, and that well-kneaded but not of finely crushed barley[2]; drink soft, white, diluted wines; take little luncheon, and only a short sleep after it; avoid as

[1] Namely, one month. See page 371.

[2] Either the text is wrong or else the dictionaries are at fault, for they give opposite meanings to *προφυρητὸς* and *ἄτριπτος*. The various readings may represent attempts to smooth away the difficulty. Perhaps *προφυρητὸς* refers to the kneading of the dough and *ἄτριπτος* to the coarseness of the flour. So apparently Littré.

ποιέεσθαι M. Littré does not record the reading of *θ*. Perhaps *τοὺς ἐμέτους* should be read; if not, Littré's punctuation must be changed.

[4] *δεῖ θ*: *χρὴ* M.

[5] M has *προφυραιτῆι*, and Littré records *προσφύραι τῆ*, *πορφύραι τῆ*, *προφυρετῆ*.

ἥκιστα τῶν σιτίων, καὶ τῷ ποτῷ ἱκανῷ[1] ἐπὶ τῷ σίτῳ χρῆσθαι· δι' ἡμέρης δὲ ὡς ἥκιστα πίνειν, ἢν μὴ ἀναγκαίῃ τινὶ ξηρασίῃ τὸ σῶμα χρήσηται· χρῆσθαι δὲ τοῖσι λαχάνοισι τοῖσιν ἑφθοῖσι, πλὴν τῶν καυσωδέων,[2] χρῆσθαι δὲ καὶ τοῖσιν ὠμοῖσι, πλὴν τῶν θερμαντικῶν[3] καὶ ξηρῶν· ἐμέτοισι δὲ μὴ χρῆσθαι, ἢν μή τις πλησμονὴ ἐγγένηται· τοῖσι δὲ ἀφροδισίοισιν ὡς ἥκιστα· λουτροῖσι δὲ χλιεροῖσι χρῆσθαι. ἡ δὲ
140 ὀπώρη ἰσχυρότερον τῆς ἀνθρωπίνης φύσιος· βέλτιον οὖν ἀπέχεσθαι· εἰ δὲ χρῷτό τις, μετὰ τῶν σίτων χρεόμενος ἥκιστ' ἂν ἐξαμαρτάνοι. τοῖσί τε πόνοισι τοῖσι τρόχοισι χρὴ γυμνάζεσθαι καὶ διαύλοισιν ὀλίγοισι μὴ πολὺν χρόνον, καὶ τοῖσι περιπάτοισιν ἐν σκιῇ, τῇ τε πάλῃ ἐν κόνει, ὅκως ἥκιστα ἐκθερμαίνηται· ἡ γὰρ ἀλίνδησις βέλτιον ἢ οἱ τρόχοι· ξηραίνουσι[4] γὰρ τὸ σῶμα κενοῦντες τοῦ ὑγροῦ· ἀπὸ δείπνου δὲ μὴ περιπατεῖν ἀλλ' ἢ ὅσον ἐξαναστῆναι· πρωῒ δὲ χρῆσθαι
150 τοῖσι περιπάτοισιν· ἡλίους δὲ φυλάσσεσθαι[5] καὶ τὰ ψύχεα τὰ πρώϊα καὶ τὰ ἐς τὴν ἑσπέρην,[6] ὅσα ποταμοὶ ἢ λίμναι ἢ χιόνες ἀποπνέουσιν. ταύτῃ δὲ τῇ διαίτῃ προσανεχέτω μέχρις ἡλίου τροπέων, ὅκως ἐν τούτῳ τῷ χρόνῳ ἀφαιρήσει πάντα ὅσα ξηρὰ καὶ θερμὰ καὶ μέλανα καὶ ἄκρητα, καὶ τοὺς ἄρτους, πλὴν εἴ τι σμικρὸν ἡδονῆς εἵνεκα. τὸν ἐχόμενον δὲ χρόνον διαιτήσεται

[1] For τῷ ποτῷ ἱκανῷ M has τῶν ποτῶν ἱκανῶς. Possibly ἱκανῶς is correct.

[2] After καυσωδέων M adds καὶ ξηρῶν.

[3] After θερμαντικῶν θ adds καὶ τῶν ξηραντικῶν· καὶ τῶν.

[4] ὀλίγοισι πουλὺν χρόνον· καὶ τοῖσι περιπάτοισιν ἐν σκιῆι· τῆι τε πάληι ἐν κόνει. ὅκως ἥκιστα διαθερμαίνοιτο· ἡ γὰρ αλινδησις

far as possible surfeits of food, and drink plentifully with food. But during the day drink as little as possible, unless the body experience an imperious dryness. Eat boiled vegetables, except those that are heating; eat also raw vegetables, except such as are warming and dry. Refrain from emetics, except in cases of surfeit. Sexual intercourse should be reduced to a minimum, and baths should be tepid. But the season's fruit is too strong for the human constitution. Accordingly, it is better to abstain from it; but if one should take it, by eating it with food the harm is reduced to a minimum. As for exercises, practice on the circular track and in the double stade should be infrequent and short, walking should be in the shade, and wrestling on dust, so as to avoid overheating as much as possible. For wrestling in the dust is preferable to circular running, as this dries the body by emptying it of its moisture. After dinner walking should be restricted to a short stroll, but in the early morning walks should be taken; one should, however, beware of the sun and of morning and evening chills, such as are given off by rivers, lakes or snow. Keep to this regimen until the solstice, so as to cut out during this period everything dry, hot, black, or undiluted, as well as wheaten bread, except just a little for pleasure's sake. During the period

---

βέλτιον· καὶ οἱ τροχοὶ δὲ βέλτιον· ψύχουσι θ: ὀλίγοισι πουλὺν χρόνον καὶ τοῖσι περιπάτοισι ἐν σκιᾶι· τῆι τε πάλῃι ἐν κόνει ὅκως ἥκιστα ἐκθερμαίνηται· ἡ γὰρ αλίνδησις βέλτιον ἢ οἱ τροχοι· ψύχουσιν M. The text is Littré's.

[5] M has *φυλάσσειν*.

[6] *τὰ πρωϊα· καὶ τὰ ἐς τὴν ἑσπέρην θ: πρωϊ καὶ τὰ ἐς τὴν ἑσπέρην.* M: *τὰ ἐν τῷ πρωὶ καὶ τὰ ἐν τῇ ἑσπέρῃ* Littré.

τοῖσι μαλθακοῖσι καὶ ὑγροῖσι καὶ ψυκτικοῖσι, λευκοῖσι καὶ καθαροῖσι, μέχρις ἀρκτούρου ἐπι-
160 τολῆς καὶ ἰσημερίης ἡμέρας ἐνενήκοντα τρεῖς. ἀπὸ δὲ ἰσημερίης ὧδε χρὴ διαιτῆσθαι, προσάγοντα πρὸς τὸν χειμῶνα ἐν τῇ φθινοπωρινῇ,[1] φυλασσόμενον τὰς μεταβολὰς τῶν ψυχέων καὶ τῆς ἀλέης ἐσθῆτι παχείῃ· χρῆσθαι δὲ ἐν τούτῳ τῷ χρόνῳ ἐν ἱματίῳ προκινήσαντα τῇ τε τρίψει καὶ τῇ πάλῃ τῇ ἐν ἐλαίῳ, ἡσυχῇ προσάγοντα· καὶ τοὺς περιπάτους ποιεῖσθαι ἐν ἀλέῃ· θερμολουσίῃ τε χρῆσθαι, καὶ τοὺς ὕπνους ἡμερινοὺς ἀφαιρεῖν, καὶ τοῖσι σιτίοισι θερμοτέροισι καὶ
170 ἧσσον ὑγροῖσι καὶ καθαροῖσι, καὶ τοῖσι πόμασι μελαντέροισι, μαλθακοῖσι δὲ καὶ μὴ ὑδαρέσι, τοῖσί τε λαχάνοισι ξηροῖσιν ἧσσόν τε, τῇ τε ἄλλῃ διαίτῃ προσάγειν πάσῃ, τῶν δὲ θερινῶν ὑφαιρέοντα, τοῖσι δὲ[2] χειμερίοισι χρῆσθαι μὴ ἐς ἄκρον, ὅπως καταστήσει ὡς ἔγγιστα τῆς χειμερινῆς διαίτης, ἐν ἡμέραις δυοῖν δεούσαιν πεντή-
177 κοντα[3] μέχρι πλειάδων δύσιος[4] ἀπὸ ἰσημερίης.

LXIX. Ταῦτα μὲν παραινέω τῷ πλήθει τῶν ἀνθρώπων, ὁκόσοισιν ἐξ ἀνάγκης εἰκῇ τὸν βίον διατελεῖν ἐστί, μηδ' ὑπάρχει αὐτοῖσι τῶν ἄλλων ἀμελήσασι τῆς ὑγιείης ἐπιμελεῖσθαι· ὅτῳ[5] δὲ

[1] τῆι φθινοπωρινῆι θ: τῶι φθινοπώρωι M.
[2] M omits this δὲ and that before θερινῶν, and has ἀφαιρεῦντα for ὑφαιρέοντα.
[3] M has ἐν ἡμέρηι δυοῖν δεούσαιν ἢ πεντήκοντα.
[4] For δύσιος θ has λύσιος.
[5] For ὅτῳ M has οἶσι.

that follows let regimen consist of things soft, moist, cooling, white and pure, for ninety-three days until the rising of Arcturus and the equinox. From the equinox regimen should be as follows, with a gradation during the autumn season to the winter, and with the use of a thick garment to guard against sudden changes of heat and cold. During this period, after some preliminary exercise in a cloak, have massage and practise wrestling with the body oiled, increasing the vigour gradually. Walks should be taken in the sun; baths should be warm; omit sleep in the day-time; food should be warmer, less moist, and pure,[1] drinks darker, soft and not diluted, vegetables dry and less in quantity; in every respect adopt a regimen departing gradually from that of summer and embracing that of winter, avoiding extremes in such a way as to take the forty-eight days from the equinox to the setting of the Pleiads in reaching the closest possible approximation to the winter regimen.

LXIX. Such is my advice to the great mass of mankind, who of necessity live a haphazard life without the chance of neglecting everything to concentrate on taking care of their health. But

[1] The present seems a suitable place to point out that it is impossible, owing to changes in habits, to find adequate English equivalents for the Greek vocabulary of foods and drinks. *σιτία*, for instance, must for convenience be rendered "food," but does not include meat, vegetables or fruit; while *καθαρὸς* does not refer to hygienic purity but to freedom from admixture. Finally, *πόμα* refers mainly to wine, practically the only drink favoured by the Greeks (they appear to have been less fond of milk than ourselves), and so *μέλας* will refer to what we call "red" wines, while *μαλθακὸς* will certainly not mean a "soft" drink, but a mild, gentle wine, not fiery, coarse or harsh.

τοῦτο παρεσκεύασται καὶ διέγνωσται, ὅτι οὐδὲν ὄφελός ἐστιν οὔτε χρημάτων οὔτε τῶν ἄλλων οὐδενὸς ἄτερ τῆς ὑγιείης, πρὸς τούτοις[1] ἔστι μοι δίαιτα ἐξευρημένη ὡς ἀνυστὸν πρὸς τὸ ἀληθέστατον τῶν δυνατῶν προσηγμένη. ταύτην μὲν
10 οὖν προϊόντος τοῦ λόγου[2] δηλώσω. τόδε δὲ τὸ ἐξεύρημα καλὸν μὲν ἐμοὶ τῷ εὑρόντι, ὠφέλιμον δὲ τοῖσι μαθοῦσιν, οὐδεὶς δέ πω τῶν πρότερον οὐδὲ ἐπεχείρησε συνεῖναι,[3] πρὸς ἅπαντα δὲ[4] τὰ ἄλλα πολλοῦ κρίνω αὐτὸ[5] εἶναι ἄξιον· ἔστι δὲ προδιάγνωσις[6] μὲν πρὸ τοῦ κάμνειν, διάγνωσις δὲ τῶν σωμάτων τί πέπονθε, πότερον τὸ σιτίον κρατεῖ τοὺς πόνους, ἢ οἱ πόνοι τὰ σιτία, ἢ μετρίως ἔχει πρὸς ἄλληλα· ἀπὸ μὲν γὰρ τοῦ κρατεῖσθαι ὁποτερονοῦν νοῦσοι ἐγγίνονται· ἀπὸ
20 δὲ τοῦ ἰσάζειν πρὸς ἄλληλα ὑγείη πρόσεστιν. ἐπὶ ταῦτα δὴ τὰ εἴδεα ἐπέξειμι, καὶ δείξω οἷά ἐστι καὶ γίνεται[7] τοῖσιν ἀνθρώποισιν ὑγιαίνειν δοκέουσι καὶ ἐσθίουσιν ἡδέως πονεῖν τε δυναμένοισι καὶ σώματος καὶ χρώματος ἱκανῶς
25 ἔχουσιν.

LXX. Αἱ ῥῖνες ἄτερ προφάσιος φανερῆς ἐμπλάσσονται ἀπό τε τοῦ δείπνου καὶ τοῦ ὕπνου, καὶ δοκέουσι μὲν πλήρεες εἶναι, μύσσονται δὲ οὐδέν· ὅταν δὲ περιπατεῖν ἄρξωνται τοῦ ὄρθρου

[1] For τούτοις M has τουτέους.
[2] προϊόντος τοῦ λόγου θ : προϊοντι τῶι χρόνωι M.
[3] For συνεῖναι M has ξυνθεῖναι. Littré has ὃ after συνεῖναι.
[4] δὲ my conjecture : δὴ θM. [5] M omits αὐτό.
[6] θ has προδιαγνώσεις and διαγνώσεις.

when a man is thus favourably situated, and is convinced that neither wealth nor anything else is of any value without health, I can add to his blessings a regimen that I have discovered, one that approximates to the truth as closely as is possible. What it is I will set forth in the sequel. This discovery reflects glory on myself its discoverer, and is useful to those who have learnt it, but no one of my predecessors has even attempted to understand[1] it, though I judge it to be of great value in respect of[2] everything else. It comprises prognosis before illness and diagnosis of what is the matter with the body, whether food overpowers exercise, whether exercise overpowers food, or whether the two are duly proportioned. For it is from the overpowering of one or the other that diseases arise, while from their being evenly balanced comes good health. Now these different conditions[3] I will set forth, and explain their nature and their arising in men who appear to be in health, eat with an appetite, can take their exercise, and are in good condition and of a healthy complexion.

LXX. The nostrils without obvious cause become blocked after dinner and after sleep, and they seem to be full without there being need to blow the nose. But when these persons have begun to walk in the

[1] Or, with the reading of M, "to set it forth in a treatise," "to compose an essay about it."

[2] Or "in comparison with."

[3] Littré translates "formes." Professor A. E. Taylor (*Varia Socratica*) maintains that in the *Corpus* εἶδος has the meaning "physical shape or appearance." Such a sense could be attributed to the word here ("healthy or unhealthy looks"), but it hardly suits οἷά ἐστι καὶ γίνεται.

[7] οἷα ἐστι· καὶ γίνεται θ: ὁκοῖα γίγνεται M.

καὶ γυμνάζεσθαι, τότε μύσσονται καὶ πτύουσι, προϊόντος δὲ τοῦ χρόνου καὶ τὰ βλέφαρα βαρέα ἴσχουσι, καὶ τὸ μέτωπον ὥσπερ ξυσμὸς[1] λαμβάνει, τῶν τε σίτων ἧσσον ἅπτονται,[2] πίνειν τε ἧσσον δύνανται, ἄχροιαι[3] τε τούτοισιν ὑπογίνονται, καὶ ἢ[4] κατάρροοι κινέονται ἢ πυρετοὶ φρικώδεες, καθ' ὅ τι ἂν τύχῃ τοῦ τόπου[5] ἡ πλησμονὴ κινηθεῖσα. ὅ τι δ' ἂν τύχῃ ποιήσας κατὰ τοῦτον τὸν καιρόν, τοῦτο αἰτιῆται οὐκ αἴτιον ἐόν· τούτῳ γὰρ κρατεῦντα τὰ σιτία τοὺς πόνους, κατὰ σμικρὸν συλλεγομένη ἡ πλησμονὴ ἐς νοῦσον προήγαγεν.[6] ἀλλ' οὐ χρὴ προΐεσθαι μέχρι τούτου, ἀλλ' ὁκόταν γνῷ τὰ πρῶτα τῶν τεκμηρίων, εἰδέναι ὅτι κρατεῖ τὰ σιτία τοὺς πόνους κατὰ σμικρὸν συλλεγόμενα, ᾗ πλησμονή ἐστι.[7] μύξα γὰρ καὶ σίαλον πλησμονῆς ἐστὶ κρίσις·[8] ἀτρεμίζοντος μὲν δὴ τοῦ σώματος, φραγνύουσι[9] τοὺς πόρους τοῦ πνεύματος, πολλῆς ἐνεούσης τῆς πλησμονῆς· θερμαινόμενον δὲ ἀπὸ[10] τοῦ πόνου, ἀποκρίνεται λεπτυνόμενον. χρὴ δὲ τὸν τοιοῦτον ἐκθεραπευθῆναι ὧδε· ἐκπονῆσαι ἐν τοῖσι γυμνασίοισι τοῖσιν εἰθισμένοισιν ἀκόπως, θερμῷ λουσάμενον ἐξεμέσαι εὐθὺς σίτοισι χρησάμενον παντοδαποῖ-

[1] ὥσπερ ξυσμὸς θ : ξυσμὴ M.
[2] ἧσσον ἅπτονται θ : ἀπέχονται M.
[3] αχροιαι θ: ἀχροιη τε (with ὑπογίνεται) M.
[4] ἢ οὖν M for καὶ ἢ.
[5] τόπου θ : χρόνου M.
[6] προήγαγεν θ : ἤγαγεν M.
[7] ἡ πλεισμονῆι ἔστι· θ : πλησμονή ἐστι M. I have translated Littré's text, but am persuaded that the words are a gloss.
[8] μύξαι γὰρ καὶ σίελα πλησμονῆς ἐστι κρίσις M.
[9] φραγνύουσι θ : φράσσουσι M. [10] ἀπὸ θ: ὑπὸ M.

morning or[1] to take exercise, then they blow the nose and spit; as time goes on the eyelids too are heavy, and as it were an itching seizes the forehead; they have less appetite for food and less capacity for drink; their complexion fades; and there come on either catarrhs or aguish fevers, according to the place occupied by the surfeit that was aroused. But the sufferer always lays the blame unjustly on the thing he may happen to do at the time of the illness. In such a case[2] food overpowers exercises, and the surfeit gathering together little by little brings on disease. One ought not, however, to let things drift to this point, but to realise, as soon as one has recognised the first of the signs, that exercises are overpowered by foods that gather together little by little, whereby comes surfeit.[3] For mucus and saliva are the crisis[4] of surfeit. Now as the body is at rest, they block up the passages of the breath, the surfeit inside being considerable; but being warmed by exercise, (the humour)[5] thins and separates itself out. Such a patient should be treated thus. He must take his usual exercise thoroughly yet without fatigue, have a warm bath, and vomit immediately after eating a

[1] Or, possibly, "and."

[2] Unless the MSS. reading be violently changed, this is an undoubted "nominative absolute." The scholar will accordingly be cautious in altering sentences containing this construction. See page 355.

[3] I have translated Littré's reading, with little belief in its correctness. The ᾗ . . . ἐστί is grammatically possible, but idiomatically strange.

[4] Or, "test."

[5] So Littré, and probably rightly. But the grammar is curious, as the natural subject of ἀποκρίνεται is σῶμα. If the middle could mean "gives off a secretion from itself," the sense would be excellent and the grammar normal. I cannot, however, find any support for this meaning of ἀποκρίνεται.

σιν·[1] ἐκ δὲ τοῦ ἐμέτου κλύσαι τὸ στόμα καὶ τὴν φάρυγγα οἴνῳ αὐστηρῷ, ὅκως ἂν στύψῃ[2] τὰ
30 στόματα τῶν φλεβῶν καὶ μηδὲν ἐπικατασπασθῇ, ὁκοῖα γίνεται ἀπὸ ἐμέτων· εἶτα ἐξαναστὰς[3] περιπατησάτω ἐν ἀλέῃ ὀλίγα· τῇ δὲ ὑστεραίῃ[4] τοῖσι μὲν περιπάτοισιν τοῖσιν αὐτοῖσι χρησάσθω,[5] τοῖσι δὲ γυμνασίοισιν ἐλάσσοσι καὶ κουφοτέροισιν ἢ πρόσθεν· καὶ ἀνάριστος διαγέτω, ἢν θέρος ᾖ· ἢν δὲ μὴ θέρος ᾖ,[6] μικρὸν ἐπιφαγέτω· καὶ τοῦ δείπνου ἀφελεῖν τὸ ἥμισυ οὗ εἴωθε δειπνεῖν· τῇ δὲ τρίτῃ τοὺς μὲν πόνους ἀποδότω τοὺς εἰθισμένους πάντας καὶ τοὺς περιπάτους, τοῖσι δὲ
40 σιτίοισι προσαγέτω ἡσυχῇ, ὅπως τῇ πέμπτῃ ἀπὸ τοῦ ἐμέτου κομιεῖται τὸ σιτίον τὸ εἰθισμένον. ἢν μὲν οὖν ἀπὸ τούτου ἱκανῶς ἔχῃ, θεραπευέσθω τὰ ἐπίλοιπα τοῖσι μὲν σιτίοισιν ἐλάσσοσι, τοῖσι δὲ πόνοισι πλείοσιν· ἢν δὲ μὴ καθεστήκῃ τὰ τεκμήρια τῆς πλησμονῆς, διαλιπὼν δύο ἡμέρας ἀφ᾽ ἧς ἐκομίσατο τὰ σιτία, ἐμεσάτω πάλιν καὶ προσαγέτω κατὰ τὰ αὐτά· ἢν δὲ[7] καὶ ἐκ τρίτου,
48 μέχρις ἂν ἀπαλλαγῇ[8] τῆς πλησμονῆς.

LXXI. Εἰσὶ δέ τινες τῶν ἀνθρώπων οἵτινες, ὅταν κρατέωνται οἱ πόνοι ὑπὸ[9] τῶν σίτων, καὶ τοιάδε[10] πάσχουσιν· ἀρχομένης τῆς πλησμονῆς

[1] ἐκπονῆσαι ἐν τοῖσι γυμνασίοισι· τοῖσιν εἰθισμένοισιν ἀκόπως. θερμῶι λουσάμενον. ἐξεμεύσαι εὐθὺς· σιτοῖσι χρησάμενον παντοδαποῖσιν· θ: διαπονήσαντα ἐν τοῖσι γυμνασίοισι τοῖσι εἰθισμένοισι ἀκόπως θερμῶι λουσάμενον παντοδάποισι· M: διαπονήσαντα . . . λουσάμενον, σιτίσαι παντοδαποῖσι καὶ ποιῆσαι ἐμέσαι. Littré, from the second hand in H.

[2] στύψηι θ: συνστυφῆι M.

[3] ἐξαναστὰς θ: ἀναστὰς M.

[4] τῆι δ᾽ ὑστεραίηι θ: ἐς δὲ τὴν ὑστεραίην M.

very varied meal.[1] After vomiting flush the mouth and throat with a harsh [2] wine, so as to contract the mouths of the veins, and prevent any result of the vomiting from being drawn down afterwards. Then one should go out for a short walk in the sun. On the next day one should take the same walks, but less and lighter exercise than before. One should take no luncheon if it be summer; if it be not summer, a light luncheon should be eaten. Reduce the usual dinner by one half. On the third day all usual exercises and walks should be resumed, and food should be gradually increased, until the usual food is restored on the fifth day from the vomiting. If as a result the patient's condition be satisfactory, let his treatment hereafter be to take less food and more exercise. But if the signs of surfeit do not disappear, let the patient wait for two days after the return to the usual diet, vomit again, and follow the same progressive increase. Even if a third vomiting be necessary, the patient should continue until he is rid of the surfeit.

LXXI. There are some men who, when exercise is overpowered by food, experience the following symptoms. At the beginning of the surfeit they

[1] The harsh asyndeton of θ's reading does not warrant our rejecting it.

[2] *i.e.* "astringent."

---

[5] χρησάσθωι· θ: χρέεσθαι M.

[6] ἢν μὴ θέρος ᾖι. θ: ἢν θέρος ᾖι· ἢν δὲ μὴ θέρος ᾖι. M.

[7] ἢν δὲ καὶ ἐκ τρίτου θ: ἢν δὲ μή. καὶ ἐκ τρίτου M.

[8] μέχρι ἀπαλλαγῆι θ: μέχρις ἀπαλλαγῆι M: μέχρις ἂν ἀπαλλαγῇ Littré.

[9] ὑπὸ M: ἀπὸ θ.

[10] καὶ τοιάδε θ: τοιάδε M: τοιαῦτα Littré (no authority given).

ὕπνοι μακροὶ καὶ ἡδέες αὐτοῖσιν ἐγγίνονται,[1] καὶ τι τῆς ἡμέρης ἐπικοιμῶνται· ὁ δὲ ὕπνος γίνεται τῆς σαρκὸς ὑγρανθείσης,[2] καὶ χεῖται τὸ αἷμα, καὶ γαληνίζεται[3] διαχεόμενον τὸ πνεῦμα. ὁκόταν δὲ μὴ δέχηται ἔτι τὸ σῶμα τὴν πλησμονήν, ἀπόκρισιν ἤδη ἀφίησιν εἴσω ὑπὸ βίης τῆς περιόδου,
10 ἥτις ὑπεναντιουμένη τῇ τροφῇ τῇ ἀπὸ τῶν σίτων ταράσσει τὴν ψυχήν. οὐκ ἔτι δὴ κατὰ τοῦτον τὸν χρόνον ἡδεῖς οἱ ὕπνοι, ἀλλ' ἀνάγκη ταράσσεσθαι τὸν ἄνθρωπον, καὶ δοκεῖν[4] μάχεσθαι· ὁκοῖα γάρ τινα πάσχει τὸ σῶμα, τοιαῦτα ὁρῇ ἡ ψυχή, κρυπτομένης τῆς ὄψιος. ὁκόταν οὖν ἐς τοῦτο ἥκῃ ὥνθρωπος, ἐγγὺς ἤδη τοῦ κάμνειν ἐστίν· ὅ τι δὲ ἥξει νόσημα, ἄδηλον· ὁκοίη γὰρ ἂν ἔλθῃ ἀπόκρισις καὶ ὅτου ἂν κρατήσῃ, τοῦτο ἐνοσοποίησεν.[5] ἀλλ' οὐ χρὴ
20 προέσθαι τὸν φρονέοντα, ἀλλ' ὁκόταν ἐπιγνῷ[6] τὰ πρῶτα, τῇσι θεραπείῃσιν ὥσπερ τὸν πρότερον ἐκθεραπευθῆναι,[7] πλείονος δὲ χρόνου καὶ λιμο-
23 κτονίης δεῖται.

LXXII. Ἔστι δὲ καὶ τὰ τοιάδε τεκμήρια πλησμονῆς· ἀλγεῖ τὸ σῶμα οἷσι μὲν ἅπαν, οἷσι δὲ μέρος τι τοῦ σώματος ὅ τι ἂν τύχῃ[8] τὸ δὲ

[1] ἐνγίγνονται θ: ἐπιγίνονται M.

[2] ὑγρανθείσης is omitted by θM, and is added by Littré from the second hand of E.

[3] γαληνίζεται θ: γαληνίζει M.

[4] δοκέειν θ: δοκέει M.

[5] ὅτι ηξει νόσημα ἄδηλον ὁκοιη γαρ ἀνἔλθηι ἀπόκρισις· καὶ ὅκου ἂν κρατήσηι τοῦτο. ἐνοσοποίησεν· θ: ὅτι δὲ ἥξει τὸ νόσημα, η μάλα δῆλον· ὁκοία γὰρ ἂν ἔλθηι ἀποκρισις· καὶ ὅτου ἦν κρατῆσαι τοῦτο, ἐνόσησεν· M.

have fall upon them long and pleasant sleeps, and they slumber for a part of the day. The sleep is the result of the flesh becoming moist[1]; the blood dissolves, and the breath, diffusing itself, is calm. But when the body can no longer contain the surfeit, it now gives out a secretion inwards through the force of the circulation,[2] which, being opposed to the nourishment from food, disturbs the soul. So at this period the sleeps are no longer pleasant, but the patient perforce is disturbed and thinks that he is struggling. For as the experiences of the body are, so are the visions of the soul when sight is cut off. Accordingly, when a man has reached this condition he is now near to an illness. What illness will come is not yet known, as it depends upon the nature of the secretion and the part that it overpowers. The wise man, however, should not let things drift, but as soon as he recognises the first signs, he should carry out a cure by the same remedies as in the first case, although more time is required and strict abstinence from food.

LXXII. The symptoms of surfeit are sometimes as follow. The body aches, in some cases all over, in others that part only of the body that happens to

[1] With the reading of θM: "The flesh goes to sleep, the blood dissolves, etc.," that is, "As the flesh, etc."

[2] For the περίοδος see pp. 241, 361, 427.

---

[6] After ἐπιγνῷ M has τῶν τεκμηρίων.

[7] τῇσι θεραπείῃσιν ὥσπερ τὸν πρῶτον ἐκθεραπευθῆναι θ: τῆς θεραπείης ἔχεσθαι, ὥσπερ τὸν πρότερον ἐκθεραπευθῆναι· M: τῆς θεραπείης ἔχεσθαι, καὶ δὴ τοῦτον ὥσπερ τὸν πρότερον ἐκθεραπευθῆναι Littré.

[8] τοῦ σώματος ὅ τι ἂν τύχῃ omitted by θ.

ἄλγος ἐστὶν οἱονεὶ[1] κόπος· δοκέοντες οὖν κοπιῆν, ῥᾳθυμίῃσί τε καὶ πλησμονῇσι θεραπεύονται, μέχρι ἂν[2] ἐς πυρετὸν ἀφικνέωνται· καὶ οὐδέπω οὐδὲ τοῦτο γινώσκουσιν, ἀλλὰ λουτροῖσί τε καὶ σίτοισι χρησάμενοι ἐς περιπλευμονίην κατέστησαν τὸ νόσημα, καὶ ἐς κίνδυνον τὸν ἔσχατον 10 ἀφικνέονται. ἀλλὰ χρὴ προμηθεῖσθαι πρὶν[3] ἐς τὰς νούσους ἀφικνέωνται, καὶ θεραπεύεσθαι τῷδε τῷ τρόπῳ· μάλιστα μὲν πυριηθέντα μαλακῇσι πυριῇσι, εἰ δὲ μή, λουσάμενον πολλῷ καὶ θερμῷ, διαλύσαντα τὸ σῶμα ὡς μάλιστα, χρησάμενον τῶν σιτίων πρῶτον μὲν τοῖσι δριμέσι καὶ πλείστοισιν, εἶτα τοῖσιν ἄλλοισιν[4] ἐξεμέσαι εὖ, καὶ ἐξαναστάντα περιπατῆσαι ὀλίγον χρόνον ἐν ἀλέῃ, ἔπειτα καταδαρθεῖν· πρωῒ δὲ τοῖσι περιπάτοισι πολλοῖσιν[5] ἐξ ὀλίγου 20 προσάγοντα χρῆσθαι καὶ τοῖσι γυμνασίοισι κούφοισι καὶ τῇσι προσαγωγῇσι καθάπερ καὶ πρότερον· ἰσχνασίης δὲ τοῦτο πλείστης δεῖται καὶ περιπάτων. ἢν δὲ μὴ προνοηθεὶς ἐς πυρετὸν ἀφίκηται, προσφέρειν μηδὲν ἄλλο[6] ἢ ὕδωρ ἡμερέων τριῶν· ἢν μὲν οὖν ἐν ταύτῃσι παύσηται· ἢν δὲ μή, πτισάνης χυλῷ θεραπεύεσθαι· ἢ γὰρ τεταρταῖος ἢ ἑβδομαῖος[7] ἐκστήσεται καὶ[8]

[1] οἱονεὶ θ : δκοῖον M. [2] ἂν is omitted by M.

[3] After πρὶν some MSS. (not θM) add ἄν.

[4] τοῖσιν ἄλλοισιν θ : τοῖσι ἄλλοισι M : Littré says "ἁλυκοῖσιν vulg." without naming MSS.

[5] πολλοῖσιν omitted by M.

[6] αλλο η θ : ἀλλ' ἢ M : ἄλλο ἀλλ' ἢ Littré (with apparently the authority of some Paris MS. or MSS.).

[7] ἢ γὰρ τεταρταῖος· ἢ ἑβδομαῖος θ : ἢ τεταρταίοις ἢ ἑβδομαίοις M : καὶ ἢ τεταρταῖος ἢ ἑβδομαῖος Littré (with apparently some authority).

be affected. The ache resembles the pain of fatigue. Accordingly, under the impression that they are suffering fatigue pains, these[3] patients adopt a treatment of rest and over-feeding, until they fall into a fever. Even then they fail to realise the true state of affairs, but indulging in baths and food they turn the illness into pneumonia, and fall into the direst peril. But what is necessary is to exercise forethought before the diseases attack, and to adopt the following treatment. Take by preference gentle vapour baths, the next best thing being copious hot baths, so as to dilate the body as much as possible, and then, after meals, at first of harsh foods and very copious, afterwards of the other kinds[1] of food, there should be a thorough emptying of the body by vomiting; after this there should be taken a short stroll in the sun, followed by sleep. In the morning walks should be long[2], though short to begin with, and gradually increased; exercises should be light, and with the same gradual increases as in the former case. Such a state requires severe reduction of flesh and plenty of walking exercise. And if through lack of forethought there is an attack of fever, nothing should be given for three days except water. If the fever go down in that time, well and good; if it does not, treat the patient with barley water, and on the fourth or the seventh day he will sweat and be quit of the

[1] Or (with the reading ἁλυκοῖσιν), "of salt foods."

[2] So Littré. It is doubtful in the *Corpus* whether πολὺς refers to quantity or to number, an ambiguity that often occurs in *Epidemics I* and *III*. "Many" is a possible meaning here, as of course the treatment is spread over several days.

---

[3] καὶ θ: ἢ M: εἰ Littré, with some Paris authority.

ἐξιδρώσει· ἀγαθὸν δὲ τοῖσι χρίσμασι χρῆσθαι τοῖσιν ἱδρωτικοῖσιν ὑπὸ τὰς κρίσιας, ἐξαναγ-
30 κάζουσι γάρ.

LXXIII. Πάσχουσι δέ τινες καὶ τοιάδε ἀπὸ πλησμονῆς· τὴν κεφαλὴν ἀλγέουσι καὶ βαρύνονται, καὶ τὰ βλέφαρα πίπτει αὐτοῖσιν ἀπὸ τοῦ δείπνου, ἔν τε τοῖς ὕπνοις ταράσσονται,[1] καὶ δοκεῖ θέρμη ἐνεῖναι, ἥ τε κοιλίη ἐφίσταται ἐνίοτε· ὁκόταν δὲ ἀφροδισιάσῃ, δοκεῖ κουφότερος εἶναι ἐς τὸ παραυτίκα, ἐξ ὑστέρου δὲ μᾶλλον βαρύνεται· τούτοισιν ἡ κεφαλὴ τὴν πλησμονὴν ἀντισπῶσα τήν τε κοιλίην ἐφίστησι, καὶ αὐτὴ
10 βαρύνεται· κίνδυνοί τε ἐπίκεινται[2] κακοί, καὶ ὅκου ἂν ῥαγῇ ἡ πλησμονή, τοῦτο διαφθείρει. ἀλλὰ χρὴ προμηθεῖσθαι ὧδε· ἢν μὲν βούληται τὴν θεραπείην ποιεῖσθαι ταχυτέρην,[3] προπυριηθέντα ἐλλεβόρῳ καθαρθῆναι, εἶτα προσάγειν τοῖσι σιτίοισι κούφοισι καὶ μαλθακοῖσιν ἐφ' ἡμέρας δέκα· τοῖσι δὲ ὄψοισι[4] διαχωρητικοῖσιν, ὅκως κρατήσει ἡ κάτω κοιλίη τὴν κεφαλὴν τῇ κάτω ἀντισπάσει· καὶ τοῖσι δρόμοισι βραδέσι καὶ τοῖσιν ὀρθρίοισι περιπάτοισιν ἱκανοῖσι, τῇ
20 τε πάλῃ ἐν ἐλαίῳ· ἀρίστῳ τε χρήσθω καὶ ὕπνῳ ἀπὸ τοῦ ἀρίστου μὴ μακρῷ· ἀπὸ τοῦ δείπνου δὲ ὅσον ἐξαναστῆναι ἱκανόν· καὶ τὸ μὲν λούεσθαι, τὸ δὲ χρίεσθαι, λούεσθαι[5] δὲ χλιερῷ, λαγνείης δὲ ἀπέχεσθαι. αὕτη μὲν ἡ ταχυτάτη[6] θεραπείη· εἰ δὲ μὴ βούλοιτο φαρμακοποτεῖν, λουσάμενον

[1] After ταράσσονται θ has τὲ (*sic*) which may (without the accent) be correct, but probably is dittography.

[2] M has ὑπόκειται, omits κακοὶ and reads ὅκηι.

[3] ταχυτερην θ: ταχείην (before ποιεῖσθαι) M.

trouble.[1] It is good to use sudorific unguents at the approach of a crisis, as they bring on sweating.

LXXIII. In certain cases the sufferers from surfeit experience the following symptoms. The head aches and feels heavy; their eyelids close after dinner; they are distressed in their sleep; they appear to be feverish, and occasionally the bowels are constipated. After sexual intercourse they seem to be for the moment more at ease, but afterwards the feeling of heaviness increases. In these cases the head, acting by revulsion on the surfeit, makes the bowels constipated and itself becomes heavy. Nasty dangers threaten, and the surfeit infects that part where it has broken out. But forethought of the following kind is required. If the quicker treatment is desired, after a vapour bath purge with hellebore, and for ten days gradually increase light and soft foods, and meats that open the bowels, that the lower belly may overpower the head by the revulsion below. Practise slow runs, longish early-morning walks, and wrestling with the body oiled. Take luncheon and a short sleep after it. After dinner a stroll is sufficient. Use baths and unguents, the baths tepid, and abstain from sexual intercourse. This is the quickest method of treatment. But if the patient wish to avoid drug-taking,[2] he should take a hot

[1] With the reading of Littré: "he will get rid of the trouble if he sweat."

[2] This refers to the hellebore mentioned earlier in the chapter. In the *Corpus* "drugs" are purges.

---

[4] *ὄψοισι* θ (not *ὀπτοῖσι* as Littré says): *σιτίοισι* M.

[5] For *λούεσθαι* θ has *λοῖσθαι*.

[6] *ταχυτάτη* θ: *ταχυτέρη* M.

θερμῷ,[1] ἐμέσαι σιτίοισι χρησάμενον τοῖσι δριμέσιν, ὑγροῖσι καὶ γλυκέσι καὶ ἁλμυροῖσιν,[2] ἐξ ἐμέτου δὲ ὅσον ἐξαναστῆναι· πρωῒ δὲ τοῖσι περιπάτοισι πρᾳέσι προσάγειν καὶ τοῖσι γυμ-
30 νασίοισι γεγραμμένοισιν ἐς ἡμέρας ἕξ· τῇ δὲ ἑβδόμῃ πλησμονὴν προσθέντα ἔμετον ποιῆσαι[3] ἀπὸ τῶν ὁμοίων σιτίων, καὶ προσάγειν κατὰ τωὐτό· χρῆσθαι δὲ τούτοισιν ἐπὶ τέσσαρας ἑβδομάδας, μάλιστα γὰρ ἐν τοσούτῳ χρόνῳ καθίσταται· εἶτα προσάγειν τοῖσί τε σίτοισι καὶ τοῖσι πόνοισι, τούς τε ἐμέτους σὺν πλείονι χρόνῳ ποιεῖσθαι, τά τε σιτία ἐν ἐλάσσονι προσάγειν, ὅκως τὸ σῶμα ἀνακομίσηται, καθιστάναι τε τὴν
39 δίαιταν ἐς τὸ σύνηθες κατὰ μικρόν.

LXXIV. Γίνεται δὲ καὶ τοιάδε ἀπὸ πλησμονῆς· ὁκόσοισιν ἡ μὲν κοιλίη καταπέσσει τὸ σιτίον, αἱ δὲ σάρκες μὴ δέχονται, ἐμμένουσα ἡ τροφὴ φῦσαν ἐμποιεῖ· ὅταν δὲ ἀριστήσῃ, καθίσταται, ὑπὸ γὰρ τοῦ ἰσχυροτέρου τὸ κουφότερον ἐξελαύνεται, καὶ δοκέουσιν ἀπηλλάχθαι· τὸ δὲ πολὺ πλέον ἐς τὴν ὑστεραίην παραγίνεται. ὅταν δὲ καθ' ἡμέρην ἑκάστην αὐξανόμενον ἰσχυρὸν γένηται, ἐκράτησε τὸ ὑπάρχον τῶν ἐπεισενεχθέντων,
10 καὶ ἐξεθέρμηνε, καὶ ἐτάραξεν ἅπαν τὸ σῶμα, καὶ ἐποίησε διαρροίην· τοῦτο γὰρ ὀνομάζεται, ἕως ἂν

[1] θερμῶι θ: πολλῶι M.
[2] ἐμεσαι σιτίοισι καὶ γλυκέσι καὶ ἁλμυροῖσι M.

bath, and then vomit after eating foods that are sharp, moist, sweet and salt[1]; after vomiting let him go for a short stroll. In the morning let him take gentle walks to begin with, and gradually increase them, and the exercises described above, for a period of six days. On the seventh day add a surfeit of like foods, and then vomit; after which make the same progressive increase. Follow this regimen for four weeks, for this is about the time required for a recovery. Then gradually increase food and exercise; increase the interval between vomitings; lessen the time taken in increasing food to the normal, so that the body may recreate itself, and restore the regimen to what is usual little by little.

LXXIV. Surfeit shows also the following symptoms. When the belly digests the food, but the flesh rejects it, the nutriment, remaining inside, causes flatulence. After luncheon, the flatulence subsides, for the lighter is expelled by the stronger, and the trouble seems to have been got rid of; but on the next day the symptoms recur much intensified. But when, owing to the daily growth, the surfeit becomes strong, what is already present overpowers the things added from without, generates heat, disturbs the whole body and causes diarrhœa. For such is the name given to

[1] Littré translates as though all four epithets applied to one food. The Greek suggests (*a*) foods sharp (acid) and moist, (*b*) sweet foods, (*c*) salt foods; perhaps (*a*) acid and moist, (*b*) sweet and moist, (*c*) salt and moist. At any rate the four qualities (which to a Greek of 400 B.C. were substances) had to be combined in one meal. As δριμὺς is generally opposed to γλυκύς, it is not surprising that the manuscript M omits two epithets.

---

[3] ποιῆσαι θ: ποιήσασθαι M.

αὐτὴ μούνη σαπεῖσα ἡ τροφὴ ὑποχωρῇ.[1] ὁκόταν δὲ θερμαινομένου τοῦ σώματος κάθαρσις δριμέα γένηται, τό τε ἔντερον ξύεται καὶ ἑλκοῦται καὶ διαχωρεῖται αἱματώδεα, τοῦτο δὲ δυσεντερίη καλεῖται, νοῦσος χαλεπὴ καὶ ἐπικίνδυνος. ἀλλὰ χρὴ προμηθεῖσθαι καὶ τὸ ἄριστον ἀφαιρεῖσθαι καὶ τοῦ δείπνου τὸ τρίτον μέρος· τοῖσι δὲ πόνοισι πλείοσι, τῇσι πάλῃσι καὶ τοῖσι δρόμοισι καὶ
20 περιπάτοισι[2] χρῆσθαι, ἀπό τε τῶν γυμνασίων καὶ ὄρθρου· ὅταν δ' ἡμέραι δέκα γένωνται, προσθέσθαι[3] τοῦ σίτου τὸ ἥμισυ τοῦ ἀφαιρεθέντος, καὶ ἔμετον ποιήσασθαι, καὶ προσάγειν[4] ἐς ἡμέρας τέσσαρας· ὁκόταν δὲ ἄλλη δεκὰς γένηται, τόν τε[5] σῖτον τὸν λοιπὸν προσθέσθαι, καὶ ἔμετον ποιήσασθαι, καὶ προσάγων πρὸς τὸν σῖτον ὑγιέα ποιήσεις ἐν τούτῳ τῷ χρόνῳ· τοῖσι δὲ
28 πόνοισι θαρρεῖν[6] τὸν τοιοῦτον πιέζων.

LXXV. Γίνεται δὲ καὶ τοιάδε· ἐς τὴν ὑστεραίην[7] τὸν σῖτον ἐρυγγάνεται ὠμὸν ἄτερ ὀξυρεγμίης, ἡ δὲ κοιλίη διαχωρεῖ, ἐλάσσω μὲν ἢ πρὸς τὰ σῖτα, ὅμως δὲ ἱκανῶς, πόνος δὲ οὐδεὶς ἐγγίνεται· τούτοισιν ἡ κοιλίη ψυχρὴ ἐοῦσα οὐ δύναται καταπέσσειν τὸν σῖτον ἐν τῇ νυκτί· ὁκόταν οὖν κινηθῇ, ἐρυγγάνεται τὸν σῖτον ὠμόν. δεῖ οὖν τούτῳ παρασκευάσαι τῇ κοιλίῃ θερμασίην ἀπό τε τῆς διαίτης καὶ[8] τῶν πόνων· πρῶτον μὲν οὖν
10 χρὴ ἄρτῳ θερμῷ[9] χρῆσθαι ζυμίτῃ, διαθρύπτοντα

[1] For ὑποχωρῇ θ has χωρέει.

[2] τοῖσι δὲ δρόμοισι πλείοσι καὶ τῇσι παλῇσι καὶ τοῖσι περιπάτοισι M.

[3] προσθέσθαι θ : προσθῆναι M.

[4] προσάγειν θ : προσαγαγεῖν M.

[5] τόν τε M : τότε θ.

the disorder so long as the waste products[1] only of food pass by stool. But when, as the body grows hot, the purging becomes harsh, the bowel is scraped, ulcers form and the stools passed are bloody; this disorder is called dysentery, a difficult and dangerous disease. Precautions must be taken, lunch omitted and dinner lessened by one-third. Use more exercises, wrestling, running and walks, both after the gymnastic practice and in the early morning. When ten days are gone, add one half of the food that has been taken away, take an emetic, and gradually increase the food for four days. When another ten days are gone, add the food that is still lacking, take an emetic, and gradually increasing the food you will effect a cure in this interval of time. Such a case as this you can without fear exercise rigorously.

LXXV. There also occurs the following kind of surfeit. On the following day the food is brought up undigested, without heartburn, copious stools are passed, but not proportionate to the food eaten, and there are no fatigue pains. In these cases the belly, being cold, cannot digest the food in the night. So when it is disturbed it brings up the food undigested. So for such a patient it is necessary to procure warmth for the belly both from regimen and from exercises. So first one should use warm, fermented bread,

[1] For σῆψις see p. 409.

---

[6] θαρρεῖν M : θαρρεῖ (θάρρει ?) θ.

[7] τοιάδε ἐς τὴν ὑστεραίην M : τοιᾶδε ἐς τὴν ὑστέραν· ἣν θ : τοιάδε πλησμονή· ἐς τὴν ὑστεραίην Littré (with some authority).

[8] After καὶ M adds ἀπό τε.

[9] θερμῶι θ : συγκομιστῶι M.

ἐς οἶνον μέλανα ἢ ἐς ζωμὸν ὕειον· τοῖσί τε ἰχθύσιν ἑφθοῖσιν ἐν ἅλμῃ δριμείῃ· χρῆσθαι δὲ καὶ τοῖσι σαρκώδεσιν, οἷον ἀκροκωλίοισί[1] τε διέφθοισι τοῖσιν ὑείοισι, τοῖσί τε πίοσιν ὑείοισιν ὀπτοῖσι, τοῖσι δὲ[2] χοιρείοισι μὴ πολλοῖσι καὶ σκυλάκων μηδὲ[3] ἐρίφων· λαχάνοισι δὲ πράσοισί τε καὶ σκορόδοισιν ἑφθοῖσι καὶ ὠμοῖσι, βλίτῳ τε ἑφθῷ καὶ κολοκύντῃ·[4] ποτοῖσί τε ἀκρήτοισιν, ἀναριστῆν τε τὴν πρώτην.[5] ὕπνοισί τε ἀπὸ τῶν γυμνασίων, τοῖσί τε δρόμοισι καμπτοῖσιν, ἐξ ὀλίγου προσάγων, πάλῃ τε μαλακῇ ἐν ἐλαίῳ, λουτροῖσί τε ὀλίγοισι, χρίσμασι πλείοσι, τοῖσι πρωῒ περιπάτοισι πλείστοισιν, ἀπὸ δείπνου δὲ[6] ὀλίγοισι· καὶ τὸ σῦκον μετὰ τῶν σίτων ἀγαθόν, ἄκρητός τε ἐπ' αὐτῷ. ἐκ δὲ ταύτης τῆς θεραπείης καθίσταται τοῖσι μὲν θᾶσσον, τοῖσι δὲ βραδύτερον.

LXXVI. Ἄλλοι δέ τινες τοιάδε πάσχουσιν· ἀχροοῦσι, καί, ὅταν φάγωσιν, ἐρυγγάνουσιν ὀλίγον ὕστερον ὀξύ,[7] καὶ ἐς τὰς ῥῖνας ἀνέρπει τὸ ὀξύ. τούτοισι τὰ σώματα οὐ καθαρά ἐστιν· ὑπὸ γὰρ τοῦ πόνου πλεῖον τὸ συντηκόμενον τῆς σαρκὸς ἢ τὸ ἀποκαθαιρόμενον[8] ὑπὸ τῆς περιόδου· ἐμμένον δὴ τοῦτο ἐναντιοῦται τῇ τροφῇ, καὶ βιάζεται, καὶ ἀποξύνει. ἡ μὲν οὖν τροφὴ ἐρυγγάνεται, αὐτὸ δὲ ὑπὸ τὸ δέρμα[9] ἐξωθεῖται, καὶ τῷ ἀνθρώπῳ ἄχροιαν ἐμποιεῖ, καὶ νούσους ὑδρω-

[1] ἀκροκωλίοισι M : ἄκροις θ.

[2] τοῖσι τὲ πίοσιν ὑείοις ὀπτοῖσι· τοῖσι δὲ θ : καὶ τοῖσι πλείοσι ὑοῖσι ἑφθοῖσι· τοῖσι τε M.

[3] μηδὲ θ : καὶ M.

[4] τῆς τε κολοκύντης θM : καὶ κολοκύντῃ Littré : possibly τῇσί τε κολοκύντῃσι.

crumbling it into dark wine or into pork broth. Also fish boiled in acrid brine. Use also fleshy meats, such as pig's feet well boiled and fat roast pork, but be sparing of sucking-pig, and the flesh of puppies and kids. Vegetables should be leeks and onions, boiled and raw, boiled blite and the pumpkin. Drink should be undiluted, and no luncheon should be taken at first. There should be sleep after exercises, running in the double course, increased gradually, gentle wrestling with the body oiled, few baths, more anointings than usual, plenty of[1] early-morning walks, but only short ones after dinner. Figs with food are good, and neat wine therewith. This treatment brings recovery, in some cases rapid, in others slower.

LXXVI. In other cases the following symptoms are experienced. There is paleness, and acid belching shortly after food, the acid matter rising into the nose. In such cases the body is impure. For the flesh melted by the fatigue is greater than that purged away by the circulation. Now this excess, remaining in the body, is antagonistic to the nourishment, forces it along, and renders it acid. So the nourishment is belched up, and the excess is pushed out under the skin, causing in the patient paleness

[1] Or, "long."

---

[5] ποτοισι τὲ ἀκρήτοισιν ἀναριστην τὲ τὴν πρώτην· θ: πόμασί τε ἀκρητεστέροισι· ὕπνοισί τε μακροῖσι ἀναριστησίῃσι τὴν πρώτην M.

[6] δὲ θ: τε M.

[7] ὀξύ θ: ὀξέα M.

[8] ἀπὸ γὰρ τοῦ πόνου πλείονος ἐόντος· συντηκομένης τῆς σαρκός· τὸ ἀπὸ καθαιρόμενον θ. The text is that of M, which, however, has τε for γὰρ.

[9] ἀυτὸ δὲ τὸ ὑπὸ τὸ δέρμα θ: ἀυτὸ δὲ ὑπὸ τοῦ δέρματος M.

ποειδέας. ἀλλὰ χρὴ προμηθεῖσθαι ὧδε· ἡ μὲν ταχυτέρη θεραπείη, ἐλλέβορον πίσαντα προσάγειν, ὥσπερ μοι πρότερον γέγραπται· ἡ δὲ ἀσφαλεστέρη[1] ὑπὸ τῆς διαιτήσιος ὧδε· πρῶτον μὲν λουσάμενον θερμῷ ἔμετον ποιήσασθαι, εἶτα προσάγειν ἐς[2] ἡμέρας ἑπτὰ τὸ σιτίον τὸ εἰθισμένον. δεκάτῃ δὲ ἡμέρῃ ἀπὸ τοῦ ἐμέτου αὖθις ἐμείτω, καὶ προσαγέτω κατὰ τωὐτό· καὶ τὸ τρίτον ὡσαύτως ποιησάτω· τοῖσι δὲ τρόχοισιν ὀλίγοισι 20 καὶ ὀξέσι καὶ ἀνακινήμασι[3] καὶ τρίψει, καὶ διατριβῇ[4] πολλῇ χρήσθω ἐν τῷ γυμνασίῳ, καὶ ἀλινδήσει χρήσθω·[5] τοῖσί τε περιπάτοισι πολλοῖσιν ἀπὸ τῶν γυμνασίων, χρῆσθαι δὲ καὶ ἀπὸ δείπνου, πλείστοισι δὲ τοῖσιν ὀρθρίοισιν· ἐγκονιόμενος[6] δὲ χριέσθω· ὅταν δὲ λούεσθαι θέλῃ, θερμῷ λουέσθω· ἀνάριστος δὲ διατελείτω τοῦτον τὸν χρόνον. καὶ ἢν μὲν ἐν μηνὶ καθιστῆται, θεραπευέσθω τὸ λοιπὸν τοῖσι προσήκουσιν· ἢν δέ τι 29 ὑπόλοιπον ᾖ, χρήσθω τῇ θεραπείῃ.

LXXVII. Εἰσὶ δέ τινες οἷσιν ἐς τὴν ὑστεραίην ὀξυρεγμίαι γίνονται· τούτοισιν ἐν τῇ νυκτὶ ἀπόκρισις ἀπὸ πλησμονῆς γίνεται·[7] ὁκόταν οὖν κινηθῇ ἐκ τοῦ ὕπνου τὸ σῶμα, πυκνοτέρῳ τῷ

[1] For ἀσφαλεστέρη θ has βραδυτέρη.
[2] ἐς is omitted by M.
[3] ἐνκινήμασι θM : ἀνακινήμασι Zwinger.
[4] For διατριβῇ M has ἐνδιατριβῆι.
[5] ἐν τῷ . . . χρήσθω omitted by θ.
[6] For ἐγκονιόμενος M has ἐκκονίων ιωένοις.
[7] M has ἡ before ἀπόκρισις, τῆς before πλησμονῆς, and ὑπὸ for ἀπὸ. θ has ἀποκρίσεις.

and dropsical diseases. The following precautions should be taken. The quicker method of treatment is to give a draught of hellebore and then to adopt the progressive diet that I have already described. The safer method, however, is by the following regimen. First a hot bath should be taken, then an emetic, and then the usual diet should be regained by a gradual increase spread over seven days. On the tenth day after the emetic another should be taken, followed by the same gradual increase of food. The treatment should be repeated a third time. Short but sharp runs should be taken in the circular course, with arm exercises, massage, long practice in the gymnasium and wrestling in dust. Plenty of walking after exercises, after dinner, but especially in the early morning. The body should be anointed when covered with dust. When the patient wishes to bathe, let the water be hot. During this time no luncheon should be taken. If recovery occur in a month, let the patient take hereafter the fitting treatment; but if the illness has not completely disappeared, let the patient continue the treatment.[1]

LXXVII. In some cases the morrow brings heartburn. When this is so, a secretion arises in the night from surfeit. Accordingly, when the body has moved after sleep, breathing more rapidly it forces

[1] The argument appears to be faulty. Why should "fitting treatment" follow complete recovery? Ermerins, seeing the difficulty, would revive an old reading (or conjecture) and add *πρότερον* before *θεραπείῃ*. This does not touch the difficulty of the clause *θεραπευέσθω . . . προσήκουσιν*. It is just possible that this clause is merely a misplaced variant of *χρῆσθω τῇ* <*πρότερον*> *θεραπείῃ*, and in the original text there was an *aposiopesis* after *καθιστῆται*. "If the patient recover in a month, well and good; if not, continue the treatment."

πνεύματι χρησάμενον, βιάζεται ἔξω σὺν τῷ πνεύματι θερμόν τε καὶ ὀξύ· ἐκ τούτου νοῦσοι γίνονται, ἢν μή τις προμηθείῃ χρήσηται. συμφέρει δὲ καὶ τούτοισιν ὥσπερ καὶ τῷ προτέρῳ[1] θεραπευθῆναι· τοῖσι δὲ πόνοισι πλείοσι τοῦτον
10 χρῆσθαι.

LXXVIII. Γίνεται δέ τισι καὶ τοιάδε· ἐν τοῖσι πυκνοσάρκοισι τῶν σωμάτων, ὅταν τὰ σιτία θερμαίνηται καὶ διαχέηται ἀπὸ πρώτου ὕπνου, θερμαινομένης τῆς σαρκὸς ὑπό τε τῶν σιτίων διά τε[2] τὸν ὕπνον, ἀπόκρισις γίνεται ἀπὸ τῆς σαρκὸς πολλὴ ὑγρῆς ἐούσης·[3] εἶτα τὴν μὲν τροφὴν ἡ σὰρξ οὐ δέχεται πυκνὴ ἐοῦσα, τὸ δὲ ἀπὸ τῆς σαρκὸς ἀποκριθὲν ἐναντιούμενον τῇ τροφῇ καὶ βιαζόμενον ἔξω[4] πνίγει τὸν ἄνθρωπον καὶ θερ-
10 μαίνει, μέχρι ἐξεμέσῃ· ἔπειτα δὲ κουφότερος ἐγένετο· πόνος δὲ οὐδεὶς ἐν τῷ σώματι φανερός· ἀχροίη δὲ ἔνεστι· προϊόντος δὲ τοῦ χρόνου πόνοι τε γίνονται καὶ νοῦσοι. πάσχουσι δὲ τούτοισι παραπλήσια καὶ ὁκόσοι ἀγύμναστοι ἐόντες, ἐξαπίνης πονήσαντες, σύντηξιν τῆς σαρκὸς πολλὴν ἐποίησαν.[5] χρὴ δὲ τοὺς τοιούτους ὧδε θεραπεύειν· ἀφελεῖν τῶν σίτων τὸ τρίτον μέρος· τοῖσι δὲ σίτοισι χρῆσθαι τοῖσι δριμέσι καὶ ξηροῖσι καὶ αὐστηροῖσι καὶ εὐώδεσι καὶ οὐρητικοῖσι, τοῖσι δὲ
20 δρόμοισι τοῖσι μὲν πλείστοισι καμπτοῖσιν ἐν ἱματίῳ, γυμνὸς δὲ[6] καὶ τοῖσι διαύλοισι καὶ τοῖσι τρόχοισι, τρίψεσι δὲ καὶ πάλῃ ὀλίγῃ,[7] ἀκρο-

[1] τὸν πρότερον θM : τῷ προτέρῳ Littré. It is possible, but rather awkward, to understand a verb to govern τὸν πρότερον.

[2] τε is omitted by M.

[3] πολλῆς ὑγρῆς ἐούσης θ : πολλὴ ὑγρασίη M.

out with the breath hot and acid matter.[1] From this come diseases, unless precautions be taken. In such cases it is beneficial to take the same treatment as that last described, but the patient must increase the amount of exercise.

LXXVIII. The following symptoms also occur. In persons of firm flesh, when the food warms and melts during first sleep, the flesh warming owing to the food and through the sleep, a copious secretion comes from the moist flesh. Then the flesh owing to its firmness will not receive the nourishment, while the secretion from the flesh, being opposed to the nourishment and forced out, warms and chokes the man until he has vomited it forth. Relief follows the vomiting, and no pain is felt in the body though the complexion is pale. In course of time, however, pain and disease occur. Similar symptoms are experienced by those who, when out of training, suddenly take violent exercise, causing a copious melting of their flesh. Such persons must be treated thus. Reduce their food by one-third. The food to be used should be acrid, dry, astringent, aromatic and diuretic. Running should be mostly on double tracks, with the cloak worn, while the double stade and circular course should be run stripped; use massage, a little wrestling, and wrestling with the

[1] Perhaps a *τι* has dropped out of the text here owing to the influence of *πνεύματι* or *τε*.

[4] *ἔξω* M: *εἴσω θ*.

[5] *σύντηξι τῆs σαρκὸs πυλλὴν ἐποιήσαντο θ*: *σύντηξιν τῆs σαρκὸs καὶ πολλὴν ἐποίησαν* M: *σύντηξιν τῆs σαρκὸs βιαίαν καὶ πολλὴν ἐποίησαν* Littré.

[6] *ἐν ἱματίοισι γυμνὸs δέ· καὶ θ*: *ἐν ἱματίωι γυμνοῖσι δὲ καὶ* M.

[7] For *ὀλίγῃ θ* has *απαληι* (*sic*).

χειρισμοῖσιν· (ἀκροχείρισις[1] καὶ κωρυκομαχίη συμφορώτερον) τοῖσι δὲ περιπάτοισιν ἀπὸ τῶν γυμνασίων πολλοῖσι καὶ τοῖσιν ὀρθρίοισι καὶ ἀπὸ δείπνου· φωνῆς δὲ πόνος ἐπιτήδειον· κένωσιν γὰρ τοῦ ὑγροῦ ποιεύμενος ἀραιοῖ τὴν σάρκα· συμφέρει δὲ ἀνάριστον διάγειν· χρῆσθαι δὲ τοῖσι τοιούτοισιν[2] ἐν ἡμέρῃσι δέκα· εἶτα προσθέσθαι
30 τὸ ἥμισυ τοῦ σίτου[3] τοῦ ἀφαιρεθέντος ἐς ἡμέρας ἕξ, καὶ ἔμετον ποιήσασθαι, ἐκ δὲ τοῦ[4] ἐμέτου προσάγειν ἐς ἡμέρας τέσσαρας τὸ σῖτον· ὅταν δὲ ἡμέραι δέκα γένωνται ἀπὸ τοῦ ἐμέτου, κομισάσθω[5] τὸ ἀφαιρεθὲν σιτίον ἅπαν· τοῖσι δὲ πόνοισι καὶ τοῖσι περιπάτοισι προσεχέτω καὶ ὑγιὴς ἔσται. ἡ δὲ τοιαύτη φύσις πόνου πλείονος
37 δεῖται ἢ σίτου.

LXXIX. Πάσχουσι δέ τινες καὶ τοιάδε· διαχωρεῖ αὐτοῖσι τὸ σιτίον ὑγρὸν ἄπεπτον οὐ διὰ νόσημα, οἷον λειεντερίην, οὐδὲ[6] πόνον οὐδένα παρέχει· πάσχουσι δὲ τοῦτο μάλιστα αἱ κοιλίαι ὅσαι ψυχραὶ καὶ ὑγραί εἰσιν· διὰ μὲν οὖν ψυχρότητα οὐ συνεψεῖ, διὰ δὲ ὑγρότητα διαχωρεῖ· τὸ οὖν σῶμα τρύχεται τροφὴν οὐ λαμβάνον τὴν προσήκουσαν, αἵ τε κοιλίαι διαφθείρονται, ἐς νούσους τε ἐμπίπτουσιν. ἀλλὰ χρὴ προμη-
10 θεῖσθαι· συμφέρει δὲ τούτῳ τῶν μὲν σίτων

[1] ἀκροχείρισις is omitted by θ. Ermerins omits ἀκροχειρισμοῖσιν.
[2] τοιούτοισιν θ : σιτίοισι M
[3] τοῦ σίτου omitted by M.
[4] ἐκ δὲ τοῦ θ : ἑκάστου M.
[5] After κομισάσθω θ adds ἀπὸ τοῦ σιτοῦ, omitting σιτίον.
[6] οὐ διανόσημα· οἷον λιεντερίην οὐδὲ θ : οἷον λιεντερίην οὐδὲ M : οἷον ἐκ λειεντερίης Littré, Ermerins.

hands (hand-wrestling and the punch-ball are more than usually valuable),[1] with long walks after exercises, in the early morning and after dinner. Voice exercises are useful, for by evacuating the moisture they rarefy the flesh. It is beneficial to abstain from luncheon. Follow this treatment for ten days; then add half the food taken away, continue thus for six days and administer an emetic. After the emetic increase the food gradually for four days. When ten days have elapsed since taking the emetic, restore food to the full original amount, keeping, however, the exercises and the walks, and the patient will recover. A constitution of such a nature needs more exercise than food.

LXXIX. The following symptoms are experienced by some patients. Their food passes watery and undigested; there is no illness like lientery to cause the trouble,[2] and no pain is felt. It is especially bowels that are cold and moist that show these symptoms. The coldness prevents digestion, and the moistness makes the bowels loose. So the body wastes away through not receiving its proper nourishment, while the bowels become diseased and illnesses occur. Precautions ought to be taken. It is beneficial in this case to reduce food by one-

[1] This sentence may be a marginal note that has crept into the text. Ermerins' emendation is probably correct. "Hand-wrestling and punch-ball are better than *πάλη*."

[2] The reading *οἷον ἐκ λιεντερίης* was probably due to a corrector who scented an inconsistency between *οὐ διὰ νόσημα* and *ἐς νούσους τε ἐμπίπτουσιν* later on. The true meaning of the passage is that, while the state of the bowels is not caused by one of the diseases that commonly do cause it, yet illnesses follow this disordered condition unless precautions be taken.

ἀφελεῖν τὸ τρίτον μέρος· ἔστω δὲ τὰ σῖτα ἄρτοι συγκομιστοὶ ἄζυμοι, κλιβανῖται, ἢ ἐγκρυφίαι, θερμοὶ ἐς οἶνον αὐστηρὸν[1] ἐμβαπτόμενοι, καὶ τῶν ἰχθύων τὰ νωτιαῖα καὶ οὐραῖα, τὰ δὲ κεφάλαια καὶ ὑπογάστρια ἐᾶν[2] ὡς ὑγρότερα· καὶ τοὺς μὲν ἑφθοὺς ἐν ἅλμῃ, τοὺς δὲ ὀπτοὺς ἐν ὄξει· καὶ τοῖσι κρέασι τεταριχευμένοισιν ἐν ἁλσὶ[3] καὶ ὄξει· καὶ τοῖσι κυνείοισιν ὀπτοῖσι·[4] καὶ φάσσης καὶ τῶν λοιπῶν τοιούτων ὀρνίθων, ἑφθοῖσι καὶ ὀπτοῖσι.
20 λαχάνοισι δὲ ὡς ἥκιστα· οἴνῳ δὲ μέλανι ἀκρητεστέρῳ αὐστηρῷ·[5] καὶ τοῖσι περιπάτοισιν ἀπό τε τοῦ δείπνου πολλοῖσι καὶ τοῖσιν ὀρθρίοισι, καὶ ἐκ τοῦ περιπάτου κοιμάσθω·[6] δρόμοισι δὲ καμπτοῖσιν ἐκ προσαγωγῆς· ἔστω δὲ καὶ τρῖψις πολλή· καὶ πάλη βραχείη καὶ ἐν τῷ ἐλαίῳ καὶ ἐν τῇ κόνει, ὅκως[7] διαθερμαινομένη ἡ σὰρξ ἀποξηραίνηταί τε καὶ τὸ ὑγρὸν ἐκ τῆς κοιλίης ἀντισπᾷ· ἀλείφεσθαι δὲ συμφέρει μᾶλλον ἢ λούεσθαι· ἀνάριστος δὲ διαγέτω· ὅταν δὲ γένωνται
30 ἡμέραι ἑπτά, προσθέσθω τὸ ἥμισυ τοῦ σίτου τοῦ ἀφαιρεθέντος, καὶ ἔμετον ποιησάσθω,[8] καὶ προσαγέτω ἐς τέσσαρας ἡμέρας τὸ σιτίον· τῇ δὲ ἄλλῃ ἑβδόμῃ κομισάσθω ἅπαν· καὶ ἔμετον πάλιν
34 ποιησάμενος προσαγέτω κατὰ τωὐτό.

LXXX. Ἄλλοισι δέ τισι γίνεται τοιάδε· τὸ διαχώρημα ἄσηπτον διαχωρεῖ, καὶ τὸ σῶμα τρύχεται τῶν σίτων οὐκ ἐπαυρισκόμενον·[9] οὗτοι

[1] For αὐστηρὸν θ has θερμὸν.
[2] ἐᾶν ὡς omitted by θ.
[3] For ἁλσὶ θ has αλει.
[4] For ὀπτοῖσι θ has ἑφθοῖσι.
[5] M omits δὲ and ἀκρητεστέρῳ αὐστηρῷ.

third. The food should consist of unleavened bread, made from unbolted meal, baked in a pot or under ashes, dipped warm into a dry wine. Of fish the parts about the back and tail; those about the head and belly are too moist and should not be taken. Fish may be boiled in brine or grilled with vinegar. Meat may be preserved in either salt or vinegar. Dog's flesh roasted; the flesh of pigeons, and of other such-like birds, boiled or roasted. Vegetables to be reduced to a minimum. Wine should be dark, dry and but little diluted. Long walks should be taken after dinner and in the early morning, with sleep after the walk. The double track should be gradually increased. Let there be plenty of massage. There should be a little wrestling, both in oil and in dust, so that the flesh may become hot and dry, and draw by revulsion the moisture from the belly. Anointing is more beneficial than bathing. The patient should not take luncheon. After seven days have passed, restore one-half of the food that has been taken away; then an emetic should be drunk, and the food increased gradually for four days. A week later restore the diet to what it was originally, administer an emetic again, and follow it by a similar gradual increase.

LXXX. In some other cases appear the following symptoms. The stools that pass are undigested, and the body wastes away, getting no profit from

---

[6] *κοιμάσθωι* θ: *κοιμᾶσθαι* M.

[7] *ὅκως* θ: *ὅπως* M. Usually M has the -*κ*- forms of the relatives and θ the others.

[8] After *ποιησάσθω* θ adds *τὸ ἥμισυ τοῦ σιτου̂*·

[9] For *οὐκ ἐπαυρισκόμενον* θ has *ἐπαυρίσκεται*.

δὲ προϊόντος τοῦ χρόνου ἐμπίπτουσιν ἐς[1] νούσους· τούτοισιν αἱ κοιλίαι ψυχραὶ καὶ ξηραί· ὁκόταν οὖν μήτε σίτοισι προσήκουσι χρέωνται μήτε γυμνασίοισι, πάσχουσι ταῦτα. συμφέρει δὴ τούτῳ[2] ἄρτοισί καθαροῖσιν ἰπνίτῃσι χρῆσθαι, καὶ τοῖσιν ἰχθύσιν ἑφθοῖσιν ἐν ὑποτρίμμασι, καὶ
10 κρέασιν ἑφθοῖσιν ὑείοισι, καὶ τοῖσιν ἀκροκωλίοισι διέφθοισι, καὶ τοῖσι πίοσιν ὀπτοῖσι,[3] καὶ τῶν δριμέων καὶ τῶν ἁλυκῶν τοῖσιν ὑγραίνουσι, καὶ τοῖσιν ἁλμυροῖσιν· οἴνοισι δὲ μέλασι μαλακοῖσι· καὶ τῶν βοτρύων καὶ τῶν σύκων ἐν τοῖσι σίτοισι·[4] χρὴ δὲ καὶ ἐναριστῆν μικρόν· τοῖσι δὲ γυμνασίοισι πλείοσι χρῆσθαι, δρόμοισι καμπτοῖσιν ἐκ προσαγωγῆς, ὑστάτοισι τε τρόχοισι, πάλῃ δὲ μετὰ τὸν δρόμον ἐν ἐλαίῳ·[5] περιπάτοισι δὲ μὴ πολλοῖσιν ἀπὸ τῶν γυμνασίων· ἀπὸ δείπνου δὲ ὅσον
20 ἐξαναστῆναι· ὄρθρου δὲ πλείοσι περιπάτοισι χρῆσθαι· λουέσθω δὲ θερμῷ· χρήσθω δὲ καὶ χρίσμασιν· ὕπνον δὲ πλείονα διδότω καὶ μαλακευνείτω· χρὴ δὲ καὶ ἀφροδισιάσαι τι· τῶν δὲ σίτων ἀφελεῖν τὸ τρίτον μέρος· ἐν ἡμέρῃσι δὲ
25 δέκα δύο[6] προσάγειν αὐτὸν πρὸς τὰ σιτία.

LXXXI. Εἰσὶ δέ τινες οἷσι τὸ διαχώρημα ὑγρὸν καὶ σεσηπὸς διαχωρεῖ, τοῖσιν ἄλλως ὑγιαίνουσι καὶ γυμναζομένοισι, καὶ πόνον οὐ παρέχει· οἱ δέ τινες ἀποκλείονται τῶν προση-

[1] Before νούσους M has τὰς.

[2] δὴ τούτωι θ: δὲ τῶι τοιούτωι M.

[3] καὶ τοῖσι μὲν ἄκροις διεφθοῖς· τοῖσι δὲ πίοσιν ἑφθοῖσι· θ: καὶ τοῖσιν ἀκροκωλίοισι διέφθοισι καὶ τοῖσι πλείστοισι ὀπτοῖσι M.

[4] ἐν τοῖσι σιτοῖσι· θ: τοῖσι δὲ σιτίοισι· M: ἐμφορεῖσθαι ἔν γε τοῖσι σιτίοισι· Littré.

[5] M has πάλη τε and τῶι before ἐλαίῳ.

[6] M has τὸ τέταρτον μέρος ἐν ἡμέρῃσι δέκα· καί.

the food. In course of time such people fall ill. In these cases the bowels are cold and dry. So when they take neither suitable food nor suitable exercises, their symptoms are those I have said. This kind of person is benefited by taking bread of bolted meal, oven-baked, boiled fish in sauce, boiled pork, extremities thoroughly boiled, fat meats roasted, of acrid, salt foods such as are moistening, and also piquant sauces.[1] Wines to be dark and soft. Some grapes and some figs to be taken with food. A little luncheon too should be eaten. Exercises should be above the average, double-track running should be gradually increased, while the last running should be on the circular track; after the running should come wrestling with the body oiled. After the exercises there should be short walks, after dinner mere strolls, but in the early morning longer walks. Let the bath be warm. Unguents should be used. Let sleep be plentiful and on a soft bed. Some sexual intercourse is necessary. Reduce food by one-third. Take twelve days to bring food back to normal.

LXXXI. In some cases the stools are watery and of waste matter;[2] the general health is good, exercise is taken and no pain is felt. Others, however,

[1] The word ἁλμυρὸς is difficult, as it is hardly to be distinguished from ἁλυκός. I suppose that it refers here to pungent dishes generally.

[2] The process whereby the digestive organs make waste matter was called σῆψις, the process of digestion πέψις. Hence both ἄπεπτος and ἄσηπτος mean "undigested," while σεσηπός means that there is plenty of waste matter, without undigested food in it; apparently the food is turned to waste without normal assimilation. Ermerins translates both ἄπεπτος and ἄσηπτος by "incoctus," σεσηπός by "concoctus." Littré has "non digéré," "non corrompu," "corrompu."

κόντων· προϊόντος δὲ τοῦ χρόνου, καὶ τὰς σάρκας ἐπισπᾶται τῇ θερμασίῃ ἡ κοιλίη, πόνον τε παρέχει, τῶν τε σίτων ἀποκλείονται, ἥ τε κοιλίη ἐξελκοῦται, στῆσαι δὲ χαλεπὸν ἤδη γίνεται αὐτήν. ἀλλὰ χρὴ πρότερον προμηθεῖσθαι γνόντα
10 τὴν κοιλίην θερμὴν καὶ ὑγρὴν παρὰ τὸ προσῆκον, πόνων τε ὑπερβολὴν ἀσυμφόρων ἐγγενομένων. τῇ οὖν διαίτῃ δεῖ ψῦξαι καὶ ξηρῆναι. πρῶτον μὲν χρὴ τὰ γυμνάσια τὰ ἡμίσεα ἀφελεῖν, τῶν τε σίτων τὸ τρίτον μέρος· χρῆσθαι δὲ μάζῃσι[1] προφυρητῇσι τριπτῇσι, καὶ τοῖσιν ἰχθύσι τοῖσι ξηροτάτοισιν ἑφθοῖσι, μήτε λιπαροῖσι μήτε ἁλμυροῖσι· χρήσθω δὲ καὶ ὀπτοῖσι· κρέασι δὲ τοῖσιν ὀρνιθίοισιν, ἑφθοῖσι μὲν φάσσης, περιστερῆς, περδίκων δὲ καὶ ἀλεκτορίδων ὀπτοῖσιν ἡδύντοισι,[2]
20 λαγῴοισι ἑφθοῖσιν ἐν ὕδατι, καὶ τοῖσιν ἀγρίοισι λαχάνοισιν ὅσα ψυκτικά, τοῖσι τεύτλοισι καθέφθοισιν ὀξηροῖσι·[3] οἴνῳ δὲ μέλανι αὐστηρῷ· γυμνασίοισι τε τρόχοισιν ὀξέσι· τρῖψις μὴ πολλὴ προσέστω,[4] ἀλλ' ὀλίγη, μηδὲ πάλη· ἀκροχειρισμὸς δὲ[5] καὶ χειρονομίη καὶ κωρυκομαχίη καὶ ἀλίνδησις ἐπιτηδείη[6] μὴ πολλή· τοῖσι δὲ περιπάτοισι καὶ ἀπὸ τοῦ γυμνασίου χρήσθω πρὸς τὸν πόνον ἱκανοῖσι, καὶ ἀπὸ δείπνου πρὸς τὰ σιτία πλείστοισι, καὶ πρωῒ πρὸς τὴν ἕξιν
30 συμμέτρως· λούσθω δὲ χλιερῷ ἀτρέμας· οὕτω δὲ διαιτηθεὶς ἡμέρας δέκα προσθέσθω τοῦ τε

[1] M has the singular, μάζῃ κ.τ.ἑ.
[2] ἀνηδύντοισι M : ηδυντοισι θ.
[3] Ermerins after ὀξηροῖσι adds καὶ τοῖσι ἀγρίοισι ἅπασι. This may be correct.
[4] προσαγέσθω M.

cannot attend to their duties. In course of time the belly by its heat draws the flesh to itself; pain is felt; there is loss of appetite; ulcers form in the belly, and hereafter the diarrhœa is difficult to arrest. Precautions should be taken early, with the knowledge that the belly is over-hot and over-watery, and that there has been excess of unsuitable exercises. Regimen, accordingly, must be such as to cool and dry the belly. First, exercise should be reduced by one-half, food by one-third. Barley cake should be eaten, the grain ground and well-kneaded. Fish of the driest kinds, that are neither rich nor salt, may be eaten boiled. They may also be grilled. As to the flesh of birds, doves and pigeons should be boiled, partridges and chickens roasted, with seasoning. Eat hares boiled in water, and such wild vegetables as are cooling; beet thoroughly boiled and with vinegar.[1] Wine should be dark and dry. Exercises to be sharp runs on the round track. Massage, but only a little, not much. No wrestling proper; but hand-wrestling, arm exercises, punch-ball and wrestling in the dust are suitable when not in excess. Walks are to be taken after exercise that are adequate considering the fatigue; after dinner they should be as long as possible considering the food; in the morning they should be proportioned to the habit of body. The bath should be tepid and taken quietly. After ten days of this regimen restore half of the food and one-

[1] The text here is very uncertain, and I have done my best to make sense of the reading of θ. It is tempting to adopt the reading of Ermerins: "vegetables that are cooling, such as beet . . ., and all wild vegetables."

---

[5] ἀκροχειρίησι M. [6] ἐπιτήδεια M : ἐπιτηδείηι θ.

σίτου τὸ ἥμισυ καὶ τῶν πόνων τὸ τρίτον μέρος· καὶ ἔμετον ποιησάσθω ἀπὸ τῶν ξηρῶν καὶ στρυφνῶν, καὶ μὴ διατριβέτω[1] ἐν τῷ σιτίῳ, ἀλλὰ ἐμείτω τὴν ταχίστην· ἐκ δὲ τοῦ ἐμέτου προσαγέτω ἐς ἡμέρας τέσσαρας[2] τὸν σῖτον καὶ τὸ ποτὸν[3] καὶ τὸν πόνον μερίζων· ὅταν δὲ δεκὰς γένηται, προσθέσθω τὸν σῖτον τὸν λοιπὸν † καὶ τῶν οἴνων τὸ πότιμον, πλὴν τῶν πόνων ἐνδεέστε-
10 ρον· †[4] καὶ ἔμετον ποιησάμενος προσαγέτω, καθάπερ γέγραπται· μονοσιτεῖν δὲ τοῦτον τὸν
12 χρόνον συμφέρει μέχρι ἂν καταστῇ.

LXXXII. Ἄλλοισι δέ τισι ξηρὸν καὶ συγκεκαυμένον τὸ διαχώρημα γίνεται, καὶ τὸ στόμα ξηρόν, προϊόντος δὲ τοῦ χρόνου καὶ πικρὸν γίνεται, καὶ ἡ κοιλίη ἵσταται καὶ οὔρησις· ὁκόταν γὰρ μὴ ἔχῃ τὸ ἔντερον ὑγρασίην, περὶ τὸν ἀπόπατον περιοιδῆσαν ἀποφράσσει τὰς διεξόδους, καὶ ὀδύνην τε παρέχει, καὶ θέρμη λαμβάνει, καὶ ὅ τι ἂν φάγῃ ἢ πίῃ ἐξεμεῖ· τελευτῶν δὲ καὶ κόπρον ἐμεῖ·[5] οὗτος οὐ βιώσιμος,
10 ὁκόταν ἐς τοῦτο ἔλθῃ. ἀλλὰ χρὴ πρότερον προμηθεῖσθαι γινώσκοντα ὅτι ξηρασίῃ θερμῇ κρατεῖται ὤνθρωπος. διαιτῆσθαι οὖν χρὴ αὐτὸν τῇ τε μάζῃ προφυρητῇ ῥαντῇ καὶ ἄρτῳ σιτανίων πυρῶν τῷ τε χυμῷ[6] τῶν πιτύρων ἐζυμωμένῳ, λαχάνοισί τε χρῆσθαι πλὴν τῶν δριμέων καὶ

[1] διατριβέτω M : διατριβε θ.
[2] τέσσαρας is omitted by θ.
[3] καὶ τὸ ποτὸν is omitted by M.
[4] καὶ τῶν σιτῶν πρὸς τὸν πόνον ἐνδεεστέρως· θ : καὶ τῶν σίτων τὸ πότιμον τὸν πόνον ἐνδεέστερον· M : καὶ τῶν οἴνων τὸ πότιμον, πλὴν τῶν πόνων ἐνδεέστερον· Littré : καὶ τὸν οἶνον, πλὴν τῶν πόνων ἐνδεέστερον· Ermerins. I have printed Littré's text

third of the exercise. An emetic should be taken after a meal of dry and astringent food, which must not remain long in the stomach; in fact the emetic should follow with all speed. After the emetic for four days increase gradually by ⟨equal⟩[1] portions food, drink and exercise. When ten days are passed, add the rest of food and drink, but not quite all the exercises.[2] After an emetic proceed progressively, as has been described. It is beneficial to take during this period one meal only a day until health is restored.

LXXXII. In some other cases the stools pass dry and burnt up, and the mouth becomes dry, in course of time becoming bitter also, while bowels and kidneys cease to act. For when the intestines have no moisture, they swell around the fæces and block up the passages, causing pain, while fever comes on and everything eaten or drunk is vomited. Finally, dung too is brought up. When this point is reached life may be despaired of. Precautions should be taken betimes, with the knowledge that the patient is overpowered by a dry heat. So his diet should consist of barley cake, well-kneaded and sprinkled, with buck-wheat bread fermented with the gruel of its bran. Vegetables should be taken except those that are acrid and dry, and they should be

[1] The word *μερίζων*, "dividing them," may merely emphasize the notion of progressive increase implied in *προσαγέτω*.

[2] The Greek admits the rendering, "but not quite enough to match the exercise." But the sense of the passage suffers.

---

between daggers and given a translation that represents the general sense.

[5] *τελευτῶν . . . ἐμεῖ* omitted by *θ*.

[6] *χυμῶι θ*: *χυλῶι* M.

ξηρῶν καὶ ἑψανοῖσι· καὶ τῶν ἰχθύων τοῖσι κουφοτάτοισιν ἑφθοῖσι· καὶ τοῖσι κεφαλαίοισι τῶν ἰχθύων καὶ καράβων· μυσὶ καὶ ἐχίνοισι καὶ τοῖσι καρκίνοισι, καὶ τῶν κογχυλίων τοῖσι
20 χυμοῖσι καὶ αὐτοῖσι τοιούτοισιν ὑγροτάτοισι· κρέασι δὲ τοῖσιν ὑείοισιν ἀκροκωλίοισιν ἐμπροσθίοισιν[1] ἑφθοῖσι καὶ ἐρίφων καὶ ἀρνῶν καὶ σκυλάκων ἑφθοῖσιν· ἰχθύων δὲ τοῖσι ποταμίοισι καὶ λιμναίοισιν ἑφθοῖσιν· οἴνῳ μαλακῷ, ὑδαρεῖ· τοῖσι δὲ πόνοισι μὴ πολλοῖσι μηδὲ ταχέσιν, ἀλλ' ἡσύχοισιν ἅπασι· τοῖσι δὲ περιπάτοισι πρωῒ μὲν χρήσθω, πρὸς τὴν ἕξιν ἱκανοῖσι καὶ ἀπὸ γυμνασίου πρὸς τὸν πόνον συμμέτροισιν· ἀπὸ δείπνου δὲ μὴ περιπατείτω· λουτροῖσι δὲ
30 χρήσθω καὶ ὕπνοισι μαλακοῖσι καὶ ἀρίστῳ· ὕπνῳ τε μετὰ τὸ ἄριστον μὴ μακρῷ· ὀπώρῃ τε τῇ ὑγραινούσῃ μετὰ τῶν σιτίων[2] χρήσθω· καὶ τοῖσιν ἐρεβίνθοισι τοῖσι χλωροῖσι, καὶ ξηροὺς δὲ βρέξας ἐν ὕδατι· ἀφελέσθω δὲ τῶν πόνων καὶ οὗτος ἐξ ἀρχῆς τοὺς ἡμίσεας τῶν πρόσθεν. καὶ ποιησάσθω[3] ἔμετον ἀπὸ γλυκέων καὶ λιπαρῶν καὶ ἁλμυρῶν καὶ πιόνων,[4] ἐνδιατριβέτω δὲ ὡς πλεῖστον χρόνον ἐν τοῖσι σιτίοισι πρὸς τοὺς ἐμέτους· εἶτα προσαγέτω τὸ σιτίον ἐς ἡμέρας
40 τρεῖς, μηδ' ἀνάριστος ἔστω· ὅταν δὲ ἡμέραι δέκα γένωνται, τῶν πόνων προσαγέσθω[5] πλείονας· ἢν μὲν οὖν ἡ πλησμονὴ ἐνῇ ἀπὸ τοῦ σίτου ἢ τῆς κοιλίης πλημμέλεια, ἐμεσάτω· ἢν δὲ μή, οὕτω
44 θεραπευέσθω τὸν ἐπίλοιπον χρόνον.

[1] ἐμπροσθίοις M : ἐμπροσθιδίοισιν θ.
[2] σιτίων M : λοιπῶν θ (perhaps rightly).
[3] For ποιησάσθω θ reads ποιησάτω.

boiled. Fish must be of the lightest and boiled. He may eat the heads of fish and of lobsters. Mussels, sea-urchins, crabs, soups from cockles, and cockles themselves of the most watery kind. Among meats, pigs' fore-feet boiled, and flesh of kids, lambs and puppies, also boiled. Fish from rivers and lakes, boiled. Soft wine, well-diluted. Exercises neither long nor sharp, but gentle in all cases. Walks are to be taken in the morning, long enough for the habit of body, and, after exercise, proportioned to the fatigue; after dinner no walk must be taken. Baths should be taken, gentle sleep, and luncheon, but the sleep after luncheon should not be long. Moistening fruit should be eaten with food. Chick-peas should be taken when fresh; if dried let them be first soaked in water. This patient too must reduce, from the very first, his former exercise by one-half. Let him also take an emetic after a meal of sweet, rich, salt, fatty[1] foods; let this meal lie in the stomach as long as possible consistently with vomiting it up. Then let the patient increase the food for three days, not forgetting to take luncheon. After ten days let him resume gradually the greater part of the exercises. If now after food there be experienced surfeit, or a disorder of the belly, let an emetic be taken. Otherwise, the same treatment should be continued for the rest of the time.

[1] The reading of M, πλειόνων, "more than usually copious," may be right. It is hard to distinguish πιόνων from λιπαρῶν. Perhaps the former is "fatty," the latter "sickly."

---

[4] For πιόνων M reads πλειόνων.
[5] For προσαγέσθω M reads προσαγέτω.

LXXXIII. Γίνεται δὲ καὶ τοιάδε· φρῖκαι ἀπὸ τῶν περιπάτων ἐγγίνονται τῶν ὀρθρίων,[1] καὶ τὴν κεφαλὴν βαρύνονται τοσούτῳ ὁκόσῳ[2] πλείονες οἱ περίπατοι τῆς συμμετρίης· κενεόμενον δὲ τὸ σῶμα καὶ ἡ κεφαλὴ τοῦ ὑγροῦ φρίσσει τε καὶ βαρύνεται· προϊόντος δὲ τοῦ χρόνου ἐς πυρετὸν ἀφικνεῖται φρικώδεα. ἀλλ' οὐ χρὴ προΐεσθαι ἐς τοῦτο, ἀλλ' ἐκθεραπεύεσθαι πρότερον ὧδε· ὅταν γένηται τάχιστα τῶν τεκ-
10 μηρίων τι, χρισάμενον καὶ ἀνατριψάμενον ὀλίγα, ἄριστον ποιήσασθαι πλέον τοῦ εἰθισμένου, καὶ πιεῖν ἱκανὸν οἶνον μαλακόν, εἶτα ὕπνῳ χρῆσθαι[3] ἀπὸ τοῦ ἀρίστου ἱκανῷ·[4] ἐς τὴν ἑσπέρην δὲ κούφοισι χρησάμενον γυμνασίοισι θερμῷ[5] λουσάμενον δειπνῆσαι τὸ εἰθισμένον· περιπάτῳ δὲ μὴ χρῆσθαι ἀπὸ δείπνου, διατρίβειν δὲ χρόνον· τῇ δὲ ὑστεραίῃ ἀφελέσθω τῶν γυμνασίων πάντων καὶ τῶν περιπάτων τὸ τρίτον μέρος, τοῖσι δὲ σίτοισι χρησάσθω ὥσπερ εἴθιστο· λουέσθω δὲ
20 χλιαρῷ, καὶ τῷ ἐλαίῳ ἀλειφέσθω[6] ἐν τῷ ὕδατι· ὕπνοισί τε μαλακοῖσι διαγέτω, ἐν ἡμέρῃσι δὲ
22 πέντε τοὺς πόνους προσαγέτω[7] κατὰ μικρόν.

LXXXIV. Εἰσὶ δέ τινες οἳ φρίσσουσιν ἐκ τῶν γυμνασίων, καὶ ἐπειδὰν ἐκδύσωνται[8] μέχρι διαπονήσωσιν· ὅταν δὲ ψύχηται, πάλιν φρίσσει·

[1] For ὀρθρίων M reads ὄρθρου.
[2] So θ. M has βαρύνεται· τούτωι πλείονες.
[3] For χρῆσθαι M has χρήσασθαι.
[4] For ἱκανῷ M has ἱκανῶς.
[5] For θερμῷ M has θερμῶς.

LXXXIII. The following symptoms also occur. Rigors come on after the early-morning walk, with heaviness of the head proportionate to the excess of the walking over the proper amount. The reason for the rigors and the heaviness is because the body and the head are emptied of their moisture. In course of time the patient falls into a fever attended by rigors. Instead of letting things slide thus far, the following treatment should be carried out before. On the first appearance of the symptoms[1] let the patient have a little unction and a little massage, take a heartier luncheon than usual, with plenty of soft wine to drink, and then a long sleep after the luncheon. In the evening light exercises should be taken, a hot bath and the usual dinner. No walk after dinner; the patient should just pass away the time. On the next day reduce all the exercises and the walks by one-third, but the usual food should be eaten. Let the patient take his bath tepid, and in the water anoint himself with oil. He must take his sleep on a soft bed, and spend five days in resuming his exercises little by little.

LXXXIV. Some have rigors as a result of[2] their exercises, that is to say, from the time they put off their clothes to the time they finish, and the rigors are renewed on cooling down. The teeth

[1] The *τι* seems to refer, not to one of the symptoms, but to their first appearance in a slight form: "as soon as the symptoms appear at all." *τῶν τεκμηρίων τι*, in fact, means, not "one symptom," but "something of the symptoms."

[2] Or "after," in which case *καὶ* means "and." I take the clauses after *καὶ* to explain *ἐκ τῶν γυμνασίων*.

---

[6] M omits *τῶν γυμνασίων . . . ἀλειφέσθω.*

[7] M omits *πόνους* and reads *προσαγέσθω.*

[8] For *ἐκδύσωνται* M has *ἐκδύηται.*

βρυγμός τε τὸ σῶμα ἔχει· ὑπνώσσει τε, ὅταν δὲ ἐξέγρηται, χασμᾶται πολλάκις· ἐκ δὲ τοῦ ὕπνου τὰ βλέφαρα βαρέα· προϊόντος δὲ τοῦ χρόνου καὶ πυρετοὶ ἐπιγίνονται ἰσχυροί,[1] καὶ φλυαρεῖ. φυλάσσεσθαι οὖν χρὴ καὶ μὴ προΐεσθαι [2] ἐς τοῦτο, ἀλλὰ ἐκδιαιτήσασθαι ὧδε· πρῶτον μὲν τῶν 10 γυμνασίων ἀφελέσθω πάντα ἢ τὰ ἡμίσεα· τοῖσι δὲ σίτοισιν πᾶσι χρήσθω ὑγροτέροισί τε καὶ ψυχροτέροισι, καὶ τοῖσι πόμασι μαλακωτέροισι καὶ ὑδαρεστέροισιν· ὁκόταν δὲ παρέλθωσιν ἡμέραι πέντε, προσθέσθω τῶν πόνων τὸ τρίτον μέρος τῶν ἀφαιρεθέντων· τοῖσι δὲ σίτοισι χρήσθω τοῖσιν αὐτοῖσι· πέμπτῃ δὲ ἄλλῃ ἡμέρῃ τοὺς ἡμίσεας τῶν λοιπῶν πόνων προσθέσθω· αὖθις δὲ πέμπτῃ μετὰ τοῦτο ἀπόδος τοὺς πόνους πάντας κουφοτέρους καὶ ἐλάσσονας, ὡς μὴ πάλιν 20 ὑπερβολὴ γένηται.

LXXXV. Τοῖσι γὰρ πάσχουσι ταῦτα τὰ τεκμήρια οἱ πόνοι κρέσσους εἰσὶ [3] τῶν σιτίων· ἀνισάζειν οὖν χρή. ἔνιοι δὲ οὐ ταῦτα πάντα πάσχουσιν, ἀλλὰ τὰ μέν, τὰ δ' οὔ. πάντων δὲ τούτων τῶν τεκμηρίων οἱ πόνοι κρατέουσι τῶν σίτων, καὶ θεραπείη ἡ αὐτή. συμφέρει δὲ τούτοισὶ θερμολουτεῖν, μαλακευνεῖν, μεθυσθῆναι ἅπαξ ἢ δίς, μὴ ἐς ὑπερβολήν· ἀφροδισιάσαι τε ὅταν ὑποπίῃ·[4] ῥᾳθυμῆσαι πρὸς τοὺς πόνους, 10 πλὴν τῶν [5] περιπάτων.

[1] For ἰσχυροί M has φαῦλοι.
[2] προσίεσθαι θ : πρόεσθαι M (which omits καὶ).
[3] θ omits τοῖσι γὰρ . . . εἰσὶ.
[4] So M. θ reads ὑποπτῆι, omitting τε.
[5] τῶν is omitted by M.

chatter.[1] The patient is sleepy, and after waking up he yawns frequently. After sleep the eyelids are heavy. In course of time high fever too comes on with delirium. So care must be taken not to let things drift so far, and the following change of regimen should be adopted. First drop all exercises or reduce them by one-half. All the food taken should be of the moister and more cooling sort, and the drink of the milder sort, well diluted. When five days are passed, let the patient add one-third of the exercises that have been dropped. The food taken should be the same. After another five days restore one-half of the remaining exercises. After another five resume all the exercises, but let them be less strenuous and less prolonged, in order that excess may not recur.

LXXXV. When patients exhibit these symptoms exercises are in excess of food. Accordingly, a due correspondence must be restored. In some cases not all the symptoms are experienced, but only some of them. But with all these symptoms exercises overpower food, and the treatment is the same. These patients ought to take their baths warm, to sleep on a soft bed, to get drunk once or twice, but not to excess, to have sexual intercourse after a moderate indulgence in wine, and to slack off their exercises, except walking.

[1] Ermerins deletes *τὸ σῶμα*, and the words are strange, although supported by all the MSS. Perhaps we should read *τὸ στόμα*. I am loth, however, to depart from the MSS., as we really know too little about Greek idioms of this type to be quite sure that the phrase *τὸ σῶμα* would be impossible in this context.

ΠΕΡΙ ΔΙΑΙΤΗΣ

ΤΟ ΤΕΤΑΡΤΟΝ

Η

ΠΕΡΙ ΕΝΥΠΝΙΩΝ

LXXXVI. Περὶ δὲ τῶν τεκμηρίων τῶν ἐν τοῖσιν ὕπνοισιν ὅστις ὀρθῶς ἔγνωκε, μεγάλην ἔχοντα δύναμιν εὑρήσει πρὸς ἅπαντα. ἡ γὰρ ψυχὴ ἐγρηγορότι μὲν τῷ σώματι ὑπηρετέουσα, ἐπὶ πολλὰ μεριζομένη, οὐ γίνεται αὐτὴ ἑωυτῆς, ἀλλ' ἀποδίδωσί τι[1] μέρος ἑκάστῳ τοῦ σώματος, ἀκοῇ, ὄψει, ψαύσει, ὁδοιπορίῃ, πρήξεσι παντὸς τοῦ σώματος·[2] αὐτὴ δὲ ἑωυτῆς ἡ διάνοια οὐ γίνεται.[3] ὅταν δὲ τὸ σῶμα ἡσυχάσῃ, ἡ ψυχὴ 10 κινεομένη καὶ ἐγρηγορέουσα[4] διοικεῖ τὸν ἑωυτῆς οἶκον, καὶ τὰς τοῦ σώματος πρήξιας ἁπάσας αὐτὴ διαπρήσσεται. τὸ μὲν γὰρ σῶμα καθεῦδον οὐκ αἰσθάνεται, ἡ δὲ ἐγρηγορέουσα γινώσκει πάντα,[5] καὶ ὁρῇ[6] τε τὰ ὁρατὰ καὶ ἀκούει τὰ ἀκουστά,[7] βαδίζει, ψαύει, λυπεῖται, ἐνθυμεῖται, ἑνὶ λόγῳ,[8] ὁκόσαι[9] τοῦ σώματος ὑπηρεσίαι ἢ τῆς ψυχῆς, πάντα ταῦτα[10] ἡ ψυχὴ ἐν τῷ ὕπνῳ

[1] τι M : τὸ θ.

[2] So θ : M has πρήξει· πάντηι τοῦ σώματος δ.ανοίη.

[3] αὕτηι δὲ ἡ διάνοια. εωυτης οὐ γίνεται θ : αὐτὴ δὲ αὐτῆς ἡ δ.ἁνοια οὐ γίνεται M.

[4] ἐγρηγορέουσα. τὰ πρήγματα θ : ἐπεξέρπουσα τὰ σώματα M : ἐπεξέρπουσα τὰ μέρη τοῦ σώματος Littré.

[5] πάντα θ : M omits.

[6] καὶ ορη θ : καθορῇ M.

REGIMEN IV

OR

DREAMS

LXXXVI. He who has learnt aright about the signs that come in sleep will find that they have an important influence upon all things. For when the body is awake the soul is its servant, and is never her own mistress, but divides her attention among many things, assigning a part of it to each faculty of the body—to hearing, to sight, to touch, to walking, and to acts of the whole body; but the mind never enjoys independence. But when the body is at rest, the soul, being set in motion and awake,[1] administers her own household, and of herself performs all the acts of the body. For the body when asleep has no perception; but the soul when awake has cognizance of all things – sees what is visible, hears what is audible, walks, touches, feels pain, ponders. In a word, all the functions of body and of soul are performed by

[1] The reading of M would mean, "pervading the body." The words τὰ πρήγματα, which θ has after ἐγρηγορέουσα, I take to be a note on τὸν ἑωυτῆς οἶκον which has crept into the text. The unusual form ἐγρηγορέουσα may possibly account for the disturbed state of the manuscript tradition.

---

[7] ἀκούει θ: διακούει M.
[8] ἑνὶ λόγῳ Mack: ἐν ὀλίγωι M: ἐν ὀλίγω ἐοῦσα θ.
[9] ὁκόσαι M: ὁκόσα θ.
[10] πάντα· ταῦτα θ: ταῦτα πάντα M.

διαπρήσσεται. ὅστις οὖν ἐπίσταται κρίνειν
19 ταῦτα ὀρθῶς μέγα μέρος ἐπίσταται σοφίης.[1]

LXXXVII. Ὁκόσα μὲν οὖν τῶν ἐνυπνίων θεῖά ἐστι καὶ προσημαίνει[2] ἢ πολέσι ἢ ἰδιώτῃσι ἢ κακὰ ἢ ἀγαθὰ † μὴ δι' αὐτῶν ἁμαρτίην,† εἰσὶ οἱ κρίνουσι περὶ τῶν τοιούτων τέχνην[3] ἔχοντες· ὁκόσα δὲ ἡ ψυχὴ τοῦ σώματος παθήματα προσημαίνει, πλησμονῆς ἢ κενώσιος ὑπερβολὴν[4] τῶν συμφυτῶν, ἢ μεταβολὴν τῶν ἀηθέων, κρίνουσι μὲν καὶ ταῦτα, καὶ τὰ μὲν τυγχάνουσι, τὰ δὲ ἁμαρτάνουσι, καὶ οὐδέτερα[5] τούτων γινώ-
10 σκουσι δι' ὅ τι[6] γίνεται, οὔθ' ὅ τι[7] ἂν ἐπιτύχωσιν οὔθ' ὅ τι ἂν ἁμάρτωσι, φυλάσσεσθαι δὲ παραινέοντες μή τι κακὸν λάβῃ. οἱ δ' οὖν[8] οὐ διδάσκουσιν ὡς χρὴ φυλάσσεσθαι,[9] ἀλλὰ θεοῖσιν εὔχεσθαι[10] κελεύουσι· καὶ τὸ μὲν εὔχεσθαι ἀγαθόν·[11] δεῖ δὲ καὶ αὐτὸν συλλαμβάνοντα τοὺς θεοὺς[12] ἐπικα-
16 λεῖσθαι.

LXXXVIII. Ἔχει δὲ περὶ τούτων ὧδε·[13] ὁκόσα τῶν ἐνυπνίων τὰς ἡμερινὰς[14] πρήξιας τοῦ ἀνθρώπου ἢ διανοίας[15] ἐς τὴν εὐφρόνην[16] ἀπο-

[1] θ omits σοφίης. [2] θ omits ἢ . . . προσημαίνει.

[3] So M. Some MSS. read ἀκριβῆ τέχνην

[4] θ has ἢ before ὑπερβολὴν, and so Diels would read προσημαίνει, ἢ ὑπερβολὴν τῶν συμφύτων κ.τ.λ., perhaps rightly. Ermerins for κενώσιος has κακώσιος, without authorities or comment.

[5] So M. θ has τυγχάνουσι. τὰ δ' οὐδετερα.

[6] διότι οὖν θ M: οὖν is omitted by the first hand in H.

[7] οὐδότι . . . οὐδότι θ.

[8] οιδων θ: οἱ δ' ὧν Diels. [9] φυλάξασθαι M.

[10] εὔξασθαι M. In θ the -ῖσιν of θεοῖσιν has been erased.

[11] The vulgate has εὔχεσθαι πρέπον καὶ λίην ἐστὶν ἀγαθόν. M has εὔχεσθαι δεῖ καὶ ἀγαθόν.

[12] In θ the -υς of τοὺς and θεοὺς has been erased.

the soul during sleep. Whoever, therefore, knows how to interpret these acts aright knows a great part of wisdom.

LXXXVII. Now such dreams as are divine, and foretell to cities or to private persons things evil or things good,[1] have interpreters in those who possess the art of dealing with such things. But all the physical symptoms foretold by the soul, excess, of surfeit or of depletion, of things natural, or change to unaccustomed things, these also the diviners interpret, sometimes with, sometimes without success. But in neither case do they know the cause, either of their success or of their failure. They recommend precautions to be taken to prevent harm,[2] yet they give no instruction how to take precautions, but only recommend prayers to the gods. Prayer indeed is good, but while calling on the gods a man should himself lend a hand.

LXXXVIII. This is the truth of the matter. Such dreams as repeat in the night a man's actions or thoughts in the day-time, representing them as

[1] The words within daggers I have omitted from my translation. Littré translates "non causés par la faute des parties intéressées." But such a meaning can apply only to κακά, not to ἀγαθά. If the words be kept, αὐτῶν must be emended to αὑτῶν or ἑωυτῶν, otherwise the order of the words is wrong.

[2] The punctuation of this passage is uncertain. I have taken παραινέοντες as a slight anacoluthon for παραινέουσι, but it might be better to put a colon or full-stop at ἁμάρτωσι and a comma at λάβῃ. So Littré and Ermerins.

---

[13] In M appears here the title Ἱπποκρατους π(ε) ἐνιπνίων ΚΓ.

[14] ἡμερινὰς M : ἑσπερινὰς θ.

[15] ἢ διανοίας θ : ἡ διάνοια M.

[16] εὐφρονην M : εὖ φρονεῖν θ. After εὐφρονην M has ἐνυπνιάζεται ἑσπέρην.

δίδωσι κατὰ τρόπον γινομένας[1] ὥσπερ[2] τῆς ἡμέρας ἐπρήχθη ἢ ἐβουλεύθη ἐπὶ[3] δικαίῳ πρήγματι, ταῦτα τῷ ἀνθρώπῳ ἀγαθά· ὑγιείην γὰρ σημαίνει, διότι ἡ ψυχὴ παραμένει τοῖσιν ἡμερινοῖσι βουλεύμασιν, οὔτε πλησμονῇ κρατηθεῖσα οὔτε κενώσει οὔτε ἄλλῳ οὐδενὶ ἔξωθεν 10 προσπεσόντι. ὅταν δὲ πρὸς τὰς ἡμερινὰς πρήξιας ὑπεναντιῶται τὰ ἐνύπνια καὶ ἐγγίνηται περὶ αὐτῶν ἢ μάχη ἢ νίκη,[4] σημαίνει τάραχον[5] ἐν τῷ σώματι· καὶ ἢν μὲν ἰσχυρὴ ᾖ, ἰσχυρὸν τὸ κακόν,[6] ἢν δὲ φαύλη, ἀσθενέστερον. περὶ μὲν οὖν τῆς πρήξιος εἴτ' ἀποτρέπειν δεῖ εἴτε μή,[7] οὐ κρίνω· τὸ δὲ σῶμα θεραπεύεσθαι συμβουλεύω· πλησμονῆς γάρ τινος ἐγγενομένης ἀπόκρισίς τις γενομένη[8] ἐτάραξε τὴν ψυχήν. ἢν μὲν οὖν ἰσχυρὸν ᾖ τὸ ἐναντιωθέν, ἔμετόν τε 20 συμφέρει ποιήσασθαι καὶ τοῖσι σίτοισι κούφοισι προσάγειν ἐς ἡμέρας πέντε, καὶ τοῖσι περιπάτοισι ὀρθρίοισι πολλοῖσι καὶ ὀξέσιν ἐκ προσαγωγῆς χρῆσθαι, καὶ τοῖσι γυμνασίοισιν, ὅστις ἐπιγυμνάζεται,[9] συμμέτροισι πρὸς τὴν προσαγωγὴν τῶν σίτων·[10] ἢν δὲ ἀσθενέστερον τὸ ὑπεναντίον[11] γένηται, ἀφελὼν τὸν ἔμετον τὸ

[1] γινομένας θ : γινόμενα M. [2] ὥσπερ M : ὅπερ θ.

[3] ἐπὶ θ : ἐν M.

[4] ἡ μάχηι· ἢ νικη θ : ἡ μάχη. ἡνίκα ἂν (with σημαίνηι) M : the text is Diels'.

[5] τάραχον θ : ταραχὴν M.

[6] θ has καὶ ἢν ἰσχυρᾶ ἰσχυρὸν τὸ σῶμα. Diels would read καὶ ἢν μὲν ἰσχυρή, ἰσχυρόν (sc. τὸν τάραχον σημαίνει); this is quite possibly correct.

[7] Both θ and M omit δεῖ, which the vulgate places after μή. M. has οὔτε for εἴτε.

occurring naturally, just as they were done or planned during the day in a normal[1] act—these are good for a man. They signify health, because the soul abides by the purposes of the day, and is overpowered neither by surfeit nor by depletion nor by any attack from without. But when dreams are contrary to the acts of the day, and there occurs about them some struggle or triumph, a disturbance in the body is indicated, a violent struggle meaning a violent mischief, a feeble struggle a less serious mischief. As to whether the act should be averted or not I do not decide, but I do advise treatment of the body. For a disturbance of the soul has been caused by a secretion arising from some surfeit that has occurred. Now if the contrast be violent, it is beneficial to take an emetic, to increase gradually a light diet for five days, to take in the early morning long, sharp walks, increasing them gradually, and to adapt exercises, when in training,[2] so as to match the gradual increase of food. If the contrast be milder, omit the emetic, reduce food

[1] The word δικαίῳ is difficult. Littré's "dans une juste affaire," and Ermerins' "in re iusta," hardly bring out the meaning, which has no reference to ethics, but only to the "sanity" of the act or thought.

[2] The reading ἐπιγυμνάζεσθαι is the easier, as few Greeks were ever "out of training." It is hard, however, to discard the reading of so good a MS. as θ, especially when we remember that "difficilior lectio potior."

---

[8] ἀποκρίσις τίς γενομένηι θ: ἀπόκρισις γέγονεν τίς. ἥτις M: ἀπόκρισις ἐγένετό τις, ἥτις Diels.

[9] ὅστις ἔτι γυμνάζεται θ: ἐπιγυμνάζεσθαι M.

[10] σιτῶν θ: σιτίων M.

[11] ὑπενάντιον θ: ὑπεναντιωθὲν M.

τρίτον μέρος ἄφελε τῶν σίτων,[1] καὶ τοῦτο[2] ἡσυχῇ προσάγου[3] πάλιν ἐπὶ πένθ' ἡμέρας· καὶ τοῖσι περιπάτοισι πιέζειν καὶ τοῖσι τῆς φωνῆς 30 πόνοισι χρῆσθαι,[4] καὶ καταστήσεται ἡ ταραχή.[5]

LXXXIX. Ἥλιον καὶ σελήνην καὶ οὐρανὸν καὶ ἄστρα[6] καθαρὰ καὶ εὐαγέα, κατὰ τρόπον ὁρεόμενα[7] ἕκαστα, ἀγαθά· ὑγιείην γὰρ τῷ σώματι σημαίνει ἀπὸ πάντων τῶν ὑπαρχόντων· ἀλλὰ χρὴ διαφυλάσσειν ταύτην τὴν ἕξιν τῇ παρεούσῃ διαίτῃ. εἰ δέ τι τούτων ὑπεναντίον γένοιτο, νοῦσόν τινα τῷ σώματι σημαίνει, ἀπὸ μὲν τῶν ἰσχυροτέρων ἰσχυροτέρην, ἀπὸ δὲ τῶν ἀσθενεστέρων κουφοτέρην. ἄστρων μὲν οὖν ἡ 10 ἡ ἔξω περίοδος, ἡλίου δὲ ἡ μέση, σελήνης δὲ ἡ πρὸς τὰ κοῖλα. ὅ τι μὲν οὖν δοκέοι[8] τῶν ἄστρων βλάπτεσθαι ἢ ἀφανίζεσθαι ἢ ἐπίσχεσθαι[9] τῆς περιόδου, ἢν μὲν ὑπ' ἠέρος ἢ νεφέλης, ἀσθενέστερον· εἰ δὲ καὶ ὕδατος ἢ χαλάζης, ἰσχυρότερον· σημαίνει δὲ ἀπόκρισιν ἐν τῷ σώματι ὑγρὴν καὶ φλεγματώδεα γενομένην ἐς τὴν ἔξω[10] περιφορὴν ἐσπεπτωκέναι. συμφέρει δὲ τούτῳ τοῖσί τε δρόμοισιν ἐν τοῖσιν ἱματίοισι

[1] τῶν σιτῶν θ: τοῦ σιτίου M. [2] τὸ M.
[3] προσαγάγου θ M. [4] χρήσθω M.
[5] After ταραχή θ has καὶ τοῖσι θεοῖσιν εὔχεσθαι with -ισι and -ισιν erased. M has καὶ τοῖσι θεοῖσι εὔχεσθαι.
[6] ἄστρα θ: ἀστέρας M.
[7] δρώμενα M: ὁραιομένα θ: ἦν before καθαρὰ and ὁρεώμενα Diels.
[8] δοκέοι θ: δοκοίη M. [9] ἐπίσχεσθαι M: ἐπίχεσθαι θ.
[10] ἔξω θ: ἔσω M.

---

[1] "Agiles" Littré; "suo motu agitata" Ermerins, as though εὐαγέα came from ἄγω.

by a third, resuming this by a gentle, gradual increase spread over five days. Insist on vigorous walks, use voice-exercises, and the disturbance will cease.

LXXXIX. To see the sun, moon, heavens and stars clear and bright,[1] each in the proper order, is good, as it indicates physical health in all its signs,[2] but this condition must be maintained by adhering to the regimen followed at the time. But if there be a contrast between the dream and reality, it indicates a physical illness, a violent contrast a violent illness, a slighter contrast a lighter illness. The stars are in the outer sphere, the sun in the middle sphere, the moon in the sphere next the hollow.[3] When any one of the heavenly bodies appears to be disfigured, to disappear, or to be arrested in its revolution, if it be through mist or cloud, the malign influence is comparatively weak; if through rain also or hail, the influence is more powerful. In any case it is indicated that a moist and phlegm like secretion, arising in the body, has fallen to the outer circuit.[4] It is beneficial for this man to make his runs long, wearing

[2] "De la part de tout ce qui y est" Littré; "omniumque eius partium" Ermerins. τὸ ὑπάρχον in this book often mean an apparition in a dream. See p. 431.

[3] The moon was supposed to be in the first and lowest of the eight concentric spheres, the sun in the fourth, the fixed stars in the eighth and outermost. τὰ κοῖλα means the concavity of the inmost sphere, by which we are surrounded. I owe this note to the kindness of Professor A. E. Housman.

[4] There is supposed to be a connexion between the spheres in which the stars move and the "circuits" or circulations in the body described in *Regimen* I.

χρῆσθαι πολλοῖσιν, ἐξ ὀλίγου προσάγοντα,
20 ὅπως ἐξιδρώσῃ[1] ὡς μάλιστα, καὶ τοῖσι περιπάτοισιν ἀπὸ τοῦ γυμνασίου πολλοῖσι, καὶ ἀνάριστον διάγειν· τῶν τε[2] σίτων[3] ἀφελόμενον τὸ τρίτον μέρος προσάγειν ἐς πένθ' ἡμέρας· εἰ δὲ δοκοίη ἰσχυρότερον εἶναι καὶ πυρίῃ[4] χρῆσθαι· τὴν γὰρ κάθαρσιν διὰ τοῦ χρωτὸς συμφέρει ποιεῖσθαι, διότι ἐν τῇ ἔξω περιφορῇ ἐστὶ τὸ βλάβος· τοῖσι δὲ σίτοισι χρῆσθαι ξηροῖσι, δριμέσιν, αὐστηροῖσιν, ἀκρήτοισι, καὶ τοῖσι πόνοισι τοῖσι ξηραίνουσι μάλιστα. εἰ δέ τι
30 τούτων ἡ σελήνη πάσχοι, εἴσω τὴν ἀντίσπασιν ποιεῖσθαι συμφέρει, ἐμέτῳ τε[5] χρῆσθαι ἀπὸ τῶν δριμέων καὶ ἁλμυρῶν καὶ μαλακῶν σίτων· τοῖσί τε τρόχοισιν[6] ὀξέσι καὶ τοῖσι περιπάτοισι· τοῖσί τε τῆς φωνῆς πόνοισι, καὶ ἀναριστίῃσι, τοῦ τε σίτου τῇ ἀφαιρέσει καὶ προσαγωγῇ ὡσαύτως. διὰ τοῦτο δὲ εἴσω ἀντισπαστέον, διότι πρὸς τὰ κοῖλα τοῦ σώματος τὸ βλαβερὸν ἐφάνη. εἰ δὲ ὁ ἥλιος τοιοῦτό τι[7] πάσχοι, ἰσχυρότερον τοῦτο ἤδη καὶ δυσεξαγωγότερον· δεῖ δὲ
40 ἀμφοτέρως τὰς ἀντισπάσιας ποιεῖσθαι, καὶ τοῖσι δρόμοισι τοῖσί τε καμπτοῖσι καὶ τοῖσι[8] τρόχοισι χρῆσθαι καὶ τοῖσι περιπάτοισι καὶ τοῖσιν ἄλλοισι πόνοισι πᾶσι, τῶν τε σίτων τῇ ἀφαιρέσει καὶ τῇ προσαγωγῇ ὡσαύτως. ἔπειτα

[1] ἐξιδρώσει Littré. Diels.
[2] τε M : δὲ θ.
[3] σίτων M : σιτῶν θ.
[4] πυριη M : πυριήσει θ.
[5] τε added by Diels.
[6] τροχοῖσι M : πόνοισιν θ.
[7] θ omits τι.
[8] M omits καὶ τοῖσι.

---

[1] "Non tempérés" Littré, that is, with their properties unmitigated by the addition of other ingredients.

his cloak the while, to increase them gradually, that he may perspire as freely as possible, and after exercise to take long walks; luncheon should be left out. Reduce food by one-third, and take five days in gradually resuming the normal quantity. Should the trouble appear to be of the more potent kind, use also the vapour-bath; for, as the mischief lies in the outer circuit, it is expedient to make the purgation through the skin. The foods employed are to be dry, acrid, astringent and unmixed;[1] the exercises such as are the most drying. But if it be the moon that shows these signs at all,[2] it is beneficial to effect the revulsion inwards, and to administer an emetic after foods that are acrid, salt and soft. There should be sharp circular runs, walks,[3] voice-exercises, omission of luncheon, the same reduction and gradual increase of food. The revulsion must be directed inwards because the harm showed itself at the hollow parts of the body. But if it be the sun that manifests the phenomena, the malady is more potent, and harder to eliminate. It is necessary to effect the revulsions in both directions, to employ running on the double track and on the round track, walks and all other exercises, the same reduction and gradual increase of food. After an emetic should come another

[2] Or "any one of the signs." See note 1, p. 417.

[3] Littré omits *τοῖσί τε τρόχοισι . . . περιπάτοισι*, on the ground that the revulsion is directed inwards. The articles I take to be generic, and the influence of *ὡσαύτως* to extend backwards only as far as *τοῦ τε σίτου*. The reading of θ looks like an attempt to extend this influence back to the beginning of the sentence, and to assimilate this prescription to the preceding, which, however, does not contain voice-exercises.

ἐξεμέσαντα αὖτις προσάγειν πρὸς τὰς πέντε· εἰ δὲ αἰθρίης ἐούσης θλίβεται,[1] καὶ ἀσθενέα δοκεῖ εἶναι καὶ[2] ὑπὸ τῆς ξηρασίης τῆς περιόδου κρατεῖσθαι, σημαίνει κίνδυνον ἐς νοῦσον ἐμπεσεῖν.[3] ἀλλὰ χρὴ τῶν πόνων ἀφαιρεῖν, τῇ τε 50 διαίτῃ τῇ ὑγροτάτῃ[4] χρῆσθαι, τοῖσί τε λουτροῖσι καὶ ῥᾳθυμίῃ πλείονι, καὶ ὕπνοῖσι, μέχρι καταστῇ. εἰ δὲ πυροειδὲς τὸ ὑπεναντιούμενον δοκοίη εἶναι καὶ θερμόν, χολῆς ἀπόκρισιν σημαίνει· εἰ μὲν οὖν κρατοίη τὰ ὑπάρχοντα, νοῦσον σημαίνει·[5] εἰ δὲ καὶ ἀφανίζοιτο τὰ κρατεύμενα, κίνδυνος ἐς θάνατον ἐκ τῆς νούσου ἐλθεῖν. εἰ δὲ τρεφθῆναι δοκοίη ἐς φυγὴν τὸ ὑπάρχον, φεύγειν δὲ ταχέως, τοὺς δὲ διώκειν, κίνδυνος μανῆναι τὸν ἄνθρωπον, ἢν μὴ θερα- 60 πευθῇ. συμφέρει δὲ τούτοισι πᾶσι μάλιστα μὲν ἐλλεβόρῳ καθαρθέντας διαιτῆσθαι· εἰ δὲ μή, τῇ πρὸς ὕδατος διαίτῃ συμφέρει χρῆσθαι, οἶνον δὲ μὴ πίνειν, εἰ μὴ[6] λευκόν, λεπτόν, μαλακόν, ὑδαρέα· ἀπέχεσθαι δὲ θερμῶν,[7] δριμέων, ξηραντικῶν, ἁλμυρῶν· πόνοισι δὲ τοῖσι κατὰ φύσιν πλείστοισι χρήσθω καὶ δρόμοισιν ἐν ἱματίῳ πλείστοισι· τρῖψις δὲ μὴ ἔστω, μηδὲ πάλη, μηδὲ ἀλίνδησις· ὕπνοισι πολλοῖσι μαλακευνείτω· ῥᾳθυμείτω πλὴν ἐκ[8] τῶν κατὰ φύσιν πόνων· 70 ἀπὸ δείπνου περιπατείτω· ἀγαθὸν δὲ καὶ πυριῆ-

[1] θ has θλίβηται and δοκῆι.
[2] καὶ omitted by θ M. First added by Zwinger.
[3] ἐνπεσεῖν θ : πεσεῖν M.
[4] ὑγροτέρηι μαλακῆι (withont τῆι) M.
[5] εἰ μὲν οὐ κρατοίη τὰ . . . σημαίνει θ. M omits, and Ermerins reads οὖν for οὐ.
[6] μὴ θ: δὲ μὴ M : δ' οὖν Littré, Ermerins.
[7] M omits θερμῶν, but has θερμαντικῶν after ξηραντικῶν.

gradual increase spread over five days. But if in a clear sky the heavenly bodies are crushed, seeming to be weak and overpowered by the dryness of the revolution,[1] it indicates a danger of falling into a disease. What is necessary is to reduce food, to employ the moistest regimen, baths and increased rest, and sleep, until there is a recovery. If the hostile influence appear to be fiery and hot, a secretion of bile is indicated. Now if the force[2] win, a disease is indicated. If the vanquished be also annihilated, there is a danger that the disease will have a fatal issue. But if the force[2] seem to be put to flight, and to flee quickly, pursued by the stars, there is a danger that the patient will become delirious, unless he be treated. In all these cases it is most beneficial to be purged with hellebore before submitting to regimen. The next best course is to adopt a watery regimen, and to abstain from wine unless it be white, thin, soft and diluted. There should be abstinence from things that are hot, acrid, drying and salt. Let there be plenty of natural exercises and long runs with the cloak worn. Let there be no massage, no ordinary wrestling, and no wrestling on dust. Long sleeps on a soft bed; rest except after[3] the natural exercises; let there be a walk after dinner. It is a good thing too to take a vapour-bath. After the

[1] If with Ermerins we transpose *καὶ* to before *σημαίνει*, and read *κρατεῖται*, we must translate: "they are overpowered, etc. and it indicates."

[2] Would the word "Thing" (capital T) represent the mysterious influence suggested by *τὸ ὑπάρχον*?

[3] Or (with *ἐκ* omitted) "from."

---

[3] *ἐκ* is omitted by M.

σθαι· καὶ ἐμεῖν ἐκ τῆς πυριῆς· τριήκοντα δὲ ἡμερέων μὴ πληρωθῇ· ὁκόταν δὲ πληρωθῇ, τρὶς ἐν τῷ μηνὶ ἐμεσάτω ἀπὸ τῶν γλυκέων καὶ ὑδαρέων καὶ κούφων. ὁκόσα δὲ τούτων πλανᾶται ἄλλοις ἄλλως,[1] ψυχῆς τάραξίν τινα σημαίνει ὑπὸ μερίμνης· συμφέρει δὲ τούτῳ ῥᾳθυμῆσαι· τὴν ψυχὴν τραπέσθαι[2] πρὸς θεωρίας, μάλιστα μὲν πρὸς τὰς γελοίας, εἰ δὲ μή, ἄλλας τινὰς ἃς[3] ὅ τι μάλιστα ἡσθήσεται θεησάμενος, ἡμέρας δυο 80 ἢ τρεῖς, καὶ καταστήσεται· εἰ δὲ μή, κίνδυνος ἐς νοῦσον πίπτειν. ὅ τι δ' ἂν ἐκ τῆς περιφορῆς ἐκπίπτειν δοκῇ τῶν ἄστρων, ὁκόσα μὲν καθαρὰ καὶ λαμπρὰ καὶ πρὸς ἕω φέρεται, ὑγείην σημαίνει· ὅ τι δ' ἂν ἐν τῷ σώματι καθαρὸν ἐνεὸν ἐκκρίνηται ἐκ τῆς περιόδου κατὰ φύσιν ἀφ' ἑσπέρας πρὸς ἠῶ,[4] ὀρθῶς ἔχει· καὶ γὰρ τὰ ἐς τὴν κοιλίην ἀποκρινόμενα καὶ τὰ ἐς τὴν σάρκα ἀπερευγόμενα πάντα ἐκ τῆς περιόδου ἐκπίπτει. ὅ τι δ' ἂν τούτων μέλαν καὶ ἀμυδρὸν καὶ πρὸς 90 ἑσπέρην δοκῇ φέρεσθαι, ἢ ἐς θάλασσαν ἢ ἐς τὴν γῆν ἢ[5] ἄνω, ταῦτα σημαίνει τὰς νούσους· τὰ μὲν ἄνω φερόμενα κεφαλῆς ῥεύματα· ὅσα δὲ ἐς θάλασσαν, κοιλίης νοσήματα· ὅσα δὲ ἐς γῆν,

[1] For the ἄλλοις ἄλλως of θ, M has ἄλλο τε ἄλλῃ μὴ ὑπο ἀνάγκης, with τινὰ after ψυχῆς.

[2] M has τραπῆναι καὶ for τραπέσθαι.

[3] M omits ἄλλας τινὰς, perhaps rightly.

[4] M reads ὅτι γάρ, ἐὸν and προσηι.

[5] θ omits ἢ and M has μᾶλλον after ἄνω.

---

[1] I take ἄλλοις ἄλλως to be an adverbial phrase independent syntactically of the rest of the sentence. I can discover no exact parallel for this, but that is no reason for rejecting the reading in a work in which a strict adherence to

vapour-bath an emetic is to be drunk. Until thirty days are gone the appetite should not be fully satisfied, and when the time has come for this full satisfaction, let an emetic be taken three times a month after a meal of sweet, watery and light foods. Whenever the heavenly bodies wander about, some in one way and others in another,[1] it indicates a disturbance of the soul arising from anxiety. Rest is beneficial in such a case. The soul should be turned to the contemplation of comic things, if possible, if not, to such other things as will bring most pleasure when looked at, for two or three days, and recovery will take place. Otherwise there is a risk of falling ill. Whenever a heavenly body appears to fall away from its orbit, should it be pure and bright, and the motion towards the east, it is a sign of health. For whenever a pure substance in the body is secreted from the circuit in the natural motion from west to east, it is right and proper. In fact secretions into the belly and substances disgorged into the flesh all fall away from the circuit. But whenever a heavenly body seems to be dark and dull, and to move towards the west, or into the sea, or into the earth, or upwards, disease is indicated. When the motion is upwards, it means fluxes of the head; when into the sea, diseases of the bowels; when

syntax is not always followed. Perhaps we should read, with the same sense, ἄλλοτε ἄλλως. M's reading would mean, "now in one direction and now in another." Perhaps ἄλλοις may refer to the dreamers: "in the way the particular dreamer may happen to see them." Professor D. S. Robertson assures me that the last interpretation is the only one consistent with ἄλλοις ἄλλως. But the Greek of *Regimen* is often abnormal.

φύματα μάλιστα σημαίνει τὰ ἐν τῇ σαρκὶ φυόμενα. τούτοισι συμφέρει τὸ τρίτον μέρος τοῦ σίτου ἀφελέσθαι, ἐμέσαντας δὲ προσάγειν ἐς ἡμέρας πέντε, ἐν ἄλλῃσι δὲ πέντε κομίσασθαι τὸ σιτίον· καὶ ἐμέσας πάλιν προσαγέσθω κατὰ τὸ αὐτό. ὅ τι δ' ἂν τῶν οὐρανίων δόξῃ σοι[1]
100 ἐφέζεσθαι καθαρὸν μὲν καὶ ὑγρὸν ἐὸν ὑγιαίνειν σημαίνει, διότι ἐκ τοῦ αἰθέρος τὸ ἐς τὸν ἄνθρωπον καθαρόν ἐστι, τοιοῦτον δὲ καὶ ἡ ψυχὴ ὁρῇ οἷόν περ ἐσῆλθεν· ὅ τι δ' ἂν μέλαν ᾖ καὶ μὴ καθαρὸν μηδὲ διαφανές, νοῦσον σημαίνει, οὔτε διὰ πλησμονὴν οὔτε διὰ κένωσιν, ἀλλ' ἔξωθεν ἐπαγωγῇ. συμφέρει δὲ τούτῳ τρόχοισιν ὀξέσι χρῆσθαι, ὅπως σύντηξις μὲν ὡς ἐλαχίστη τοῦ σώματος γένηται, πνεύματι δὲ ὡς πυκνοτάτῳ χρησάμενος ἐκκρίνῃ τὸ παρελθόν· ἀπὸ δὲ τῶν
110 τρόχων περιπάτοισιν ὀξέσιν. ἡ δίαιτα μαλακὴ καὶ κούφη[2] προσαχθήτω ἐς ἡμέρας τέσσαρας. ὅ τι δ' ἂν παρὰ θεοῦ δοκῇ λαμβάνειν καθαροῦ καθαρόν, ἀγαθὸν πρὸς ὑγείην· σημαίνει γὰρ τὰ ἐς τὸ σῶμα ἐσιόντα εἶναι καθαρά. ὅ τι δ' ἂν τούτου ἐναντίον δοκῇ ὁρῆν, οὐκ ἀγαθόν· νοσηρὸν γάρ τι[3] σημαίνει ἐς τὸ σῶμα ἐσεληλυθέναι· χρὴ οὖν ὥσπερ τὸν πρότερον θεραπευθῆναι καὶ τοῦτον. εἰ δὲ δοκοίη ὕεσθαι ὕδατι μαλθακῷ ἐν εὐδίῃ, καὶ μὴ σφόδρα βρέχεσθαι μηδὲ δεινῶς

[1] σοι is not in θ, which has, however, θιεφίζεσθαι.
[2] Both θ and M have datives τῇ . . . κούφῃ. Either read nominatives, or omit προσαχθήτω (understanding χρῆσθαι). Possibly, however, προσαχθήτω could take a dative.

into the earth, most usually tumours growing in the flesh. In such cases it is beneficial to reduce food by one-third and to take an emetic, to be followed by a gradual increase of food for five days, the normal diet being resumed in another five. Another emetic should be followed by the same gradual increase. Whenever a heavenly body seems to settle on you, if it be pure[1] and moist, it indicates health, because what descends from the ether on to the person is pure, and the soul too sees it in its true character as it entered the body. But should the heavenly body be dark, impure and not transparent, it indicates disease caused neither by surfeit nor by depletion, but by the entrance of something from without. It is beneficial in this case to take sharp runs on the round track, that there may be as little melting of the body as possible, and that by breathing as rapidly as possible the patient may secrete the foreign body. After these runs let there be sharp walks. Diet to be soft and light[2] for four days. Whatsoever a man seems to receive pure from a pure god is good for health; for it indicates that the matter is pure that enters the body. But whatever he seems to see that is the opposite thereof is not good; for it indicates that something diseased has entered the body. Accordingly the treatment in this case should be the same as the former. Should it seem to rain with a gentle shower from a clear sky, with neither a violent

[1] That is, "clear."

[2] Perhaps we should add "gradually increased" (προσαχθήτω). It is often uncertain whether προσάγω carries this meaning or not.

---

[3] νόσον γὰρ M.

120 χειμάζειν, ἀγαθόν· σημαίνει γὰρ σύμμετρον καὶ καθαρὸν τὸ πνεῦμα ἐκ τοῦ ἠέρος ἐληλυθέναι. εἰ δὲ τούτων τἀναντία, σφόδρα ὕεσθαι καὶ χειμῶνα καὶ ζάλην εἶναι, ὕδατί τε μὴ καθαρῷ, νοῦσον σημαίνει ἀπὸ τοῦ πνεύματος τοῦ ἐπακτοῦ· ἀλλὰ χρὴ καὶ τοῦτον ὡσαύτως διαιτηθῆναι, σίτοισι δὲ ὀλίγοισι παντελῶς τοῦτον.[1] περὶ μὲν οὖν τῶν οὐρανίων σημείων οὕτω γινώσκοντα χρὴ προμηθεῖσθαι καὶ ἐκδιαιτῆσθαι καὶ τοῖσι θεοῖσιν εὔχεσθαι, ἐπὶ μὲν τοῖσι ἀγαθοῖσι Ἡλίῳ, Διὶ 130 οὐρανίῳ, Διὶ κτησίῳ, Ἀθηνᾷ κτησίῃ, Ἑρμῇ, Ἀπόλλωνι, ἐπὶ δὲ τοῖσι ἐναντίοισι τοῖσι ἀποτροπαίοισι, καὶ Γῇ καὶ ἥρωσιν, ἀποτρόπαια τὰ 133 χαλεπὰ εἶναι πάντα.[2]

XC. Προσημαίνει δὲ καὶ τάδε ἐς ὑγείην· τῶν ἐπὶ γῆς ὀξὺ ὁρῆν καὶ ὀξὺ[3] ἀκούειν, ὁδοιπορεῖν τε ἀσφαλῶς καὶ τρέχειν ἀσφαλῶς καὶ ταχὺ[4] ἄτερ φόβου, καὶ τὴν γῆν ὁρῆν λείην καὶ καλῶς εἰργασμένην, καὶ τὰ δένδρεα θαλέοντα καὶ πολύκαρπα καὶ ἥμερα, καὶ ποταμοὺς ῥέοντας κατὰ τρόπον καὶ ὕδατι καθαρῷ μήτε πλέονι μήτε ἐλάσσονι τοῦ προσήκοντος, καὶ[5] τὰς κρήνας καὶ τὰ φρέατα ὡσαύτως. ταῦτα πάντα σημαίνει 10 ὑγείην τῷ ἀνθρώπῳ, καὶ τὸ σῶμα κατὰ τρόπον πάσας τε τὰς περιόδους καὶ τὰς προσαγωγὰς καὶ τὰς ἀποκρίσεις εἶναι. εἰ δέ τι τούτων ὑπεναντίον ὁρῷτο, βλάβος σημαίνει τι ἐν τῷ σώματι· ὄψιος μὲν καὶ ἀκοῆς βλαπτομένων, περὶ τὴν κεφαλὴν νοῦσον σημαίνει· τοῖσιν οὖν ὀρθρίοισι περιπάτοισι

[1] M has σιτίοισί τε ὀλίγοισι πάντας τούτους.

[2] I have followed M in this passage. θ has been "bowdlerized" by some Christian enthusiast, who has

downpour nor a terrible storm, it is a good sign; for it indicates that the breath has come from the air in just measure and pure. If the reverse occur, violent rain, storm and tempest, and the water be foul, it indicates disease from the breath that comes from without. In this case also the same regimen must be employed, and diet must be very strictly limited. So with this knowledge about the heavenly bodies, precautions must be taken, with change of regimen and prayers to the gods; in the case of good signs, to the Sun, to Heavenly Zeus, to Zeus, Protector of Home, to Athena, Protectress of Home, to Hermes and to Apollo; in the case of adverse signs, to the Averters of evil, to Earth and to the Heroes, that all dangers may be averted.

XC. The following too are signs that foretell health. To see and hear clearly the things on the earth, to walk surely, to run surely, quickly and without fear, to see the earth level and well tilled, trees that are luxuriant, covered with fruit and cultivated, rivers flowing naturally, with water that is pure, and neither higher nor lower than it should be, and springs and wells that are similar. All these indicate health for the dreamer, and that the body with all its circuits, diet and secretions are proper and normal. But if anything be seen that is the reverse of these things, it indicates some harm in the body. If sight or hearing be impaired, it indicates disease in the region of the head. In addition to the preceding regimen the dreamer

erased the -σιν of θεοῖσιν, and also about a line and a quarter (ἡλίῳ . . . Ἀπόλλωνι) to avoid the heathen deities.

[3] ὀξὺ omitted by M, which has τε after τὰς.

[4] ταχὺ omitted by M. [5] καὶ omitted by M.

καὶ τοῖσιν ἀπὸ δείπνου πλείοσι χρηστέον πρὸς τῇ προτέρῃ διαίτῃ. τῶν σκελέων δὲ βλαπτομένων, ἐμέτοισιν ἀντισπαστέον, καὶ τῇ πάλῃ πλείονι χρηστέον πρὸς τῇ προτέρῃ διαίτῃ.[1] γῆ δὲ
20 τραχείη[2] οὐ καθαρὴν τὴν σάρκα σημαίνει· τοῖσιν οὖν ἀπὸ τῶν γυμνασίων περιπάτοισι πλείοσι χρηστέον.[3] δένδρων ἀκαρπία σπέρματος τοῦ ἀνθρωπίνου διαφθορὴν δηλοῖ· ἢν μὲν οὖν φυλλορροοῦντα ᾖ τὰ δένδρα, ὑπὸ τῶν ὑγρῶν καὶ ψυχρῶν βλάπτεται· ἢν δὲ τεθήλῃ μέν, ἄκαρπα δὲ ᾖ, ὑπὸ τῶν θερμῶν καὶ ξηρῶν· τὰ μὲν οὖν θερμαίνειν καὶ ξηραίνειν τοῖσι διαιτήμασι χρή, τὰ δὲ ψύχειν τε καὶ ὑγραίνειν. ποταμοὶ δὲ κατὰ τρόπον μὴ γινόμενοι αἵματος περίοδον
30 σημαίνουσι, πλέον μὲν ῥέοντες ὑπερβολήν, ἔλασσον δὲ ῥέοντες ἔλλειψιν· δεῖ δὲ τῇ διαίτῃ τὸ μὲν αὐξῆσαι, τὸ δὲ μειῶσαι. μὴ καθαρῷ[4] δὲ ῥέοντες ταραχὴν σημαίνουσι·[5] καθαίρεται δὲ ὑπὸ τῶν τρόχων καὶ τῶν περιπάτων πνεύματι πυκνῷ διακινεόμενα.[6] κρῆναι καὶ φρέατα περὶ τὴν κύστιν τι σημαίνει·[7] ἀλλὰ χρὴ τοῖσιν οὐρητικοῖσιν ἐκκαθαίρειν. θάλασσα δὲ ταρασσομένη κοιλίης νοῦσον σημαίνει· ἀλλὰ χρὴ τοῖσι διαχωρητικοῖσι καὶ κούφοισι καὶ μαλακοῖσιν ἐκκαθαίρειν. γῆ
40 κινευμένη ἢ οἰκίη ὑγιαίνοντι μὲν ἀσθενείην σημαίνει, νοσέοντι δὲ ὑγείην καὶ μετακίνησιν τοῦ ὑπάρχοντος. τῷ μὲν οὖν[8] ὑγιαίνοντι μεταστῆσαι τὴν δίαιταν συμφέρει· ἐμεσάτω δὲ πρῶτον,

[1] θ omits τῶν σκελέων . . . διαίτῃ.
[2] τῇ δὲ ταχεια θ: τῆι δὲ τραχείη M.
[3] χρηστέον θ: πονητέον M.
[4] καθαροὶ M. [5] σημαίνει θ.

should take longer walks in the early morning and after dinner. If it be the legs that are injured, the revulsion should be made with emetics, and in addition to the preceding regimen there should be more wrestling. For the earth to be rough indicates that the flesh is impure. So the walks after exercises must be made longer. Fruitless trees signify corruption of the human seed. Now if the trees are shedding their leaves, the harm is caused by moist, cold influences; if leaves abound without any fruit, by hot, dry influences. In the former case regimen must be directed towards warming and drying; in the latter towards cooling and moistening. When rivers are abnormal they indicate a circulation of the blood; high water excess of blood, low water defect of blood. Regimen should be made to increase the latter and lessen the former. Impure streams indicate disturbance of the bowels. The impurities are removed by running on the round track and by walks, which stir them up by accelerated respiration. Springs and cisterns indicate some trouble of the bladder; it should be thoroughly purged by diuretics. A troubled sea indicates disease of the belly; it should be thoroughly purged by light, soft aperients. Trembling of the earth or of a house indicates illness when the dreamer is in health, and a change from disease to health when he is sick. So it is beneficial to change the regimen of a healthy dreamer. Let him first take an emetic, that he may resume nourish-

---

[6] *διακινούμενα* θ : *ἀνακινεύμενα* M

[7] M has *κρῆναι δὲ καὶ φρέατα πνεύματα περὶ τὴν κύστιν τι σημαίνει.*

[8] *οὖν* M : *νῦν* θ.

ἵνα προσδέξηται αὖτις κατὰ μικρόν· ἀπὸ γὰρ τῆς ὑπαρχούσης κινεῖται[1] πᾶν τὸ σῶμα. τῷ δὲ ἀσθενέοντι συμφέρει χρῆσθαι τῇ αὐτῇ διαίτῃ· μεθίσταται γὰρ ἤδη τὸ σῶμα ἐκ τοῦ παρεόντος. κατακλυζομένην γῆν ἀπὸ ὕδατος ἢ θαλάσσης ὁρῆν νοῦσον σημαίνει, ὑγρασίης πολλῆς ἐνεούσης
50 ἐν τῷ σώματι· ἀλλὰ χρὴ τοῖσιν ἐμέτοισι καὶ τῇσιν ἀναριστίῃσι[2] καὶ τοῖσι πόνοισι καὶ τοῖσι διαιτήμασι ξηροῖσι· ἔπειτα προσάγειν ἐξ ὀλίγων καὶ ὀλίγοισιν.[3] οὐδὲ μέλαιναν ὁρῆν τὴν γῆν οὐδὲ κατακεκαυμένην ἀγαθόν, ἀλλὰ κίνδυνος ἰσχυροῦ νοσήματος ἀντιτυχεῖν καὶ θανασίμου· ξηρασίης γὰρ ὑπερβολὴν σημαίνει ἐν τῇ σαρκί· ἀλλὰ χρὴ τούς τε πόνους ἀφελεῖν, τοῦ τε σίτου ὅσα τε ξηρὰ καὶ[4] δριμέα καὶ οὐρητικά· διαιτῆσθαί τε τῆς τε πτισάνης καθέφθῳ τῷ χυλῷ, καὶ[5] σίτοισι
60 κούφοισιν ὀλίγοισι, ποτῷ δὲ πλέονι ὕδαρεῖ λευκῷ, λουτροῖσι πολλοῖσι·[6] μὴ ἄσιτος λουέσθω, μαλακευνείτω, ῥᾳθυμείτω, ψῦχος καὶ ἥλιον φυλασσέσθω· εὔχεσθαι δὲ Γῇ καὶ Ἑρμῇ καὶ ἥρωσιν.[7] εἰ δὲ κολυμβῆν ἐν λίμνῃ ἢ ἐν θαλάσσῃ ἢ ἐν ποταμοῖσι δοκεῖ[8] οὐκ ἀγαθόν· ὑπερβολὴν γὰρ ὑγρασίης σημαίνει· συμφέρει δὲ καὶ τούτῳ ξηραίνειν τῇ διαίτῃ, τοῖσί τε[9] πόνοισι πλείοσι· πυρέσσοντι δὲ ἀγαθόν· σβέννυται γὰρ τὸ θερμὸν
69 ὑπὸ τῶν ὑγρῶν.

XCI. Ὅ τι δ' ἂν τις περὶ αὐτοῦ ὁρῇ κατὰ τρόπον

[1] κρίνεται θ.
[2] τῆι ἀναριστήσει M: τῇσιν ἀναριστηισι θ.
[3] καὶ ὀλίγοισιν is omitted by θ.
[4] After καὶ M has θερμὰ καὶ.
[5] After καὶ M has πᾶσι τοῖσι μαλακοῖσι καὶ instead of σίτοισι.

ment again little by little, for it is the present nourishment that is troubling all the body. A sick dreamer benefits by continuing the same regimen, for the body is already changing from its present condition. To see the earth flooded by water or sea signifies a disease, as there is much moisture in the body. What is necessary is to take emetics, to avoid luncheon, to exercise and to adopt a dry diet. Then there should be a gradual increase of food, little by little, and little to begin with. It is not good either to see the earth black or scorched, but there is a danger of catching a violent, or even a fatal disease, for it indicates excess of dryness in the flesh. What is necessary is to give up exercises and such food as is dry and acrid and diuretic. Regimen should consist of barley-water well boiled, light and scanty meals, copious white wine well diluted, and numerous baths. No bath should be taken on an empty stomach, the bed should be soft and rest abundant. Chill and the sun should be avoided. Pray to Earth, Hermes and the Heroes. If the dreamer thinks that he is diving in a lake, in the sea, or in a river, it is not a good sign, for it indicates excess of moisture. In this case also benefit comes from a drying regimen and increased exercises. But for a fever patient these dreams are a good sign, for the heat is being suppressed by the moisture.

XCI. The sight of something connected with the

---

[6] After πολλοῖσι M has θερμοῖσι.

[7] The "Christian" corrector of θ has struck out the words Γῆ . . . ἥρωσιν.

[8] δοκοιη θ : δοκέειν M.

[9] M has τοῖσι τε πόνοισι χρῆσθαι. θ omits τε.

γινόμενον πρὸς τὴν φύσιν τὴν ἑωυτοῦ μήτε μέζω μήτε ἐλάσσω, ἀγαθὸν πρὸς ὑγείην σημαίνει· καὶ ἐσθῆτα λευκὴν τὴν ὑπάρχουσαν[1] καὶ ὑπόδεσιν τὴν καλλίστην, ἀγαθόν. ὅ τι δ' ἂν ᾖ μεῖζον τῶν μελέων ἢ ἔλασσον, οὐκ ἀγαθόν· ἀλλὰ χρὴ τὸ μὲν[2] αὔξειν τῇ διαίτῃ, τὸ δὲ μειοῦν. τὰ δὲ μέλανα νοσερώτερα καὶ ἐπικινδυνώτερα·[3] ἀλλὰ χρὴ μαλάσσειν καὶ ὑγραίνειν. καὶ τὰ καινὰ
10 μεταλλαγὴν σημαίνει.

XCII. Τοὺς δὲ ἀποθανόντας ὁρῆν καθαροὺς ἐν ἱματίοισι λευκοῖσιν ἀγαθόν, καὶ λαμβάνειν τι παρ' αὐτῶν καθαρὸν ὑγείην σημαίνει καὶ τῶν σωμάτων καὶ τῶν ἐσιόντων· ἀπὸ γὰρ τῶν ἀποθανόντων αἱ τροφαὶ καὶ αὐξήσιες καὶ σπέρματα γίνεται· ταῦτα δὲ καθαρὰ ἐσέρπειν ἐς τὸ σῶμα ὑγείην σημαίνει. εἰ δὲ τοὐναντίον τις ὁρῴη γυμνοὺς ἢ μελανοειμονας ἢ μὴ καθαροὺς ἢ λαμβάνοντάς τι ἢ φέροντας ἐκ τῆς οἰκίης, οὐκ ἐπιτήδειον· σημαίνει γὰρ
10 νοῦσον· τὰ γὰρ ἐσιόντα ἐς τὸ σῶμα βλαβερά· ἀλλὰ χρὴ τοῖσι τρόχοισι καὶ τοῖσι περιπάτοισιν ἀποκαθαίρεσθαι, καὶ τῇ τροφῇ τῇ μαλακῇ τε καὶ
13 κούφῃ προσάγειν ἐμέσαντα.

XCIII. Ὁκόσα δὲ ἀλλόμορφα σώματα φαίνεται ἐν τοῖσιν ὕπνοισι καὶ φοβεῖ τὸν ἄνθρωπον, σιτίων ἀσυνήθων σημαίνει πλησμονὴν καὶ ἀπόκρισιν καὶ χολέραν καὶ νοῦσον κινδυνώδεα· ἀλλὰ χρὴ ἔμετον ποιήσασθαι καὶ προσάγειν ἐς ἡμέρας

[1] τὴν ὑπάρχουσαν θM : ἐνδεδύσθαι Littré, Ermerins.
[2] θ has τὰ μὲν followed by τὰ δὲ.
[3] M has τε before καὶ and reads ἐπικίνδυνα.

---

[1] It is tempting to think that Ermerins is right in reading μέζον and ἔλασσον. The sentence thus becomes far more

person that is normal, and for which the physique is neither too large nor too small,[1] is a good sign for the health. To be wearing white clothes, and the most beautiful shoes, is also a good sign. But anything too large or too small for the limbs is not good. What is necessary is in the latter case to increase by regimen, in the former to diminish. Black objects indicate a worse and more dangerous disease; what is necessary is to soften and to moisten. New objects indicate a change.

XCII. To see the dead clean[2] and in white cloaks is a good sign, and to receive something clean from them indicates both health of body and the healthiness of the things that enter it. For from the dead come nourishment, growth and seed, and for these to enter the body clean[2] indicates health. But if, on the contrary, one should see them naked, or clothed in black, or not clean, or taking something, or bringing something out of the house, the sign is unfavourable, as it indicates disease, the things entering the body being harmful. What is necessary is to purge them away by runs on the round track and by walks, and after an emetic gradually to increase a soft and light diet.

XCIII. Monstrous bodies that are seen in sleep and frighten a man indicate a surfeit of unaccustomed food, a secretion, a bilious flux and a dangerous disease. What is necessary is an emetic, followed

idiomatic, though the sense is not materially altered: "neither too large nor too small for the physique."

[2] The word καθαρός is difficult, and to render it consistently by one English word is impossible. Littré uses "pur" in both these cases; Ermerins has "nitidus" and "purus." "Neat" or "tidy" seems to be the meaning in the first case, "pure" in the other.

πέντε σίτοισιν ὡς κουφοτάτοισι, μὴ πολλοῖσι μηδὲ δριμέσι, μήτε τοῖσι ξηροῖσι μήτε τοῖσι θερμοῖσι, καὶ τῶν πόνων τοῖσι κατὰ φύσιν μάλιστα, πλὴν τῶν ἀπὸ δείπνου περιπάτων· χρῆσθαι δὲ καὶ 10 θερμολουσίῃ καὶ ῥᾳθυμίῃσιν· ἥλιον δὲ καὶ ψῦχος φυλασσέσθω. ὁκόταν[1] δὲ ἐν τῷ ὕπνῳ ἐσθίειν δοκῇ ἢ πίνειν τῶν συνήθων ποτῶν ἢ σιτίων,[2] ἔνδειαν σημαίνει τροφῆς καὶ ψυχῆς ἀθυμίην·†[3] κρέα δὲ τὰ μὲν ἰσχυρότατα, μεγίστης ὑπερβολῆς, τὰ δὲ ἀσθενέστερα ἧσσον· ὥσπερ γὰρ ἐσθιόμενον ἀγαθόν, οὕτω καὶ ὁρεόμενον· ἀφαιρεῖν οὖν τῶν σιτίων συμφέρει· τροφῆς γὰρ ὑπερβολὴν σημαίνει·†[4] καὶ ἄρτοι τυρῷ καὶ μέλιτι πεποιημένοι ὡσαύτως σημαίνουσιν. ὕδωρ πινόμενον 20 καθαρὸν οὐ βλάπτει· τὰ δὲ ἄλλα πάντα βλάπτει. ὁκόσα δὲ δοκεῖ ἄνθρωπος θεωρεῖν τῶν συνήθων, ψυχῆς ἐπιθυμίην σημαίνει. ὅσα δὲ φεύγει πεφοβημένος, ἐπίστασιν τοῦ αἵματος σημαίνει ὑπὸ ξηρασίης· συμφέρει δὲ ψῦξαι καὶ ὑγρῆναι τὸ σῶμα. ὅσα δὲ μάχεται ἢ κεντεῖται ἢ συνδεῖται ὑπ᾽ ἄλλου, ἀπόκρισιν σημαίνει ὑπεναντίην τῇ περιόδῳ γεγονέναι ἐν τῷ σώματι· ἐμεῖν συμφέρει καὶ ἰσχναίνειν καὶ περιπατεῖν· σίτοισι κούφοισι χρῆσθαι, καὶ προσάγειν ἐκ τοῦ

[1] ὁκόταν θ: ἢν M.

[2] ποτῶν· ἢ σιτίων θ: σιτίων ἢ πομάτων M.

[3] M has ἔνδειαν σημαίνει ψυχῆς καὶ τροφῆς ἀθυμίην. Littré and Ermerins read ψυχῆς ἐπιθυμίην.

[4] Littré would rewrite the passage between daggers. For μεγίστης ὑπερβολῆς he reads ἐνδείας ὑπερβολήν; he adds οὐ before συμφέρει and ἐνδείας before ὑπερβολήν.

by a gradual increase, for five days, of the lightest food possible, neither abundant nor acrid, neither dry nor hot, with such exercises as are most natural, excepting walks after dinner. The dreamer should take hot baths and rest, and avoid the sun and cold. Whenever in his sleep a man thinks he is eating or drinking his usual food or drink, it indicates a want of nourishment and depression of the soul. †Meats if they be very strong show a very great excess; if they be weaker, a less excess. For just as eating is good, so eating in a dream is a good sign. So it is beneficial to reduce the quantity of food, for an excess of nourishment is indicated.†[1] The meaning is the same when bread is eaten, prepared with cheese and honey. To drink clean water in dreams is no sign of harm, but it is to drink any other kind. Whenever a man thinks that he beholds familiar objects, it indicates a desire of the soul. Whenever he runs away in fear, it indicates that the blood is arrested by dryness. It is in this case beneficial to cool and moisten the body. Fighting, or to be pierced or bound by another, indicates that there has occurred in the body a secretion opposed to the circuit. It is beneficial to take an emetic, to reduce the flesh, to walk, to eat light foods, and after the

[1] It is easy to see that the passage within daggers, which is a translation of θ, cannot represent the original. But the bold emendations of Littré, although they yield a possible sense, are most unlikely to be correct. I cannot solve the difficulties satisfactorily, but a great many are removed by transposing the sentence κρέα . . . ἧσσον· to after ὀρεόμενον· We then get the following sequence of ideas. "To eat in dreams one's usual food is a good sign; but to dream one is eating strong meat indicates excess, and diet should be reduced."

30 ἐμέτου πρὸς ἡμέρας τέσσαρας.[1] καὶ πλάνοι καὶ ἀναβάσιες χαλεπαὶ ταῦτα σημαίνουσιν. ποταμῶν διαβάσιες καὶ ὁπλῖται πολέμιοι καὶ[2] τέρατα ἀλλόμορφα νοῦσον σημαίνει ἢ μανίην. συμφέρει σίτοισιν ὀλίγοισι κούφοισι μαλακοῖσι χρῆσθαι καὶ ἐμέτοισι, προσάγειν ἡσυχῇ ἐς ἡμέρας πέντε,[3] καὶ πόνοισι τοῖσι κατὰ φύσιν πολλοῖσι πλὴν ἀπὸ τοῦ δείπνου, θερμολουσίην, ῥᾳθυμίην, ψῦχος, ἥλιον φυλάσσεσθαι. τούτοισι χρώμενος ὡς γέγραπται, ὑγιανεῖ τὸν βίον, καὶ εὕρηταί μοι
40 δίαιτα ὡς δυνατὸν εὑρεῖν ἄνθρωπον ἐόντα σὺν
41 τοῖσι θεοῖσιν.

[1] M has ἐς ἡμέρας πέντε.
[2] With Littré I insert καὶ here. θM omit.
[3] Before καὶ θ has καὶ σιτοῖσι.

emetic to increase food gradually for four days. Wanderings and difficult ascents have the same meaning. Crossing rivers, enemy men-at-arms and strange monsters indicate disease or raving. It is beneficial to take small meals of light, soft food, and emetics, and gently to increase food for five days, with plenty of natural exercise except after dinner; but hot baths, rest, cold and sun are to be avoided. Using these means in the way I have described a man will live a healthy life: in fact I have discovered regimen, with the gods' help, as far as it is possible for mere man to discover it.

HERACLEITUS

ON THE UNIVERSE

WITH AN ENGLISH TRANSLATION BY

W. H. S. JONES

INTRODUCTION

Greek philosophy began in wonder at the repeated miracle of motion and change, and first manifested itself in an effort to discover the material (φύσις) out of which the universe is made, phenomena being regarded as the transient modifications of this permanent reality. It differed from earlier thought in that it discarded the myth, or fairy story, as an explanation, and substituted rational causation; it differed from later science in that it proceeded from an unproved postulate,[1] upon which it built deductively,[2] attaching little importance to observation of phenomena, and still less to experiment.

In considering the history of early philosophy we must remember that the age of mythology did not pass away suddenly and completely. Mythological figures, indeed, disappear, but the artistic spirit of the romancer, which demands a complete picture, led the Greek philosopher to indulge his imagination in supplying details for which he had no warrant from experience and observation.[3] Another fact to be borne in mind is that the conception of im-

[1] Called later on ὑπόθεσις.

[2] Deductive science preceded inductive, probably because of the influence of mathematics, the first science to reach a high state of development.

[3] Heracleitus seems freer from this fault than many other early philosophers.

material existence was as yet unformed; soul and mind were looked upon as matter. The sciences, too, of logic and grammar were still to be born, and consequently men were often deceived by false analogies and verbal fallacies.

The first impulse to philosophic thought came, not unnaturally,[1] from a contemplation of the earth and sky; cosmologies succeeded cosmogonies. Thales of Miletus (*floruit* 585 B.C.) looked upon the world as water modifying itself; Anaximander[2] (560 B.C.) as "the Boundless" modifying itself in two opposite directions; Anaximenes[3] (546 B.C.) as air modifying itself in two directions by thickening and thinning.[4] In Western Greece the Pythagorean brotherhood, founded in the latter part of the sixth century, began under the influence of mathematical studies to lay stress upon the dualities apparent in the world.[5]

The Ionian school of material monists had their

[1] Observation of the sky was more common in days when there were no almanacs, no clocks, and no compass.

[2] Also of Miletus. His "Boundless" (τὸ ἄπειρον) may have been a kind of mist or cloud.

[3] Also of Miletus. Pre-Socratic philosophy bears many traces of its Eastern birth, notably the religious tinge in its phraseology.

[4] In other words, Anaximenes took a quantitative view of change.

[5] The Pythagoreans apparently began with the pair even)(odd. See Aristotle, *Metaphysics*, A 986*a*. Other (perhaps later) members of the brotherhood increased the number of pairs :—

limit)(unlimited,	rest)(motion,
odd)(even,	straight)(bent,
one)(multitude,	light)(darkness,
right)(left,	good)(bad,
male)(female,	square)(oblong.

last representative in Heracleitus of Ephesus. He is said to have flourished in the sixty-ninth Olympiad (504–500 B.C.). We know practically nothing about his life, and the title of his writings, which have come down to us only in fragments, has not been preserved.

Heracleitus was called "the dark" by the ancients, who had all his work before them; to the moderns, who possess only isolated sentences, he is darker still. It is both confusing and depressing to read the treatises of Lassalle, Teichmüller and Pfleiderer, and to see how the most opposite and inconsistent conclusions can be drawn by learned and intelligent men from exactly the same evidence. But in spite of all this diversity of opinion there is gradually shaping itself a more stable view of the doctrine of Heracleitus in its main outlines, although the details are still obscure, and may, in fact, in some cases never be elucidated.

It seems reasonable to suppose, when we consider the period in which he lived, that the phenomenon of change was the primary interest of his researches. His contribution to the problem was to point out that change is constant and perpetual. For no two seconds together is a thing ever the same. There is no pause in change; it is as much a *continuum* as is time. All things are for ever passing into something else.

In this eternal flux the only really constant thing is the principle of change itself, yet in some way or other fire, according to Heracleitus, has an individuality of its own which gives it precedence over all other things. The world "was ever, is now, and ever shall be an ever-living Fire, in measures being kindled and in measures going out." Nothing could be plainer than this declaration of the eternal nature of fire, and nothing could be more logically inconsistent with the doctrine of perpetual flux. Hence several scholars have held that the fire of Heracleitus is not the fire which burns and crackles, but warm vital force or something even more abstract still. Such a conception seems alien from the thought of the period, and the most recent research regards the Heracleitean fire as the ordinary fire of the every-day world. It is perhaps rash to hazard a guess when so many scholars have been baffled, but it may be that Heracleitus consciously or unconsciously identified fire and change. If so, there is less inconsistency in regarding fire as an eternal reality, though it is bad interpretation to twist facts in order to make a Greek philosopher self-consistent; we are not warranted in assuming that all early philosophy *was* consistent. Perhaps the fragments of Heracleitus do not support my guess, but the Heracleitean treatise *Regimen I* expressly states that the δύναμις of fire is to cause motion.[1] In any case, symbolically or actually, fire is a good example of physical transformation. Fuel is supplied from below, the flames quickly alter its nature, and finally it rises as smoke and fumes. The most obvious and the most rapid changes with which we are familiar are all connected with fire; it destroys, it cleanses and it renews. The sun seems to be a great mass of the very best fire, and it is the sun that transforms, by its alternate advance and retreat, the face of the earth from

[1] *Regimen I*, ch. iii. In this treatise δύναμις often means essence, and the sentence referred to virtually identifies change and fire.

season to season and from day to day. The world is an ever-living fire; it is always becoming all things, and all things are always returning into it.

There is thus a twofold way in nature, to fire and from fire, and this leads us to the most fundamental thought of Heracleitus, the "attunement" or harmonious unity resulting from the strife of opposites.[1] There is a "road up" to fire and a "road down" from fire, and these two roads are "one and the same." If they are one and the same, there must be a perpetual strain resulting from two, as it were, opposite forces. The way up fights with the way down. It is like the tension in a bow-string or in the cord of a harp. The flight of the arrow, the note of the string, are due solely to opposite tension (παλίντονος ἁρμονίη). This conception of universal strife dominated the theory of Heracleitus to such an extent that it is sometimes pushed to illogical extremes.[2] Each opposite is tending to turn into its opposite, and so in a sense each is the same as the other. "God is day and night, winter and summer, war and peace, surfeit and hunger." What Heracleitus really meant, and should have said, is that day and night, with all other opposites, are two sides of the same process, inseparably conjoined like concavity and convexity. Neither is possible without the other. Any ex-

[1] See in particular Philo, *Rer. Div. Her.* 43: ἓν γὰρ τὸ ἐξ ἀμφοῖν τῶν ἐναντίων, οὗ τμηθέντος γνώριμα τὰ ἐναντία. οὐ γὰρ τοῦτ' ἔστιν ὅ φασιν Ἕλληνες τὸν μέγαν καὶ ἀοίδιμον παρ' αὐτοῖς Ἡράκλειτον κεφάλαιον τῆς αὑτοῦ προστησάμενον φιλοσοφίας αὐχεῖν ὡς ἐφ' εὑρέσει καινῇ;

[2] Strictly speaking, the two opposites should produce a third thing, as male and female produce the offspring, but there is no third thing produced by (say) night and day.

planation of one will be the explanation of the other. It is "the common" that we should seek to know, that which manifests itself now as one thing and now as its opposite.

We are told by Diogenes Laertius that the book of Heracleitus was divided into three parts, one dealing with the universe, one with politics and one with theology.[1] Bywater has attempted with fair success to arrange the fragments under these three heads, his sections being Nos. 1–90, 91–97, 98–130.

We have only a few fragments dealing with ethics and politics, and it is difficult to extract from them a definite ethical standpoint, but this was certainly dependent on the physical theory. Heracleitus lays great stress on "the common." By this he meant, in the case of the State, the law, but it is harder to conjecture what meaning he attached to it in the case of the individual. The most attractive explanation hitherto given is that of Patrick.[2] He holds that Heracleitus pleaded for unity with nature through obedience to the law of "the common." Communion with the fields and trees could teach men more than discussing virtue and justice. Heracleitus stood for the instinctive, the unconscious, the naïve. "The philosophy and ethics of Heracleitus, as we have seen, stood in vital opposition to"[3] over self-consciousness, too much inwardness and painful self-inspection, absence of trust in our instincts and of the healthful study of nature. We may be sure,

[1] Diogenes Laertius, IX. 5.

[2] *The Fragments of the Work of Heraclitus of Ephesus on Nature*, by G. T. W. Patrick, Baltimore, 1889. See especially pp. 73–83.

[3] *Op. cit.* p. 77.

too, that Heracleitus warned his readers not to expect too much. Perfect bliss is unattainable, for satisfaction is impossible without want, health implies disease, and rest implies painful effort.

The religious teaching of Heracleitus appears to have been directed against customs and ritual rather than against the immoral legends of Homer and Hesiod. He attacks idolatry, mystery-mongers and purification through blood. There is thus no evidence that he was a prophet of Orphism and the mysteries connected with that way of belief. His God must have been the "ever-living Fire," but he appears to have believed that heroic men, who died through excess of fire (*i.e.* in battle or other brave struggle), and not through excess of water (*i.e.* through sottish habits or decay), became the guardians of the living and of the dead. So gods and men are in a sense one. "They live each others' life and die each others' death."

Patrick lays stress, and rightly, upon the stern, prophetic character of many of the fragments. Heracleitus is like a Hebrew seer. He despised all his contemporaries, both the common people and their would-be teachers. Hesiod, Pythagoras, Xenophanes and Hecataeus, all are attacked and condemned. As for the vulgar many, they are spoken of with contempt for their blindness, stupidity and grossness. "Thus the content of Heracleitus' message to his countrymen was *ethical*. It was a call to men everywhere to *wake up*, to purify their βαρβάρους ψυχάς, and to see things in their reality."[1]

It was to this message, in all probability, that he

[1] *Op. cit.*, p. 59.

refers in the word λόγος. Many commentators think that λόγος means "reason" or "law." This was certainly the meaning attached to the word in the ethical system of the Stoics, but although this school borrowed largely from Heracleitus, they developed and indeed transformed his thought, adapting it to the more advanced conceptions of their own day. We are, in fact, tempted to look at Heracleitus through Stoic eyes, and so it is necessary to guard against this danger whenever we are dealing with an ancient statement about Heracleitus that comes from or through a Stoic source.

Our evidence for the doctrines of Heracleitus falls into two classes. We have first the fragments quoted by later writers, with their comments thereon. Then we have the so-called doxographies, or summaries of the views of philosophers. Several of these exist, but they are all derived, directly or indirectly, from a lost work of Theophrastus called Φυσικαὶ δόξαι. In the case of Heracleitus our chief doxographical evidence is contained in the ninth book of the scrappy series of lives of philosophers that goes by the name of Diogenes Laertius. The compiler, whoever he was, probably lived in the third century A.D.

I have followed Bywater in numbering the fragments, though occasionally I do not adopt his readings. Sincere thanks are due to the Delegates of the Oxford University Press for allowing me to use Bywater's numbering and references.

BIBLIOGRAPHY

1807. Schleiermacher, *Herakleitos.*

1848. Bernays (Jac.). *Heraclitea.* Also *Heraklitische Studien* in *Mus. Rh.* 1850, and *Die Heraklitischen Briefe*, 1869.

1858. Lassalle (Ferd.), *Die Philosophie Herakleitos des Dunkeln von Ephesos.*

1873. Schuster (Paul), *Heraklit von Ephesus* in *Actis* soc. phil. Lips., ed. Fr. Ritschelius.

1876. Teichmüller, *Neue Stud. z. Gesch der Begriffe* (Heft II, 1878).

1877. Bywater (I.), *Heracliti Ephesii Reliquiae.*

1886. Pfleiderer, *Die Philosophie des Heraklit von Ephesus im Lichte der Mystericnidee.*

1889. Patrick (G. T. W.), *The Fragments of the Work of Heraclitus of Ephesus.*

1909. Diels (H.), *Herakleitos von Ephesos.*

See also Eduard Zeller, *Die Philosophie der Griechen*, Bd. 1, and John Burnet, *Early Greek Philosophy.*

LIFE OF HERACLITUS[1]

Heraclitus, son of Bloson or, according to some, of Heracon, was a native of Ephesus. He flourished in the 69th Olympiad.[2] He was lofty-minded beyond all other men,[3] and over-weening, as is clear from his book in which he says: "Much learning does not teach understanding; else would it have taught Hesiod and Pythagoras, or, again, Xenophanes and Hecataeus."[4] For "this one thing is wisdom, to understand thought, as that which guides all the world everywhere."[5] And he used to say that "Homer deserved to be chased out of the lists and beaten with rods, and Archilochus likewise."[6]

Again he would say: "There is more need to extinguish insolence than an outbreak of fire,"[7] and "The people must fight for the law as for city-

[1] Taken from R. D. Hicks' translation of Diogenes Laertius in the Loeb Classical Library. The spelling "Heraclitus" is retained. "D." = Diels and "B." = Bywater.

[2] 504–500 B.C.

[3] The biographers used by our author laid evident stress on this characteristic of the Ephesian, for §§ 1–3 (excepting two fragments cited in § 2) dwell on this single theme. As to the criticism of Pythagoras *cf.* Clem. Alex. *Strom.* i. 129 *s. f.*, who, dealing with chronology, says that Heraclitus was later than Pythagoras, for Pythagoras is mentioned by him.

[4] Fr. 40 D., 16 B.

[5] Fr. 41 D., 19 B.

[6] Fr. 42 D., 119 B.

[7] Fr. 43 D., 103 B.

walls."[1] He attacks the Ephesians, too, for banishing his friend Hermodorus: he says: "The Ephesians would do well to end their lives, every grown man of them, and leave the city to beardless boys, for that they have driven out Hermodorus, the worthiest man among them, saying, 'We will have none who is worthiest among us; or if there be any such let him go elsewhere and consort with others.'"[2] And when he was requested by them to make laws, he scorned the request because the state was already in the grip of a bad constitution. He would retire to the temple of Artemis and play at knuckle-bones with the boys; and when the Ephesians stood round him and looked on, "Why, you rascals," he said, "are you astonished? Is it not better to do this than to take part in your civil life?"

Finally, he became a hater of his kind and wandered on the mountains, and there he continued to live, making his diet of grass and herbs. However, when this gave him dropsy, he made his way back to the city and put this riddle to the physicians, whether they were competent to create a drought after heavy rain. They could make nothing of this, whereupon he buried himself in a cowshed, expecting that the noxious damp humour would be drawn out of him by the warmth of the manure. But, as even this was of no avail, he died at the age of sixty.

There is a piece of my own about him as follows[3]:

[1] Fr. 44 D., 100 B. [2] Fr. 121 D., 114 B.

[3] *Anth. Pal.* vii. 127.

Often have I wondered how it came about that Heraclitus endured to live in this miserable fashion and then to die. For a fell disease flooded his body with water, quenched the light in his eyes and brought on darkness.

Hermippus, too, says that he asked the doctors whether anyone could by emptying the intestines draw off the moisture; and when they said it was impossible, he put himself in the sun and bade his servants plaster him over with cow-dung. Being thus stretched and prone, he died the next day and was buried in the market-place. Neanthes of Cyzicus states that, being unable to tear off the dung, he remained as he was and, being unrecognisable when so transformed, he was devoured by dogs.

He was exceptional from his boyhood; for when a youth he used to say that he knew nothing, although when he was grown up he claimed that he knew everything. He was nobody's pupil, but he declared that he "inquired of himself,"[1] and learned everything from himself. Some, however, had said that he had been a pupil of Xenophanes, as we learn from Sotion who also tells us that Ariston in his book *On Heraclitus* declares that he was cured of the dropsy and died of another disease. And Hippobotus has the same story.

As to the work which passes as his, it is a continuous treatise *On Nature*, but is divided into three discourses, one on the universe, another on politics, and a third on theology. This book he deposited in the temple of Artemis and, according to some, he deliberately made it the more obscure in order that none but adepts should approach it, and lest familiarity should breed contempt. Of our philosopher Timon[2] gives a sketch in these words:[3]

[1] Fr. 101 D., 80 B.
[2] Fr. 43 D.
[3] *Cf. Il.* i. 247, 248.

In their midst uprose shrill, cuckoo-like, a mob-reviler, riddling Heraclitus.

Theophrastus puts it down to melancholy that some parts of his work are half-finished, while other parts make a strange medley. As a proof of his magnanimity Antisthenes in his *Successions of Philosophers* cites the fact that he renounced his claim to the kingship in favour of his brother. So great fame did his book win that a sect was founded and called the Heracliteans, after him.

Here is a general summary of his doctrines. All things are composed of fire, and into fire they are again resolved; further, all things come about by destiny, and existent things are brought into harmony by the clash of opposing currents; again, all things are filled with souls and divinities. He has also given an account of all the orderly happenings in the universe, and declares the sun to be no larger than it appears. Another of his sayings is: "Of soul thou shalt never find boundaries, not if thou trackest it on every path; so deep is its cause."[1] Self-conceit he used to call a falling sickness (epilepsy) and eyesight a lying sense.[2] Sometimes, however, his utterances are clear and distinct, so that even the dullest can easily understand and derive therefrom elevation of soul. For brevity and weightiness his exposition is incomparable.

Coming now to his particular tenets, we may state them as follows: fire is the element, all things are exchange for fire and come into being by rarefaction and condensation[3]; but of this he gives no clear explanation. All things come into being by conflict of opposites, and the sum of things flows like a stream. Further, all that is is limited and forms one world.

[1] Fr. 45 D., 71 B. [2] F. 46 D., 132 B.
[3] *Cf.* Fr. 90 D., 22 B.

And it is alternately born from fire and again resolved into fire in fixed cycles to all eternity, and this is determined by destiny. Of the opposites that which tends to birth or creation is called war and strife, and that which tends to destruction by fire is called concord and peace.[1] Change he called a pathway up and down, and this determines the birth of the world.

For fire by contracting turns into moisture, and this condensing turns into water; water again when congealed turns into earth. This process he calls the downward path. Then again earth is liquefied, and thus gives rise to water, and from water the rest of the series is derived. He reduces nearly everything to exhalation from the sea. This process is the upward path. Exhalations arise from earth as well as from sea; those from sea are bright and pure, those from earth dark. Fire is fed by the bright exhalations, the moist element by the others. He does not make clear the nature of the surrounding element. He says, however, that there are in it bowls with their concavities turned towards us, in which the bright exhalations collect and produce flames. These are the stars. The flame of the sun is the brightest and the hottest; the other stars are further from the earth and for that reason give it less light and heat. The moon, which is nearer to the earth, traverses a region which is not pure. The sun, however, moves in a clear and untroubled region, and keeps a proportionate distance from us. That is why it gives us more heat and light. Eclipses of the sun and moon occur when the bowls are turned

[1] *Cf.* Fr. 80 D., 62 B.

upwards; the monthly phases of the moon are due to the bowl turning round in its place little by little. Day and night, months, seasons and years, rains and winds and other similar phenomema are accounted for by the various exhalations. Thus the bright exhalation, set aflame in the hollow orb of the sun, produces day, the opposite exhalation when it has got the mastery causes night; the increase of warmth due to the bright exhalation produces summer, whereas the preponderance of moisture due to the dark exhalation brings about winter. His explanations of other phenomena are in harmony with this. He gives no account of the nature of the earth, nor even of the bowls. These, then, were his opinions.

The story told by Ariston of Socrates, and his remarks when he came upon the book of Heraclitus, which Euripides brought him, I have mentioned in my Life of Socrates.[1] However, Seleucus the grammarian says that a certain Croton relates in his book called *The Diver* that the said work of Heraclitus was first brought into Greece by one Crates, who further said it required a Delian diver not to be drowned in it. The title given to it by some is *The Muses*,[2] by others *Concerning Nature*; but Diodotus calls it [3]

A helm unerring for the rule of life;

others "a guide of conduct, the keel of the whole

[1] ii. 22.

[2] Plato, alluding to Heraclitus, speaks of "Ionian Muses" (*Soph*. 242 E). He is followed by Clement of Alexandria (*Strom*. v. 9, 682 P. *αἱ γοῦν Ἰάδες Μοῦσαι διαρρήδην λέγουσι*), and possibly, as M. Ernout thinks, by Lucretius, i. 657, where "Musae" is the MS. reading. But *cf*. Lachmann, *ad loc*.

[3] Nauck, *T.G.F.*[2], *Adesp*. 287.

world, for one and all alike." We are told that, when asked why he kept silence, he replied, "Why, to let you chatter." Darius, too, was eager to make his acquaintance, and wrote to him as follows[1]:

"King Darius, son of Hystaspes, to Heraclitus the wise man of Ephesus, greeting.

"You are the author of a treatise *On Nature* which is hard to understand and hard to interpret. In certain parts, if it be interpreted word for word, it seems to contain a power of speculation on the whole universe and all that goes on within it, which depends upon motion most divine; but for the most part judgement is suspended, so that even those who are the most conversant with literature are at a loss to know what is the right interpretation of your work. Accordingly King Darius, son of Hystaspes, wishes to enjoy your instruction and Greek culture. Come then with all speed to see me at my palace. For the Greeks as a rule are not prone to mark their wise men; nay, they neglect their excellent precepts which make for good hearing and learning. But at my court there is secured for you every privilege and daily conversation of a good and worthy kind, and a life in keeping with your counsels."

"Heraclitus of Ephesus to King Darius, son of Hystaspes, greeting.

"All men upon earth hold aloof from truth and justice, while, by reason of wicked folly, they devote themselves to avarice and thirst for popularity. But

[1] The request of Darius is mentioned by Clem. Alex. *Strom.* i. 65 οὗτος βασιλέα Δαρεῖον παρακαλοῦντα ἥκειν εἰς Πέρσας ὑπερεῖδεν. The story is not made more plausible by the two forged letters to which it must have given rise.

I, being forgetful of all wickedness, shunning the general satiety which is closely joined with envy, and because I have a horror of splendour, could not come to Persia, being content with little, when that little is to my mind."

So independent was he even when dealing with a king.

Demetrius, in his book on *Men of the Same Name*, says that he despised even the Athenians, although held by them in the highest estimation; and, notwithstanding that the Ephesians thought little of him, he preferred his own home the more. Demetrius of Phalerum, too, mentions him in his *Defence of Socrates* [1]; and the commentators on his work are very numerous, including as they do Antisthenes and Heraclides of Pontus, Cleanthes and Sphaerus the Stoic, and again Pausanius who was called the imitator of Heraclitus, Nicomedes, Dionysius, and among the grammarians, Diodotus. The latter affirms that it is not a treatise upon nature, but upon government, the physical part serving merely for illustration.[2]

Hieronymus tells us that Scythinus, the satirical poet, undertook to put the discourse of Heraclitus into verse. He is the subject of many epigrams, and amongst them of this one [3]:

Heraclitus am I. Why do ye drag me up and down, ye illiterate? It was not for you I toiled, but for such as

[1] This work is again quoted in ix. 37 and ix. 57, and is perhaps the source of the first sentence of § 52 also.

[2] Apparently D. L. is using through another of his sources, the very same citation from Diodotus which he has given verbatim in § 12.

[3] *Anth. Pal.* vii. 128.

understand me. One man in my sight is a match for thirty thousand, but the countless hosts do not make a single one. This I proclaim, yea in the halls of Persephone.

Another runs as follows[1]:

Do not be in too great a hurry to get to the end of Heraclitus the Ephesian's book: the path is hard to travel. Gloom is there and darkness devoid of light. But if an initiate be your guide the path shines brighter than sunlight.

Five men have borne the name of Heraclitus: (1) our philosopher; (2) a lyric poet, who wrote a hymn of praise to the twelve gods; (3) an elegiac poet of Halicarnassus, on whom Callimachus wrote the following epitaph[2]:

They told me, Heraclitus, they told me you were dead,
They brought me bitter news to hear and bitter tears to shed.
I wept as I remembered how often you and I
Had tired the sun with talking and sent him down the sky.

And now that thou art lying, my dear old Carian guest,
A handful of grey ashes, long, long ago at rest,
Still are thy pleasant voices, thy nightingales, awake;
For Death, he taketh all away, but them he cannot take;[3]

(4) a Lesbian who wrote a history of Macedonia; (5) a jester who adopted this profession after having been a musician.

---

[1] *Anth. Pal.* ix. 540. [2] *Anth. Pal.* vii. 80.

[3] From Cory's *Ionica*, p. 7. In bare prose: "One told me of thy death, Heraclitus, and moved me to tears, when I remembered how often we two watched the sun go down upon our talk. But though thou, I ween, my Halicarnassian friend, art dust long, long ago, yet do thy 'Nightingales' live on, and Death, that insatiate ravisher, shall lay no hand on them." Perhaps "Nightingales" was the title of a work. Laertius deserves our gratitude for inserting this little poem, especially on so slight a pretext.

HERACLEITUS
ON THE UNIVERSE

ΗΡΑΚΛΕΙΤΟΥ ΕΦΕΣΙΟΥ

ΠΕΡΙ ΤΟΥ ΠΑΝΤΟΣ.

The order of the fragments is that of Bywater.

I. Οὐκ ἐμεῦ ἀλλὰ τοῦ λόγου ἀκούσαντας
2 ὁμολογέειν σοφόν ἐστι, ἓν πάντα εἶναι.

II. Τοῦ δὲ λόγου τοῦδ' ἐόντος αἰεὶ ἀξύνετοι γίνονται ἄνθρωποι καὶ πρόσθεν ἢ ἀκοῦσαι καὶ ἀκούσαντες τὸ πρῶτον. γινομένων γὰρ πάντων κατὰ τὸν λόγον τόνδε ἀπείροισι ἐοίκασι πειρώμενοι καὶ ἐπέων καὶ ἔργων τοιουτέων ὁκοίων ἐγὼ διηγεῦμαι, διαιρέων ἕκαστον κατὰ φύσιν καὶ φράζων ὅκως ἔχει. τοὺς δὲ ἄλλους ἀνθρώπους λανθάνει ὁκόσα ἐγερθέντες ποιέουσι, ὅκωσπερ
9 ὁκόσα εὕδοντες ἐπιλανθάνονται.

I. Hippolytus *Ref. Haer.* ix. 9: Ἡράκλειτος μὲν οὖν ⟨ἓν⟩ φησιν εἶναι τὸ πᾶν, διαιρετὸν ἀδιαίρετον, γενητὸν ἀγένητον, θνητὸν ἀθάνατον, λόγον αἰῶνα, πατέρα υἱόν, θεὸν δίκαιον. Οὐκ ἐμοῦ ἀλλὰ τοῦ δόγματος ἀκούσαντας ὁμολογεῖν σοφόν ἐστιν, ἓν πάντα εἰδέναι, ὁ Ἡράκλειτός φησι· καὶ ὅτι τοῦτο οὐκ ἴσασι πάντες οὐδὲ ὁμολογοῦσιν, ἐπιμέμφεται ὧδέ πως· Οὐ ξυνίασιν ὅκως διαφερόμενον ἑωυτῷ ὁμολογέει· παλίντροπος ἁρμονίη ὅκωσπερ τόξου καὶ λύρης.

λόγου is a conjecture of Bernays, εἶναι a conjecture of Miller. Bergk would reconstruct thus: δίκαιον οὐκ ἐμοῦ ἀλλὰ τοῦ δόγματος ἀκούσαντας ὁμολογέειν ὅτι ἓν τὸ σοφόν ἐστιν, ἓν πάντα εἰδέναι. The conjectures in the text do not arouse any strong confidence, though δόγματος might well be a gloss on λόγου. But if εἶναι be correct, why should it have been corrupted to εἰδέναι? I am on the whole inclined to think that Bergk's restoration is nearer to the actual words of Heracleitus.

HERACLEITUS

ON THE UNIVERSE

I. It is wise to listen, not to me but to the Word, and to confess that all things are one.

For λόγος see Heinze, *Lehre vom Logos*, 1873 ; Zeller, i. 630 ; Aall, *Gesch. d. Logosidee* 1896. "All things are one" because they are all resolved into fire and come from fire.

II. This Word, which is ever true, men prove as incapable of understanding when they hear it for the first time as before they have heard it at all. For although all things happen in accordance with this Word, men seem as though they had no experience thereof, when they make experiment with such words and works as I relate, dividing each thing according to its nature and setting forth how it really is. The rest of men know not what they do when awake, just as they forget what they do when asleep.

Aristotle was in doubt whether αἰεὶ should be taken with ἐόντος or with ἀξύνετοι γίνονται. See *Rhetoric*, III. 5, 1407, *b* 14. ἐόντος means "true" in Ionic with words like λόγος. See Burnet, *E. G. Ph.* note on Fragment II. I have tried in my translation to bring out the play on words in ἀπείροισι ἐοίκασι πειρώμενοι.

II Hipp. *Ref. Haer.* ix. 9 ; Aristotle *Rhetoric* iii. 5 ; Sextus Empiricus *adversus Mathematicos* vii. 132 ; Clement of Alex. *Strom.* v. 14, p. 716 ; Eusebius *Praep. Ev*. xiii 13, p. 680. The MSS. (except those of Sextus) read τοῦ δεόντος.

III. Ἀξύνετοι ἀκούσαντες κωφοῖσι ἐοίκασι·
2 φάτις αὐτοῖσι μαρτυρέει παρεόντας ἀπεῖναι.

IV. Κακοὶ μάρτυρες ἀνθρώποισι ὀφθαλμοὶ καὶ
2 ὦτα, βαρβάρους ψυχὰς ἐχόντων.

V. Οὐ φρονέουσι τοιαῦτα πολλοὶ ὁκόσοισι
ἐγκυρέουσι οὐδὲ μαθόντες γινώσκουσι, ἑωυτοῖσι
3 δὲ δοκέουσι.

VI. Ἀκοῦσαι οὐκ ἐπιστάμενοι οὐδ' εἰπεῖν.

VII. Ἐὰν μὴ ἔλπηαι, ἀνέλπιστον οὐκ ἐξευρή-
2 σει, ἀνεξερεύνητον ἐὸν καὶ ἄπορον.

VIII. Χρυσὸν οἱ διζήμενοι γῆν πολλὴν ὀρύσ-
2 σουσι καὶ εὑρίσκουσι ὀλίγον.

IX. Ἀγχιβασίην.

X. Φύσις κρύπτεσθαι φιλεῖ.

XI. Ὁ ἄναξ οὗ τὸ μαντεῖόν ἐστι τὸ ἐν
Δελφοῖς, οὔτε λέγει οὔτε κρύπτει, ἀλλὰ
3 σημαίνει.

III. Clem. Alex. *Strom.* v. 14, p. 718; Euseb. *P.E.* xiii. 13, p. 681.

IV. Sextus Emp. *adv. Math.* vii. 126; Stobaeus *Florilegium* iv. 56. βορβόρου ψυχὰς ἔχοντος Bernays.

V. Clem. Alex. *Strom.* ii. 2, p. 432; Marcus Antoninus iv. 46.

VI. Clem. Alex. *Strom.* ii. 5, p. 442.

VII. Clem. Alex. *Strom.* ii. 4, p. 437. Theodoretus *Therap.* i. p. 15, 51. The sources have ἔλπηται and ἐλπίζητε. ἔλπηαι Schuster and Bywater. Some would put the comma after ἀνέλπιστον instead of before it.

VIII. Clem. Alex. *Strom.* iv. 2, p. 565; Theodoretus *Therap.* i. p. 15, 52.

IX. Suidas *s.v.*

X. Themistius *Or.* v. p. 69.

XI. Plutarch *de Pyth. Orac.* 21, p. 404; Iamblichus *de Myst.* iii. 15; Stobaeus *Flor.* v. 72 and lxxxi. 17.

III. The stupid when they have heard are like the deaf; of them does the proverb bear witness that when present they are absent.

IV. Bad witnesses are eyes and ears to men, if they have souls that understand not their language.

This passage is not a general attack on the senses; it merely lays stress on the need of an intelligent soul to interpret the sense-impressions. The clever emendation of Bernays would mean: "when mud holds the soul," *i.e.* when the soul is moist, and therefore (on Heracleitean principles) dull and stupid.

V. Many do not interpret aright such things as they encounter, nor do they have knowledge of them when they have learned, though they seem to themselves so to do.

H. seems to be referring to (*a*) the correct apprehension of phenomena and (*b*) the difference between unintelligent learning and understanding.

VI. Knowing neither how to listen nor how to speak.

VII. If you do not expect it, you will not find out the unexpected, as it is hard to be sought out and difficult.

Heracleitus is laying stress upon the importance of the constructive imagination in scientific enquiry—what the early Christians might have called "faith."

VIII. Gold-seekers dig much earth to find a little gold.

IX. Critical discussion.

X. Nature is wont to hide herself.

φύσις is not necessarily an abstraction here, but merely the truth about the Universe. It is easy, however, to see why the Stoics could maintain that their pantheism was founded on Heracleitus. See Fragments XIX, XCI, XCII.

XI. The Lord whose is the oracle in Delphi neither declares nor hides, but sets forth by signs.

XII. Σίβυλλα δὲ μαινομένῳ στόματι ἀγέλαστα καὶ ἀκαλλώπιστα καὶ ἀμύριστα φθεγγομένη χιλίων ἐτέων ἐξικνέεται τῇ φωνῇ διὰ 4 τὸν θεόν.

XIII. Ὅσων ὄψις ἀκοὴ μάθησις, ταῦτα ἐγὼ 2 προτιμέω.

XIV. Τοῦτο γὰρ ἴδιόν ἐστι τῶν νῦν καιρῶν, ἐν οἷς πάντων πλωτῶν καὶ πορευτῶν γεγονότων οὐκ ἂν ἔτι πρέπον εἴη ποιηταῖς καὶ μυθογράφοις χρῆσθαι μάρτυσι περὶ τῶν ἀγνοουμένων, ὅπερ οἱ πρὸ ἡμῶν περὶ τῶν πλείστων, ἀπίστους ἀμφισβητουμένων 6 παρεχόμενοι βεβαιωτὰς κατὰ τὸν Ἡράκλειτον.[1]

XV. Ὀφθαλμοὶ τῶν ὤτων ἀκριβέστεροι μάρ- 2 τυρες.

XVI. Πολυμαθίη νόον ἔχειν οὐ διδάσκει· Ἡσίοδον γὰρ ἂν ἐδίδαξε καὶ Πυθαγόρην· αὖτίς τε 3 Ξενοφάνεα καὶ Ἑκαταῖον.

XII. Plutarch *de Pyth. Orac.* 6, p. 397.

XIII. Hipp. *Ref. Haer.* ix. 9.

Bywater prints this fragment with a question mark at the end.

XV. Polybius xii. 27.

XVI. Diogenes Laertius ix. 1; cf. Clem. Alex. *Strom.* i. 19, p. 373; Athenaeus xiii. p. 610 B; Aulus Gellius *praef.* 12.

[1] Polybius iv. 40.

XII. The Sibyl with raving mouth utters things mirthless, unadorned and unperfumed, but with her voice she extends over a thousand years because of the God.

In this and the preceding H. seems to be calling attention to his oracular style, which was in part due to the strong religious emotion of his age. There is much that is oracular in Aeschylus and Pindar.

XIII. The things that can be seen, heard and learnt, these I honour especially.

This and the following two fragments emphasise the importance of personal research, as contrasted with learning from authority. Bywater's punctuation would make the meaning to be: "Am I to value highly those things that are learnt by sight or hearing?"—an attack upon the accuracy and value of the senses. But H. does not distrust the senses, but only sense-impressions interpreted in a stupid way.

XIV. Particularly at the present time, when all places can be reached by water or by land, it would not be right to use as evidence for the unknown the works of poets and mythologists, as in most things our predecessors did, proving themselves, as Heracleitus has it, unreliable supporters of disputed points.

XV. Eyes are more accurate witnesses than ears.

First-hand information is better than hearsay.

XVI. Much learning does not teach understanding, or it would have taught Hesiod and Pythagoras, as well as Xenophanes and Hecataeus.

As is plain from the following fragment, this is an attack on confusing second-hand information with true understanding and education. It is unfair to the mathematical achievements of Pythagoras and scarcely does justice to the theological acumen of Xenophanes, to say nothing of his wonderful

XVII. Πυθαγόρης Μνησάρχου ἱστορίην ἤσκησε ἀνθρώπων μάλιστα πάντων. καὶ ἐκλεξάμενος ταύτας τὰς συγγραφὰς ἐποιήσατο ἑωυτοῦ
4 σοφίην, πολυμαθίην, κακοτεχνίην.

XVIII. Ὁκόσων λόγους ἤκουσα οὐδεὶς ἀφικνέεται ἐς τοῦτο, ὥστε γινώσκειν ὅτι σοφόν ἐστι
3 πάντων κεχωρισμένον.

XIX. Ἓν τὸ σοφόν, ἐπίστασθαι γνώμην ᾗ
2 κυβερνᾶται πάντα διὰ πάντων.

XX. Κόσμον τόνδε τὸν αὐτὸν ἁπάντων οὔτε τις θεῶν οὔτε ἀνθρώπων ἐποίησε, ἀλλ' ἦν αἰεὶ καὶ ἔστι καὶ ἔσται πῦρ ἀείζωον, ἁπτόμενον μέτρα
4 καὶ ἀποσβεννύμενον μέτρα.

XXI. Πυρὸς τροπαὶ πρῶτον θάλασσα· θαλάσσης δὲ τὸ μὲν ἥμισυ γῆ, τὸ δὲ ἥμισυ
3 πρηστήρ.

XVII. Diogenes Laertius viii. 6. One MS. has ἐποίησεν and one ἐποίησατο. Bywater reads ἐποίησε and Burnet ἐποιήσατο.

XVIII. Stobaeus *Flor*. iii. 81.

XIX. Diogenes Laertius ix. 1.

XX. Clem. Alex. *Strom*. v. 14, p 711; cf. Simplicius in Aristotle *de Caelo*, p. 132; Plutarch *de Anim. Procreatione* 5, p. 1014.

XXI. Clem. Alex. *Strom*. v. 14, p. 712.

anticipation of the modern doctrine of scientific progress. See Fragment XVI. (In Stob. *Flor.* 29, 41):

οὔ τοι ἀπ' ἀρχῆς πάντα θεοὶ θνητοῖς παρέδειξαν,
ἀλλὰ χρόνῳ ζητοῦντες ἐφευρίσκουσιν ἄμεινον.

XVII. Pythagoras, son of Mnesarchus, practised research more than any other man, and choosing out these writings claimed as his own a wisdom that was only much learning, a mischievous art.

An attack on book-learning that is merely the acquisition of second-hand information. Diels rejects the fragment as spurious, chiefly because it makes Pythagoras a writer of books. But the reading *ἐποιήσατο* for *ἐποίησεν* does away with this objection.

XVIII. Of all those whose discourses I have heard, not one attains to this, to realise that wisdom is a thing apart from all.

This has been interpreted to mean that true wisdom is attained by none, or that general opinions do not contain real wisdom.

XIX. Wisdom is one thing—to know the thought whereby all things are steered through all things.

That is, to understand the doctrine of opposites and of perpetual change.

XX. This world, which is the same for all, was made neither by a god nor by man, but it ever was, and is, and shall be, ever-living Fire, in measures being kindled and in measures going out.

The use of *κόσμος* to mean "world" is Pythagorean. *μέτρα* refers to the approximate correspondence between the things that are becoming fire and the things that are coming out of fire. The balance of nature is not disturbed by perpetual flux.

XXI. The transformations of Fire are, first, sea; of sea half is earth and half fiery storm-cloud.

This is the famous "road up and down" (or at any rate the best illustration of it) with its three stages—earth, water,

XXII. Πυρὸς ἀνταμείβεται πάντα καὶ πῦρ ἁπάντων, ὥσπερ χρυσοῦ χρήματα καὶ χρημάτων 3 χρυσός.

XXIII. Θάλασσα διαχέεται καὶ μετρέεται ἐς 2 τὸν αὐτὸν λόγον ὁκοῖος πρόσθεν ἦν ἢ γενέσθαι.

XXIV. Χρησμοσύνη . . . κόρος.

XXV. Ζῇ πῦρ τὸν ἀέρος θάνατον, καὶ ἀὴρ ζῇ τὸν πυρὸς θάνατον· ὕδωρ ζῇ τὸν γῆς θάνατον, γῆ 3 τὸν ὕδατος.

XXVI. Πάντα τὸ πῦρ ἐπελθὸν κρινέει καὶ 2 καταλήψεται.

XXVII. Τὸ μὴ δῦνόν ποτε πῶς ἄν τις λάθοι;

XXVIII. Τὰ δὲ πάντα οἰακίζει κεραυνός.

XXIX. Ἥλιος οὐχ ὑπερβήσεται μέτρα· εἰ δὲ 2 μή, Ἐρινύες μιν δίκης ἐπίκουροι ἐξευρήσουσι.

XXII. Plutarch *de* EI 8, p. 388; Diog. Laert. ix. 8; Eusebius *Praep. Evang.* xiv. 3, p. 720.

XXIII. Clem. Alex. *Strom.* v. 14, p. 712; Euseb. *P. E.* xiii. 13, p. 676.

The MSS. of Clement read γῆ after γενέσθαι, whence Schuster reads γῆν. In any case earth is referred to, and γῆ is probably the subject of διαχέεται. See Burnet.

XXIV. Hipp. *Ref. Haer.* ix. 10; Philo *de Victim.* 6, p. 242; Plutarch *de* EI 9, p. 389.

XXV. Maximus Tyr. xli. 4, p. 489. See also Plutarch *de* EI 18, p. 392, and M. Anton. iv. 46.

In the texts ἀέρος and γῆς are transposed. Diels reads as above; Bywater retains the old order.

XXVI. Hipp. *Ref. Haer.* ix. 10.

XXVII. Clem. Alex. *Paedag.* ii. 10, p. 229.

XXVIII. Hipp. *Ref. Haer.* ix. 10.

XXIX. Plutarch *de Exil.* 11, p. 604.

fire. On the earth is the sea, above the sea is the sun. Sea is half composed of earth transforming itself to water and half of fiery cloud, the latter representing water on its way to become fire. This explanation of πρηστήρ I owe to Burnet.

XXII. All things are exchanged for Fire and Fire for all things, even as goods for gold and gold for goods.

XXIII. It is melted into sea, and is measured to the same proportion as before it became earth.

The subject is γῆ, and the whole fragment means that along the "road up" the proportion of the "measures" remains constant. The amount of earth in the universe remains approximately the same, because the "measures" of water turning to earth equal the "measures" of earth turning to water.

XXIV. Want . . . surfeit.

E.g. the "want" of earth for water to increase it equals the "surfeit" of earth which makes some of it turn to water.

XXV. Fire lives the death of air, and air lives the death of Fire; water lives the death of earth, earth that of water.

XXVI. Fire when it has advanced will judge and convict all things.

For the "advances" of fire see περὶ διαίτης I, Chap. III. Such statements as the one above led the Stoics to develop their theory of ἐκπύρωσις, the destruction of all things periodically by fire, to be followed by a re-birth and restoration of all things.

XXVII. How can you hide from that which never sets?

XXVIII. The thunderbolt steers all things.

XXIX. The sun will not overstep his measures; otherwise the Erinyes, helpers of Justice, will find him out.

See the notes to XX and XXIII.

XXX. Ἠοῦς καὶ ἑσπέρης τέρματα ἡ ἄρκτος,
2 καὶ ἀντίον τῆς ἄρκτου οὖρος αἰθρίου Διός.

XXXI. Εἰ μὴ ἥλιος ἦν, ἕνεκα τῶν ἄλλων
2 ἄστρων εὐφρόνη ἂν ἦν.

XXXII. Νέος ἐφ' ἡμέρῃ ἥλιος.

XXXIII. Δοκεῖ δὲ (scil. Θαλῆς) κατά τινας πρῶτος ἀστρολογῆσαι καὶ ἡλιακὰς ἐκλείψεις καὶ τροπὰς προειπεῖν, ὥς φησιν Εὔδημος ἐν τῇ περὶ τῶν ἀστρολογουμένων ἱστορίᾳ· ὅθεν αὐτὸν καὶ Ξενοφάνης καὶ Ἡρόδοτος θαυμάζει· μαρτυρεῖ δ'
5 αὐτῷ καὶ Ἡράκλειτος καὶ Δημόκριτος.[1]

XXXIV. Οὕτως οὖν ἀναγκαίαν πρὸς τὸν οὐρανὸν ἔχων συμπλοκὴν καὶ συναρμογὴν ὁ χρόνος οὐχ ἁπλῶς ἐστι κίνησις ἀλλ', ὥσπερ εἴρηται, κίνησις ἐν τάξει μέτρον ἐχούσῃ καὶ πέρατα καὶ περιόδους. ὧν ὁ ἥλιος ἐπιστάτης ὢν καὶ σκοπός, ὁρίζειν καὶ βραβεύειν καὶ ἀναδεικνύναι καὶ ἀναφαίνειν μεταβολὰς καὶ ὥρας αἳ πάντα φέρουσι, καθ' Ἡράκλειτον, οὐδὲ φαύλων οὐδὲ μικρῶν, ἀλλὰ τῶν μεγίστων καὶ
10 κυριωτάτων τῷ ἡγεμόνι καὶ πρώτῳ θεῷ γίνεται συνεργός.[2]

XXX. Strabo i. 6, p 3.

XXXI. Plutarch *Aquae et Ignis Comp.* 7, p. 957, and *de Fortuna* 3, p. 98. Cf. Clem. Alex. *Protrept.* ii. p. 87.

Bywater does not include the words ἕνεκα . . . ἄστρων in the text, but considers them to be a part of the narrator's explanation.

XXXII. Aristotle *Meteor.* ii. 2, p. 355, *a* 9. See the comments of Alex. *Aphrod.* and of Olympiodorus. Also Proclus *in Timaeum*, p. 334 B.

[1] Diogenes Laert. i. 23.
[2] Plutarch *Qu. Plat.* viii. 4, p. 1007.

XXX. The limits of the East and West are the Bear, and opposite the Bear is the boundary of bright Zeus.

The "boundary of bright Zeus" is, according to Diels, the South Pole. Burnet takes it to be the horizon, and the whole passage a protest against the Pythagorean view of a southern hemisphere.

XXXI. If there were no sun, there would be night, in spite of the other stars.

XXXII. The sun is new every day.

This is because of the perpetual flux. One sun is extinguished at sunset; another is kindled at sunrise.

XXXIII. Thales is supposed by some to have been the first astronomer and the first to foretell the eclipses and turnings of the sun, as Eudemus declares in his account of astronomical discoveries. For this reason both Xenophanes and Herodotus pay him respectful honour, and both Heracleitus and Democritus bear witness to him.

XXXIV. So time, having a necessary connection and union with the firmament, is not motion merely, but, as I have said, motion in an order having measure, limits and periods. Of which the sun, being overseer and warder, to determine, judge, appoint and declare the changes and seasons, which, according to Heracleitus, bring all things, is a helper of the leader and first God, not in trivial or small things, but in the greatest and most important.

XXXV. Διδάσκαλος δὲ πλείστων Ἡσίοδος· τοῦτον ἐπίστανται πλεῖστα εἰδέναι, ὅστις ἡμέρην
3 καὶ εὐφρόνην οὐκ ἐγίνωσκε· ἔστι γὰρ ἕν.

XXXVI. Ὁ θεὸς ἡμέρη εὐφρόνη, χειμὼν θέρος, πόλεμος εἰρήνη, κόρος λιμός· ἀλλοιοῦται δὲ ὅκωσπερ πῦρ, ὁκόταν συμμιγῇ θυώμασι, ὀνομάζε-
4 ται καθ' ἡδονὴν ἑκάστου.

XXXVII. Εἰ πάντα τὰ ὄντα καπνὸς γένοιτο,
2 ῥῖνες ἂν διαγνοῖεν.

XXXVIII. Αἱ ψυχαὶ ὀσμῶνται καθ' ᾅδην.

XXXIX. Τὰ ψυχρὰ θέρεται, θερμὸν ψύχεται,
2 ὑγρὸν αὐαίνεται, καρφαλέον νοτίζεται.

XL. Σκίδνησι καὶ συνάγει, πρόσεισι καὶ
2 ἄπεισι.

XLI. Ποταμοῖσι δὶς τοῖσι αὐτοῖσι οὐκ ἂν
2 ἐμβαίης· ἕτερα γὰρ <καὶ ἕτερα> ἐπιρρέει ὕδατα.

XXXV. Hipp. *Ref. Haer.* ix. 10.

XXXVI. Hipp. *Ref. Haer.* ix. 10. Diels reads ὅκωσπερ ⟨πῦρ⟩ :
Bywater adds θύωμα after συμμιγῇ, with Bernays, and Zeller adds ἀήρ in the same place.

XXXVII. Aristotle *de Sensu* 5, p. 443, *a* 21.

XXXVIII. Plutarch *de Fac. in Orbe Lunae* 28, p. 943.

XXXIX. Scholiast, Tzetzes *ad Exeg. in Iliada*, p. 126.

XL. Plutarch *de* EI 18, p. 392.

XLI. Plutarch *Quaest. nat.* 2, p. 912; *de sera Num. Vind.* 15, p. 559; *de* EI 18, p. 392. See Plato *Cratylus* 402 A, and Aristotle *Meta.* iv. 5, p. 1010 *a* 13.

XLII. I omit this, as being obviously a corrupt form of XLI.

XXXV. The teacher of most men is Hesiod. They think that he knew very many things, though he did not understand day and night. For they are one.

In *Theogony* 124 Hesiod calls day the daughter of night. According to Heracleitus day and night, two opposites, are really one, or, as we should say, two aspects of the same thing.

XXXVI. God is day and night, winter and summer, war and peace, surfeit and hunger. But he undergoes transformations, just as fire, when it is mixed with spices, is named after the savour of each.

"Unity of opposites" again. Burnet renders ἡδονὴ "savour," and I have followed him, though with some hesitation, especially as the reading of the second sentence is dubious. καθ' ἡδονὴν ἑκάστου could mean: "according to individual caprice," and I am not certain that this is not the meaning here.

XXXVII. If all existing things were to become smoke, the nostrils would distinguish them.

XXXVIII. Souls smell in Hades.

It is difficult to see what sense can be given to this fragment except that in Hades souls are a smoky exhalation, and so come under the sense of smell. Pfleiderer suggested ὁσιοῦνται, "are made holy," a thought foreign to Heracleitus.

XXXIX. Cold things become warm, warmth cools, moisture dries, the parched gets wet.

XL. It scatters and gathers, it comes and goes.

XLI. You could not step twice into the same rivers; for other waters are ever flowing on to you.

XLIII. Καὶ Ἡράκλειτος ἐπιτιμᾷ τῷ ποιήσαντι· ὡς ἔρις ἔκ τε θεῶν καὶ ἀνθρώπων ἀπόλοιτο· οὐ γὰρ ἂν εἶναι ἁρμονίαν μὴ ὄντος ὀξέος καὶ βαρέος, οὐδὲ τὰ ζῷα ἄνευ θήλεος καὶ ἄρρενος,
5 ἐναντίων ὄντων.[1]

XLIV. Πόλεμος πάντων μὲν πατήρ ἐστι πάντων δὲ βασιλεύς, καὶ τοὺς μὲν θεοὺς ἔδειξε τοὺς δὲ ἀνθρώπους, τοὺς μὲν δούλους ἐποίησε
4 τοὺς δὲ ἐλευθέρους.

XLV. Οὐ ξυνίασι ὅκως διαφερόμενον ἑωυτῷ ὁμολογέει· παλίντονος ἁρμονίη ὅκωσπερ τόξου
3 καὶ λύρης.

XLVI. Καὶ περὶ αὐτῶν τούτων ἀνώτερον ἐπιζητοῦσι καὶ φυσικώτερον· Εὐριπίδης μὲν φάσκων ἐρᾶν μὲν ὄμβρου γαῖαν ξηρανθεῖσαν, ἐρᾶν δὲ σεμνὸν οὐρανὸν πληρούμενον ὄμβρου πεσεῖν ἐς γαῖαν· καὶ Ἡράκλειτος τὸ ἀντίξουν συμφέρον, καὶ ἐκ τῶν διαφερόντων καλλίστην
7 ἁρμονίαν, καὶ πάντα κατ' ἔριν γίνεσθαι.[2]

XLIII. See also Simplicius in Arist. *Categ.* p. 104 Δ. Eustathius on *Iliad* xviii. p. 107, and the Ven. A, Scholiast.

XLIV. Hipp. *Ref. Haer.* ix. 9; Plutarch *de Iside.* 48, p 370.

XLV. Plato *Symposium* 187 A, *Sophist* 242 D; Plutarch *de Anim. Procreatione* 27, p. 1026, *de Iside* 45, p. 369, παλίντονος γὰρ ἁρμονίη κόσμου ὅκωσπερ λύρης καὶ τόξου καθ' Ἡράκλειτον. Burnet thinks (rightly) that Heracleitus could not have said both παλίντροπος and παλίντονος; he prefers the latter and Diels the former. The one refers to the shape of the bow, the latter to the tension in the bow-string. Bywater reads παλίντροπος (as in Plut. *de An. Pr.* and Hipp. *Ref. Haer.* ix. 9).

XLIII. And Heracleitus rebukes the poet who says, "would that strife might perish from among gods and men." For there could be (he said) no attunement without the opposites high and low, and no animals without the opposites male and female.

XLIV. War is the father of all and the king of all; some he has marked out to be gods and some to be men, some he has made slaves and some free.

XLV. They understand not how that which is at variance with itself agrees with itself. There is attunement of opposite tensions, like that of the bow and of the harp.

With the reading παλίντροπος the meaning is: "a harmony from opposite shapes."

XLVI. In reference to these very things they look for deeper and more natural principles. Euripides says that "the parched earth is in love with rain," and that "high heaven, with rain fulfilled, loves to fall to earth." And Heracleitus says that "the opposite is beneficial," and that "from things that differ comes the fairest attunement," and that "all things are born through strife."

Burnet thinks that there is a reference to the medical theory of "like is cured by unlike" in the first of these quotations from Heracleitus (τὸ ἀντίξουν συμφέρον). See also Stewart on Aristotle, *Eth. Nic.* 1104, *b*16.

---

[1] Aristotle, *Eth. Eud.* vii. 1, p. 1235a, 26.
[2] Aristotle, *Eth. Nic.* viii. 2, p. 1151b1.

XLVII. Ἁρμονίη ἀφανὴς φανερῆς κρείσσων.

XLVIII. Μὴ εἰκῆ περὶ τῶν μεγίστων συμβα-
2 λώμεθα.

XLIX. Χρὴ εὖ μάλα πολλῶν ἵστορας φιλο-
2 σόφους ἄνδρας εἶναι.

L. Γναφέων ὁδὸς εὐθεῖα καὶ σκολιὴ μία ἐστὶ
2 καὶ ἡ αὐτή.

LI. Ὄνοι σύρματ' ἂν ἕλοιντο μᾶλλον ἢ χρυσόν.

LIa. Heraclitus dixit quod si felicitas esset in delectationibus corporis boves felices diceremus,
3 cum inveniant orobum ad comedendum.[1]

LII. Θάλασσα ὕδωρ καθαρώτατον καὶ μιαρώτατον, ἰχθύσι μὲν πότιμον καὶ σωτήριον,
3 ἀνθρώποις δὲ ἄποτον καὶ ὀλέθριον.

LIII. Siccus etiam pulvis et cinis, ubicunque cohortem porticus vel tectum protegit, iuxta parietes reponendus est, ut sit quo aves se perfundant: nam his rebus plumam pinnasque emendant, si modo credimus Ephesio Heraclito qui ait: sues coeno,
6 cohortales aves pulvere (vel cinere) lavari.[2]

LIV. Βορβόρῳ χαίρειν.

XLVII. Plutarch *de Anim. Procreatione* 27, p. 1026; Hipp. *Ref. Haer.* ix. 9.

XLVIII. Diog. Laert ix. 73.

XLIX. Clem. Alex. *Strom.* v. 14, p. 733.

L. Hipp. *Ref. Haer.* ix. 10. γραφέων MSS.; γναφέων Bywater; γναφείῳ Bernays.

LI. Aristotle *Eth. Nic.* x. 5, p. 1176 *a* 6. LI.*a* is Bywater's discovery. See *Journal of Philology*, ix. (1880), p. 230.

LII. Hipp. *Ref. Haer.* ix. 10.

LIV. Athenaeus v. p. 178 F. Cf. Clem. Alex. *Protrept.* 10, p. 75.

[1] Albertus Magnus *de Veget.* vi. 401, p. 545 Meyer.

[2] Columella *de R. R.* viii. 4.

XLVII. The invisible attunement is superior to the visible.

This apparently means that the attunement of opposites in the natural world is a superior "harmony" to that which we hear from musical instruments. ἁρμονία means "tune" rather than "harmony."

XLVIII. Let us not make random guesses about the greatest things.

XLIX. Men who love wisdom must have knowledge of very many things.

This is not inconsistent with πολυμαθίη νόον ἔχειν οὐ διδάσκει. Though πολυμαθίη is not enough, yet the true philosopher will have it.

L. The straight and the crooked way of the cloth-carders is one and the same.

This is a reference to the motion of the fuller's comb, which both revolved and also moved in a straight line.

LI. Asses would prefer straw to gold.

LI*a*. Heracleitus said that if happiness consisted in bodily delights we should call oxen happy when they find bitter vetches to eat.

LII. Sea-water is both very pure and very foul; to fishes it is drinkable and healthful, to men it is undrinkable and deadly.

Here we have the "unity of opposites" in a slightly different form.

LIII. Dry dust also and ashes must be placed near the walls wherever the porch or roof protects the chicken-run, that the birds may have a place to sprinkle themselves; for with these things they improve their plumage and wings, if only we believe Heracleitus the Ephesian, who says: "pigs wash in mud and barnyard fowls in dust (or ash)."

LIV. To delight in mud.

LV. Πᾶν ἑρπετὸν πληγῇ νέμεται.

LVI. Παλίντονος ἁρμονίη κόσμου ὅκωσπερ
2 λύρης καὶ τόξου.

LVII. Ἀγαθὸν καὶ κακὸν ταὐτόν.

LVIII. Καὶ ἀγαθὸν καὶ κακόν (scil. ἕν ἐστι)· οἱ γοῦν ἰατροί, φησὶν ὁ Ἡράκλειτος, τέμνοντες καίοντες πάντη βασανίζοντες κακῶς τοὺς ἀρρωστοῦντας ἐπαιτιέονται μηδέν' ἄξιον μισθὸν λαμβάνειν παρὰ τῶν ἀρρωστούντων, ταῦτα ἐργα-
6 ζόμενοι τὰ ἀγαθὰ καὶ †τὰς νόσους†.[1]

LIX. Συνάψιες οὖλα καὶ οὐχὶ οὖλα, συμφερόμενον διαφερόμενον, συνᾷδον διᾷδον· ἐκ πάντων
3 ἓν καὶ ἐξ ἑνὸς πάντα.

LX. Δίκης οὔνομα οὐκ ἂν ᾔδεσαν, εἰ ταῦτα
2 μὴ ἦν.

LV. Aristotle *de Mundo* 6, p. 401 *a* 8 (with the reading τὴν γῆν); Stobaeus *Ecl.* i. 2, p. 86 (with the reading πληγῇ). Zeller retains τὴν γῆν.

LVI. See Plutarch *de Tranquill.* 15, p. 473; *de Iside* 45, p. 369; Porphyrius *de Antro Nymph.* 29. It is unlikely that the aphorism occurred with both παλίντονος and παλίντροπος. See XLV.

LVII. Aristotle *Phys.* i. 2, p. 185 *b* 20, and Hipp. *Ref. Haer.* ix. 10.

LVIII. Many readings have been suggested for the corrupt τὰς νόσους—καὶ ⟨τὰ κακὰ⟩ τὰς νόσους, κατὰ τὰς νόσους and καὶ βασάνους. See Bywater's note. ἐπαιτέονται Bernays for the MS. reading ἐπαιτιῶνται.

LIX. Aristotle *de Mundo* 5, p. 396 *b* 12; Stobaeus *Ecl.* i. 34 p. 690. συνάψιες Diels: συνάψειας MSS.

LX. Clem. Alex. *Strom.* iv. 3, p. 568.

[1] Hippolytus *Ref. Haer.* ix. 10.

LV. Every creature is driven to pasture with blows.

The reading τὴν γῆν, preferred by Zeller and Pfleiderer, will refer to the "crawling creatures" (worms) which feed on earth. But cf. Aeschylus, *Agamemnon* 358 and Plato, *Critias* 109 B, καθάπερ ποιμένες κτήνη πληγῇ νέμοντες. See Diels in *Berl. Sitzb.* 1901, p. 188. Men do not know what is good for them, and have to be forced to it.

LVI. The attunement of the world is of opposite tensions, as is that of the harp or bow.

See Fragment XLV.

LVII. Good and bad are the same.

This refers (*a*) to a thing being good for some and bad for others; (*b*) to goodness and badness being two aspects of the same thing.

LVIII. Goodness and badness are one. At any rate doctors, as Heracleitus says, cut, burn, and cruelly rack the sick, asking to get from the sick a fee that is not their deserts, in that they effect such benefits † in sickness.†

With ἐπαιτιῶνται the meaning is: "complain that the patients do not give them an adequate return." See Plato, *Republic* VI, 497B.

LIX. Couples are wholes and not wholes, what agrees disagrees, the concordant is discordant. From all things one and from one all things.

The reading συνάψειας could be taken as a potential optative without ἄν. Burnet renders συμφερόμενον διαφερόμενον "what is drawn together and what is drawn asunder," and takes all three pairs to be explanatory of συνάψιες.

LX. Men would not have known the name of Justice were it not for these things.

That is, justice is known only through injustice.

LXI. Ἀπρεπές φασιν, εἰ τέρπει τοὺς θεοὺς πολέμων θέα. ἀλλ' οὐκ ἀπρεπές· τὰ γὰρ γενναῖα ἔργα τέρπει. ἄλλως τε πόλεμοι καὶ μάχαι ἡμῖν μὲν δεινὰ δοκεῖ, τῷ δὲ θεῷ οὐδὲ ταῦτα δεινά. συντελεῖ γὰρ ἅπαντα ὁ θεὸς πρὸς ἁρμονίαν τῶν ὅλων, οἰκονομῶν τὰ συμφέροντα, ὅπερ καὶ Ἡράκλειτος λέγει, ὡς τῷ μὲν θεῷ καλὰ πάντα καὶ ἀγαθὰ καὶ δίκαια, ἄνθρωποι δὲ ἃ μὲν 9 ἄδικα ὑπειλήφασιν, ἃ δὲ δίκαια.[1]

LXII. Εἰδέναι χρὴ τὸν πόλεμον ἐόντα ξυνόν, καὶ δίκην ἔριν· καὶ γινόμενα πάντα κατ' ἔριν καὶ 3 †χρεώμενα†.

LXIII. Ἔστι γὰρ εἱμαρμένα πάντως * * * *.

LXIV. Θάνατός ἐστι ὁκόσα ἐγερθέντες ὁρέομεν, 2 ὁκόσα δὲ εὕδοντες ὕπνος.

LXV. Ἓν τὸ σοφὸν μοῦνον λέγεσθαι οὐκ ἐθέλει 2 καὶ ἐθέλει Ζηνὸς οὔνομα.

LXVI. Τοῦ βιοῦ οὔνομα βίος, ἔργον δὲ 2 θάνατος.

LXII. Origen *contra Celsum* vi. 42, p. 312.
LXIII. Stobaeus *Ecl.* i. 5, p. 178.
LXIV. Clem. Alex. *Strom.* iii. 3, p. 520.
LXV. Clem. Alex. *Strom.* v. 14, p. 718.
LXVI. Eustathius *in Iliad.* i. 49; *Etymol. magnum* s.v. βιός; Schol. *in Iliad.* i. 49 ap. Cramer *A. P.* iii. p. 122.

[1] Schol. B. *in Il.* iv. 4, p. 120 Bekk.

LXI. They say that it is unseemly that the sight of wars delights the gods. But it is not unseemly, for noble deeds delight them. Wars and fighting seem to our thoughtlessness (?) terrible, but in the sight of God even these things are not terrible. For God makes everything contribute to the attunement of wholes, as he dispenses the things that benefit, even as Heracleitus says that to God all things are fair and good and just, but men have supposed that some things are unjust, other things just.

LXII. We must know that war is common to all and that strife is justice, and that everything comes into being by strife and . . .

The corrupt *χρεώμενα* has been emended to *καταχρεώμενα*, to *φθειρόμενα* and *κρινόμενα*, but no reading commends itself as really probable.

LXIII. For there are things foreordained wholly.

LXIV. Whatsoever things we see when awake are death, just as those we see in sleep are slumber.

Diels thinks that the original went on to say that "what we see when dead is life." The road up and down has three stages, Fire, Water, Earth, or, Life, Sleep, Death.

LXV. The one and only wisdom is both unwilling and willing to be spoken of under the name of Zeus.

"Unum illud principium mundi est materia causa lex regimen. *Ζεύς, Δίκη, σοφόν, λόγος*: varia nomina, res non diversa. Idem significat illud . . . *πῦρ αἰείζωον*, unde manat omnis motus, omnis vita, omnis intellectus." Ritter and Preller, *Hist. Philos. Gr.* § 40, note *a*. This is admirably said, and puts a great deal of Heracleitus' teaching into three sentences.

LXVI. The name of the bow is life, but its work is death.

A pun on *βιός* (bow) and *βίος* (life).

LXVII. Ἀθάνατοι θνητοί, θνητοὶ ἀθάνατοι, ζῶντες τὸν ἐκείνων θάνατον τὸν δὲ ἐκείνων βίον
3 τεθνεῶτες.

LXVIII. Ψυχῇσι γὰρ θάνατος ὕδωρ γενέσθαι, ὕδατι δὲ θάνατος γῆν γενέσθαι· ἐκ γῆς δὲ ὕδωρ
3 γίνεται, ἐξ ὕδατος δὲ ψυχή.

LXIX. Ὁδὸς ἄνω κάτω μία καὶ ὡυτή.

LXX. Ξυνὸν ἀρχὴ καὶ πέρας.

LXXI. Ψυχῆς πείρατα οὐκ ἂν ἐξεύροιο πᾶσαν
2 ἐπιπορευόμενος ὁδόν· οὕτω βαθὺν λόγον ἔχει.

LXXII. Ψυχῇσι τέρψις ὑγρῇσι γενέσθαι.

LXXIII. Ἀνὴρ ὁκότ' ἂν μεθυσθῇ, ἄγεται ὑπὸ παιδὸς ἀνήβου σφαλλόμενος, οὐκ ἐπαΐων ὅκη
3 βαίνει, ὑγρὴν τὴν ψυχὴν ἔχων.

LXXIV. Αὔη ψυχὴ σοφωτάτη καὶ ἀρίστη

LXVII. Hipp. *Ref. Haer.* ix. 10. The fragment (or parts of it) are quoted by many authors. See Bywater, Patrick or Diels.

LXVIII. Hipp *Ref. Haer.* v. 16; Clem. Alex. *Strom.* vi. 2, p. 746; Philo *de Incorr. Mundi* 21, p 509; Proclus *in Tim.* 36 c.

LXIX. Hipp. *Ref. Haer.* ix. 10; Diog. Laert ix 8; Max. Tyr. xli. 4, p. 489; Cleomedes περὶ μετεώρων i. p. 75; Stobaeus *Ecl.* i. 41.

LXX. Porphyry ap. Schol. B. *Il.* xiv. 200, p. 392 Bekk.

LXXI. Diog. Laert. ix. 7.

LXXII. Numenius ap. Porphyr. *de Antro Nymph.* 10.

LXXIII. Stobaeus *Flor.* v. 120.

LXXIV. Plutarch *Romulus* 28; Stobaeus *Flor.* v. 120 (in the form αὔη ξηρὴ ψυχὴ σοφωτάτη καὶ ἀρίστη, where ξηρὴ is a gloss). In several cases (*e.g.* Plutarch *de Carn. Esu* i. 6, p. 995; *de Defectu Orac.* 41, p. 432; Hermeias in Plato *Phaedr.* p. 73, Ast) the fragment occurs in the form αὐγὴ ξηρὴ ψυχὴ σοφωτάτη καὶ ἀρίστη. Another very old form, going back at least to Philo, is οὗ γῆ ξηρή, ψυχὴ σοφωτάτη

LXVII. Immortal mortals, mortal immortals, one living the others' death and dying the others' life.

For the sake of symmetry in English I have translated τεθνεῶτες rather inaccurately. Being perfect in tense it strictly means "being dead," *i.e.* their being dead is the others' life.

LXVIII. For it is death to souls to become water, and death to water to become earth. But from earth comes water, and from water, soul.

The best commentary on this is Aristotle, *de Anima* I. 2, 405 *a*, 25: καὶ Ἡράκλειτος δὲ τὴν ἀρχὴν εἶναί φησι ψυχήν, εἴπερ τὴν ἀναθυμίασιν, ἐξ ἧς τἆλλα συνίστησιν.

LXIX. The road up and the road down is one and the same.

LXX. The beginning and end are common.

Heracleitus is referring to a point on the circumference of a circle.

LXXI. The limits of soul you could not discover though you journeyed the whole way, so deep a measure it has.

Burnet renders λόγον "measure," as in Fragment XXIII.

LXXII. It is delight to souls to become moist.

Perhaps because the change to moisture means death, and the rest of death is pleasant. Or, the way down to death is really a way to the joy of a new life. Or (finally), the passage cannot be altogether without a reference to the τέρψις of intoxication. See the next fragment.

LXXIII. A man when he has become drunk is led by a mere stripling, stumbling, not knowing where he walks, having his soul moist.

LXXIV. A dry soul is wisest and best.

καὶ ἀρίστη. The steps in the corruption seem to be αὔη—αὔη ξηρὴ—αὐγὴ ξηρὴ—οὗ γῆ ξηρή. See Bywater's notes on LXXV and LXXVI.

LXXV. †Αὐγὴ ξηρὴ ψυχὴ σοφωτάτη καὶ
2 ἀρίστη.†

LXXVI. †Οὗ γῆ ξηρή, ψυχὴ σοφωτάτη καὶ
2 ἀρίστη.†

LXXVII. Ἄνθρωπος, ὅκως ἐν εὐφρόνῃ φάος,
2 ἅπτεται ἀποσβέννυται.

LXXVIII. Πότε γὰρ ἐν ἡμῖν αὐτοῖς οὐκ ἔστιν
ὁ θάνατος; καὶ ᾗ φησιν Ἡράκλειτος, ταὐτ' εἶναι
ζῶν καὶ τεθνηκός, καὶ τὸ ἐγρηγορὸς καὶ τὸ
καθεῦδον, καὶ νέον καὶ γηραιόν· τάδε γὰρ μεταπεσόντα ἐκεῖνά ἐστι κἀκεῖνα πάλιν μεταπεσόντα
6 ταῦτα.[1]

LXXIX. Αἰὼν παῖς ἐστι παίζων πεσσεύων·
2 παιδὸς ἡ βασιληίη.

LXXX. Ἐδιζησάμην ἐμεωυτόν.

LXXXI. Ποταμοῖσι τοῖσι αὐτοῖσι ἐμβαίνομέν
2 τε καὶ οὐκ ἐμβαίνομεν, εἶμέν τε καὶ οὐκ εἶμεν.

LXXVII. Clem Alex. *Strom.* iv. 22, p. 628.
LXXIX. Clem. Alex. *Paedag.* i. 5 p. 111; Hipp. *Ref. Haer.* ix. 9; Proclus *in Tim.* 101 F.
LXXX. Plutarch *adv. Colot.* 20, p. 1118; Dio Chrysost. *Or.* 55, p. 282; Suidas *s.v.* **Ποστοῦμος.**
LXXXI. Heraclitus *Alleg. Hom.* 24 and Seneca *Epp.* 58.

[1] Plutarch, *Consol. ad Apoll.* 10, p. 106.

LXXV. Dry light is the wisest and best soul.

LXXVI. Where earth is dry, the soul is wisest and best.

For LXXV and LXXVI see notes on the text.

LXXVII. Man, like a light in the night, is kindled and put out.

LXXVIII. For when is death not within our selves? And as Heracleitus says: "Living and dead are the same, and so are awake and asleep, young and old. The former when shifted are the latter, and again the latter when shifted are the former."

Burnet takes the metaphor in μεταπέσοντα to be the moving of pieces from one γραμμὴ of the draught-board to another.

LXXIX. Time is a child playing draughts; the kingship is a child's.

Cf. Homer, *Iliad* XV. 362:

> ὡς ὅτε τις ψάμαθον πάϊς ἄγχι θαλάσσης,
> ὅς τ' ἐπεὶ οὖν ποιήσῃ ἀθύρματα νηπιέῃσιν,
> ἂψ αὖτις συνέχευε ποσὶν καὶ χερσὶν ἀθύρων.

The changes of time are like the changes of the child's game.

LXXX. I searched my self.

See Ritter and Preller, § 48. Possibly it means: "I inquired of myself, and did not trust others." See Fragments XV–XVIII. Some see a reference to γνῶθι σεαυτόν, and it is possible that Heracleitus gave a new meaning to this old saying. But Pfleiderer's theory, that H. sought for the τέλος in introspection, is a strangely distorted view.

LXXXI. Into the same rivers we step and do not step; we are and we are not.

LXXXII. Κάματός ἐστι τοῖς αὐτοῖς μοχθεῖν
2 καὶ ἄρχεσθαι.

LXXXIII. Μεταβάλλον ἀναπαύεται.

LXXXIV. Καὶ ὁ κυκεὼν διίσταται μὴ κινεό-
2 μενος.

LXXXV. Νέκυες κοπρίων ἐκβλητότεροι.

LXXXVI. Γενόμενοι ζώειν ἐθέλουσι μόρους τ' ἔχειν· μᾶλλον δὲ ἀναπαύεσθαι, καὶ παῖδας κατα-
3 λείπουσι μόρους γενέσθαι.

LXXXVII. Οἱ μὲν "ἡβῶντος" ἀναγινώσκοντες[1] ἔτη τριάκοντα ποιοῦσι τὴν γενέαν καθ' Ἡράκλειτον· ἐν ᾧ χρόνῳ γεννῶντα παρέχει τὸν ἐξ αὐτοῦ
4 γεγεννημένον ὁ γεννήσας.[2]

LXXXVIII. Ὁ τριάκοντα ἀριθμὸς φυσικώτατός ἐστιν· ὃ γὰρ ἐν μονάσι τριάς, τοῦτο ἐν δεκάσι τριακοντάς. ἐπεὶ καὶ ὁ τοῦ μηνὸς κύκλος συνέστηκεν ἐκ τεσσάρων τῶν ἀπὸ μονάδος ἑξῆς τετραγώνων α', δ', θ', ις'. ὅθεν οὐκ ἀπὸ σκοποῦ
6 Ἡράκλειτος γενεὰν τὸν μῆνα καλεῖ.[3]

LXXXIX. Ex homine in tricennio potest avus
2 haberi.

LXXXII. Plotinus *Enn.* iv. 8, p. 468; Iamblichus *ap.* Stob. *Ecl.* i. 41, p. 906

LXXXIII. Same as for LXXXII.

LXXXIV. Theophrastus περὶ ἰλίγγων 9, p 138.

LXXXV. Strabo xvi. 26, p. 784; Plutarch *Qu. conviv.* iv. 4, p. 669; Pollux *Onom.* v. 163; Origen *contra Cels.* v. 14, p. 247; Julianus *Or.* vii. p 226 *c.* The scholiast V on *Iliad* xxiv. 54, p. 630 Bekk. assigns the fragment to Empedocles.

LXXXVI. Clem. Alex. *Strom.* iii. 3, p. 516.

LXXXVII. Cf. Censorinus *de D. N.* 17.

LXXXIX. Philo Qu. in Gen. ii. 5, p. 82, Aucher.

LXXXII. It is toil to labour for the same masters and to be ruled by them.

I.e. change is restful. Cf. the next fragment.

LXXXIII. By changing it rests.

LXXXIV. The posset too separates if it be not stirred.

An example of change and motion giving existence and reality.

LXXXV. Corpses are more fit to be thrown out than is dung.

LXXXVI. When born they wish to live and to have dooms—or rather to rest, and they leave children after them to become dooms.

LXXXVII. Some reading ἡβῶντος in this passage make a generation to consist of thirty years, as Heracleitus has it, this being the time it takes a father to have a son who is himself a father.

LXXXVIII. The number thirty is one most intimately bound up with nature, as it bears the same relation to tens as three does to units. Then again the cycle of the moon is composed of the numbers 1, 4, 9, 16, which are the squares of the first four numbers. Wherefore Heracleitus hit the mark when he called the month (or moon) a generation.

LXXXIX. In thirty years a man may become a grandfather.

The Fragments LXXXVI–LXXXIX refer to the "cycle of life." The circle is complete when the son himself becomes a father.

---

[1] Apud Hesiod *fr.* 163 Goettling.
[2] Plutarch *de Orac. Def.* 11, p. 415.
[3] Io. Lydus *de Mensibus*, iii. 10, p. 37 ed. Bonn.

XC. Πάντες εἰς ἓν ἀποτέλεσμα συνεργοῦμεν, οἱ μὲν εἰδότως καὶ παρακολουθητικῶς, οἱ δὲ ἀνεπιστάτως· ὥσπερ καὶ τοὺς καθεύδοντας, οἶμαι, ὁ Ἡράκλειτος ἐργάτας εἶναι λέγει καὶ
5 συνεργοὺς τῶν ἐν τῷ κόσμῳ γινομένων.[1]

XCI. Ξυνόν ἐστι πᾶσι τὸ φρονέειν. ξὺν νόῳ λέγοντας ἰσχυρίζεσθαι χρὴ τῷ ξυνῷ πάντων, ὅκωσπερ νόμῳ πόλις καὶ πολὺ ἰσχυροτέρως. τρέφονται γὰρ πάντες οἱ ἀνθρώπειοι νόμοι ὑπὸ ἑνὸς τοῦ θείου· κρατέει γὰρ τοσοῦτον ὁκόσον
6 ἐθέλει καὶ ἐξαρκέει πᾶσι καὶ περιγίνεται.

XCII. Διὸ δεῖ ἕπεσθαι τῷ ξυνῷ. τοῦ λόγου δ' ἐόντος ξυνοῦ, ζώουσι οἱ πολλοὶ ὡς ἰδίην ἔχοντες
3 φρόνησιν.

XCI. Stobaeus *Flor.* iii. 84. Cf. *Hymn* of Cleanthes 24, *οὔτ' ἐσορῶσι θεοῦ κοινὸν νόμον οὔτε κλύουσιν, ᾧ κεν πειθόμενοι σὺν νῷ βίον ἐσθλὸν ἔχοιεν.*

XCII. Sext. Emp. *adv. Math.* vii. 133. Bywater does not regard Διὸ . . . *ξυνῷ* as Heracleitean and Burnet rejects *τοῦ . . . ξυνοῦ.*

[1] M. Antoninus vi. 42.

ON POLITICS AND ETHICS

XC. We all work together to one end, some wittingly and with understanding, others unconsciously. In this sense, I think, Heracleitus says that even sleepers are workers and co-operators in the things that take place in the world.

XCI. Thought is common to all. Men must speak with understanding and hold fast to that which is common to all, as a city holds fast to its law, and much more strongly still. For all human laws are nourished by the one divine law. For it prevails as far as it wills, suffices for all, and there is something to spare.

"The common" will be fire, which is the one true wisdom. So men who have understanding must "keep their souls dry" and refuse to cut themselves off from the great principle of the universe by letting their souls grow moist. See Introduction, p. 457. Passages like this were eagerly seized upon by the Stoics when they elaborated their theory of a great *κοινὸς λόγος* animating the universe. True virtue, they held, was for a man consciously and lovingly to follow this *λόγος*, which is really the will of God, and to try to associate himself with it. What is crude and imperfect in Heracleitus became mature and complete in Stoicism. Christianity seized upon this thought, and developed the *λόγος*-doctrine of St. John and the early Fathers.

XCII. Therefore one must follow the common. But though the Word is common, the many live as though they had a wisdom of their own.

Burnet thinks that *τοῦ λόγου δ' ἐόντος ξυνοῦ* does not belong to Heracleitus, appealing to the MSS. reading *δὲ ὄντος* in support of his contention. He is chiefly influenced by his conviction that *λόγος* can mean only the message or gospel of Heracleitus. But at this early stage in the history of thought there could be no distinction made between (*a*) the message and (*b*) the truth which the message tries to explain. It is the latter meaning that I think *λόγος* has in this passage.

XCIII. Ὧι μάλιστα διηνεκέως ὁμιλέουσι, τούτῳ
2 διαφέρονται.

XCIV. Οὐ δεῖ ὥσπερ καθεύδοντας ποιεῖν καὶ
2 λέγειν.

XCV. Ὁ Ἡράκλειτός φησι, τοῖς ἐγρηγορόσιν
ἕνα καὶ κοινὸν κόσμον εἶναι, τῶν δὲ κοιμωμένων
3 ἕκαστον εἰς ἴδιον ἀποστρέφεσθαι.[1]

XCVI. Ἦθος γὰρ ἀνθρώπειον μὲν οὐκ ἔχει
2 γνώμας, θεῖον δὲ ἔχει.

XCVII. Ἀνὴρ νήπιος ἤκουσε πρὸς δαίμονος
2 ὅκωσπερ παῖς πρὸς ἀνδρός.

XCVIII. Ἢ οὐ καὶ Ἡράκλειτος ταὐτὸν τοῦτο
λέγει, ὃν σὺ ἐπάγει, ὅτι ἀνθρώπων ὁ σοφώτατος
πρὸς θεὸν πίθηκος φανεῖται καὶ σοφίᾳ καὶ κάλλει
4 καὶ τοῖς ἄλλοις πᾶσιν ;[2]

XCIX. Ὦ ἄνθρωπε, ἀγνοεῖς ὅτι τὸ τοῦ Ἡρα-
κλείτου εὖ ἔχει, ὡς ἄρα πιθήκων ὁ κάλλιστος
αἰσχρὸς ἄλλῳ γένει συμβάλλειν, καὶ χυτρῶν ἡ
καλλίστη αἰσχρὰ παρθένων γένει συμβάλλειν, ὥς
5 φησιν Ἱππίας ὁ σοφός.[3]

XCIII and XCIV. M. Antoninus iv. 46. Diels adds λόγῳ τῷ τὰ ὅλα διοικοῦντι, which Burnet rejects as belonging to M. Aurelius (Stoic idea).

XCVI and XCVII. Origen *contra Cels.* vi. 12, p. 291.

[1] Plutarch *de Superst.* 3, p. 166.

XCIII. They are at variance with that with which they have most continuous intercourse.

XCIV. We ought not to act and to speak as though we were asleep.

XCV. Heracleitus says that there is one world in common for those who are awake, but that when men are asleep each turns away into a world of his own.

Sleepiness to Heracleitus was the state of a man who allowed his soul to sink on the downward path into moisture or mud. See Fragments XCI and XCII. To be awake was to have one's soul dry, and to be in close connection with "the ever-living fire" of the universe.

XCVI. Human nature has no understanding, but that of God has.

This fragment expresses in another way the thought that τὸ ξυνὸν is good, τὸ ἴδιον evil.

XCVII. Man is called a baby by the deity as a child is by a man.

ON RELIGION

XCVIII. And does not Heracleitus too, whom you bring forward, say this very same thing, that the wisest of men compared with God will appear as an ape in wisdom, in beauty and in everything else?

XCIX. Sir, you do not know that the remark of Heracleitus is a sound one, to the effect that the most beautiful of apes is ugly in comparison with another species, and that the most beautiful of pots is ugly in comparison with maidenhood, as says Hippias the wise.

---

[2] Plato *Hipp. mai.* 289 B.
[3] Plato *Hipp. mai.* 289 A.

C. Μάχεσθαι χρὴ τὸν δῆμον ὑπὲρ τοῦ νόμου
2 ὅκως ὑπὲρ τείχεος.

CI. Μόροι γὰρ μέζονες μέζονας μοίρας λαγχά-
2 νουσι.

CII. Ἀρηιφάτους θεοὶ τιμῶσι καὶ ἄνθρωποι.

CIII. Ὕβριν χρὴ σβεννύειν μᾶλλον ἢ πυρ-
2 καϊήν.

CIV. Ἀνθρώποισι γίνεσθαι ὁκόσα θέλουσι
οὐκ ἄμεινον. νοῦσος ὑγίειαν ἐποίησε ἡδύ, κακὸν
3 ἀγαθόν, λιμὸς κόρον, κάματος ἀνάπαυσιν.

CV. Θυμῷ μάχεσθαι χαλεπόν· ὅ τι γὰρ ἂν
2 χρηίζῃ γίνεσθαι, ψυχῆς ὠνέεται.

CVI. †Ἀνθρώποισι πᾶσι μέτεστι γιγνώσκειν
2 ἑαυτοὺς καὶ σωφρονεῖν.†

CVII. †Σωφρονεῖν ἀρετὴ μεγίστη· καὶ σοφίη
2 ἀληθέα λέγειν καὶ ποιεῖν κατὰ φύσιν ἐπαΐοντας.†

CVIII. Ἀμαθίην ἄμεινον κρύπτειν· ἔργον δὲ ἐν
2 ἀνέσει καὶ παρ᾽ οἶνον.

C. Diogenes Laertius ix. 2.

CI Clem. Alex. *Strom.* iv. 7, p. 586.

CII. Clem. Alex. *Strom.* iv. 4, p. 571; Theodoretus *Therap.* viii. p. 117, 33.

CIII Diogenes Laertius ix. 2.

CIV. Stobaeus *Flor.* iii. 83. Cf. Clem. Alex. *Strom.* ii. 21, p. 497. I accept (with some hesitation) κακὸν for the MS. reading καί (Heitz, Diels, Burnet).

CV. Iamblichus *Protrept.* p. 140; Aristotle *Eth. Nic.* 1105*a* 8, *Eth. Eud.* 1223*b* 22, and *Pol.* 1315*a* 29; Plutarch *de cohibenda Ira* 9, p. 457 and *Coriol.* 22.

CVI. Stobaeus *Flor.* v. 119.

CVII. Stobaeus *Flor.* iii. 84.

CVIII. Plutarch *Qu. conviv.* iii. *prooem.* p. 644; *de Audiendo* 12, p. 43 and *Virt. doc. posse* 2, p. 439; Stob. *Flor.* xviii. 32.

C. The people should fight for their law as for a wall.

This is because the law is ξυνόν, is, in fact, but a reflection of the great ξυνὸν of the natural world.

CI. For greater dooms win greater destinies.

This refers to the "fiery deaths" of heroic men. See Introduction, p. 457, and also the following fragment.

CII. Gods and men honour those who are killed in battle.

CIII. You should put out insolence even more than a fire.

CIV. For men to get all they wish is not the better thing. It is disease that makes health a pleasant thing; evil, good; hunger, surfeit; and toil, rest.

CV. It is hard to contend against one's heart's desire; for whatever it wishes to have it buys at the cost of soul.

Burnet so translates θυμός; the word covers a wider area than any English equivalent, but includes much of what we include under "instinct," "urge," "passionate craving." Aristotle understood θυμὸς to mean anger (*Ethic. Nicom.* II. 2, 1105 *a* 8). To gratify θυμὸς is to allow one's soul "to become moist."

CVI. It is the concern of all men to know themselves and to be sober-minded.

CVII. To be sober-minded is the greatest virtue, and wisdom is to speak the truth and to act it, listening to the voice of nature.

These two fragments (both are of doubtful authenticity) express positively what is stated in Fragment CV in a quasi-negative form.

CVIII. It is better to hide ignorance, but it is hard to do this when we relax over wine.

CIX. †Κρύπτειν ἀμαθίην κρέσσον ἢ ἐς τὸ
2 μέσον φέρειν.†

CX. Νόμος καὶ βουλῇ πείθεσθαι ἑνός.

CXI. Τίς γὰρ αὐτῶν νόος ἢ φρήν; [δήμων] ἀοιδοῖσι ἕπονται καὶ διδασκάλῳ χρέωνται ὁμίλῳ, οὐκ εἰδότες ὅτι πολλοὶ κακοὶ ὀλίγοι δὲ ἀγαθοί. αἱρεῦνται γὰρ ἓν ἀντία πάντων οἱ ἄριστοι, κλέος ἀέναον θνητῶν, οἱ δὲ πολλοὶ κεκόρηνται ὅκωσπερ
6 κτήνεα.

CXII. Ἐν Πριήνῃ Βίας ἐγένετο ὁ Τευτάμεω, οὗ
2 πλέων λόγος ἢ τῶν ἄλλων.

CXIII. Εἷς ἐμοὶ μύριοι, ἐὰν ἄριστος ᾖ.

CXIV. Ἄξιον Ἐφεσίοις ἡβηδὸν ἀπάγξασθαι πᾶσι καὶ τοῖς ἀνήβοις τὴν πόλιν καταλιπεῖν, οἵτινες Ἑρμόδωρον ἄνδρα ἑωυτῶν ὀνήιστον ἐξέβαλον, φάντες· ἡμέων μηδὲ εἷς ὀνήιστος ἔστω,
5 εἰ δὲ μή, ἄλλῃ τε καὶ μετ' ἄλλων.

CXV. Κύνες καὶ βαΰζουσι ὃν ἂν μὴ γινώ-
2 σκωσι.

CXVI. Ἀπιστίῃ διαφυγγάνει μὴ γινώσκεσθαι.

CXVII. Βλὰξ ἄνθρωπος ἐπὶ παντὶ λόγῳ
2 ἐπτοῆσθαι φιλέει.

CIX. Stobaeus *Flor.* iii. 82.

CX. Clem Alex. *Strom.* v. 14, p. 718.

CXI. Clem. Alex *Strom.* v. 9, p. 682 and iv. 7, p. 586; Proclus *in Alcib.* p. 255, Creuzer.

CXII. Diogenes Laertius i. 88.

CXIII. Galen περὶ διαγνώσεως σφυγμῶν i. 1; Theodorus Prodromus in Lazerii *Miscell.* i. p. 20; Seneca *Epp.* 7.

CXIV. Strabo xiv. 25, p. 642; Cicero *Tusc. Disp.* v 105; Musonius *ap.* Stob. *Flor.* xl. 9; Diog. Laert. ix. 2; Iamblichus *de Vit. Pyth.* 30, p. 154 Arcer.

CXV. Plutarch *an Seni sit ger. Resp.* vii. p. 787.

CIX. To hide ignorance is preferable to bringing it to light.

CX. It is law too to obey the advice of one.

CXI. For what mind or sense have they? They follow the bards and use the multitude as their teacher, not realising that there are many bad but few good. For the best choose one thing over all others, immortal glory among mortals, while the many are glutted like beasts.

CXII. In Priene lived Bias, son of Teutamas, who is of more account than the others.

CXIII. One man to me is as ten thousand, if he be the best.

Fragments CXI–CXIII show the aristocratic tendencies of the mind of Heracleitus. His "common," of course, has nothing to do with "common-sense" or with general opinions. It refers to the law or principle of nature, which each man must apprehend for himself. He who can do so best is a natural leader and lawgiver.

CXIV. All the Ephesians from the youths up would do well to hang themselves and leave their city to the boys. For they banished Hermodorus, the best man of them, saying, "We would have none among us who is best; if there be such an one, let him be so elsewhere among other people."

CXV. Dogs also bark at him they know not.

CXVI. He escapes being known because of men's unbelief.

"A prophet is not without honour save in his own city."

CXVII. A fool is wont to be in a flutter at every word.

CXVI. Plutarch *Coriolanus* 38; Clem. Alex. *Strom.* v. 13, p. 699.

CXVII. Plutarch *de Audiendo* 7, p. 41 and *de aud. Poet.* 9, p. 28.

CXVIII. Δοκεόντα ὁ δοκιμώτατος γινώσκει †φυλάσσειν·† καὶ μέντοι καὶ δίκη καταλήψεται
3 ψευδέων τέκτονας καὶ μάρτυρας.

CXIX. Τόν θ' Ὅμηρον ἔφασκεν ἄξιον ἐκ τῶν ἀγώνων ἐκβάλλεσθαι καὶ ῥαπίζεσθαι, καὶ Ἀρχί-
3 λοχον ὁμοίως.[1]

CXX. Unus dies par omni est.

CXXI. Ἦθος ἀνθρώπῳ δαίμων.

CXXII. Ἀνθρώπους μένει τελευτήσαντας ἅσσα
2 οὐκ ἔλπονται οὐδὲ δοκέουσι.

CXXIII. Ἐπανίστασθαι καὶ φύλακας γίνεσθαι
2 ἐγερτὶ ζώντων καὶ νεκρῶν.

CXXIV. Νυκτιπόλοι, μάγοι, βάκχοι, λῆναι,
2 μύσται.

CXXV. Τὰ γὰρ νομιζόμενα κατ' ἀνθρώπους
2 μυστήρια ἀνιερωστὶ μυεῦνται.

CXXVI. Καὶ τοῖς ἀγάλμασι τουτέοισι εὔχονται, ὁκοῖον εἴ τις τοῖς δόμοισι λεσχηνεύοιτο, οὔ τι
3 γινώσκων θεοὺς οὐδ' ἥρωας, οἵτινές εἰσι.

CXVIII. Clem. Alex. *Strom.* v. 1, p. 649. The MS. reading is δοκεόντων; Schleiermacher suggested δοκέοντα and Diels δοκέοντ' ὦν. The MS. φυλάσσειν has been emended to φυλάσσει (Schleiermacher), φλυάσσειν (Bergk), πλάσσειν (Bernays and Bywater).

CXX. Seneca *Epp.* 12; Plutarch *Camillus* 19.

CXXI. Plutarch *Qu. Plat.* i. 2, p. 999; Alex. Aphrod. *de Fato* 6, p. 16; Stob *Flor.* civ. 23.

CXXII. Clem. Alex. *Strom.* iv. 22, p 630; Theodoretus *Therap.* viii. p. 118, 1; Themistius in Stob. *Flor.* cxx 28.

CXXIII. Hipp. *Ref. Haer.* ix. 10. The MS. has before ἐπανίστασθαι the words ἔνθα δεόντι. Various emendations have been suggested: ἐνθάδε ἐόντας Bernays; ἔνθα θεὸν δεῖ Sauppe; ἐνθάδε ἔστι Petersen. So the MS. also has ἐγερτιζόντων. The text is that of Bernays.

CXVIII. The one most in repute knows only what is reputed. And yet justice will overtake the makers of lies and the false witnesses.

Of all the emendations of the corrupt φυλάσσειν I prefer Bergk's φλυάσσειν, but I follow Burnet in deleting the word.

CXIX. He said that Homer deserved to be expelled from the lists and beaten, and Archilochus likewise.

CXX. One day is like any other.

CXXI. A man's character is his fate.

CXXII. There await men after death such things as they neither expect nor look for.

CXXIII. To rise up and become wakeful guards of the living and of the dead.

CXXIV. Night - walkers, Magians, priests of Bacchus and priestesses of the vat, the initiated.

CXXV. The mysteries that are celebrated among men it is unholy to take part in.

CXXVI. And to these images they pray, as if one were to talk to one's house, knowing not the nature of gods and heroes.

CXXIV. Clem. Alex. *Protrept.* 2, p. 18 = Eusebius *P. E.* ii. 3, p. 66.

CXXV. Clem. Alex. *Protrept.* 2, p. 19 = Eusebius *P. E.* ii. 3, p. 67.

CXXVI. Clem. Alex. *Protrept.* 4, p. 44; Origen *contra Cels.* i. 5, p. 6, and vii. 62, p. 384.

[1] Diogenes Laert. ix. 1.

CXXVII. Εἰ μὴ γὰρ Διονύσῳ πομπὴν ἐποιεῦντο καὶ ὕμνεον ᾆσμα αἰδοίοισι, ἀναιδέστατα εἴργαστ' ἄν· ὡυτὸς δὲ Ἀΐδης καὶ Διόνυσος, ὅτεῳ μαίνονται
4 καὶ ληναΐζουσι.

CXXVIII. Θυσιῶν τοίνυν τίθημι διττὰ εἴδη· τὰ μὲν τῶν ἀποκεκαθαρμένων παντάπασιν ἀνθρώπων, οἷα ἐφ' ἑνὸς ἄν ποτε γένοιτο σπανίως, ὥς φησιν Ἡράκλειτος, ἤ τινων ὀλίγων εὐαριθμήτων ἀνδρῶν· τὰ δ' ἔνυλα καὶ σωματοειδῆ καὶ διὰ μεταβολῆς συνιστάμενα, οἷα τοῖς ἔτι κατεχομένοις
7 ὑπὸ τοῦ σώματος ἁρμόζει.[1]

CXXIX. Ἄκεα.

CXXX. Καθαίρονται δὲ αἵματι μιαινόμενοι ὥσπερ ἂν εἴ τις ἐς πηλὸν ἐμβὰς πηλῷ ἀπο-
3 νίζοιτο.

CXXVII. Plutarch *de Iside* 28, p. 362; Clem. Alex. *Protrept.* 2, p. 30.

CXXIX. Iamblichus *de Myst.* i. 11.

CXXX. Gregorius Naz. *Or.* xxv. (xxiii.) 15, p. 466 with Elias Cretensis *in loc.* See Apollonius *Epp.* 27. Professor D. S. Robertson inserts αἷμα before αἵματι.

[1] Iamblichus *de Myst.* v. 15.

CXXVII. For if it were not to Dionysus that they made procession and sang the phallic hymn, it would be a most disgraceful action. But Hades is the same as Dionysus, in whose honour they rave and keep the feast of the vat.

CXXVIII. I distinguish, therefore, two kinds of sacrifices. First, that of men wholly cleansed, such as would rarely take place in the case of a single individual, as Heracleitus says, or in the case of very few men. Second, material and corporeal sacrifices, arising from change, such as befit those who are still fettered by the body.

CXXIX. Cures (atonements).

CXXX. When defiled they purify themselves with blood, just as if one who had stepped in mud were to wash himself in mud.

INDEX

OF CHIEF NAMES AND SUBJECTS

INDEX

INDEX

INDEX

INDEX

INDEX

INDEX

INDEX

INDEX

SUPPLEMENTARY INDEX

INDEX OF DISEASES

A

B

C

D

SUPPLEMENTARY INDEX

SUPPLEMENTARY INDEX